The Sally Series
Pure and Applied
UNDERGRADUATE TEXTS · 22

Early Fourier Analysis

Hugh L. Montgomery

American Mathematical Society
Providence, Rhode Island

2010 *Mathematics Subject Classification*. Primary 42-01.

Cover photo credit:
White Wall Mural by Joe St. Jean
Photographed by Leslie Patron
with permission of Chef Chris Nixon of Element 112, Sylvania, OH.

Page 267: "Young Heisenberg" by cartoonist/anthropolgist Emily Holt.
Reprinted with permission.

For additional information and updates on this book, visit
www.ams.org/bookpages/amstext-22

Library of Congress Cataloging-in-Publication Data

Montgomery, Hugh L., author.
Early Fourier analysis / Hugh L. Montgomery
pages cm. — (Pure and applied undergraduate texts ; volume 22)
Includes bibliographical references and index.
ISBN 978-1-4704-1560-0 (alk. paper)
1. Harmonic analysis. 2. Fourier transformations. I. Title.

QA403.M66 2015
515′.2433—dc23

2014035196

∞ The paper used in this book is acid-free and falls within the guidelines
established to ensure permanence and durability.
Visit the AMS home page at http://www.ams.org/

10 9 8 7 6 5 4 3 2 1 19 18 17 16 15 14

Dedicated with love to my children —

Lowell and Peter,

Hal and Eve,

Philip and Katharina

Contents

Preface

While such subjects as number theory and probability theory are commonly offered to undergraduates, it seems that Fourier analysis is rarely found, which is shocking when one considers the value of this subject not just within mathematics but also in the physical sciences and engineering. The author hopes that this book will encourage the view that Fourier analysis can be fruitfully presented not just to undergraduates, but even to younger undergraduates with no more experience than three or four terms of calculus. Such students will find a gentle introduction to the art of writing proofs and will be better prepared for advanced calculus and complex variables.

A student who has taken a course in advanced calculus may wonder what can be done with that machinery. The answer is: harmonic analysis (among other things). Paul Halmos is reported to have said words to the effect that the tragedy of Fourier series is that they were invented (in 1807) before convergence. The wonderful thing is that analysts such as Cauchy, Dirichlet, Riemann, and Weierstrass were motivated to develop the foundations of real analysis in order to make sense of Fourier series. In particular, Riemann defined his integral in order to provide a more rigorous basis for the discussion of Fourier series.

This book could be used for a capstone course of an undergraduate program or for beginning graduate students as a way to motivate the study of the Lebesgue integral. Since it is hoped that this book will be useful at a wide range of levels, it contains far more material than would ever be used in a single one term course. The author will be happy to provide suggestions adjusted to the instuctor's purpose.

We study Fourier analysis in three important settings. First we consider the Discrete Fourier Transform, which has to do with the use of roots of unity to describe periodic sequences. The results in this setting are easily obtained, and they form a framework for our endeavors in the more difficult subsequent settings. The point is that in the discrete setting there is no issue of convergence, but with Fourier Series we discover that convergence is a delicate matter. With Fourier Transforms of functions defined on the real line, matters are similar, but with additional difficulties. In the two latter situations we encounter points in our arguments where a detail is needed from advanced calculus or Lebegue measure theory. On such occasions, we simply quote the needed result and move on.

On the subject of Fourier Series, some authors use $\cos nx$ and $\sin nx$, so that all functions have period 2π. The consequence of this prescription is that most formulas have a $1/(2\pi)$ or $1/\pi$. Our contention is that the subject is more elegant when one works with functions with period 1, so that the basic building blocks are $\cos 2\pi nx$ and $\sin 2\pi nx$. But $\cos 2\pi nx + i \sin 2\pi nx = e^{2\pi inx}$ (a fact that will be a subject of discussion in Chapter 1), and it is more elegant still to use the complex exponential rather than sines and cosines. Of course, to proceed in this way, one must first become more comfortable with complex numbers. Hence that is the topic of Chapter 1. In general, when we are faced with a function with some strange period, we make a linear change of variable so that everything is translated into issues of functions with period 1. If sines and cosines are involved, we may convert to complex exponentials. When we resolve whatever is at issue, we may convert back to sines and cosines, if we wish. This is a little reminiscent of a problem expressed in terms of pounds and feet, which we would convert to grams and meters, and then convert back after the calculation is done.

Fourier analysis has links to many other branches of mathematics. We occasionally make remarks relating to such topics as linear algebra, probability theory, or number theory. Such digressions may be safely ignored by readers who are unfamiliar with the related subject in question.

Among the following chapters, sections, and appendices are found several valuable topics that are rarely found in the undergraduate (and sometimes even the graduate) curriculum. These include linear recurrences (in §F.4), summability theory (in §4.3), Bernoulli polynomials and Euler–Maclaurin summation (in §9.5), uniform distribution (in §9.6), Chebyshev polynomials (in Appendix C), and inequalities (in Appendix I).

The author is indebted to colleagues Al Taylor and Jack Goldberg for initiating this educational experiment and to the late Curtis Huntington for his unwavering support. In addition, the author is happy to thank Dick Askey, John Benedetto, Edward Crane, Peter Duren, Emily Holt, Alex Iosevich, Michael Kelly, Harsh Mehta, Kristen Moore, Michael Mossinghoff, Chris Nixon, Olivier Ramaré, Elmer Rees, Babar Saffari, and Jeff Vaaler for their valuable contributions. It has been a pleasure to work with editor Sergei Gelfand and his competent and attentive support staff at the AMS. Finally, the author thanks Michele MacFarlane, who cheerfully accepted a double dose of domestic chores in order that the author would have more time to write.

Hugh L. Montgomery
Ann Arbor, Michigan
August 31, 2014

Background

In the first section below, terms are defined and proofs are included. In the second section, most terms are defined, but proofs are omitted. In the third section, not only are proofs omitted, but some terms are undefined. The point is that everything we do can be made fully rigorous, but neither lack of rigor nor absence of advanced training in analysis should interfere with the acquisition of Fourier analysis in its most classical settings.

0.1. Elementary mathematics

An *arithmetic progression* (sometimes abbreviated AP) is a set of the form $\{nq+a : \in \mathbb{Z}\}$. Hence a sequence $\{u_n\}$ is said to be in arithmetic progression if $u_{n+1} - u_n$ is the same for all n. That means that $u_n = nq + a$ for some q and a. We frequently sum such numbers.

Theorem 0.1. *If $u_1, u_2, \ldots, u_N$ are consecutive members of an arithmetic progression, then*

$$u_1 + u_2 + \cdots + u_N = N \cdot \frac{u_1 + u_N}{2}. \tag{0.1}$$

For example,

$$1 + 2 + \cdots + N = \frac{N(N+1)}{2}.$$

Proof. Let d be determined so that $u_{n+1} - u_n = d$ for all n. We write the sum twice, first in its natural order and then in reverse order. Thus if S is the sum, then

$$\begin{aligned} S &= u_1 + u_2 + u_3 + \cdots + u_{N-1} + u_N, \\ S &= u_N + u_{N-1} + u_{N-2} + \cdots + u_2 + u_1. \end{aligned}$$

We now sum in columns. On the left hand side we have $2S$. In the first column on the right we have $u_1 + u_N$. In the second column on the right we have $u_2 + u_{N-1} = (u_1 + d) + (u_N - d) = u_1 + u_N$. In the third column we have $u_3 + u_{N-2} =$

$(u_2+d)+(u_{N-1}-d) = u_2+u_{N-1} = u_1+u_N$. Continuing in this way, we find that $u_k+u_{N+1-k} = u_1+u_N$ for all k. Hence the right hand side above is $N(u_1+u_N)$, so we have the result. □

A sequence $\{u_n\}$ is said to be a *geometric progression* if u_{n+1}/u_n is the same for all n. That means that $u_n = ar^n$ for some a and r.

Theorem 0.2. *The sum of a geometric progression is*

$$1+r+r^2+\cdots+r^{N-1} = \begin{cases} \dfrac{1-r^N}{1-r} & (r \neq 1), \\ N & (r=1). \end{cases} \tag{0.2}$$

If the first term to be summed is a power of r, we can simply factor out that amount. Thus for example,

$$\sum_{n=M}^{M+N-1} r^n = \frac{r^M - r^{M+N}}{1-r}$$

if $r \neq 1$. If $|r| < 1$, then $r^N \to 0$ as $N \to \infty$, so

$$\sum_{n=0}^{\infty} r^n = \frac{1}{1-r}. \tag{0.3}$$

Proof. If $r = 1$, then each of the N terms on the left is 1, so the sum is N. Suppose that $r \neq 1$, and let S denote the sum. Then

$$rS = r + r^2 + r^3 + \cdots + r^N.$$

On subtracting this from S we see that most terms cancel, leaving

$$S - rS = 1 - r^N.$$

The stated formula now follows on dividing both sides by $1-r$. □

If $p(x) = ax^2+bx+c$ is a quadratic polynomial, then

$$4ap(x) = 4a^2x^2 + 4abx + 4ac = (2ax+b)^2 + 4ac - b^2.$$

This manipulation is called *completing the square.* If $p(x) = 0$, then $(2ax+b)^2 = d$ where $d = b^2-4ac$ is the *discriminant* of the polynomial. If $d > 0$, then the equation $p(x) = 0$ has two distinct real roots, namely

$$r_1 = \frac{-b+\sqrt{d}}{2a}, \qquad r_2 = \frac{-b-\sqrt{d}}{2a}.$$

If $d = 0$, then $p(x)$ has a double root, $r_1 = r_2 = -b/(2a)$. If $d < 0$, then $p(x)$ has no real root, but it has two complex roots,

$$\frac{-b \pm i\sqrt{-d}}{2a}.$$

In all three cases the sum of the roots is $-b/a$, the product of the roots is c/a, and the polynomial factors, $p(x) = a(x-r_1)(x-r_2)$.

Let $a_1, a_2, \ldots, a_N$ and $b_1, b_2, \ldots, b_N$ be real numbers. *Cauchy's Inequality* asserts that

$$\Big|\sum_{n=1}^{N} a_n b_n\Big| \le \Big(\sum_{n=1}^{N} a_n^2\Big)^{1/2}\Big(\sum_{n=1}^{N} b_n^2\Big)^{1/2}. \tag{0.4}$$

That this is so follows immediately from the algebraic identity

$$\Big(\sum_{n=1}^{N} a_n^2\Big)\Big(\sum_{n=1}^{N} b_n^2\Big) - \Big(\sum_{n=1}^{N} a_n b_n\Big)^2 = \frac{1}{2}\sum_{m=1}^{N}\sum_{n=1}^{N}(a_m b_n - a_n b_m)^2,$$

since a sum of squares of real numbers is nonnegative. Moreover, from the above identity it is clear that equality holds in (0.4) if and only if the two sequences $\{a_n\}$, $\{b_n\}$ are proportional.

The *Principle of Mathematical Induction* is one of the axioms that define the integers. It can be formulated in a number of (equivalent) ways.

(1) *Weak induction*: If S is a set of positive integers, if $1 \in S$, and if $n+1 \in S$ whenever $n \in S$, then S is the set of all positive integers.

(2) *Strong induction*: If S is a set of positive integers, if $1 \in S$, and if $k \in S$ for all positive integers $k < n$ implies that $n \in S$, then S is the set of all positive integers.

(3) *Well ordering*: If S is a set of positive integers, and S is non-empty, then S contains a least member.

In all three cases we are inducting from 1, but of course one could instead induct from 0 or any other convenient point.

The Binomial Theorem is treated in Appendix B. A catalogue of trigonometric formulæ is provided in Appendix T, for convenience. The manner in which we express $\cos n\theta$ as a polynomial in $\cos\theta$ is the subject of Appendix C.

0.2. Real analysis

It is not our purpose to summarize all of real analysis. We mention only specific items that we need, and these are largely concerned with such issues as conditions that ensure that (a) one can exchange two limiting operations; and (b) a sequence that appears to tend to a limit does so.

Theorem 0.3. *A bounded monotonic sequence of real numbers has a limit.*

A sequence $\{x_n\}$ is said to be a *Cauchy sequence*if

$$\lim_{\substack{m\to\infty\\ n\to\infty}} (x_m - x_n) = 0. \tag{0.5}$$

Clearly any sequence that tends to a finite limit is a Cauchy sequence. What is important is that the converse is also true:

Theorem 0.4. *If $\{x_n\}$ is a Cauchy sequence, then $\lim_{n\to\infty} x_n$ exists and is finite.*

Note that a sequence of rational numbers tending to $\sqrt{2}$ is a Cauchy sequence, but does not not have a limit within the system of rational numbers, because $\sqrt{2}$ is irrational. In a set-theoretic sense, the real numbers are constructed by filling in the holes found among the rational numbers. Because all Cauchy sequences have a limit, we say that the real numbers are *complete.*

A function f is continuous at a if $\lim_{x\to a} f(x) = f(a)$. That is, for every $\varepsilon > 0$ there is a $\delta > 0$ such that $|f(x) - f(a)| < \varepsilon$ if $|x - a| < \delta$. Here the choice of δ depends on both ε and a. However, if in a domain $\mathcal{D}$ we have $|f(x) - f(y)| < \varepsilon$ whenever $x \in \mathcal{D}$, $y \in \mathcal{D}$ and $|x - y| < \delta$, then we say that f is *uniformly continuous* on $\mathcal{D}$.

Theorem 0.5. *If a real-valued function $f(x)$ is continuous on a closed bounded interval $[a, b]$, then it is uniformly continuous on that interval, and attains its maximum and minimum values.*

The same theorem also holds for real-valued continuous functions defined on closed bounded sets in the plane $\mathbb{R}^2$.

Theorem 0.6. (The Squeeze Theorem) *Suppose that $\delta > 0$, that f_-, f, and f_+ are functions such that $f_-(x) \le f(x) \le f_+(x)$ for $a-\delta < x < a+\delta$. If $\lim_{x\to a} f_-(x) = c$ and $\lim_{x\to a} f_+(x) = c$, then $\lim_{x\to a} f(x) = c$.*

Theorem 0.7. (Rolle's Theorem) *If $f(x)$ is a continuous real-valued function on the interval $a \le x \le b$ with $f(x)$ differentiable for $a < x < b$, and if $f(a) = f(b)$, then there exists a $\xi \in (a, b)$ such that $f'(\xi) = 0$.*

Theorem 0.8. (The Mean Value Theorem of Differential Calculus) *If $f(x)$ is a continuous real-valued function on the interval $a \le x \le b$ and $f(x)$ is differentiable for $a < x < b$, then there exists a $\xi \in (a, b)$ such that*

$$f'(\xi) = \frac{f(b) - f(a)}{b - a}.$$

Theorem 0.9. *If $\sum_{n=1}^{\infty} |a_n| < \infty$, then the sum $\sum_{n=1}^{\infty} a_n$ converges.*

Theorem 0.10. *Suppose that $a_{mn} \ge 0$ for all m and n. We form two sums:*

$$\sum_{m=1}^{\infty}\Big(\sum_{n=1}^{\infty} a_{mn}\Big), \qquad \sum_{n=1}^{\infty}\Big(\sum_{m=1}^{\infty} a_{mn}\Big). \tag{0.6}$$

If either of these sums is finite, then the other one is also finite, and they are equal.

Theorem 0.11. *If $|a_{mn}| \le A_{mn}$ for all m and n, and if*

$$\sum_{m=1}^{\infty}\Big(\sum_{n=1}^{\infty} A_{mn}\Big) < \infty,$$

then the sums (0.6) *converge and are equal.*

Theorem 0.12. *Let $f(z) = \sum_{k=0}^{\infty} a_k z^k$ be a power series, and let R be defined by the relation*

$$\frac{1}{R} = \limsup_{n\to\infty} |a_n|^{1/n}.$$

Then the power series is convergent for $|z| < R$, and is divergent for $|z| > R$. For $|z| < R$ a power series may be differentiated term-by-term, and the differentiated power series has the same radius of convergence R.

To define what it means to say that a function is Riemann–integrable on an interval $[a, b]$, we start with a partition $\boldsymbol{\pi}$, which is to say a sequence $\{x_n\}$ such that

$$a = x_0 < x_1 < \cdots < x_J = b,$$

and choose interspersing numbers ξ_j so that

$$a = x_0 \le \xi_1 \le x_1 \le \xi_2 \le x_2 \le \cdots \le \xi_{J-1} \le x_{J-1} \le \xi_J \le x_J = b. \tag{0.7}$$

A *Riemann sum* for $\int_a^b f(x)\,dx$ is then a sum of the form

$$S(\boldsymbol{\pi}, \boldsymbol{\xi}) = \sum_{j=1}^{J} f(\xi_j)(x_j - x_{j-1})\,.$$

The *mesh* of $\boldsymbol{\pi}$ is defined to be

$$\text{mesh}(\boldsymbol{\pi}) = \max_{1\le j\le J}(x_j - x_{j-1})\,. \tag{0.8}$$

That is, the mesh is the length of the longest subinterval defined by $\boldsymbol{\pi}$. We say that the integral exists and has the value I if for every $\varepsilon > 0$ there is a $\delta > 0$ such that if $\text{mesh}(\boldsymbol{\pi}) < \delta$, then $|S(\boldsymbol{\pi}, \boldsymbol{\xi}) - I| < \varepsilon$ for any choice of the interspersing points $\boldsymbol{\xi}$.

As to sums and integrals such as

$$\sum_{a\le n\le b} u_n \quad \text{and} \quad \int_a^b f(x)\,dx,$$

we have two different conventions. In a sum, we sum over all n that satisfy the indicated constraints. Thus if $b < a$, then there is no such n, and the value of the sum is 0. However, for integrals, if $b < a$, we simply say that the value of the integral is $-\int_b^a f(x)\,dx$.

The arc length of a parameterized curve $(x(t), y(t))$ for $a \le t \le b$ is the supremum of all sums of the form

$$\sum_{j=1}^{J} \sqrt{(x(t_j) - x(t_{j-1}))^2 + (y(t_j) - y(t_{j-1}))^2} \tag{0.9}$$

where $a = t_0 < t_1 < \cdots < t_J = b$.

Theorem 0.13. (Fundamental Theorem of Calculus, First Form) *Suppose that $f(x)$ is Riemann-integrable on the interval $[a, b]$. For $a \le x \le b$, put*

$$F(x) = \int_a^x f(u)\,du\,.$$

Then $F(x)$ is continuous on the interval $[a, b]$. If $a < c < b$ and if $f(x)$ is continuous at $x = c$, then $F(x)$ is differentiable at $x = c$, and $F'(c) = f(c)$.

Theorem 0.14. (Fundamental Theorem of Calculus, Second Form) *Suppose that $f(x)$ is Riemann-integrable on the interval $[a,b]$. Suppose further that $F(x)$ is a differentiable function on the interval $[a,b]$ such that $F'(x) = f(x)$ for $a \le x \le b$. Then*

$$\int_a^b f(x)\,dx = F(b) - F(a)\,.$$

The Second Form of the Fundamental Theorem follows from the First Form if f is continuous. The point of the Second Form is that it holds under the weaker assumption that f is Riemann-integrable.

Theorem 0.15. (Integration by parts) *Suppose that f is Riemann-integrable on $[a,b]$, that F is a function such that $F'(x) = f(x)$, and that g is a differentiable function such that $g'(x)$ is Riemann-integrable on $[a,b]$. Then*

$$\int_a^b f(x)g(x)\,dx = F(b)g(b) - F(a)g(a) - \int_a^b F(x)g'(x)\,dx\,.$$

This follows immediately from the Second Form of the Fundamental Theorem, in view of the differentiation formula $(Fg)' = F'g + Fg' = fg + Fg'$.

Theorem 0.16. (The triangle inequality for integrals) *If f is Riemann-integrable, then*

$$\left|\int_a^b f(x)\,dx\right| \le \int_a^b |f(x)|\,dx\,.$$

Theorem 0.17. (Leibniz's Rule) *If $f(x,y)$ and $\frac{\partial}{\partial x}f(x,y)$ exist and are continuous on the closed rectangle $a \le x \le b$, $c \le y \le d$, then the function*

$$F(x) = \int_c^d f(x,y)\,dy$$

is differentiable for $a \le x \le b$, and

$$F'(x) = \int_c^d \frac{\partial}{\partial x} f(x,y)\,dy\,.$$

Theorem 0.18. *Suppose that $f(x) = \sum_{n=1}^{\infty} f_n(x)$. If the functions f_n are differentiable, and if the series $\sum_{n=1}^{\infty} f_n'(x)$ is uniformly convergent, then f is differentiable, and*

$$f'(x) = \sum_{n=1}^{\infty} f_n'(x)\,.$$

Theorem 0.19. (Dominated Convergence) *If a_{mn} is a double sequence, and A_n is such that $\lim_{m\to\infty} a_{mn} = A_n$ exists, and if there is a sequence M_n such that $|a_{mn}| \le M_n$ for all m, and $\sum_{n=1}^{\infty} M_n < \infty$, then*

$$\lim_{m\to\infty} \sum_{n=1}^{\infty} a_{mn} = \sum_{n=1}^{\infty} A_n\,.$$

Theorem 0.20. (Green's Theorem) *Suppose that* $P(x,y)$, $Q(x,y)$, $\frac{\partial}{\partial y}P(x,y)$ *and* $\frac{\partial}{\partial x}Q(x,y)$ *are continuous on and inside a simple closed curve* $\mathcal{C}$ *with interior* R. *Then*

$$\int_{\mathcal{C}} P\,dx + Q\,dy = \iint_R \left(\frac{\partial Q}{\partial x} - \frac{\partial P}{\partial y}\right) dA\,.$$

A discussion of summation by parts, and its application to Abel's Test and Dirichlet's Test, is found in §4.2.

0.3. Lebesgue measure theory

We begin by noting Littlewood's three principles:

- Every [measurable] set of real numbers is nearly a finite union of intervals.
- Every [measurable] function is nearly continuous.
- Every convergent sequence of [measurable] functions is nearly uniformly convergent.

Our first task is to formulate precise theorems that embody these principles.

Theorem 0.21. *Let* E *be a measurable set of real numbers. Then for every* $\varepsilon > 0$ *there is an open set* O *such that* $E \subseteq O$ *and* $meas(O \setminus E) < \varepsilon$. *Also, there is a closed set* F *such that* $F \subseteq E$ *and* $meas(E \setminus F) < \varepsilon$.

Theorem 0.22. *Continuous functions are dense in* $L^1([a,b])$ *in the sense that if* $f \in L^1([a,b])$ *and* $\varepsilon > 0$ *are given, then there is a continuous function* $g(x)$ *such that* $\int_a^b |f(x) - g(x)|\,dx < \varepsilon$.

For $1 \le p < \infty$, the same theorem holds for $L^p([a,b])$, for $L^p(\mathbb{T})$, and for $L^p(\mathbb{R})$.

Theorem 0.23. (Egorov's Theorem) *Let* f *and a sequence* $\{f_n\}$ *of functions all be defined on* $I = [a,b]$, *and suppose that* $\lim_{n\to\infty} f_n(x) = f(x)$ *for almost all* $x \in [a,b]$. *Then for any* $\varepsilon > 0$ *there is a set* $A \subseteq I$ *with* $meas(A) < \varepsilon$ *such that* $f_n(x)$ *tends uniformly to* $f(x)$ *as* $n \to \infty$, *for all* $x \in I \setminus A$.

Theorem 0.24. (Monotone Convergence Theorem) *Suppose that* $0 \le f_1(x) \le f_2(x) \le \cdots$ *for all* $x \in [a,b]$, *and that there is a function* f *such that* $\lim_{n\to\infty} f_n(x) = f(x)$ *for all* $x \in [a,b]$. *We assume that the functions* f_n *are measurable. Then*

$$\lim_{n\to\infty} \int_a^b f_n(x)\,dx = \int_a^b f(x)\,.$$

Theorem 0.25. (The Principle of Dominated Convergence) *Let* $f_1, f_2, \ldots$ *be measurable functions on* $[a,b]$ *with* $\lim_{n\to\infty} f_n(x) = f(x)$ *for all* $x \in [a,b]$. *Suppose that there is a function* $F(x)$ *such that* $F(x) \ge 0$ *for all* $x \in [a,b]$, *with* $\int_a^b F(x)\,dx < \infty$, *and such that* $|f_n(x)| \le F(x)$ *and* $|f(x)| \le F(x)$ *for all* n *and all* $x \in [a,b]$. *Then*

$$\int_a^b f(x)\,dx = \int_a^b \lim_{n\to\infty} f_n(x)\,dx = \lim_{n\to\infty} \int_a^b f_n(x)\,dx\,.$$

In the above, if the function $F(x)$ is taken to be a constant, then the result is known as the Bounded Convergence Theorem.

Just as the real numbers are complete, the spaces $L^p([a,b])$ are complete in the following sense.

Theorem 0.26. *Suppose that $1 \le p < \infty$, and that $f_1, f_2, \dots$ is a sequence of functions in $L^p([a,b])$ such that*

$$\lim_{\substack{m\to\infty \\ n\to\infty}} \|f_m - f_n\|_p \to 0\,.$$

Then there is an $f \in L^p([a,b])$ such that $\lim_{n\to\infty} \|f_n - f\|_p = 0$.

The above also holds for $L^p(\mathbb{T})$ and for $L^p(\mathbb{R})$.

Theorem 0.27. (Fubini's Theorem) *Let $I = [a,b]$, $J = [c,d]$, and let R be the rectangle $R = I \times J$. If $f(x,y)$ is a measurable function on R such that*

$$\int_R |f(x,y)|\, dA < \infty,$$

then

$$\int_R f(x,y)\, dA = \int_a^b \Big(\int_c^d f(x,y)\, dy \Big)\, dx = \int_c^d \Big(\int_a^b f(x,y)\, dx \Big)\, dy\,.$$

The full definition of the measure of a set is a little complicated, but sets of measure zero are easily described: If S is a set of real numbers, then S has measure 0 if for every $\varepsilon > 0$ there exist intervals $I_1, I_2, \dots$ such that

$$S \subseteq \bigcup_{n=1}^{\infty} I_n, \qquad \sum_{n=1}^{\infty} |I_n| < \varepsilon\,.$$

Theorem 0.28. *If f is a bounded function on a finite interval $[a,b]$, then f is Riemann-integrable on this interval if and only if the set of points $x \in [a,b]$ at which f is discontinuous is a set of Lebesgue measure 0.*

Theorem 0.29. *If f is Lebesgue integrable and $F(x) = \int_a^x f(u)\, du$, then F is continuous, and $F'(x) = f(x)$ for almost all x.*

In the above situation, we say that F is *absolutely continuous*. A continuous function may have derivative 0 almost everywhere and yet be nonconstant. Such a function is said to be *singular*.

If f is a Lebesgue integrable function, then we say that x is a *Lebesgue point* of f if

$$\lim_{h\to 0^+} \frac{1}{h} \int_{x-h}^{x+h} |f(u) - f(x)|\, du = 0\,.$$

Roughly, x is a Lebesgue point if $f(u)$ is on average near $f(x)$ when u is near x.

Theorem 0.30. *If f is a Lebesgue integrable function, then almost all x are Lebesgue points of f.*

Chapter 1

Complex Numbers

1.1. Basics

If x is a real number, then $x^2 \geq 0$. Since $-1 < 0$, it follows that -1 has no square root among the real numbers. However, the real numbers are contained in a larger system of numbers, the *complex numbers*, and among the complex numbers there is one, called i, with the property that $i^2 = -1$. The most general complex number may be written $x + iy$ where x and y are real. We visualize the complex numbers as forming a plane; thus the complex number $x + iy$ is plotted as the point (x, y) in the Cartesian plane. The sum of two complex numbers is

$$(x + iy) + (u + iv) = (x + u) + i(y + v)\,.$$

This is the same as vector addition:

$$\begin{bmatrix} x \\ y \end{bmatrix} + \begin{bmatrix} u \\ v \end{bmatrix} = \begin{bmatrix} x + u \\ y + v \end{bmatrix}.$$

The product of two complex numbers is

$$\begin{aligned}(x + iy)(u + iv) &= x(u + iv) + iy(u + iv) = xu + ixv + iyu + i^2 yv \\ &= (xu - yv) + i(xv + yu)\,.\end{aligned}$$

Finally, the reciprocal of a complex number is

$$\frac{1}{x + iy} = \frac{x - iy}{(x + iy)(x - iy)} = \frac{x - iy}{x^2 + y^2} = \frac{x}{x^2 + y^2} + i\frac{-y}{x^2 + y^2}$$

provided that x and y are not both zero. The complex number $0 + i0$ is the zero (i.e., additive identity) for the complex number system, and the number $1 = 1 + 0i$ is the multiplicative identity. It is easy to verify that with the basic operations formed this way, the familiar algebraic rules of commutativity, associativity, distributivity, etc. hold. The real numbers form a subset of the complex numbers, since $x + iy$ is real if $y = 0$. Thus for complex numbers, the x-axis is the real axis, and the y-axis is the imaginary axis.

One might well ask how we can be sure that the complex numbers really exist. In fact it is easy to construct the complex numbers, provided that one already knows that the real numbers exist. On pairs of real numbers we define

$$\begin{aligned}(x, y) + (u, v) &= (x + u, y + v),\\ (x, y) \cdot (u, v) &= (xu - yv, xv + yu),\\ (x, y)^{-1} &= \left(\frac{x}{x^2 + y^2}, \frac{-y}{x^2 + y^2}\right).\end{aligned}$$

The pair $(0, 0)$ is the additive identity, and the pair $(1, 0)$ is the multiplicative identity, and these pairs (x, y) *are* the complex numbers, although the notation is somewhat cumbersome.

If $z = x + iy$ is a complex number, then the *complex conjugate* of z, denoted $\overline{z}$, is $\overline{z} = x - iy$. In geometric terms, the complex conjugate is obtained by reflecting z across the real axis. Of course, $\overline{\overline{z}} = z$. The real number x is called the *real part* of z, and y is the *imaginary part* of z. In symbols, $x = \operatorname{Re} z$, $y = \operatorname{Im} z$. These two quantities can be expressed in terms of z and $\overline{z}$, since

$$\operatorname{Re} z = \frac{z + \overline{z}}{2}, \qquad \operatorname{Im} z = \frac{z - \overline{z}}{2i}. \tag{1.1}$$

In our calculation of the reciprocal of z, we already saw that $z\overline{z} = x^2 + y^2$. From our definition of the sum of two complex numbers it is obvious that $\overline{z + w} = \overline{z} + \overline{w}$. Perhaps a little less obviously, if $z = x + iy$ and $w = u + iv$, then

$$\overline{zw} = \overline{(xu - yv) + i(xv + yu)} = (xu - yv) - i(xv + yu) = (x - iy)(u - iv) = \overline{z}\,\overline{w}.$$

For real numbers we have a concept of absolute value. This can be extended to complex numbers by setting $|z| = \sqrt{x^2 + y^2}$. Thus $|z|$ is the distance of the point z in the complex plane from the origin. Note that this is consistent with the absolute value for real numbers, since if z is real (i.e., $y = 0$), then $|z| = \sqrt{x^2} = |x|$. Since $z\overline{z} = x^2 + y^2$, we see that $|z| = \sqrt{z\overline{z}}$. We note also that $|\overline{z}| = |z|$. The *modulus* of a complex number is the same as its absolute value. Thus if $|z| = 1$, then we say that z is "unimodular".

For real numbers we know that $|ab| = |a||b|$. This holds also for complex numbers, since

$$|zw| = \sqrt{zw\overline{zw}} = \sqrt{zw\overline{z}\,\overline{w}} = \sqrt{z\overline{z}\,w\overline{w}} = \sqrt{z\overline{z}}\sqrt{w\overline{w}} = |z||w|.$$

There is a second way of deriving this identity. If $z = x + iy$ and $w = u + iv$, then

$$\begin{aligned}|zw|^2 &= |(xu - yv) + i(xv + yu)|^2 = (xu - yv)^2 + (xv + yu)^2\\ &= x^2u^2 + y^2v^2 + x^2v^2 + y^2u^2 = (x^2 + y^2)(u^2 + v^2) = |z|^2|w|^2.\end{aligned}$$

We also note that $|x| = \sqrt{x^2} \le \sqrt{x^2 + y^2} = |z|$, and similarly for $|y|$. That is,

$$|\operatorname{Re} z| \le |z|, \qquad |\operatorname{Im} z| \le |z|. \tag{1.2}$$

For real numbers we have a triangle inequality, $|a + b| \le |a| + |b|$. There is also a triangle inequality for the complex numbers,

$$|z + w| \le |z| + |w|. \tag{1.3}$$

Indeed, the three points 0, z, and $w+z$ are the vertices of a triangle whose side-lengths are $|z|$, $|w|$, and $|z+w|$, so the triangle inequality for complex numbers is exactly the triangle inequality in plane Euclidean geometry. However, we can also prove the triangle inequality for complex numbers by direct calculation. If a and b are nonnegative real numbers, then $a \le b$ if and only if $a^2 \le b^2$. Hence the inequality $|z+w| \le |z| + |w|$ is equivalent to

$$|z+w|^2 \le (|z|+|w|)^2 = |z|^2 + 2|z||w| + |w|^2 .$$

Here the left hand side is $(z+w)(\overline{z+w}) = z\overline{z} + z\overline{w} + w\overline{z} + w\overline{w}$. But $z\overline{z} = |z|^2$, and $w\overline{w} = |w|^2$, and these quantities appear also on the right hand side above. Thus our problem is to show that

$$z\overline{w} + w\overline{z} \le 2|z||w| .$$

Let $t = z\overline{w}$. Then $\overline{t} = \overline{z\overline{w}} = \overline{z}\,\overline{\overline{w}} = \overline{z}w$. Hence the left hand side above is $t + \overline{t} = 2\operatorname{Re} t$, by (1.1). But by (1.2), $\operatorname{Re} t \le |t| = |z\overline{w}| = |z||\overline{w}| = |z||w|$, so our proof of (1.3) is complete.

By the triangle inequality we see that $|z| = |z+w-w| \le |z+w| + |w|$. On rearranging, this gives

$$|z+w| \ge |z| - |w|, \tag{1.4}$$

which is an alternate presentation of the triangle inequality. Which variant we use depends on the context—we make frequent use of both forms.

We have seen that the addition of complex numbers can be interpreted geometrically. The multiplication of complex numbers also has a geometric interpretation, as follows. Consider the line segment from 0 to $z = x + iy$ in the complex plane, and let θ denote the angle that this segment makes with the positive real axis. This angle θ is known as the *argument* of z; in symbols we write $\theta = \arg z$. Let $r = |z|$. Then

$$x = r\cos\theta, \qquad y = r\sin\theta .$$

Similarly, let $w = u + iv$ make an angle ϕ with the positive real axis, and put $R = |w|$, so that

$$u = R\cos\phi, \qquad v = R\sin\phi .$$

Then

$$\begin{aligned} zw &= (r\cos\theta + ir\sin\theta)(R\cos\phi + iR\sin\phi) \\ &= rR(\cos\theta\cos\phi - \sin\theta\sin\phi) + irR(\cos\theta\sin\phi + \sin\theta\cos\phi) . \end{aligned}$$

But, as noted in Appendix T,

$$\cos(\theta+\phi) = \cos\theta\cos\phi - \sin\theta\sin\phi, \tag{1.5}$$

$$\sin(\theta+\phi) = \cos\theta\sin\phi + \sin\theta\cos\phi, \tag{1.6}$$

so

$$zw = rR\cos(\theta+\phi) + irR\sin(\theta+\phi) .$$

That is, zw is a complex number with absolute value rR and argument $\theta+\phi$. Hence we have a third proof that $|zw| = |z||w|$, and the new information that

$$\arg(zw) = \arg z + \arg w .$$

Of course, in the special case that $w = 1/z$, the product is 1, and $\arg 1 = 0$, so $\arg(1/z) = -\theta$. Also, $|1/z| = 1/|z|$, and indeed

$$\frac{1}{z} = \frac{1}{r}\cos(-\theta) + \frac{i}{r}\sin(-\theta)\,.$$

The argument of a complex number is not uniquely determined, because an angle of 2π is the same as the angle 0. Indeed, if $\theta = \arg z$, then $\theta + 2k\pi$ is also the argument of z, for any integer k. There is no ambiguity beyond this, for if θ_1 and θ_2 are two different arguments of the same number z, then $\theta_2 - \theta_1$ must be an integral multiple of 2π. We express this by saying, "The argument of z is uniquely determined modulo 2π." For example, if $\arg z = \frac{3}{2}\pi$ and $\arg w = \frac{4}{3}\pi$, then $\arg zw = \frac{17}{6}\pi$, which is the same as saying that $\arg zw = \frac{5}{6}\pi$. The argument of 0 is undefined.

The number $\cos\theta + i\sin\theta$ has modulus 1 and argument θ. Since the modulus of a product is the product of the moduli, and the argument of a product is the sum of the arguments, it follows that $(\cos\theta + i\sin\theta)^n$ has modulus 1 and argument $n\theta$. That is,

$$(\cos\theta + i\sin\theta)^n = \cos n\theta + i\sin n\theta\,. \tag{1.7}$$

This is *de Moivre's Formula*, which dates from 1730.

In the first exercise below we encounter the concept of the sum of two sets. If $\mathcal{A}$ and $\mathcal{B}$ are two sets of objects from a domain on which a summation is defined, we define the *sum set* of $\mathcal{A}$ and $\mathcal{B}$, written $\mathcal{A} + \mathcal{B}$, to be the set of all numbers that can be written in the form $a + b$ with $a \in \mathcal{A}$ and $b \in \mathcal{B}$. In symbols,

$$\mathcal{A} + \mathcal{B} = \{a + b\,:\, a \in \mathcal{A}, b \in \mathcal{B}\}. \tag{1.8}$$

Exercises

1. Let z and w be complex numbers, put $R = |z|$ and $r = |w|$, and suppose that $0 < r < R$.
   (a) Use the triangle inequality in the two forms (1.3), (1.4) to show that $R - r \le |z + w| \le R + r$.
   (b) Let $\mathcal{A} = \{z \in \mathbb{C} : |z| = R\}$, $\mathcal{B} = \{w \in \mathbb{C} : |w| = r\}$. Show that $\mathcal{A} + \mathcal{B} = \{z \in \mathbb{C} : R - r \le |z| \le R + r\}$. (A set of this latter sort, namely the set of all points between two concentric circles, is known as an *annulus*.)
2. Suppose that $z = x + iy \ne 0$. Show that
$$\arg z = \begin{cases} \operatorname{Arctan} y/x & \text{if } x > 0, \\ \operatorname{Arctan} y/x \pm \pi & \text{if } x < 0, \\ \pi/2 & \text{if } x = 0 \text{ and } y > 0, \\ -\pi/2 & \text{if } x = 0 \text{ and } y < 0. \end{cases}$$
3. Suppose that z and w are complex numbers such that $|z| = |w|$. Show that $|z+s|^2 + |z-s|^2 = |w+s|^2 + |w-s|^2$ for all complex numbers s.
4. Express $(2+i)/(3+4i)$ in the form $x + iy$.

5. Suppose that $z = x + iy$ where x and y are real.
   (a) Express $\mathrm{Re}(z^2)$ and $\mathrm{Im}(z^2)$ in terms of x and y.
   (b) Explain why $\left(\mathrm{Re}(z^2)\right)^2 + \left(\mathrm{Im}(z^2)\right)^2 = (x^2 + y^2)^2$.
   (c) Note that if $z = 2 + i$, the above reads $3^2 + 4^2 = 5^2$. Make a different choice of z to obtain another Pythagorean triangle.

1.2. Euclidean geometry via complex numbers

We have already used properties of complex numbers to prove the triangle inequality (1.3), which is equivalent to the triangle inequality of plane geometry. Many further theorems of plane geometry can be easily derived by systematic use of complex numbers. We now consider a few examples to illustrate this point.

Example 1.1. To Show: The diagonals of a parallelogram intersect at their midpoints. To see why this is so, let a and b be complex numbers such that $0, a, b$ are not colinear in the complex plane. Thus $0, a, b, a + b$ form the vertices of a parallelogram, as depicted in Figure 1.1. For real t, the points $(1 - t)a + tb$ form a line. (Think of t as denoting time.) At time $t = 0$ we are at a. At time $t = 1$ we are at b. Indeed, with this parameterization of the line through a and b we are traveling with the uniform velocity $b - a$. We obtain the midpoint of the segment from a to b by taking $t = 1/2$. Thus the midpoint is $(a + b)/2$. We can similarly parameterize the line from 0 to $a + b$ by writing $(a + b)t$. We are at 0 when $t = 0$, and at $a + b$ when $t = 1$. The midpoint between is obtained by taking $t = 1/2$, and we see that it is also $(a + b)/2$. Thus the midpoint of the diagonal from a to b is also the midpoint of the diagonal from 0 to $a + b$.

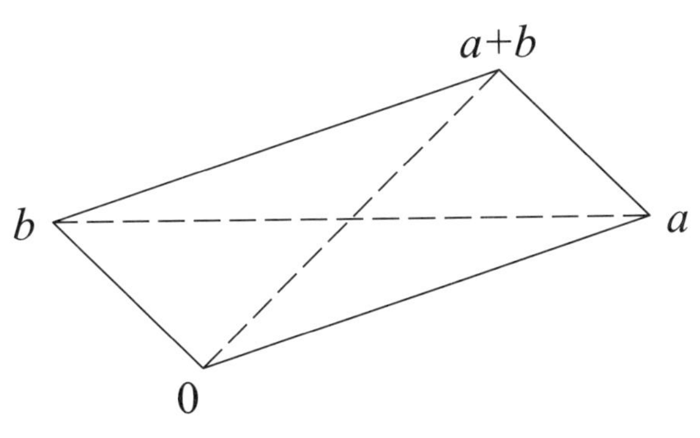

Figure 1.1. The parallelogram with vertices $0, a, a + b, b$, and its diagonals.

Example 1.2. The *Parallelogram Law* asserts that the sum of the squares of the lengths of the sides of a parallelogram is equal to the sum of the squares of the lengths of the two diagonals. To see why this is so, consider a parallelogram in the complex plane whose vertices are 0, a, $a + b$, and b, as in Figure 1.1. The side from 0 to a is parallel to the side from b to $a + b$; they have length $|a|$. The side from 0 to b is parallel to the side from a to $a+b$; they have length $|b|$. The diagonals run from 0 to $a+b$ and from a to b; they have lengths $|a+b|$ and $|a-b|$ respectively. Thus

$$\begin{aligned}|a+b|^2+|a-b|^2 &= (a+b)(\overline{a+b})+(a-b)(\overline{a-b})\\ &= (a+b)(\overline{a}+\overline{b})+(a-b)(\overline{a}-\overline{b})\\ &= a\overline{a}+a\overline{b}+b\overline{a}+b\overline{b}+a\overline{a}-a\overline{b}-b\overline{a}+b\overline{b}\\ &= 2a\overline{a}+2b\overline{b}=2|a|^2+2|b|^2,\end{aligned}$$

as claimed.

Example 1.3. To Show: The diagonals of a rhombus are perpendicular. We consider a rhombus with vertices $0, a, a+b, b$ where a is real and $b = a(\cos\theta + i\sin\theta)$, as in Figure 1.2. Thus the angle at 0 is θ. The ray from 0 through $a+b$ makes an angle $\arg(a+b)$ with the positive real axis. The ray from a through b makes an angle $\arg(b-a)$ with the positive real axis. We recall that $\arg z/w = \arg z - \arg w$. Thus to determine the difference between our two angles we divide:

$$\begin{aligned}\frac{a+b}{b-a} &= \frac{a(1+\cos\theta+i\sin\theta)}{a(\cos\theta+i\sin\theta-1)}\\ &= \frac{1+\cos\theta+i\sin\theta}{\cos\theta+i\sin\theta-1}\cdot\frac{\cos\theta-1-i\sin\theta}{\cos\theta-1-i\sin\theta}\\ &= \frac{\cos^2\theta-1+\sin^2\theta-i2\sin\theta}{(\cos\theta-1)^2+\sin^2\theta}\\ &= \frac{-2i\sin\theta}{2-2\cos\theta} = -i\frac{\sin\theta}{1-\cos\theta}\end{aligned}$$

since $\cos^2\theta+\sin^2\theta=1$. Since this quantity is purely imaginary, its argument is $\pm\pi/2$, which is to say that the two diagonals are perpendicular.

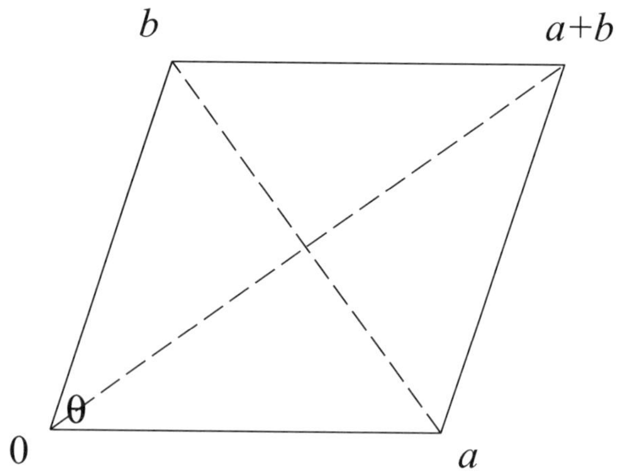

Figure 1.2. The rhombus with vertices $0, a, a+b, b$, and its diagonals.

We conclude this section with a much more ambitious undertaking.

Example 1.4. Napoleon's Theorem: Let $\triangle abc$ be an arbitrary triangle. Let the point C be chosen so that the triangle $\triangle abC$ is an equilateral triangle with c and C on opposite sides of the line through a and b. Similarly, let A be chosen so that $\triangle bcA$ is an equilateral triangle with a and A on opposite sides of the line through b and c. Also, let B be chosen so that $\triangle caB$ is an equilateral triangle with b and B on opposite sides of the line through a and c. Finally, let α be the centroid of triangle $\triangle bcA$, β be the centroid of $\triangle caB$, and γ the centroid of $\triangle abC$. Then $\triangle\alpha\beta\gamma$ is an equilateral triangle, as depicted in Figure 1.3.

While this is known as Napoleon's Theorem, and Napoleon *was* a geometry buff, historians have determined that this theorem was proved first not by Napoleon, but rather by one of his generals. Proving Napoleon's theorem by tools of classical geometry would seem to be quite a formidable problem, but when we use complex numbers it becomes just a matter of a simple algebraic calculation—no cleverness required.

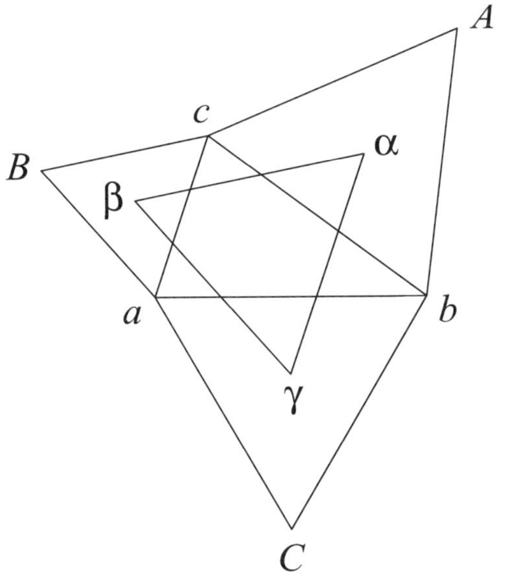

Figure 1.3. The triangle $\triangle abc$, and equilateral triangles $\triangle abC$, $\triangle bcA$, $\triangle caB$, $\triangle\alpha\beta\gamma$.

We first determine α, β, γ in terms of a, b, c. The midpoint of the segment from a to b is $(a+b)/2$. To construct a perpendicular line through this point we multiply $b-a$ by $-i$. Since $\arg -i = -\pi/2$, the line $(a+b)/2 - i(b-a)t$ is the desired line, with positive values of t corresponding to moving away from $\triangle abc$. We recall that an equilateral triangle with side length s has height $\frac{1}{2}\sqrt{3}s$. Thus

$$C = \frac{a+b}{2} - i\frac{\sqrt{3}}{2}(b-a)\,.$$

We also recall that the centroid of an equilateral triangle lies one-third of the way from the midpoint of the base, $(a+b)/2$ to the top, C of the triangle. Thus

$$\gamma = \frac{a+b}{2} - i\frac{\sqrt{3}}{6}(b-a) = \Big(\frac{1}{2} + \frac{i\sqrt{3}}{6}\Big)a + \Big(\frac{1}{2} - \frac{i\sqrt{3}}{6}\Big)b\,.$$

Similarly,

$$\begin{aligned} \alpha &= \Big(\frac{1}{2} + \frac{i\sqrt{3}}{6}\Big)b + \Big(\frac{1}{2} - \frac{i\sqrt{3}}{6}\Big)c, \\ \beta &= \Big(\frac{1}{2} + \frac{i\sqrt{3}}{6}\Big)c + \Big(\frac{1}{2} - \frac{i\sqrt{3}}{6}\Big)a\,. \end{aligned}$$

In $\triangle\alpha\beta\gamma$, the (directed) edge from α to β corresponds to the complex number $\beta - \alpha$. From the above equations we see that

$$\beta - \alpha = \Big(\frac{1}{2} - \frac{i\sqrt{3}}{6}\Big)a - \Big(\frac{1}{2} + \frac{i\sqrt{3}}{6}\Big)b + \frac{i\sqrt{3}}{3}c\,.$$

Similarly, for the other two edges,

$$\begin{aligned} \gamma - \beta &= \frac{i\sqrt{3}}{3}a + \Big(\frac{1}{2} - \frac{i\sqrt{3}}{6}\Big)b - \Big(\frac{1}{2} + \frac{i\sqrt{3}}{6}\Big)c, \\ \alpha - \gamma &= -\Big(\frac{1}{2} + \frac{i\sqrt{3}}{6}\Big)a + \frac{i\sqrt{3}}{3}b + \Big(\frac{1}{2} - \frac{i\sqrt{3}}{6}\Big)c\,. \end{aligned}$$

We note that the coefficients in these equations are the same, apart from a cyclic permutation from one equation to the next. We proceed from α to β in the direction $\beta-\alpha$. If we are going to create an equilateral triangle, then at β we need to turn in the counter-clockwise direction through an angle of 120° ($= 2\pi/3$). To accomplish this, we multiply $\beta - \alpha$ by a unimodular number with argument $2\pi/3$, which is to say by the number $\omega = (-1 + i\sqrt{3})/2$. To prepare for this, we first compute the product of ω times each of the coefficients. By direct computation (i.e., by using the definition of the product of two complex numbers) we find that

$$\Big(\frac{1}{2} - \frac{i\sqrt{3}}{6}\Big)\omega = \frac{i\sqrt{3}}{3}, \qquad -\Big(\frac{1}{2} + \frac{i\sqrt{3}}{6}\Big)\omega = \frac{1}{2} - \frac{i\sqrt{3}}{6}, \qquad \frac{i\sqrt{3}}{3}\omega = -\Big(\frac{1}{2} + \frac{i\sqrt{3}}{6}\Big)\,.$$

Hence

$$(\beta-\alpha)\omega = \gamma - \beta, \qquad (\gamma-\beta)\omega = \alpha - \gamma, \qquad (\alpha-\gamma)\omega = \beta - \alpha\,.$$

Since $|\omega| = 1$, it follows that $|\beta-\alpha| = |\gamma-\beta| = |\alpha-\gamma|$. As these are the side-lengths of $\triangle\alpha\beta\gamma$, we conclude that the triangle is equilateral. While the proof is complete, it is still instructive to note that

$$\begin{aligned} \beta - \alpha &= \frac{i\sqrt{3}}{3}\big(\omega^2 a + \omega b + c\big), \\ \gamma - \beta &= \frac{i\sqrt{3}}{3}\omega^2\big(\omega^2 a + \omega b + c\big), \\ \alpha - \gamma &= \frac{i\sqrt{3}}{3}\omega\big(\omega^2 a + \omega b + c\big)\,. \end{aligned}$$

Thus the sides of $\triangle\alpha\beta\gamma$ are all of length $|\omega^2 a + \omega b + c|/\sqrt{3}$.

Exercises

1. Let ω be defined as in Example 1.4. Suppose that x, y, z are real numbers such that $x + \omega y + \omega^2 z = 0$. Show that $x = y = z$.
2. In the context of Example 1.4, define C' in the same way that C is defined, but with C' on the *same* side of the line through a and b as c. Define A' and B' similarly. Let α' be the centroid of the equilateral triangle $\triangle bcA'$, and define β' and γ' similarly. Show that $\triangle\alpha'\beta'\gamma'$ is equilateral.

1.3. Polynomials

Polynomials with complex coefficients work much like those with real coefficients. We can add and multiply polynomials as before, and we can divide one polynomial into another to obtain a quotient and remainder. Thus if $f(z)$ and $g(z)$ are given polynomials, and f is not identically 0, then there exists a unique quotient polynomial $q(z)$ and remainder polynomial $r(z)$ such that

$$\frac{g(z)}{f(z)} = q(z) + \frac{r(z)}{f(z)}$$

where r has degree strictly less than that of f. On multiplying through by $f(z)$, we see that the above is equivalent to the identity

$$g(z) = q(z)f(z) + r(z)\,.$$

Take $f(z) = z - z_1$ where z_1 is some complex number. Since f has degree 1, and r must have a strictly smaller degree, it follows that $r(z)$ is a constant. That is,

$$g(z) = q(z)(z - z_1) + r$$

for some number r. This equation holds for all z, and in particular when $z = z_1$. Thus $g(z_1) = q(z_1)(z_1 - z_1) + r = r$. In particular, $g(z_1) = 0$ if and only if $r = 0$. To summarize: If

$$g(z) = q(z)(z - z_1) \tag{1.9}$$

for all z, then $g(z_1) = 0$, and conversely, if $g(z_1) = 0$, then g can be written in the form (1.9). All this is true for polynomials with real coefficients and real z, but it is just as true for polynomials with complex coefficients and complex z. (Note that a complex coefficient may actually be a real number.) Thus roots of a polynomial correspond precisely to its linear factors. Of course a linear factor may be repeated, in which case we have a multiple root. Since each linear factor has degree 1, and each such factor increases the degree of f by 1, the total number of linear factors cannot exceed the degree of g. That is, the number of roots of a polynomial, counting multiplicity, does not exceed the degree of that polynomial.

A polynomial such as $z^2 - 5z + 6$ has two real roots, and factors completely, $z^2 - 5z + 6 = (z - 2)(z - 3)$, but $z^2 + 1$ has no real roots and no linear factors if one is using only real numbers. However, if complex numbers are allowed, then we may write $z^2 + 1 = (z - i)(z + i)$, and we see that this polynomial has two roots, namely i and $-i$. Here we encounter a major difference between polynomials over the reals and over the complex numbers. A polynomial of degree n, with real or complex coefficients, *always* factors into n linear factors $z - z_j$, where the roots z_j are complex numbers. Some of the z_j may be real, possibly even all of them. This important result is known as

Theorem 1.1. (The Fundamental Theorem of Algebra) *Let*

$$f(z) = a_n z^n + a_{n-1} z^{n-1} + \cdots + a_1 z + a_0$$

be a polynomial with real or complex coefficients and $a_n \neq 0$. *Then there exist complex numbers* $z_1, z_2, \ldots, z_n$ *such that*

$$f(z) = a_n (z - z_1)(z - z_2) \cdots (z - z_n)$$

identically.

This may be proved in many ways; we provide two proofs.

Analytic Proof. If f has a root, say z_1, then we can write $f(z) = q(z)(z - z_1)$, and then turn our attention to $q(z)$. If $q(z)$ has a root, we can repeat this process. As long as we have roots, we continue. The only way that a problem can arise would be to encounter a polynomial of degree ≥ 1 with no root. Suppose that $f(z)$ is such a polynomial. We note that $|a_n| r^n$ tends to infinity as $r \to \infty$ faster than

$$|a_{n-1}| r^{n-1} + |a_{n-2}| r^{n-2} + \cdots + |a_1| r + 2|a_0| .$$

Thus there is an R such that

$$|a_n| r^n > |a_{n-1}| r^{n-1} + |a_{n-2}| r^{n-2} + \cdots + |a_1| r + 2|a_0|$$

for all $r \geq R$. By the triangle inequality in the form (1.4) it follows that if $|z| = r \geq R$, then $|f(z)| \geq |a_n| r^n - (|a_{n-1}| r^{n-1} + |a_{n-2}| r^{n-2} + \cdots + |a_0|) > |a_0|$. From elementary calculus we know that a continuous function on a closed bounded interval is not only bounded, but actually attains its maximum and minimum values. Now $|f(z)|$ is a continuous function in the complex plane, and the disc $|z| \leq R$ is closed and bounded, so it follows that there is a point z_0 in this disk at which $|f(z)|$ attains its minimum value. We note that $|f(0)| = |a_0|$, so the minimum value in question is $\leq |a_0|$. Since $|f(z)| > |a_0|$ when $|z| = R$, the minimizing point z_0 does not lie on the bounding circle $|z| = R$, but rather in its interior. We note that $|f(z_0)| > 0$, since $f(z)$ has no root. By expanding the monomials $a_k(z_0 + w)^k$, we may write $f(z_0 + w)$ in terms of powers of w,

$$f(z_0 + w) = c_0 + c_1 w + \cdots + c_n w^n .$$

Suppose that $|w|$ is small, say $w = \delta \cos\theta + i\delta \sin\theta$. If δ is sufficiently small, then $|c_1|\delta$ is much larger than $|c_2|\delta^2 + \cdots + |c_n|\delta^n$. As θ runs from 0 to 2π, the point $c_0 + c_1 w$ describes a circle of radius $|c_1|\delta$, centered at $c_0 = f(z_0)$. Thus there is a θ for which $|c_0 + c_1 w| = |c_0| - |c_1|\delta$. Hence $|f(z_0 + w)| < |f(z_0)|$, a contradiction. Thus the assumption that f has no root is untenable, and the proof is complete. □

The above argument overlooks a small detail: It might be the case that $c_1 = 0$. If this were to happen, then we would work instead with $c_2 w^2$, and note that as θ runs from 0 to 2π, the quantity $c_0 + c_2 w^2$ moves two times around a circle of radius $|c_2|\delta^2$ centered at c_0. Thus there is a θ for which $|c_0 + c_2 w^2| = |c_0| - |c_2|\delta^2$.

If $c_1 = c_2 = 0$, then we would work with c_3, and so on. The numbers $c_1, c_2, \ldots, c_n$ cannot all be 0, since $c_n = a_n \neq 0$.

In §0.1 we described the roots of a quadratic polynomial. Similar formulas exist for cubic and quartic polynomials, but for polynomials of degree 5 and higher there is in general no simple formula for the roots. The roots certainly exist, and we can compute good numerical approximations of their values, but in general they cannot be expressed in closed form by means of radicals.

Put $z = r\cos\theta + ir\sin\theta$. As θ runs from 0 to 2π, the number $z + c$ traverses one complete circuit of the circle of radius r centered at c. If $|c| < r$, then $\arg(z+c)$ increases by 2π as this happens. If $|c| > r$, then $\arg(z + c)$ oscillates, but returns to its original value, as θ runs from 0 to 2π. If $f(z) = a_n \prod_{j=1}^{n}(z - z_j)$, then

$$\arg f(z) = \arg a_n + \sum_{j=1}^{n} \arg(z - z_j)\,.$$

This is true for arbitrary z, but if we put $z = r\cos\theta + ir\sin\theta$ and let θ run from 0 to 2π, we find that $\arg f(z)$ increases by $2\pi k$ where k is the number of roots z_j for which $|z_j| < r$. By calculating this change in argument for various r we can determine (approximately) the absolute values of the roots of f.

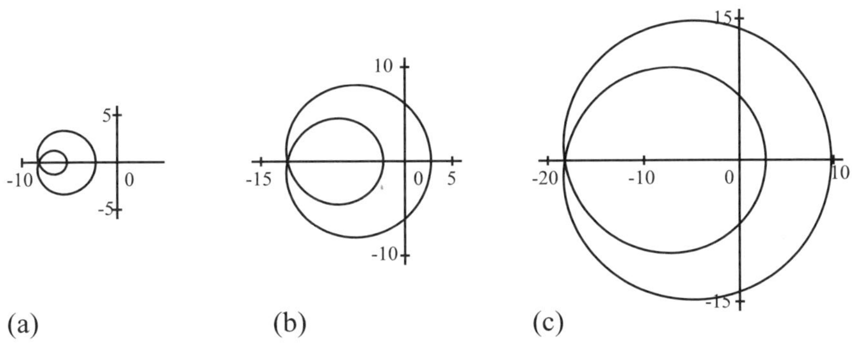

Figure 1.4. Graph of $f(z) = z^2 - z - 6$ for $z = r\cos\theta + ir\sin\theta$, $0 \le \theta \le 2\pi$; (a) $r = 1.5$, (b) $r = 2.5$, (c) $r = 3.5$.

Let $f(z) = z^2 - z - 6$. In Figure 1.4 we see that $f(z)$ has no root of modulus < 1.5, one root with modulus < 2.5, and two roots with modulus < 3.5. In fact the roots are at 3 and -2. This way of counting zeros by noting the increase in argument of a polynomial suggests that the Fundamental Theorem of Algebra can be established by means of a

Topological proof. As in the first proof, we need only to establish the existence of at least one root. Let $\mathcal{C}(r)$ denote the closed curve $\mathcal{C}(r) = \{f(r(\cos\theta + i\sin\theta)) : 0 \le \theta \le 2\pi\}$, and let

$$\delta(r) = \min_{|z|=r} |f(z)|.$$

Thus $\delta(r) = 0$ if and only if f has a root z_j such that $|z_j| = r$. The function $\delta(r)$ is a continuous function of r, and (as in the first proof), $\delta(r) \asymp r^n$ for all sufficiently large r, say $r \geq R$. Now $\delta(r)$ must attain its minimum in the interval $[0, R]$, and if f has no root, then this number is positive; call it $m > 0$. For a given r, let $\arg f(r(\cos\theta + i\sin\theta))$ be a continuous function of θ. The total change in argument will be of the form $2\pi k(r)$ where $k(r)$ is an integer. Since the curves $\mathcal{C}(r)$ are bounded away from the origin, a small change in r produces a small change in argument, and hence a small change in $k(r)$. That is, $k(r)$ is a continuous function of r. But $k(r)$ is an integer, and so $k(r)$ is constant. When r is small, the constant term a_0 dominates, so $k(r) = 0$ for all sufficiently small r. (As in the first proof, we may assume that $a_0 \neq 0$.) If r is large, then the leading term dominates, and so $k(r) = n$ for all sufficiently large r. We have reached a contradiction, based on the assumption that f has no root. □

For a collection of applications of the Fundamental Theorem of Algebra see Appendix F.

Exercises

1. Let

$$P(z) = a_n z^n + a_{n-1} z^{n-1} + \cdots + a_0,$$
$$Q(z) = b_n z^n + b_{n-1} z^{n-1} + \cdots + b_0.$$

Also let $z_1, z_2, \ldots, z_{n+1}$ denote $n+1$ distinct real or complex numbers. Suppose that $P(z_j) = Q(z_j)$ for $j = 1, 2, \ldots, n+1$. Show that $a_k = b_k$ for all k.

2. Let

$$f(z) = a_n z^n + \cdots + a_0$$

be a polynomial with arbitrary real or complex coefficients. Suppose that f has the property that $f(x)$ is real for every real number x. Show that all the coefficients of f are real.
Hint: Put

$$F(z) = \sum_{j=0}^{n} (\operatorname{Re} a_j) z^j,$$
$$G(z) = \sum_{j=0}^{n} (\operatorname{Im} a_j) z^j,$$

so that $f(z) = F(z) + iG(z)$ identically.

3. Throughout this exercise, assume that $f(z)$ is a polynomial with real coefficients.
   (a) Show that $\overline{f(z)} = f(\overline{z})$.
   (b) Deduce that if $f(z_1) = 0$, then $f(\overline{z_1}) = 0$.
   (c) Let $x_1, \ldots, x_r$ be the real roots of f, and write $n = r + 2s$. Show that there exist real numbers b_k and c_k with $b_k^2 - 4c_k < 0$ such that

$$f(z) = a_n \prod_{j=1}^{r} (z - x_j) \prod_{k=1}^{s} (z^2 + b_k z + c_k)$$

identically. (This is the factorization of f over the real number system.)

4. Throughout this exercise, assume that $f(z) = a_n z^n + a_{n-1} z^{n-1} + \cdots + a_0$ is a polynomial of degree n such that $f(x) \geq 0$ for all real x.
   (a) Show that $a_n > 0$.
   (b) Show that n is even.
   (c) Show that there is a polynomial g of degree $n/2$ with complex coefficients such that $f(x) = |g(x)|^2$ for all real x.
   (d) Show that there exist polynomials $h(z)$ and $j(z)$ with real coefficients and degrees not exceeding $n/2$ such that $f(x) = h(x)^2 + j(x)^2$ for all real x.
5. Let $f(z) = 2z^4 - 2z^3 + z^2 + 2z + 2$.
   (a) Plot $f(x)$ for $-2 \leq x \leq 2$, to confirm that f has no real roots. In Maple, this could be done by typing the commands `with(plots);` [Return]

   `f := x -> 2*x^4 - 2*x^3 + x^2 + 2*x + 2;` [Return]
   `plot(f(x),x=-2..2);` [Return]

   (b) For $r = .5$, plot $f(z)$ where $z = r\cos t + ir\sin t$. In Maple, this can be done by typing `r := 0.5;` [Return]

   `plot([Re(f(r*cos(t)+I*r*sin(t))),Im(f(r*cos(t)+I*r*sin(t))),`
   `t=0..2*Pi], scaling=CONSTRAINED);` [Return]

   Thus confirm that f has no root in the disk $|z| \leq .5$.
   (c) Repeat part (b) with $r = 1$. How many roots does f have with modulus < 1?
   (d) Repeat part (b) with $r = 1.5$. How many roots does f have with modulus < 1.5?
   (e) By repeatedly adjusting the value of r and replotting, determine (approximately) the moduli of the roots of f.
   (f) With r set to a value that is approximately the modulus of a root, list the values of $f(r\cos t + ir\sin t)$ for $t = 2\pi n/100$ with $n = 1, 2, \ldots, 100$. In Maple: `for n from 1 to 100 do`

   `print(n, evalf(f(r*cos(2*Pi*n/100)+I*r*sin(2*Pi*n/100)))) od;`
   [Return]

   (g) By noting the value of n for which $|f(z)|$ is smallest, derive approximations to the roots of f.

1.4. Power series

Let

$$f(z) = \sum_{n=0}^{\infty} a_n z^n \tag{1.10}$$

be a given power series, and let R be a number chosen so that $|a_n|/(R+\varepsilon)^n$ is a bounded sequence but $|a_n|/(R-\varepsilon)^n$ is not, for all $\varepsilon > 0$. There always exists such an R, and it is unique. By comparison with a geometric series we deduce that the power series is absolutely convergent if $|z| < R$, and that it is divergent when $|z| > R$, because the terms do not tend to 0. The region of convergence is thus an interval from $-R$ to R, possibly including one or the other of the endpoints. This is the reasoning used on the real line in calculus, but it works also in the complex plane. The region of convergence consists of an open disc $|z| < R$, where the series is absolutely convergent, together with possibly some of the points on the circle of convergence $|z| = R$. Of course, as in the calculus context, we may have $R = 0$ or $R = \infty$, depending on the size of the coefficients.

Let D denote an open set in the complex plane. If $a \in D$ and if

$$\lim_{z\to a} \frac{f(z) - f(a)}{z - a} \tag{1.11}$$

exists and is finite, then we call this limit $f'(a)$. If this limit exists for all $a \in D$, then we say that f is *analytic* in D. Certainly a polynomial is analytic throughout the complex plane, but furthermore, if f is defined by the power series (1.10) with radius of convergence R, then by a routine application of Theorem 0.18 the power series may be differentiated term-by-term, so that

$$f'(z) = \sum_{n=1}^{\infty} n a_n z^{n-1}$$

for $|z| < R$ and hence f is analytic in this disc. If $R = \infty$, then f is analytic throughout the complex plane; such a function is said to be *entire*. It is often the case that a function f defined by a power series of finite radius of convergence R can be extended analytically to a larger domain than the disc $|z| < R$. For example, if $f(z) = 1 + z + z^2 + z^3 + \cdots$ for $|z| < 1$, then $f(z) = 1/(1-z)$ in this disc. However, this latter formula defines f for all $z \neq 1$, and it is easy to show (see Exercise 1) that $1/(1-z)$ is analytic whenever $z \neq 1$.

The further discussion of power series in the complex plane is a matter of complex analysis, and therefore outside our scope, but there is one particular power series of great interest to us, namely the exponential function

$$e^z = \sum_{n=0}^{\infty} \frac{z^n}{n!}. \tag{1.12}$$

This power series converges for all z, because $n!$ tends to infinity faster than R^n, no matter how large R is. Thus e^z is an entire function. This power series can be used to define the exponential function on the real line, but it can likewise be used to extend this definition to the complex plane. One of the main properties of the exponential function for real variables is that $e^{a+b} = e^a e^b$. We show now that this also is true for complex arguments. First,

$$e^{z+w} = \sum_{n=0}^{\infty} \frac{(z+w)^n}{n!}.$$

We expand $(z+w)^n$ by means of the binomial theorem, in the form given in (B.8). Thus the above is

$$= \sum_{n=0}^{\infty} \frac{1}{n!} \sum_{k=0}^{n} \binom{n}{k} z^k w^{n-k}.$$

By (B.1) we know that

$$\binom{n}{k} = \frac{n!}{k!(n-k)!}\,.$$

We invert the order of summation (which is justified by the absolute convergence of everything in sight—recall Theorem 0.11) to see that

$$e^{z+w} = \sum_{k=0}^{\infty} \frac{z^k}{k!} \sum_{n=k}^{\infty} \frac{w^{n-k}}{(n-k)!}\,.$$

Put $m = n - k$, and reindex with m in place of n. Thus the inner sum above is

$$\sum_{m=0}^{\infty} \frac{w^m}{m!} = e^w$$

which does not depend on k, so we have the desired result:

$$e^{z+w} = e^z e^w\,. \tag{1.13}$$

In particular, $e^{x+iy} = e^x e^{iy}$. Here the second factor is of particular interest. If n is even, then i^n is ± 1, while if n is odd, then i^n is $\pm i$. Thus

$$\begin{aligned} e^{iy} &= \sum_{n=0}^{\infty} \frac{(iy)^n}{n!} = \sum_{m=0}^{\infty} \frac{(iy)^{2m}}{(2m)!} + \sum_{m=0}^{\infty} \frac{(iy)^{2m+1}}{(2m+1)!} \\ &= \sum_{m=0}^{\infty} \frac{(-1)^m y^{2m}}{(2m)!} + i \sum_{m=0}^{\infty} \frac{(-1)^m y^{2m+1}}{(2m+1)!}\,. \end{aligned}$$

Here we recognize the power series of $\cos y$ and of $\sin y$. Thus

$$e^{iy} = \cos y + i \sin y\,. \tag{1.14}$$

Euler proved this in 1747. He drew particular attention to the special case $y = \pi$:

$$1 + e^{i\pi} = 0\,. \tag{1.15}$$

This is an identity linking the five most fundamental constants of mathematics. From Euler's identity (1.14) with y replaced by $-y$ we see that

$$e^{-iy} = \cos y - i \sin y\,. \tag{1.16}$$

We add (1.14) and (1.16) and divide by 2 to solve for $\cos y$, and subtract (1.16) from (1.14) and divide by $2i$ to solve for $\sin y$. Thus

$$\cos y = \frac{e^{iy} + e^{-iy}}{2}, \qquad \sin y = \frac{e^{iy} - e^{-iy}}{2i}\,. \tag{1.17}$$

By comparing (1.14) and (1.16) we see that $\overline{e^{iy}} = e^{-iy}$ when y is real. Thus

$$|e^{iy}|^2 = e^{iy}\overline{e^{iy}} = e^{iy}e^{-iy}$$

which by (1.13) is

$$= e^{iy-iy} = e^0 = 1\,.$$

From (1.14) we deduce that $\left|e^{iy}\right|^2 = (\cos y)^2 + (\sin y)^2$, and so $(\sin y)^2 + (\cos y)^2 = 1$. In trigonometry this is just Pythagoras's theorem, but it is reassuring that we can derive this from the definition (1.17) of sine and cosine.

With trigonometry in mind, we see from (1.14) that the segment from 0 to e^{iy} makes an angle y with the positive real axis, which is to say that $\arg e^{iy} = y$. Since $\cos y$ and $\sin y$ both have period 2π, it follows that e^{iy} also has period 2π. Indeed, if $p(t) = \cos t + i \sin t$ is our position at time t, then the velocity of this motion is $v(t) = p'(t) = -\sin t + i \cos t = ip(t)$. Hence the speed of this motion is $|v(t)| = 1$ for all t, and the direction of the motion is counter-clockwise. Since the arc length of a unit circle is 2π, the quantity e^{it} runs once around the unit circle, at unit speed, as t runs through an interval of length 2π, say $[a, a + 2\pi]$. In particular,

$$e^{2\pi ix} = 1 \text{ if and only if } x \text{ is an integer.} \tag{1.18}$$

de Moivre's Formula (1.7) is especially transparent in the present context, since by (1.14),

$$(\cos\theta + i\sin\theta)^n = \left(e^{i\theta}\right)^n = e^{i\theta} \times e^{i\theta} \times \cdots \times e^{i\theta}.$$

By repeated use of (1.13), this is $e^{in\theta}$, which by (1.14) is $\cos n\theta + i \sin n\theta$.

Complex numbers z for which $z^n = 1$ are of great importance to us, especially in the next chapter. They are called n^{th} roots of unity. Write $z = re^{i\theta}$. Then $z^n = r^n e^{in\theta}$. Now $|z^n| = r^n = 1$ only if $r = 1$. Also, $e^{in\theta} = 1$ only if $n\theta$ is an integral multiple of 2π, which happens precisely when θ is of the form $\theta = 2\pi k/n$ for some integer k. Thus the nth roots of unity are the numbers $e^{2\pi ik/n}$. If k_1 differs from k_2 by an integral multiple of n, then k_1/n differs from k_2/n by an integer, and then $e^{2\pi ik_1/n} = e^{2\pi ik_2/n}$. On the other hand, if $0 \le k_1 < k_2 < n$, then $e^{2\pi ik_1/n} \ne e^{2\pi ik_2/n}$ since the first number has argument $2\pi k_1/n$ and the second number has argument $2\pi k_2/n$. To summarize, there are exactly n n^{th} roots of unity. They form the vertices of a regular n-gon on the unit circle in the complex plane. By the Fundamental Theorem of Algebra (Theorem 1.1), the polynomial $z^n - 1$ can be written as a product over its zeros. Since we have identified these zeros, it follows that

$$z^n - 1 = \prod_{k=0}^{n-1} \left(z - e^{2\pi ik/n}\right). \tag{1.19}$$

It is worth noting that an nth root of unity may already have been a dth root of unity for some $d < n$. For example, $z = 1$ is an nth root of unity for all n. Also $z = -1$ is an nth root of unity whenever n is even. In general, a dth root of unity is an nth root of unity if and only if d divides n. Of course n/d exists as a rational number for any pair of integers with $d \ne 0$, but when we say that 'd divides n' we mean that n is a multiple of d, so that n/d is an integer. In symbols, we express this by writing $d|n$.

In §1.1 we used the familiar formulæ (1.5) for the cosine and sine of the sum of two angles. These formulæ follow easily from (1.13) and (1.14), since

$$\begin{aligned}\cos(\theta+\phi) + i\sin(\theta+\phi) &= e^{i(\theta+\phi)} = e^{i\theta}e^{i\phi} = (\cos\theta + i\sin\theta)(\cos\phi + i\sin\phi)\\ &= (\cos\theta\cos\phi - \sin\theta\sin\phi) + i(\sin\theta\cos\phi + \cos\theta\sin\phi).\end{aligned}$$

The formulæ (1.5) now follow by equating real and imaginary parts.

We note that the factorization (1.19) has the following nice application to trigonometry.

Theorem 1.2. *Let n be a positive integer. Then the identity*

$$\prod_{k=0}^{n-1} \sin \pi(x + k/n) = \frac{\sin \pi n x}{2^{n-1}}$$

holds for all x.

Proof. By (1.17) the left hand side is

$$\prod_{k=0}^{n-1} \frac{e^{i\pi(x+k/n)} - e^{-i\pi(x+k/n)}}{2i}\,.$$

To separate $e^{i\pi x}$ from $e^{i\pi k/n}$, we simultaneously multiply and divide by $e^{i\pi(x-k/n)}$. Thus the above is

$$= \prod_{k=0}^{n-1} e^{-\pi i(x-k/n)} \frac{e^{2\pi i x} - e^{-2\pi i k/n}}{2i}\,.$$

From (0.1) we know that $\sum_{k=0}^{n-1} k = n(n-1)/2$, so it follows that $\prod_{k=0}^{n-1} e^{\pi i k/n} = e^{\pi i(n-1)/2} = i^{n-1}$. Hence the above is

$$= \frac{i^{n-1} e^{-\pi i n x}}{(2i)^n} \prod_{k=0}^{n-1} \left(e^{2\pi i x} - e^{-2\pi i k/n}\right).$$

On taking $z = e^{2\pi i x}$ in (1.19), we see that the product above is $z^n - 1 = e^{2\pi i n x} - 1$. Hence the above is

$$= \frac{1}{2^{n-1}} \cdot \frac{e^{\pi i n x} - e^{-\pi i n x}}{2i} = \frac{\sin \pi n x}{2^{n-1}}$$

by (1.17), and the proof is complete. □

For real numbers, the exponential has an inverse, the logarithm, which is defined for positive real numbers. For complex numbers, we might start with a number $z = x + iy$ and ask whether there is a number $w = u + iv$ such that $e^w = z$. Since $e^w = e^{u+iv} = e^u e^{iv}$, we deduce that $e^u = |z|$ and $v = \arg z$. The first of these equations has a unique solution, $u = \log |z|$ provided that $z \ne 0$. Thus if $z \ne 0$, then

$$\log z = \log |z| + i \arg z\,.$$

Since the argument is determined only modulo 2π, it follows that the logarithm is defined only up to integral multiples of $2\pi i$. One could solve this problem by insisting that $0 \le \arg z < 2\pi$ or all $z \ne 0$, but this gives rise to jump discontinuities when z traverses a path that wraps around the origin. In such situations it is preferable to make $\log z$ locally continuous, even though this may mean that $\operatorname{Im} \log z$ may wander outside the interval $[0, 2\pi)$.

The quantity z^n is simply z multiplied by itself n times if n is a positive integer, and by taking reciprocals we can define z^{-n}. We can extend this definition to z^a for arbitrary complex a by putting

$$z^a = e^{a \log z}\,.$$

Since there are multiple possible values of $\log z$, there are correspondingly multiple values for z^a, unless a is an integer.

Example 1.5. If a and b are nonnegative real numbers, then $\sqrt{a}\sqrt{b} = \sqrt{ab}$. However, a problem arises if a and b are complex. In particular, we note that $\sqrt{-1}\sqrt{-1} = i \cdot i = -1$, but $\sqrt{-1 \cdot -1} = \sqrt{1} = 1$. The explanation for this is that when dealing with complex numbers, $\sqrt{a}$ refers to one of the square roots of a, with neither of them being favored over the other. Thus $\sqrt{-1}$ refers to i or $-i$, and $(\pm i)(\pm i) = \pm 1$, which are the two square roots of 1. The identity $\sqrt{a}\sqrt{b} = \sqrt{ab}$ is still valid, but each square root has two values, not one. Gauss (1831) introduced the symbol i to stand for the square root of -1 that has positive imaginary part, precisely to remove the ambiguity in the notation $\sqrt{-1}$.

Suppose that $w(x) = u(x) + iv(x)$ is complex-valued function of a real variable. If we want to differentiate w, we can form the difference-quotient in the usual way, separate the real and imaginary parts, to find that

$$w'(x) = u'(x) + iv'(x)\,.$$

Similarly, if we want to integrate w, we can form Riemann sums, and separate real and imaginary parts, to see that

$$\int_a^b w(x)\,dx = \int_a^b u(x)\,dx + i\int_a^b v(x)\,dx\,.$$

If f is a real-valued function on $[a, b]$, then

$$\left|\int_a^b f(x)\,dx\right| \le \int_a^b |f(x)|\,dx\,. \tag{1.20}$$

To understand why this is so, note that the integral on the left is a limit of Riemann sums

$$\sum_{k=1}^{n} f(\xi_k)(x_k - x_{k-1}) \tag{1.21}$$

where $a = x_0 \le x_1 \le \cdots \le x_n = b$ is a partitioning of $[a, b]$ and $x_{k-1} \le \xi_k \le x_k$ for all k. Similarly, the integral on the right in (1.20) is the limit of Riemann sums

$$\sum_{k=1}^{n} |f(\xi_k)|(x_k - x_{k-1})\,. \tag{1.22}$$

The expression (1.21) has absolute value not exceeding the quantity (1.22), by the triangle inequality for real numbers, and so the inequality (1.20) follows by taking limits. However, we have shown that the triangle inequality (1.3) also holds for complex numbers. Hence by the above reasoning we deduce that the triangle inequality (1.20) for integrals holds also when f is complex-valued.

Consider the calculus formula

$$\frac{d}{dx}e^{ax} = ae^{ax}, \tag{1.23}$$

which is familiar when a is real. Now suppose that a is complex, say $a = \alpha + i\beta$. Then

$$\begin{aligned}
\frac{d}{dx}e^{ax} &= \frac{d}{dx}\left(e^{\alpha x}\cos\beta x + ie^{\alpha x}\sin\beta x\right) = \frac{d}{dx}\left(e^{\alpha x}\cos\beta x\right) + i\frac{d}{dx}\left(e^{\alpha x}\sin\beta x\right) \\
&= \alpha e^{\alpha x}\cos\beta x - \beta e^{\alpha x}\sin\beta x + i(\alpha e^{\alpha x}\sin\beta x + \beta e^{\alpha x}\cos\beta x) \\
&= (\alpha + i\beta)e^{\alpha x}\cos\beta x + i(\alpha + i\beta)e^{\alpha x}\sin\beta x \\
&= ae^{\alpha x}(\cos\beta x + i\sin\beta x) = ae^{\alpha x}e^{i\beta x} \\
&= ae^{ax}\,.
\end{aligned}$$

Thus (1.23) holds also when a is complex. Similarly,

$$\begin{aligned}
\int e^{ax}\,dx &= \int e^{\alpha x}\cos\beta x\,dx + i\int e^{\alpha x}\sin\beta x\,dx \\
&= \frac{\alpha}{\alpha^2+\beta^2}e^{\alpha x}\cos\beta x + \frac{\beta}{\alpha^2+\beta^2}e^{\alpha x}\sin\beta x \\
&\quad - \frac{i\beta}{\alpha^2+\beta^2}e^{\alpha x}\cos\beta x + \frac{i\alpha}{\alpha^2+\beta^2}e^{\alpha x}\sin\beta x \ + C \\
&= \frac{\alpha - i\beta}{\alpha^2+\beta^2}e^{\alpha x}\cos\beta x + i\frac{\alpha - i\beta}{\alpha^2+\beta^2}e^{\alpha x}\sin\beta x \ + C \\
&= \frac{1}{\alpha + i\beta}e^{\alpha x}e^{i\beta x} \ + C = \frac{1}{a}e^{ax} \ + C
\end{aligned} \tag{1.24}$$

provided that $a \neq 0$.

Suppose we have a definite integral of a real variable, as in calculus, but that the integrand is complex, say

$$\int_a^b f(x) + ig(x)\,dx\,.$$

Suppose also that F is an antiderivative of f, and that G is an antiderivative of g. Then the above is

$$= \int_a^b f(x)\,dx + i\int_a^b g(x)\,dx = [F(x)|_a^b + i[G(x)|_a^b = [F(x) + iG(x)|_a^b\,.$$

Here $F(x) + iG(x)$ is an antiderivative of $f(x) + ig(x)$, and so the Fundamental Theorem of Calculus holds also for complex-valued functions of a real variable.

By combining (1.24) with the above remarks, we find in particular that

$$\int_a^b e^{2\pi ix}\,dx = \left[\frac{e^{2\pi ix}}{2\pi i}\right|_a^b = \frac{e^{2\pi ib} - e^{2\pi ia}}{2\pi i} = \frac{e^{2\pi ia}\left(e^{2\pi i(b-a)} - 1\right)}{2\pi i} = 0$$

if and only if $b - a$ is an integer. This has the following pleasing—and unexpected—application to geometry.

Theorem 1.3. (The Rectangle Theorem) *Suppose that a rectangle $\mathcal{R} = [a,b]\times[c,d]$ is partitioned into finitely many subrectangles $\mathcal{R}_j = [a_j,b_j]\times[c_j,d_j]$. Suppose that each of the subrectangles has the property that at least one of its side-lengths is an integer. Then $\mathcal{R}$ also has this property.*

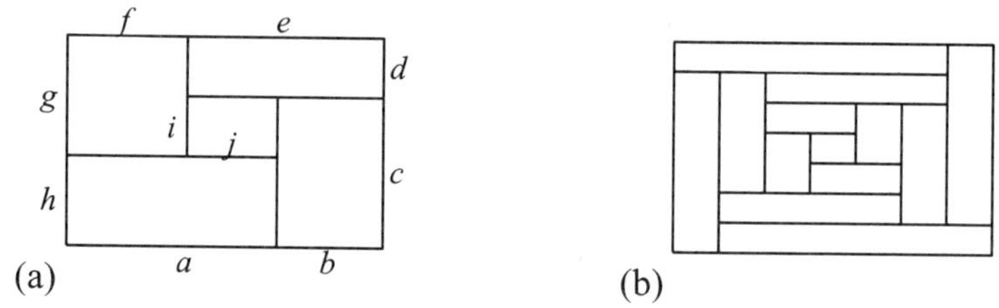

Figure 1.5. Examples of subdivided rectangles

Consider the situation in Figure 1.5(a) with the indicated side-lengths. Suppose that a is an integer. If b is an integer, then $a+b$ is an integer, and we are done. So suppose that instead c is an integer. If d is an integer, then we are done; so suppose instead that e is an integer. If f is an integer, then we are done; so suppose instead that g is an integer. The height of the rectangle is $c-i+g$, and its width is $a-j+e$. Thus if i is an integer, then the height of the rectangle is an integer, while if j is an integer, then the width is an integer. This involves quite a few cases, and we are only half done, since we started by assuming that a is an integer, and we have yet to consider the alternative case in which h is an integer. Such a case-by-case analysis is unsuited for a more complicated configuration such as in Figure 1.5(b), so we need a better approach.

Proof. Clearly

$$\iint_{\mathcal{R}} e^{2\pi i(x+y)}\,dx\,dy = \sum_j \iint_{\mathcal{R}_j} e^{2\pi i(x+y)}\,dx\,dy = \sum_j \Big(\int_{a_j}^{b_j} e^{2\pi i x}\,dx\Big)\Big(\int_{c_j}^{d_j} e^{2\pi i y}\,dy\Big).$$

For each j, at least one of the last two integrals vanishes, since at least one of $b_j - a_j$ or $d_j - c_j$ is an integer. Hence each summand is 0, so the sum is 0. But the left hand side is

$$= \Big(\int_a^b e^{2\pi i x}\,dx\Big)\Big(\int_c^d e^{2\pi i y}\,dy\Big).$$

In order for this product to vanish, one of the factors must vanish. But the first integral is 0 only when $b-a$ is an integer, and the second integral is 0 only when $d-c$ is an integer. Since at least one of these must hold, at least one of the side-lengths of $\mathcal{R}$ is an integer. □

Exercises

1. At the beginning of this section it is stated without proof that $f(z) = 1/(1-z)$ is analytic throughout the complex plane except at $z = 1$. For $a \neq 1$, form an appropriate difference quotient, simplify the expression so that the limit as $z \to a$ can be determined, and thus verify that $f(z)$ is indeed analytic.
2. Use the identities (1.5) to show that
   (a) $\sin(\theta + \pi/2) = \cos\theta$.
   (b) $\sin(\theta + \pi) = -\sin\theta$.

(c) $\sin(\theta - \pi/2) = -\cos\theta$.

3. (a) Use Exercise 2(a) and Theorem 1.2 to show that

$$\prod_{k=0}^{n-1} \cos\pi(x + k/n) = \frac{\sin\pi n(x + 1/2)}{2^{n-1}}.$$

(b) Use Exercise 2 to show that the above is

$$= \begin{cases} \dfrac{\sin\pi nx}{2^{n-1}} & (n \equiv 0 \pmod 4), \\ \dfrac{\cos\pi nx}{2^{n-1}} & (n \equiv 1 \pmod 4), \\ -\dfrac{\sin\pi nx}{2^{n-1}} & (n \equiv 2 \pmod 4), \\ -\dfrac{\cos\pi nx}{2^{n-1}} & (n \equiv 3 \pmod 4). \end{cases}$$

4. (a) By taking logarithmic derivatives in Theorem 1.2, show that if nx is not an integer, then

$$\sum_{k=0}^{n-1} \cot\pi(x + k/n) = n\cot\pi nx\,.$$

(b) By differentiating both sides of the above, show that if nx is not an integer, then

$$\sum_{k=0}^{n-1} \csc^2\pi(x + k/n) = n^2\csc^2\pi nx\,.$$

(c) Show that

$$\lim_{x\to 0}(\csc^2 x - x^{-2}) = \frac{1}{3}\,.$$

(d) Deduce that

$$\sum_{k=1}^{n-1} \csc^2\pi k/n = \frac{n^2 - 1}{3}\,.$$

(e) Deduce that

$$\sum_{k=1}^{n-1} \cot^2\pi k/n = \frac{(n-1)(n-2)}{3}\,.$$

5. Find all possible correct values for the following:
(a) $\log(-1)$;
(b) $\log(2i)$;
(c) $(-1)^i$.

6. In the derivation of (1.24) we used the classical integration formulæ

$$\int e^{\alpha x}\cos\beta x\,dx = \frac{e^{\alpha x}}{\alpha^2 + \beta^2}(\alpha\cos\beta x + \beta\sin\beta x) + C, \tag{1.25}$$

$$\int e^{\alpha x}\sin\beta x\,dx = \frac{e^{\alpha x}}{\alpha^2 + \beta^2}(-\beta\cos\beta x + \alpha\sin\beta x) + C\,. \tag{1.26}$$

These formulæ can be proved by standard calculus methods, using integration by parts. However, in our new state of knowledge, if one were presented with such an integral, one could start by writing

$$\int e^{\alpha x}\cos\beta x\,dx = \frac{1}{2}\int e^{\alpha x}e^{i\beta x}\,dx + \frac{1}{2}\int e^{\alpha x}e^{-i\beta x}\,dx$$
$$= \frac{1}{2(\alpha+i\beta)}e^{(\alpha+i\beta)x} + \frac{1}{2(\alpha-i\beta)}e^{(\alpha-i\beta)x} + C\,.$$

(a) Complete the proof of (1.25).
(b) Similarly prove (1.26).

7. Suppose that k is an integer.
(a) Explain why
$$\int_0^{2\pi} e^{ikx}\,dx = \begin{cases} 2\pi & (k=0), \\ 0 & (k\neq 0). \end{cases}$$
(b) Show that if n is a nonnegative integer, then
$$(\cos x)^{2n} = \frac{1}{2^{2n}}\sum_{k=0}^{2n}\binom{2n}{k}e^{2(k-n)ix}\,.$$
(c) Show that
$$\int_0^{2\pi}(\cos x)^{2n}\,dx = \frac{\pi}{2^{2n-1}}\binom{2n}{n} = \frac{\pi(2n)!}{2^{2n-1}n!^2} = 2\pi\frac{1\cdot 3\cdot 5\cdots(2n-1)}{2\cdot 4\cdot 6\cdots(2n)}\,.$$

8. Complete the argument that was begun following Figure 1.5, by treating the case in which h is an integer.

9. We wish to tile an $A\times B$ rectangle with $a\times b$ tiles. Here a and b are integers. Of course if $A = ma$ and $B = nb$, then this can be done with mn parallel tiles. We call this a *trivial tiling*. The object of this exercise is to show that if $a|b$, then a $A\times B$ rectangle can be tiled only if it can be trivially tiled. That is, if a divides both A and B, and b divides at least one of A and B.
(a) Scale the tiles and the rectangle by a factor a, both horizontally and vertically. Thus the tiles are now $1\times(b/a)$ and the rectangle is now $(A/a)\times(B/a)$. Note that both 1 and b/a are integers. Argue that if the shrunken rectangle can be tiled with shrunken tiles, then both A/a and B/a are integers.
(b) Now scale the original tiles and rectangle by a factor b. Thus the new shrunken tiles are $(a/b)\times 1$ and the new shrunken rectangle has dimensions $(A/b)\times(B/b)$. Apply the Rectangle Theorem to show that at least one of A/b and B/b is an integer.

10. (Euler, 1744) The object of this exercise is to show that the number e is irrational.
(a) Show that if a/q and a'/q' are rational numbers, and $a/q\neq a'/q'$, then $|a/q - a'/q'|\geq 1/(qq')$.
(b) Set
$$\frac{a}{q} = \sum_{n=0}^{N}\frac{1}{n!}\,.$$

Show that

$$0 < e - \frac{a}{q} < \frac{2}{(N+1)!}$$

(c) Show that $q = N!$ or at least $q \mid N!$.

(d) Deduce that

$$0 < e - \frac{a}{q} < \frac{2}{q(N+1)}\,.$$

(e) Conclude that e is irrational.

11. Suppose that there are integers a, b, c, not all 0, such that $ae^2 + be + c = 0$.

(a) Explain why

$$b + \sum_{n=0}^{\infty} \frac{a + (-1)^n c}{n!} = 0\,.$$

(b) Show that if the above is multiplied by $N!$, the result is an integer plus

$$\frac{a + (-1)^{N+1} c}{N+1} + O(1/N^2)\,.$$

(c) Deduce that $a + c = 0$ and that $a - c = 0$.

(d) Deduce that $a = c = 0$.

(e) Deduce that $b = 0$.

(f) Conclude that the number e is neither rational nor a quadratic irrationality.

Notes

1.1. The term "complex number" was coined by Gauss in 1831. The term "conjugate" had been already coined by Cauchy in 1821.

1.2. For a more extended account of the ways that complex numbers can be used in Euclidean geometry see Hahn (1994). The linear-fractional functions (sometimes called Möbius functions)

$$f(z) = \frac{az + b}{cz + d}$$

where $ad - bc \neq 0$, also have interesting geometric properties: They map lines and circles to lines and circles. In this context, a line may be thought of as a circle that happens to pass through infinity. The composition of one such map after another is again linear-fractional, with coefficients given by the matrix product of the two sets of coefficients.

1.3. In the eighteenth century it was generally expected that a polynomial of degree n would have n roots. In some cases it was understood that some of the roots would be complex numbers, which is to say numbers of the form $a + bi$ where a and b are real. However, it was speculated that some roots might not be of this form, but would be more "imaginary", and that a polynomial whose coefficients were "imaginary" might have roots that are even more "imaginary", and so on, creating a hierarchy of ever more imaginary roots. Fortunately, the truth is much simpler.

The gifted amateur mathematician Jean-Robert Argand, who lived and worked in Paris, published (at his own expense) a small pamphlet in 1806, with important contents. First he observed that it can be fruitful to plot a complex number $a + bi$ in a Cartesean plane where a is the x coordinate and b is the y coordinate. For centuries this has been known as an "Argand diagram", although today we just think of it as the complex plane. Secondly, Argand coined the term "modulus" for the absolute value of a complex number. Finally, and most spectacularly, he stated and proved the Fundamental Theorem of Algebra for arbitrary polynomials with real or complex coefficients. Our first proof of this is essentially that of Argand.

Gauss attempted a proof of the Fundamental Theorem of Algebra in his doctoral thesis of 1799, but it contained a major gap that was only filled by Ostrowski in 1921. Gauss gave two correct proofs in 1816. Argand's proof also contained a gap, since in 1806 it had not yet been proved that a continuous function on a closed bounded interval attains its maximum, but this was established within a few years of Argand's initial work. While myriads of proofs of the Fundamental Theorem of Algebra have now been published, Argand's remains one of the most popular. Other writers have observed that the Fundamental Theorem of Algebra is neither fundamental nor a theorem of algebra. Rather, it is a theorem of analysis that is a corollary of the fact that the real numbers are complete. (This latter fact is considered to be fundamental.)

1.4. It seems to be unknown who first proved the "Rectangle Theorem", although it can be proved in remarkably many different ways; see Wagon (1987).

Chapter 2

The Discrete Fourier Transform

2.1. Sums of roots of unity

Suppose we are given a polynomial in factored form, say

$$P(z) = \sum_{k=0}^{n} a_k z^k = a_n \prod_{j=1}^{n} (z - z_j) \tag{2.1}$$

with $a_n \neq 0$. When the product is expanded, the coefficient of z^{n-1} is

$$-a_n \sum_{j=1}^{n} z_j \,.$$

Thus the sum of the n roots of P is

$$z_1 + z_2 + \cdots + z_n = -\frac{a_{n-1}}{a_n} \,.$$

In the case of the polynomial $P(z) = z^n - 1$, whose factorization is given in (1.19), we have $a_{n-1} = 0$ when $n > 1$, and hence

$$\sum_{j=1}^{n} e^{2\pi i j/n} = 0 \,. \tag{2.2}$$

This is not surprising, since the roots $e^{2\pi i j/n}$ are the vertices of a regular n-gon inscribed in the unit circle $|z| = 1$, and it might be expected that their center of gravity would lie at the center. In what follows, we need to understand not just the sum of the roots of unity, but also the sum of the kth powers of these roots, for various k.

Starting now, and continuing through the rest of our work, many formulas will include an expression of the form $e^{2\pi i x}$ where the quantity x may itself involve several terms. To ease our typography we now adopt a notation introduced many decades ago by the great Russian mathematician I. M. Vinogradov:

$$e(x) = e^{2\pi i x}. \tag{2.3}$$

We refer to this as *the complex exponential with period one.*

Theorem 2.1. *Let q be a positive integer. Then for any integer k,*

$$\sum_{n=1}^{q} e(kn/q) = \begin{cases} q & (q|k), \\ 0 & (\textit{otherwise}). \end{cases}$$

Proof. If q divides k evenly among the integers (in symbols $q|k$), then k/q is an integer, so $e(kn/q) = 1$ for all n by (1.18), and hence the q terms sum to q. Suppose that q does not divide k (in symbols $q \nmid k$). The terms $e(kn/q)$ form a geometric progression with common ratio $e(k/q) \neq 1$ from one term to the next, so by the formula (0.2) for the sum of a geometric progression we see that

$$\sum_{n=1}^{q} e(kn/q) = \frac{e(k(q+1)/q) - e(k/q)}{e(k/q) - 1} = e(k/q)\frac{e(kq/q) - 1}{e(k/n) - 1}.$$

Here $e(kq/q) = e(k) = 1$, so the numerator is 0, and the denominator is nonzero, so the sum is 0 in this case. □

Example 2.1. Let $\zeta = e(1/5)$, and put

$$a = 2\cos 2\pi/5 = \zeta + \zeta^4, \qquad b = 2\cos 4\pi/5 = \zeta^2 + \zeta^3.$$

Then $a + b = \zeta + \zeta^2 + \zeta^3 + \zeta^4 = -1$ by (2.2), and $ab = (\zeta + \zeta^4)(\zeta^2 + \zeta^3) = \zeta + \zeta^2 + \zeta^3 + \zeta^4 = -1$. Thus

$$(z - a)(z - b) = z^2 - (a + b)z + ab = z^2 + z - 1.$$

Hence by the formula for the roots of a quadratic polynomial, a and b are $(-1 \pm \sqrt{5})/2$ in some order. But clearly $a > 0$ and $b < 0$, so we conclude that

$$\cos\frac{2\pi}{5} = \frac{-1 + \sqrt{5}}{4}, \qquad \cos\frac{4\pi}{5} = -\frac{1 + \sqrt{5}}{4}. \tag{2.4}$$

Exercises

Let $z_1, z_2, \ldots, z_n$ be given complex numbers. For an integer k, $1 \le k \le n$, multiply k of them together, in all $\binom{n}{k}$ different possible ways, and sum the results. This quantity is denoted σ_k, and is called the kth *elementary symmetric function* of the z_j. By convention, we put $\sigma_0 = 1$.

1. (a) Show that $\cos(\pi - \theta) = -\cos\theta$.
   (b) Deduce that
   $$\cos\frac{\pi}{5} = \frac{1 + \sqrt{5}}{4}.$$
2. Let $\zeta = e(1/7)$, and put $a = \zeta + \zeta^6 = 2\cos 2\pi/7$, $b = \zeta^2 + \zeta^5 = 2\cos 4\pi/7$, and $c = \zeta^3 + \zeta^4 = 2\cos 6\pi/7$.
   (a) Show that $a + b + c = -1$.

(b) Show that $ab + bc + ca = -2$.
(c) Show that $abc = 1$.
(d) Conclude that $(z-a)(z-b)(z-c) = z^3 + z^2 - 2z - 1$.

3. Suppose that $P(z)$ is given as in (2.1). Show that $\sigma_k = (-1)^k a_{n-k}/a_n$.

4. Suppose that
$$Q(z) = \prod_{j=1}^{n} (1 - z_j z) .$$
(a) Show that
$$Q(z) = 1 - \sigma_1 z + \sigma_2 z^2 - \cdots + (-1)^n \sigma_n z^n .$$
(b) Show that
$$Q'(z) = -\sigma_1 + 2\sigma_2 z - \cdots + (-1)^n n \sigma_n z^{n-1} .$$
(c) Show that the logarithmic derivative of $1 - z_j z$ is $-z_j/(1 - z_j z)$.
(d) Show that
$$\frac{Q'(z)}{Q(z)} = -\sum_{j=1}^{n} \frac{z_j}{1 - z_j z} .$$
(e) Suppose that z is so small that $|z_j z| < 1$. Explain why
$$\frac{1}{1 - z_j z} = \sum_{m=0}^{\infty} (z_j z)^m .$$
(f) Let $s_m = z_1^m + z_2^m + \cdots + z_n^m$. This is called a *power sum*. (Guess why.) Deduce that
$$Q'(z) = -Q(z) \sum_{m=0}^{\infty} s_{m+1} z^m$$
when $|z|$ is sufficiently small. By comparing the constant terms on both sides, deduce that $\sigma_1 = s_1$. By comparing the coefficients of z on the two sides above, deduce that $2\sigma_2 = \sigma_1 s_1 - s_2$. In general, by comparing the coefficient of z^{m-1} on each side, deduce that

$$\sum_{k=1}^{m} (-1)^{k-1} \sigma_{m-k} s_k = \begin{cases} m\sigma_m & (1 \le m \le n), \\ 0 & (m > n). \end{cases} \tag{2.5}$$

These are the *Newton–Girard formulæ*. From the theory of linear recurrences (see §F.4) we know that the s_m must satisfy a linear recurrence. By these formulæ the linear recurrence is made explicit. Given the σ_k, we can use these formulæ to compute the s_m. In the reverse direction, if A is an $n \times n$ matrix and the z_j are its eigenvalues, then $\operatorname{tr} A^m = s_m$, so by computing $A^2, A^3, \ldots, A^n$ one can compute the s_m, and hence the σ_k, and consequently the characteristic polynomial of A.

5. We apply the above to $Q(z) = 1 - z^q$.
(a) Explain why in this case
$$s_m = \sum_{n=1}^{q} e(mn/q) .$$
(b) Show that $\sigma_k = 0$ for $0 < k < q$, and that $\sigma_q = (-1)^{q-1} q$.

(c) Use the Newton–Girard formulæ to show that $s_m = 0$ for $m = 1, 2, ..., q-1$ and that $s_q = q$.
(d) Observe that in the present situation, $s_{m+q} = s_m$.
(e) Use these results to complete a second proof of Theorem 2.1.

2.2. The Transform

We say that $f(n)$ is an *arithmetic function* if it is defined for all integers. We also say that an arithmetic function is *periodic with period* q if $f(n+mq) = f(n)$ for all integers m. For an arithmetic function f with period q, we define the *Discrete Fourier Transform* (DFT) to be

$$\widehat{f}(k) = \sum_{n=1}^{q} f(n)e(-nk/q) \tag{2.6}$$

for each integer k. The numbers $\widehat{f}(k)$ also have period q, since $e(-n(k+mq)/q) = e(-nk/q)e(-nm) = e(-nk/q)$ for all n. The $\widehat{f}(k)$ are defined in terms of the $f(n)$, but this can be reversed:

Theorem 2.2. *Suppose that $f(n)$ has period q, and let $\widehat{f}(k)$ be defined as in* (2.6). *Then*

$$f(n) = \frac{1}{q}\sum_{k=1}^{q} \widehat{f}(k)e(kn/q)\,. \tag{2.7}$$

This is the Fourier expansion of f, by which we see that any arithmetic function with period q can be expressed as a linear combination of the basic functions $e(kn/q)$. Conventions vary among authors as to where the factor $1/q$ should go. Some writers put q on the right hand side of (2.6) rather than in the denominator in (2.7), while others put a factor $1/\sqrt{q}$ in both formulæ.

Proof. From the definition (2.6), we see that the right hand side above is

$$= \frac{1}{q}\sum_{k=1}^{q} e(kn/q)\sum_{a=1}^{q} f(a)e(-ak/q)\,.$$

In (2.6) we used n to index the sum, but in the above we have used a new letter, a, since n is now in use. By inverting the order of summation we see that the above is

$$= \frac{1}{q}\sum_{a=1}^{q} f(a)\sum_{k=1}^{q} e(k(n-a)/q)\,.$$

By Theorem 2.1, the inner sum is q if $q|(n-a)$. Otherwise the inner sum is 0. For $1 \le a \le q$ there is exactly one a such that $q|(n-a)$, and for this a we have $f(a) = f(n)$ since f has period q. Hence the above is $f(n)$, as desired. □

When $q|(n-a)$, as we encountered above, we say that a is *congruent to* n *modulo* q, and in symbols we write $a \equiv n \pmod{q}$.

The functions $f_k(n) = e(kn/q)$ have period q, and the import of Theorem 2.2 is that they form a basis for the vector space of all functions with period q. When considering vectors in n-dimensional real space $\mathbb{R}^n$, the vectors $\boldsymbol{u} = [u_j]$ and $[\boldsymbol{v}] = [v_j]$ have the inner product (or 'dot product') $\langle \boldsymbol{u}, \boldsymbol{v}\rangle = \sum_{j=1}^n u_j v_j$. When dealing with n-dimensional vectors with complex elements, this is modified as follows:

$$\langle \boldsymbol{u}, \boldsymbol{v}\rangle = \sum_{j=1}^{n} u_j \overline{v_j}\,. \tag{2.8}$$

Thus, while in the real case $\langle \boldsymbol{u}, \boldsymbol{v}\rangle = \langle \boldsymbol{v}, \boldsymbol{u}\rangle$, in the complex case we have

$$\langle \boldsymbol{v}, \boldsymbol{u}\rangle = \overline{\langle \boldsymbol{u}, \boldsymbol{v}\rangle}\,. \tag{2.9}$$

With this understanding, we see by Theorem 2.1 that the functions f_k are not merely linearly independent; they are orthogonal. Indeed, in (2.6) and (2.7) we witness the usual computation of the coordinates of a vector with respect to an orthogonal basis.

In (2.7) we have expressed f as a linear combination of the basis vectors f_k. It follows by principles of linear algebra that this is the unique linear combination that yields f, but we give a direct computational proof of this result.

Theorem 2.3. *Suppose that f has period q, and that $c(1), c(2), \ldots, c(q)$ are numbers such that*

$$f(n) = \frac{1}{q}\sum_{k=1}^{q} c(k)e(kn/q) \tag{2.10}$$

for all n. Then $c(k) = \widehat{f}(k)$ for all k.

Proof. On subtracting (2.7) from (2.10) we find that

$$\sum_{k=1}^{q}(c(k) - \widehat{f}(k))e(kn/q) = 0$$

for all n. Let $g(k) = c(k) - \widehat{f}(k)$. From the definition (2.6), with the roles of k and n reversed, we see that $\widehat{g}(n) = 0$ for all n. From (2.7) we know that

$$g(k) = \frac{1}{q}\sum_{n=1} \widehat{g}(n)e(kn/q)\,.$$

Since $\widehat{g}(n) = 0$ for all n, it follows that $g(k) = 0$ for all k. That is, $c(k) = \widehat{f}(k)$. □

If f has period q and a is an integer, then $f(a+1), f(a+2), \ldots, f(a+q)$ are the numbers $f(1), f(2), \ldots, f(q)$ in a permuted order. Hence

$$\sum_{n=a+1}^{a+q} f(n) = \sum_{n=1}^{q} f(n)\,. \tag{2.11}$$

This applies to such sums as the one in (2.6). Since $f(n)$ has period q and $e(-kn/q)$ has period q, it follows that $f(n)e(-kn/q)$ has period q, and hence

$$\widehat{f}(k) = \sum_{n=a+1}^{a+q} f(n)e(-kn/q) \tag{2.12}$$

for any integer a. Similarly, from (2.7) it follows that

$$f(n) = \frac{1}{q} \sum_{k=a+1}^{a+q} \widehat{f}(k) e(kn/q) \tag{2.13}$$

for any a.

From (2.6), by the triangle inequality we see that

$$|\widehat{f}(k)| \le \sum_{n=1}^{q} |f(n)| \tag{2.14}$$

for all k. Similarly, from (2.7) we deduce that

$$|f(n)| \le \frac{1}{q} \sum_{k=1}^{q} |\widehat{f}(k)| \tag{2.15}$$

for all n.

The DFT is linear in the sense that if f and g have period q and $h(n) = \alpha f(n) + \beta g(n)$ for all n, then from (2.6) we see that $\widehat{h}(k) = \alpha \widehat{f}(k) + \beta \widehat{g}(k)$. Further properties of the DFT are described below.

Theorem 2.4. *Suppose that $f(n)$ has period q.*

(a) *If a is an integer and $g(n) = f(n+a)$ for all n, then $\widehat{g}(k) = \widehat{f}(k)e(ka/q)$ for all k.*

(b) *If b is an integer and $h(n) = f(n)e(bn/q)$ for all n, then $\widehat{h}(k) = \widehat{f}(k-b)$ for all k.*

(c) *If $j(n) = f(-n)$ for all n, then $\widehat{j}(k) = \widehat{f}(-k)$ for all k.*

(d) *If $\ell(n) = \overline{f(n)}$ for all n, then $\widehat{\ell}(k) = \overline{\widehat{f}(-k)}$ for all k.*

Proof. (a) From (2.6) we know that

$$\begin{aligned}
\widehat{g}(k) &= \sum_{n=1}^{q} g(n) e(-kn/q) = \sum_{n=1}^{q} f(n+a) e(-kn/q) \\
&= e(ka/q) \sum_{n=1}^{q} f(n+a) e(-k(n+a)/q) \\
&= e(ka/q) \sum_{n=a+1}^{a+q} f(n) e(-kn/q) = e(ka/q) \widehat{f}(k)
\end{aligned}$$

by (2.12).

(b) From (2.6) we know that

$$\widehat{h}(k) = \sum_{n=1}^{q} h(n) e(-kn/q) = \sum_{n=1}^{q} f(n) e((b-k)n/q) = \widehat{f}(k-b).$$

(c) $\widehat{j}(k) = \sum_{n=1}^{q} f(-n)e(-kn/q)$. Put $m = -n$. Then this is

$$\sum_{m=-q}^{-1} f(m)e(mk/q) = \widehat{f}(-k)$$

by (2.12).

(d) We first replace k by $-k$ in (2.6), and then take complex conjugates of both sides to see that

$$\overline{\widehat{f}(-k)} = \sum_{n=1}^{q} \overline{f(n)}e(-kn/q) = \widehat{\ell}(k)\,.$$

□

Example 2.2. Suppose that $2N < q$. Define $f(n)$ with period q by setting

$$f(n) = \begin{cases} 1 & (-N \le n \le N), \\ 0 & (N < n < q - N). \end{cases}$$

Clearly $\widehat{f}(0) = 2N + 1$. Suppose that $q \nmid k$. Then

$$\widehat{f}(k) = \sum_{n=-N}^{N} e(-kn/q) = e(kN/q)\sum_{n=0}^{2N} e(-kn/q).$$

Here we have a geometric series, so by (0.2) the above is

$$\begin{aligned} &= e(kN/q)\frac{1 - e(-k(2N+1)/q)}{1 - e(-k/q)} \\ &= e(kN/q)\frac{e(-\tfrac{1}{2}k(2N+1)/q)(e(\tfrac{1}{2}k(2N+1)/q) - e(-\tfrac{1}{2}k(2N+1)/1))}{e(-\tfrac{1}{2}k/q)(e(\tfrac{1}{2}k/q) - e(-\tfrac{1}{2}k/q))}\,. \end{aligned}$$

But $e(kN/q)e(-\tfrac{1}{2}k(2N+1)/q) = e(-\tfrac{1}{2}k/q)$. We divide the numerator and denominator by $2i$ to see that the above is

$$= \frac{\sin \pi k(2N+1)/q}{\sin \pi k/q}\,.$$

Suppose that $f(n)$ and $g(n)$ are arithmetic functions with period q. The *convolution* of f and g is

$$(f * g)(n) = \sum_{a=1}^{q} f(n-a)g(a), \tag{2.16}$$

which is also an arithmetic function with period q. This is the first of several different convolutions that we shall encounter. The precise definition of a convolution depends on the context, but in all cases one forms a product of the two functions at arguments that sum to one fixed value, and then sums (or integrates) over all such pairs of arguments. Some authors would put a $1/q$ on the right hand side in (2.16). Thus one needs to be aware of the conventions being followed when reading from other sources.

Theorem 2.5. *Suppose that $f(n)$ and $g(n)$ are arithmetic functions with period q, and that $h = f * g$. Then*
$$\widehat{h}(k) = \widehat{f}(k)\widehat{g}(k)\,.$$

Proof. From the definition (2.6) we know that

$$\begin{aligned}
\widehat{h}(k) &= \sum_{n=1}^{q} h(n)e(-kn/q) = \sum_{n=1}^{q} e(-kn/q) \sum_{a=1}^{q} f(n-a)g(a) \\
&= \sum_{a=1}^{q} g(a) \sum_{n=1}^{q} f(n-a)e(-kn/q) \\
&= \sum_{a=1}^{q} g(a)e(-ka/q) \sum_{n=1}^{q} f(n-a)e(-k(n-a)/q)\,. \qquad (2.17)
\end{aligned}$$

In the last sum above, as n runs from 1 to q, $n-a$ runs from $1-a$ to $q-a$. Thus by (2.12), this sum is

$$= \sum_{n=1-a}^{q-a} f(n)e(-kn/q) = \sum_{n=1}^{q} f(n)e(-kn/q) = \widehat{f}(k).$$

This is independent of a, so when we insert this value in (2.17) the remaining sum is just $\widehat{g}(k)$, and the proof is complete. □

Example 2.3. Suppose that f has period q, that $2N < q$, and that

$$f(n) = \begin{cases} 1 & (0 \le n < N), \\ 0 & (N \le n < q). \end{cases}$$

By arguing as in Example 2.2, we find that

$$\widehat{f}(k) = \begin{cases} N & (q|k), \\ e(-\frac{1}{2}k(N-1)/q)\dfrac{\sin \pi kN/q}{\sin \pi k/q} & (q \nmid k). \end{cases}$$

Let $g(n) = f(-n)$, and set $h(n) = f * g$. Then

$$h(n) = \begin{cases} N - |n| & (-N < n < N), \\ 0 & (N \le n \le q - N). \end{cases}$$

By Theorem 2.4(d) we know that $\widehat{g}(k) = \widehat{f}(-k)$. Thus by Theorem 2.5 we deduce that

$$\widehat{h}(k) = \begin{cases} N^2 & (q|k), \\ \left(\dfrac{\sin \pi kN/q}{\sin \pi k/q}\right)^2 & (q \nmid k)\,. \end{cases}$$

Let f be an arithmetic function with period q. The average of f over a full period is $\widehat{f}(0)/q$. If the numbers $\widehat{(k)}$ are not too large for $0 < k < q$, we may hope that f has approximately the same average in shorter intervals.

Theorem 2.6. *Let f be an arithmetic function with period q, and suppose that $|\widehat{f}(k)| \le M$ for $0 < k < q$. If a and b are integers with $a < b$, then*

$$\Big|\sum_{n=a}^{b-1} f(n) - \frac{b-a}{q}\widehat{f}(0)\Big| \le \frac{2}{\pi} M \log\frac{4q}{\pi}\,.$$

Proof. By the Discrete Fourier expansion of f we see that

$$\sum_{n=a}^{b-1} f(n) = \sum_{n=a}^{b-1} \frac{1}{q}\sum_{k=1}^{q} \widehat{f}(k)e(kn/q) = \frac{1}{q}\sum_{k=1}^{q} \widehat{f}(k)\sum_{n=a}^{b-1} e(kn/q)\,.$$

On separating $k = q$ from the other terms and using the formula for the sum of a geometric progression we deduce that

$$\sum_{n=a}^{b-1} f(n) - \frac{b-a}{q}\widehat{f}(0) = \frac{1}{q}\sum_{k=1}^{q-1} \widehat{f}(k)\frac{e(kb/q) - e(ka/q)}{e(k/q) - 1}\,. \tag{2.18}$$

Here the right hand side has absolute value not exceeding

$$\frac{2M}{q}\sum_{k=1}^{q-1}\frac{1}{|e(k/q) - 1|}\,,$$

and $|e(k/q) - 1| = 2\sin \pi k/q$ for $0 < k < q$, so to complete the proof it suffices to show that

$$\sum_{k=1}^{q-1} \csc\frac{\pi k}{q} < \frac{2}{\pi} q \log\frac{4q}{\pi}\,. \tag{2.19}$$

Since $\csc \pi u$ is convex, it follows that

$$\csc \pi k/q \le q\int_{\frac{k-1/2}{q}}^{\frac{k+1/2}{q}} \csc \pi u\, du\,.$$

On summing this over k we see that the left hand side of (2.19) is

$$\begin{aligned}
&\le q\int_{1/(2q)}^{1-1/(2q)} \csc \pi u\, du = \frac{q}{\pi}\Big[\log(\csc \pi u - \cot \pi u)\Big|_{1/(2q)}^{1-1/(2q)}\\
&= \frac{2q}{\pi}\log(\csc \pi/(2q) + \cot \pi/(2q)) = \frac{2q}{\pi}\log(\cot \pi/(4q)) < \frac{2}{\pi} q \log(4q/\pi)\,.
\end{aligned}$$

This gives (2.19), so the proof is complete. □

In the notation of the proof just completed, let

$$F(n) = \frac{1}{q}\sum_{k=1}^{q-1} \widehat{f}(k)\frac{e(kn/q)}{e(k/q) - 1}\,. \tag{2.20}$$

This acts as a discrete analogue of an antiderivative of $f(n) - \widehat{f}(0)/q$, since the right hand side of (2.18) is $F(b) - F(a)$. The distribution of the values of $F(n)$ is of great interest to us.

Example 2.4. Suppose that S has period q, that $S(0) = 0$, and that $S(n) = 1/2 - n/q$ for $n = 1, 2, \ldots, q-1$. This is a discrete sawtooth function. We determine the Discrete Fourier Transform of S. First we note that

$$\widehat{S}(0) = \sum_{n=1}^{q-1}\Big(\frac{1}{2} - \frac{n}{q}\Big) = \frac{q-1}{2} - \frac{1}{q}\cdot\frac{q(q-1)}{2} = 0,$$

which is to be expected, since $S(n)$ is an odd function. Now suppose that $0 < k < q$. We find that

$$\sum_{n=1}^{q-1} ne(-kn/q) = \sum_{n=1}^{q-1}\Big(\sum_{m=1}^{n} 1\Big)e(-kn/q) = \sum_{m=1}^{q-1}\sum_{n=m}^{q-1} e(-kn/q)\,.$$

By the formula (0.2) for the sum of a geometric progression we see that the above is

$$= \sum_{m=1}^{q-1}\frac{e(-km/q) - 1}{1 - e(-k/q)} = \frac{1}{1 - e(-k/q)}\sum_{m=1}^{q-1}\Big(e(-km/q) - 1\Big)\,.$$

By Theorem 2.1 we know that $\sum_{m=1}^{q} e(-km/q) = 0$. In the above, the term $e(-kq/q)$ arising from $m = q$ is missing, and so the sum has value -1. Thus the above is

$$= \frac{1}{1 - e(-k/q)}\big(-1 - (q-1)\big) = \frac{-q}{1 - e(-kd/q)}\,.$$

Hence

$$\widehat{S}(k) = \sum_{n=1}^{q-1}\Big(\frac{1}{2} - \frac{n}{q}\Big)e(-kn/q) = -\frac{1}{2} + \frac{1}{1 - e(-k/q)} = \frac{1 + e(-k/q)}{2(1 - e(-k/q))}\,.$$

We multiply the numerator and denominator by $e(k/(2q))$ to see that this is

$$= \frac{e(k/(2q)) - e(-k/(2q))}{2\big(e(k/(2q)) - e(-k/(2q))\big)} = \frac{\cos\pi k/q}{2i\sin\pi k/q}\,.$$

To summarize, if $S(0) = 0$ and $S(n) = 1/2 - n/q$ for $0 < n < q$, then

$$\widehat{S}(k) = \begin{cases} -\dfrac{i}{2}\cot\pi k/q & (0 < k < q), \\ 0 & (k = 0). \end{cases} \tag{2.21}$$

When we convolve S with a function f we find that if $a < b$, then

$$\frac{1}{2}f(a) + \sum_{a<n<b} f(n) + \frac{1}{2}f(b) - \frac{b-a}{q}\widehat{f}(0) = (f * S)(b) - (f * S)(a)\,. \tag{2.22}$$

Thus $(f * S)(n)$ is a close relative of the function $F(n)$ in (2.20), but now the sum of $f(n)$ is symmetrically placed on the interval $[a, b]$, with the first and last terms counted with weight $1/2$.

Example 2.5. Let X be a random variable that takes values $1, 2, \ldots, q$ with probabilities $p_1, p_2, \ldots, p_q$ respectively. Let Y be a second random variable, independent of X, that takes the values $1, 2, \ldots, q$ with probabilities $r_1, r_2, \ldots, r_q$. Here the p_n and r_n are nonnegative and

$$\sum_{n=1}^{q} p_n = 1, \qquad \sum_{n=1}^{q} r_n = 1\,.$$

We extend the definitions of the sequences p_n and r_n to all integers by making them periodic with period q. We form $X+Y$, and if the result is greater than q, then we subtract q in order to obtain a value in the range $1, 2, \ldots, q$. The probability that $X + Y \equiv n \pmod q$ is

$$(p * r)(n) = \sum_{a=1}^{q} p_a r_{n-a}\,.$$

Now suppose that $q_n = 1/q$ for all n. We call a variable Y with this distribution *uniform*. In the above sum, the $1/q$ factors out, and the p_a sum to 1, so we deduce that $(p * q)(n) = 1/q$ for all n. That is, $X + Y$ is uniform (mod q) if Y is. This has an application to cryptography: Suppose we have a sequence of characters that form our plaintext. These may be thought of as integers in a range $1, 2, \ldots, q$. For example, we could take $q = 26$, and have each integer refer to a letter of the alphabet. More realistically, each character is associated via its ASCII code to a sequence of 8 bits. Thus we have a bit stream, and we work with $q = 2$. We add a quasi-random sequence to this (mod q) to obtain our cryptotext. The quasi-random sequence is the output of a random-number generator, which starts from a known initial state. If the recipient has an identical random generator and knows the appropriate initial state (the key), then the recipient can generate the same quasi-random sequence and by subtraction recover the original message. If the quasi-random sequence were truly random, then the cryptotext would also be random, and hence the code would be impossible to break. A cryptanalyst seeking to break the code would need to identify and exploit imperfections in the design of the random number generator.

Theorem 2.7. *If $f(n)$ and $g(n)$ are arithmetic functions with period q, then*

$$\sum_{n=1}^{q} f(n)\overline{g(n)} \;=\; \frac{1}{q}\sum_{k=1}^{q} \widehat{f}(k)\overline{\widehat{g}(k)}\,.$$

We refer to the above as "Parseval's Identity for the DFT." The original Parseval Identity pertains to Fourier Series; see Theorems 5.7 and 5.8.

Proof. By the Discrete Fourier Expansion of $g(n)$ found in Theorem 2.2 we see that the left hand side above is

$$\sum_{n=1}^{q} f(n)\overline{\frac{1}{q}\sum_{k=1}^{q}\widehat{g}(k)e(kn/q)} = \frac{1}{q}\sum_{n=1}^{q} f(n)\sum_{k=1}^{q}\overline{\widehat{g}(k)}e(-kn/q)\,.$$

On reversing the order of summation we see that this is

$$= \frac{1}{q}\sum_{k=1}^{q}\overline{\widehat{g}(k)}\sum_{n=1}^{q} f(n)e(-kn/q) = \frac{1}{q}\sum_{k=1}^{q}\widehat{f}(k)\overline{\widehat{g}(k)},$$

as desired. □

The case $g = f$ of this is of special interest.

Corollary 2.8. *If f is an arithmetic function with period q, then*

$$q\sum_{n=1}^{q}|f(n)|^2 = \sum_{k=1}^{q}|\widehat{f}(k)|^2 .$$

Example 2.6. Suppose we have a random number generator that produces the integers $1, 2, \dots, q$ with probabilities $p_1, p_2, \dots, p_q$. If the random number generator were perfect, then we would have $p_n = 1/q$ for all n. However, the "quasi random" generator is imperfect, so the p_n are only close to $1/q$, say $p_n = (1 + \delta_n)/q$ where the δ_n are small. One prescription for improving such a quasi random generator is to take two samples of it, X and Y, add them, and if the sum is greater than q, then subtract q so that the result lies between 1 and q. The question is to what extent we should believe that this operation gives rise to a distribution that is closer to uniform. Since $\sum_{n=1}^{q} p_n = 1$, it follows that $\sum_{n=1}^{q} \delta_n = 0$. The distribution of $X + Y$ modulo q is

$$\begin{aligned}(p * p)(n) &= \frac{1}{q^2}\sum_{a=1}^{q}(1+\delta_a)(1+\delta_{n-a}) \\ &= \frac{1}{q} + \frac{1}{q^2}\sum_{a=1}^{q}\delta_a\delta_{n-a} \\ &= \frac{1}{q}\Big(1 + \frac{1}{q}\sum_{a=1}^{q}\delta_a\delta_{n-a}\Big).\end{aligned}$$

By Theorems 2.2 and 2.5 we see that

$$\sum_{a=1}^{q}\delta_a\delta_{n-a} = \frac{1}{q}\sum_{k=1}^{q}\widehat{\delta}(k)^2 e(kn/q) .$$

By the triangle inequality, the left hand side above has absolute value not exceeding

$$\frac{1}{q}\sum_{k=1}^{q}\big|\widehat{\delta}(k)\big|^2 = \sum_{n=1}^{q}|\delta_n|^2$$

by Corollary 2.8. Thus the original relative error δ_n is replaced by a new relative error that does not exceed

$$\frac{1}{q}\sum_{n=1}^{q}|\delta_n|^2 .$$

In particular, if δ is a number such that $|\delta_n| \le \delta$ for all n, then the new relative error does not exceed δ^2. It may actually be even smaller, since in applying the triangle inequality we may be throwing away some cancelation.

Theorem 2.9. *Let p be an odd prime number, and put*

$$S_{k,p} = \sum_{n=1}^{p} e(kn^2/p) .$$

If $p|k$, then $S_{k,p} = p$, but if $p \nmid k$, then $|S_{k,p}| = \sqrt{p}$.

Proof. The first assertion is obvious. Suppose the $p \nmid k$. Then

$$\Big|\sum_{n=1}^{q} e(kn^2/p)\Big|^2 = \sum_{m=1}^{p}\sum_{n=1}^{p} e((m^2-n^2)/p)\,.$$

The change of variable $m = n+h$ gives $m^2 - n^2 = 2nh + h^2$, so the above is

$$= \sum_{h=1}^{p} e(kh^2/p) \sum_{n=1}^{p} e(2khn/p)\,.$$

Now $(2k,p) = 1$, so $p|2kh$ if and only if $p|h$. Thus the above sum over n vanishes unless $h = p$, which gives the result. □

Example 2.7. For an odd prime p, let N_p denote the number of solutions of the congruence $x^2+y^2+z^2 \equiv 0 \pmod p$. That is, among the triples of integers (x,y,z) with $1 \le x,y,z \le p$, the number N_p counts those triples for which $p|(x^2+y^2+z^2)$. Since we start with p^3 triples, and any one of them has a chance $1/p$ of producing a quantity that is divisible by p, we anticipate that N_p should be approximately p^2. We show that this is so, in the following precise sense: $|N_p - p^2| < p^{3/2}$. To see this, observe that

$$\sum_{k=1}^{p} e(k(x^2+y^2+z^2)/p) = \begin{cases} p & (p|(x^2+y^2+z^2)),\\ 0 & (\text{otherwise}).\end{cases}$$

Thus

$$\begin{aligned} pN_p &= \sum_{x,y,z}\sum_{k=1}^{p} e(k(x^2+y^2+z^2)/p)\\ &= \sum_{k=1}^{p}\Big(\sum_{x=1}^{p} e(kx^2/p)\Big)\Big(\sum_{y=1}^{p} e(ky^2/p)\Big)\Big(\sum_{z=1}^{p} e(kz^2/p)\Big) = \sum_{k=1}^{p} S_{k,p}^3 \end{aligned}$$

in the notation of Theorem 2.9. The contribution of $k = p$ on the right hand side is p^3. We subtract this from both sides to see that

$$|pN_p - p^3| = \Big|\sum_{k=1}^{p-1} S_{k,p}^3\Big| \le \sum_{k=1}^{p-1} |S_{k,p}|^3 < p^{5/2} \tag{2.23}$$

since $|S_{k,p}| = \sqrt{p}$ when $p \nmid k$. Thus we have the stated estimate.

Exercises

1. Suppose that f has period q. Use Theorem 2.4(c) to show the following:
   (a) If $f(n) = f(-n)$ for all n, then $\widehat{f}(k) = \widehat{f}(-k)$ for all k.
   (b) If $\widehat{f}(k) = \widehat{f}(-k)$ for all k, then $f(n) = f(-n)$ for all n.
2. Suppose that f has period q, and again use Theorem 2.4(c) to show that $f(n) = -f(-n)$ for all n if and only if $\widehat{f}(k) = -\widehat{f}(-k)$ for all k. (As in the preceding exercise, this requires two arguments.)
3. Suppose that f has period q.

(a) Use Theorem 2.4(d) to show that $\overline{f(n)} = f(n)$ for all n (i.e., f is real-valued) if and only if $\widehat{f}(-k) = \overline{\widehat{f}(k)}$ for all k.
(b) Show that if f is real-valued, then $\widehat{f}(k)e(kn/q) + \widehat{f}(-k)e(-kn/q)$ is real for all k.
(c) Suppose that f is real-valued and that q is even. When $k = q/2$, the two terms above are equal, and therefore the common value must be real. Show that $\widehat{f}(q/2)$ is real, and that $e((q/2)n/q)$ is real.

4. Suppose that f is an arithmetic function with period q. Put $g(k) = \widehat{f}(k)$. Then g has period q. Show that $\widehat{g}(n) = qf(-n)$.

5. Suppose that f is an arithmetic function with period q, and put $g(n) = f(n) + \widehat{f}(-n)/q$. Show that $\widehat{g}(k) = g(k)$.

6. For an arithmetic function with period q, let $\|f\|_1 = \sum_{n=1}^{q} |f(n)|$.
(a) Show that if f and g are arithmetic functions with period q, then $\|f+g\|_1 \le \|f\|_1 + \|g\|_1$.
(b) Show that if f and g are arithmetic functions with period q, then $\|f*g\|_1 \le \|f\|_1\|g\|_1$.

7. Let $f(n)$ be an arithmetic function with period 3, with $f(0) = 0$, $f(1) = 1$, and $f(2) = -1$.
(a) Find $\widehat{f}(k)$ for $k = 0, 1, 2$, and use the fact that $e(\pm 1/3) = -1/2 \pm i\sqrt{3}/2$ to simplify your answers.
(b) There is a constant c such that $\widehat{f}(k) = cf(k)$ for all k. What is this c?
(c) Write $f(n)$ in its discrete Fourier expansion, so that it is expressed as a linear combination of the three basic functions 1, $e(n/3)$, and $e(2n/3)$.

8. Let $f(n)$ be an arithmetic function with period 4, with $f(0) = f(2) = 0$, $f(1) = 1$, and $f(3) = -1$.
(a) Find $\widehat{f}(k)$ for $k = 0, 1, 2, 3$.
(b) There is a constant c such that $\widehat{f}(k) = cf(k)$ for all k. What is this c?
(c) Write $f(n)$ in its discrete Fourier expansion, so that it is expressed as a linear combination of the four functions 1, i^n, $(-1)^n$, $(-i)^n$.

9. Suppose that f is an arithmetic function with period q, that $|f(n)| = 1$ for N values of n in a period, that otherwise $f(n) = 0$, and that $\widehat{f}(k)$ is nonzero for exactly N values of k in a period. Show that $\max_k |\widehat{f}(k)| \ge \sqrt{q}$.

10. Suppose that δ is an arithmetic function with period q. Show that the following two assertions are equivalent:
(i) $\delta(0) = 1$ and $\delta(n) = 0$ for $0 < n < q$;
(ii) $\widehat{\delta}(k) = 1$ for all k.
The function $\delta(n)$ as in (i) above is the *DFT Dirac delta function.*

11. Suppose that f is an arithmetic function with period q. Show that the following two assertions are equivalent:
(i) There is a constant c such that $f(n) = c$ for all n;
(ii) $\widehat{f}(k) = 0$ for all k in the interval $0 < k < q$.

12. Let X and Y be independent random variables as in Example 2.5 with $q = 2$. Show that if $X+Y \pmod 2$ is uniform, then at least one of X or Y is uniform.

13. Let X and Y be independent random variables as in Example 2.5 with $q = 4$, and $p_1 = 1/8$, $p_2 = 3/8$, $p_3 = 3/8$, $p_4 = 1/8$, $r_1 = 1/8$, $r_2 = 3/8$, $r_3 = 1/8$, $r_4 = 3/8$.
   (a) Show that $\widehat{p}(2) = 0$.
   (b) Show that $\widehat{r}(1) = \widehat{r}(3) = 0$.
   (c) Show that $\widehat{p}(0) = \widehat{r}(0) = 1$.
   (d) Deduce that $(p * q)(n) = 1/4$ for all n. That is, $X + Y \pmod 4$ is uniform, even though neither X nor Y are. (Hint: Use Exercise 11.)

14. Let f be an arithmetic function with period q. In the context of Exercise 11, if f is not quite constant, we might measure the extent to which it deviates from its mean $\mu = \widehat{f}(0)/q$, in terms of the sizes of the $\widehat{f}(k)$.
   (a) Let $g(n) = \mu$ for all n. Explain why $\widehat{g}(k) = 0$ for $0 < k < q$.
   (b) Let $h(n) = f(n) - g(n)$. Show that $\widehat{h}(0) = 0$ and that $\widehat{h}(k) = \widehat{f}(k)$ for $0 < k < q$.
   (c) Show that
$$\sum_{n=1}^{q} |f(n) - \mu|^2 = \frac{1}{q}\sum_{k=1}^{q-1} |\widehat{f}(k)|^2 .$$

15. Let $c_1, c_2, \ldots, c_N$ be real or complex numbers, and put
$$S(x) = \sum_{n=1}^{N} c_n e(nx) .$$
Set $Z = S(0)$ and put
$$Z(q,k) = \sum_{\substack{n=1 \\ n \equiv k \pmod q}}^{N} a_n .$$
   (a) Show that
$$S(a/q) = \sum_{k=1}^{q} Z(q,k)e(ka/q)$$
   for $a = 1, 2, \ldots, q$.
   (b) Show that the average of the numbers $Z(q,k)$ is Z/q.
   (c) Use the preceding exercise to show that
$$\sum_{k=1}^{q} |Z(q,k) - Z/q|^2 = \frac{1}{q}\sum_{a=1}^{q-1} |S(a/q)|^2 .$$
   Let $\mathcal{N}$ be a collection of Z integers in the interval $[1, N]$, and put $a_n = 1$ if $n \in \mathcal{N}$, $a_n = 0$ otherwise. Then $Z(q,k)$ counts the members of $\mathcal{N}$ that are congruent to k modulo q. From the above we see that the members of $\mathcal{N}$ are evenly distributed into residue classes modulo q (in the mean square sense) if the sum on the right is small.

16. Use Example 2.4 and Corollary 2.8 to show that
$$\sum_{k=1}^{q-1} \cot^2 \frac{\pi k}{q} = \frac{(q-1)(q-2)}{3} .$$

(This was established in Exercise 4 by a different method.) Note: One of the formulas in (9.27) may prove to be useful.

17. Suppose that f and g are arithmetic functions with period q. Show that

$$\widehat{fg}(k) = \frac{1}{q}(\widehat{f} * \widehat{g})(k)\,.$$

18. Suppose that f has period q. Show that

$$\sum_{k=1}^{q} e(hk/q)|\widehat{f}(k)|^2 = q\sum_{a=1}^{q} f(a)\overline{f(a-h)}\,.$$

19. Let $g(x) = ax^2 + bx + c$ where a, b, c are integers. Suppose that p is an odd prime and that $p \nmid a$. Put

$$S_k(g) = \sum_{n=1}^{p} e(kg(n)/p)\,.$$

Show that $S_0(g) = p$ and that $|S_k(g)| = \sqrt{p}$ if $p \nmid k$.

20. Let $f(x) = ax^3 + bx^2 + cx + d$ be a cubic polynomial with integer coefficients. Let p be an odd prime, and suppose that $p \nmid a$. Let $S_k(f)$ be defined in the manner of the preceding exercise.
   (a) Let $g_h(x) = f(x+h) - f(x)$. Show that $g_h(x) = 3ahx^2 + (3ah^2 + 2bh)x + (ah^3 + bh^2 + ch)$.
   (b) Show that $|S_k(f)|^2 = \sum_{h=1}^{p} S_k(g_h)$.
   (c) Deduce that $|S_k(f)| \le 2p^{3/4}$ if $p \nmid k$.

2.3. The Fast Fourier Transform

Suppose that f is an arithmetic function with period q. The computation of $\widehat{f}(k)$ involves q multiplications and $q-1$ additions. If we do this for all q values of k, then we perform $\asymp q^2$ arithmetic operations. However, there is a faster way. We confine ourselves to the Fast Fourier Transform (FFT) as it was originally invented when q is a power of 2, but there are other FFTs, including some that work for general q, even prime q.

Theorem 2.10. *Let f have period q, and suppose that all q^{th} roots of unity have already been computed. Let $N(q)$ denote the least number of arithmetic operations required to compute $\widehat{f}(k)$ for all k $(1 \le k \le q)$. Then $N(2q) \le 2N(q) + 3q$.*

Proof. Suppose that f has period $2q$. Put $g(n) = f(2n)$ and $h(n) = f(2n+1)$. Thus g and h have period q, and

$$\begin{aligned}\widehat{f}(k) &= \sum_{n=1}^{2q} f(n)e(-kn/(2q)) \\ &= \sum_{n=1}^{q} f(2n)e(-kn/q) + \sum_{n=0}^{q-1} f(2n+1)e(-k(2n+1)/(2q)) \\ &= \sum_{n=1}^{q} g(n)e(-kn/q) + e(-k/(2q))\sum_{n=0}^{q-1} h(n)e(-kn/q) \\ &= \widehat{g}(k) + e(-k/(2q))\widehat{h}(k)\,.\end{aligned}$$

We compute $\widehat{g}(k)$ and $\widehat{h}(k)$ for all q values of k. This requires at most $2N(q)$ operations. The computation of $\widehat{f}(k)$ then requires two additional arithmetic operations. When this is done for all $2q$ values of k, we perform $4q$ operations. However, a further economy can be made. We note that $e(-(k+q)/(2q)) = -e(-k/(2q))$. Thus

$$\begin{aligned}\widehat{f}(k) &= \widehat{g}(k) + e(-k/(2q))\widehat{h}(k), \\ \widehat{f}(k+q) &= \widehat{g}(k) - e(-k/(2q))\widehat{h}(k).\end{aligned}$$

The multiplication of $e(-k/(2q))$ by $\widehat{h}(k)$ is performed only once, and the result of this multiplication is either added to $\widehat{g}(k)$ or subtracted from it. Thus we determine both $\widehat{f}(k)$ and $\widehat{f}(k+q)$ in three arithmetic operations. This is done q times to obtain $\widehat{f}(k)$ for all k. Thus $N(2q) \le 2N(q) + 3q$. □

Corollary 2.11. *Let $N(q)$ be defined as in Theorem 2.10. Then $N(2^r) \le \frac{3}{2}r2^r$.*

Proof. We note that $N(1) = 0$, since $\widehat{f}(k) = f(n)$ when f is an arithmetic function with period 1. Suppose that $N(2^r) \le \frac{3}{2}r2^r$. By Theorem 7 it follows that

$$N(2^{r+1}) \le 2N(2^r) + 3\cdot 2^r \le 3r2^r + 3\cdot 2^r = 3(r+1)2^r = \frac{3}{2}(r+1)2^{r+1}\,.$$

Thus the result follows by induction. □

At this stage of our discussion, we do not have a precise algorithm, but it is fairly plausible that by making systematic use of binary expansions we should be able to compute all q Discrete Fourier Transforms in $O(q\log q)$ operations, when q is a power of 2. Let us consider how the calculation would proceed when $q = 8$. For brevity we let $\omega = e(-1/8)$. We start with a first column of eight numbers in the order shown below left, and form linear combinations to create a second column, shown below right.

(2.24)

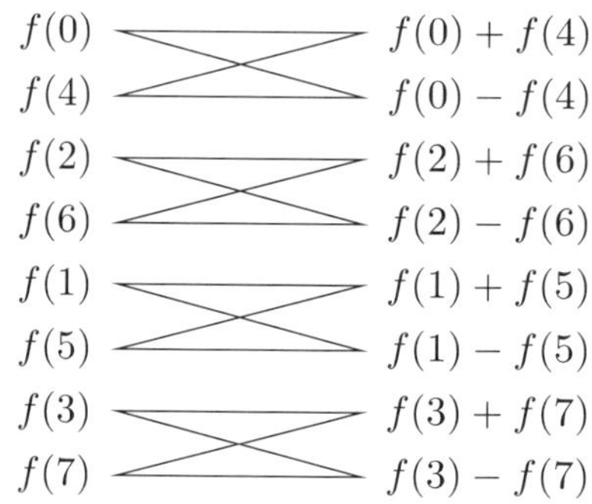

Number the rows 0 through 7. We combine rows $2j$ and $2j+1$ first with a $+$ sign and then with a $-$ sign to form rows $2j$ and $2j+1$ in the second column. The sequence of mulipliers $1, -1, \ldots, 1, -1$ applied to the second member in row j on the right can be thought of as $(\omega^4)^j$. Since we have no further use for the numbers in first column, the numbers in the second column can be stored in the same memory locations that had been used to store the first column. We next use the second column to form a third column

(2.25)

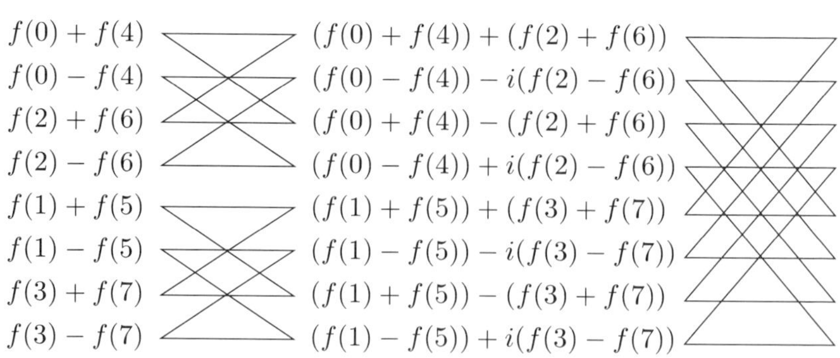

Here row $4j$ is combined with row $4j+2$ in two different ways to form new rows $4j$ and $4j+2$, and row $4j+1$ is combined with row $4j+3$ in two different ways to form new rows $4j+1$ and $4j+3$. Here the multiplier $1, -i, -1, i, \ldots$ applied to the second term in row j is $(\omega^2)^j$. Finally, we use this third column to generate a fourth column

$$
\begin{aligned}
\widehat{f}(0) &= ((f(0)+f(4))+(f(2)+f(6)))+((f(1)+f(5))+(f(3)+f(7)))\\
\widehat{f}(1) &= ((f(0)-f(4))-i(f(2)-f(6)))+\omega((f(1)-f(5))-i(f(3)-f(7)))\\
\widehat{f}(2) &= ((f(0)+f(4))-(f(2)+f(6)))+\omega^2((f(1)+f(5))-(f(3)+f(7)))\\
\widehat{f}(3) &= (((f(0)-f(4))+i(f(2)-f(6)))+\omega^3((f(1)-f(5))+i(f(3)-f(7)))\\
\widehat{f}(4) &= ((f(0)+f(4))+(f(2)+f(6)))+\omega^4((f(1)+f(5))+(f(3)+f(7)))\\
\widehat{f}(5) &= ((f(0)-f(4))-i(f(2)-f(6)))+\omega^5((f(1)-f(5))-i(f(3)-f(7)))\\
\widehat{f}(6) &= ((f(0)+f(4))-(f(2)+f(6)))+\omega^6((f(1)+f(5))-(f(3)+f(7)))\\
\widehat{f}(7) &= ((f(0)-f(4))+i(f(2)-f(6)))+\omega^7((f(1)-f(5))+i(f(3)-f(7)))
\end{aligned}
$$

For $0 \le j < 4$ we combine row j with row $j+4$ in two different ways to form a new row j and row $j+4$. The multiplier attached to the second term in row j is ω^j. One may wonder at the peculiar order in which we listed the values of $f(n)$ to start this calculation. However, all becomes clear when one observes that the numbers 0 through 7 in binary are 000, 001, 010, 011, 100, 101, 110, 111. When the binary digits are written in reversed order, we have 000, 100, 010, 110, 001, 101, 011, 111. These are the binary expansions of $0, 4, 2, 6, 1, 5, 3, 7$, respectively. We could go on to construct an inductive proof that these observed patterns continue indefinitely. The calculation can be coded efficiently, so that the nominal saving of $O(q \log q)$ versus $O(q^2)$ is achieved without an overly severe penalty in the constants.

For many applications we can choose the q we want to work with, so we are free to take q to be a power of 2. For purposes of digitizing music, we want q to be of the order of 10^4 or 10^5, so $O(q \log q)$ is a huge savings on the naïve $O(q^2)$.

Suppose that f and g have period q and that we want to compute the convolution $(f * g)(n)$. This requires $\asymp q$ arithmetic operations for one particular n. If we repeat this for all q values of n, then we perform $\asymp q^2$ arithmetic operations. However, by the FFT we can compute $\widehat{f}(k)$ and $\widehat{g}(k)$ in $\asymp q \log q$ operations, the products $\widehat{f}(k)\widehat{g}(k)$ in q multiplications, and by a third application of FFT we recover $(f * g)(n)$ for all n in an additional $\asymp q \log q$ operations. Thus FFT can be useful even when the Fourier transform is not the object of our interest. For example, when computing the coefficients of the product of two polynomials we are calculating a convolution. Thus the FFT can be used for rapid multiplication of polynomials of high degree.

Notes

2.2. The machinery of linear algebra, as discussed in Appendix L, enables an instructive insight concerning the Discrete Fourier Transform. Let $U = [u_{jk}]$ be a $q \times q$ matrix with entries

$$u_{jk} = \frac{e(-jk/q)}{\sqrt{q}}\,.$$

The inner product of the kth column of this matrix with the ℓth column is

$$\sum_{j=1}^{q} \frac{e(-jk/q)}{\sqrt{q}} \cdot \frac{e(j\ell/q)}{\sqrt{q}} = \frac{1}{q} \sum_{j=1}^{q} e(j(\ell - k)/q) = \begin{cases} 1 & (k = \ell), \\ 0 & (\text{otherwise}). \end{cases}$$

That is, the columns of this matrix are orthonormal. Let $\boldsymbol{f} \in \mathbb{C}^q$ be a vector with coordinates $f(j)$. From the definition (2.6) of the Discrete Fourier Transform, we see that

$$U\boldsymbol{f} = \frac{1}{\sqrt{q}} \widehat{\boldsymbol{f}}.$$

But $\|U\boldsymbol{f}\| = \|\boldsymbol{f}\|$, so we obtain again Parseval's Identity (Corollary 2.8) for the Discrete Fourier transform. If we had put a factor $1/\sqrt{q}$ in our definition of the Discrete Fourier transform, then the act of taking the transform would be a unitary transformation, and all distances would be preserved (and there would be no factor q in Theorem 2.7). Some authors do this, and one can see the advantage of this. We instead adopted a convention by which the FFT is easier to discuss. As things stand, all distances are rescaled by the same factor.

In Appendix L.3 we use the DFT to diagonalize circulant matrices.

2.3. The Fast Fourier Transform was introduced by Cooley and Tukey (1965). Cooley and Tukey had both been involved with John von Neumann's project in the 1940s to build an early digital computer at the Institute for Advanced Study in Princeton. At that time, Tukey invented the term "bit", short for "binary digit". In the 1960s, for the purposes of a treaty with the Soviet Union, the U. S. needed to develop means to detect atomic explosions from a distance. Tukey, a statistician at Princeton University, became involved in this effort to analyze offshore seismic data. He had the critical insight to realize the computation could be greatly reduced, and joined forces with Cooley, a researcher at IBM, who was the first to program the algorithm and thus verified its feasibility. A secondary application at that time was for the acoustic detection of nuclear submarines. Other applications soon followed. The Cooley–Tukey paper revolutionized the theory of mathematical algorithms. It has been said that this paper has been cited more times than any other paper in the mathematical literature. The FFT is used routinely in optics, acoustics, quantum physics, telecommunications, signal processing, and image processing. Thus it is incorporated on a DVD, in JPEG and MP3 formats, and in the tomography of MRI and CAT scans. Our world would be very different indeed without this algorithm.

Over time, it has become recognized that the FFT was discovered, in whole or in part, many times. Gauss had a form of it in 1805, as did Runge in 1803, and it seems that Lagrange may have had it earlier still. It's a tangled (but interesting) tale. It's safe to say that Cooley and Tukey discovered it for the last time. For a more detailed treatment of the FFT, see Benedetto (1997, pp. 239–248).

Chapter 3

Fourier Coefficients and First Fourier Series

3.1. Definitions and basic properties

We now turn to functions of period 1. In analogy with the functions $e(kn/q)$ for the Discrete Fourier Transform, we have functions $e(nx)$ with period 1. As in the former situation, the main question is whether we can express an arbitrary function as a linear combination of these basic functions, say

$$f(x) = \sum_{n=-\infty}^{\infty} c_n e(nx)\,. \tag{3.1}$$

Since in general infinitely many of the c_n are non-zero, questions of convergence arise, and this turns out to be rather tricky. We group $c_n e(nx)$ with $c_{-n}e(-nx)$, and note that

$$\begin{aligned} c_n e(nx) + c_{-n}e(-nx) &= c_n(\cos 2\pi nx + i\sin 2\pi nx) + c_{-n}(\cos 2\pi nx - i\sin 2\pi nx) \\ &= (c_n + c_{-n})\cos 2\pi nx + i(c_n - c_{-n})\sin 2\pi nx\,. \end{aligned}$$

Thus (3.1) is equivalent to writing

$$f(x) = \frac{1}{2}a_0 + \sum_{n=1}^{\infty}(a_n \cos 2\pi nx + b_n \sin 2\pi nx) \tag{3.2}$$

where

$$a_n = c_n + c_{-n}, \qquad b_n = ic_n - ic_{-n}\,. \tag{3.3}$$

Conversely, if we start with f written in the form (3.2), then we obtain (3.1) by taking

$$c_n = \frac{1}{2}(a_n - ib_n), \qquad c_{-n} = \frac{1}{2}(a_n + ib_n) \tag{3.4}$$

for nonnegative integers n. We shall work less with sine and cosine and more with the complex exponential, for which the formulæ are simpler, but we can always switch from one to the other whenever we please.

The audacious assertion of Fourier was that for any f with period 1 there is a choice of the c_n so that (3.1) holds. This was greeted with considerable skepticism, but later Dirichlet proved that he was essentially correct. Our most immediate concern is to determine how the c_n should be chosen in terms of f, if we are to have any hope that (3.1) should be true. We note that the functions $e(nx)$ are orthonormal on the interval $[0, 1]$ in the sense that if m and n are distinct integers, then

$$\int_0^1 e(nx)e(-mx)\,dx = \int_0^1 e((n-m)x)\,dx = \left[\frac{e((n-m)x)}{2\pi i(n-m)}\right|_0^1 = 0\,. \tag{3.5}$$

Thus if we multiply both sides of (3.1) by $e(-mx)$ and integrate from 0 to 1, then we find that

$$c_m = \int_0^1 f(x)e(-mx)\,dx\,.$$

Although this reasoning is not rigorous because of the unjustified exchange of integration and an infinite sum, it is nevertheless a useful guide.

Definition 3.1. If f has period 1 and $\int_0^1 |f(x)|\,dx < \infty$, then for each integer n we put

$$\widehat{f}(n) = \int_0^1 f(x)e(-nx)\,dx. \tag{3.6}$$

We say that the numbers $\widehat{f}(n)$ are the *Fourier coefficients* of f, and that the *Fourier Series* of f is

$$\sum_{n=-\infty}^{\infty} \widehat{f}(n)e(nx)\,. \tag{3.7}$$

Orally, the symbol $\widehat{f}$ is called 'eff-hat'. As we learn more, it will emerge that the only likely formula for $f(x)$ of the type (3.1) has $c_n = \widehat{f}(n)$ for all n. The Fourier Series (FS) of f can also be written in the form (3.2), where by (3.3) we see that

$$a_n = \widehat{f}(n) + \widehat{f}(-n) = 2\int_0^1 f(x)\cos 2\pi nx\,dx, \qquad (n \ge 0), \tag{3.8}$$

$$b_n = i\widehat{f}(n) - i\widehat{f}(-n) = 2\int_0^1 f(x)\sin 2\pi nx\,dx \qquad (n > 0). \tag{3.9}$$

Conversely, if we start with the Fourier Series of f in the form (3.2), we can recover the Fourier coefficients $\widehat{f}(n)$ by the formulæ

$$\widehat{f}(n) = \begin{cases} \frac{1}{2}(a_n - ib_n) & (n > 0), \\ \frac{1}{2}(a_{-n} + ib_{-n}) & (n < 0), \\ \frac{1}{2}a_0 & (n = 0). \end{cases} \tag{3.10}$$

In either of these systems of notation, the constant term is the average value of f,

$$\widehat{f}(0) = \frac{1}{2}a_0 = \int_0^1 f(x)\,dx\,. \tag{3.11}$$

We note also that if $n > 0$, then

$$\widehat{f}(n)e(nx) + \widehat{f}(-n)e(-nx) = a_n \cos 2\pi nx + b_n \sin 2\pi nx\,. \tag{3.12}$$

Thus the nth term in (3.2) is obtained by combining the nth and $(-n)^{\text{th}}$ terms in (3.7).

To avoid repeating the words "f has period 1 and $\int_0^1 |f(x)|\,dx < \infty$" many times, we introduce a notational shorthand. If S is a set over which one might integrate, then we define $L^p(S)$ to be the set of functions f for which

$$\int_S |f(x)|^p\,dx < \infty\,. \tag{3.13}$$

In the present situation, the set S of interest is the set of real numbers modulo 1. In this setting, x and y are interchangeable if they have the same fractional part, $\{x\} = \{y\}$. Equivalently, $x - y$ is an integer. In the same way that we say that two integers m and n are congruent modulo q if $q|(m-n)$, we say that two real numbers x and y are congruent modulo 1 if $x - y$ is an integer, and we write $x \equiv y \pmod 1$. The set of real numbers is denoted $\mathbb{R}$, the integers $\mathbb{Z}$, so $\mathbb{R}/\mathbb{Z}$ denotes the set of real numbers modulo 1 (which is to say modulo integers). We give this an even shorter name: $\mathbb{R}/\mathbb{Z} = \mathbb{T}$, and refer to this as the "circle group" because of the way that $e(x)$ moves around the unit circle in the complex plane. Thus, instead of saying "f has period 1 and $\int_0^1 |f(x)|\,dx < \infty$", we shall simply say "$f \in L^1(\mathbb{T})$". In these definitions, the letter L refers to the Lebesgue integral (a generalization of the Riemann integral), whose basic properties are recorded in §0.3. For the most part the reader can think of our integrals as if they were Riemann integrals. The symbol $\mathbb{T}$ might also seem puzzling. Suppose that $f(x, y)$ has period 1 with respect to both x and y. Thus the values of f are completely determined by its values in the unit square $[0, 1) \times [0, 1)$. Here the top edge is identified with the bottom, and the left edge is identified with the right. This is topologically a torus (aka doughnut). Hence the letter $\mathbb{T}$, even though $\mathbb{T}$ is only a one-dimensional torus (i.e., a circle). For real numbers we have an absolute value $|x|$ that measures size, and $|x - y|$ is the distance between x and y. Correspondingly, for $x \in \mathbb{T}$ we define

$$\|x\| = \min_{n\in\mathbb{Z}} |x - n|\,. \tag{3.14}$$

Thus if $x, y \in \mathbb{T}$, then the distance between them is $\|x - y\|$. For example, the distance between 0.1 and 0.9 is $\|0.1 - 0.9\| = \|-0.8\| = 0.2$.

If f has period 1, then we might integrate it from 0 to 1, but we now show that integrating over any other interval of length 1 will produce the same result. This is analogous to our remark (2.11) concerning sums of arithmetic functions with period q.

Lemma 3.1. *If f has period 1 and $\int_0^1 f(x)\,dx$ exists, then for any real number a,*

$$\int_a^{a+1} f(x)\,dx = \int_0^1 f(x)\,dx\,.$$

Proof. Suppose that $a = n + b$ where n is an integer and $0 \le b < 1$. Then

$$\int_a^{a+1} f(x)\,dx = \int_{n+b}^{n+b+1} f(x)\,dx = \int_b^{b+1} f(n+x)\,dx = \int_b^{b+1} f(x)\,dx$$

because f has period 1. To complete the proof we observe that

$$\begin{aligned}\int_b^{b+1} f(x)\,dx &= \int_b^1 f(x)\,dx + \int_1^{b+1} f(x)\,dx = \int_b^1 f(x)\,dx + \int_0^b f(x+1)\,dx \\ &= \int_b^1 f(x)\,dx + \int_0^b f(x)\,dx = \int_0^1 f(x)\,dx,\end{aligned}$$

as desired. □

From Definition 3.1, the triangle inequality (1.20) for integrals, and the fact that $|e(-nx)| = 1$ we deduce that

$$|\widehat{f}(n)| \le \int_0^1 |f(x)|\,dx \tag{3.15}$$

for all n. This is analogous to (2.14) for the Discrete Fourier Transform. Also like the DFT, Fourier coefficients are linear in the sense that if $f \in L^1(\mathbb{T})$, $g \in L^1(\mathbb{T})$ and $h(x) = \alpha f(x) + \beta g(x)$ for all x, then $h \in L^1(\mathbb{T})$ and $\widehat{h}(n) = \alpha\widehat{f}(n) + \beta\widehat{g}(n)$. Some further analogues are as follows.

Theorem 3.2. *Suppose that $f \in L^1(\mathbb{T})$.*

(a) *If a is a real number and $g(x) = f(x+a)$ for all x, then $\widehat{g}(n) = \widehat{f}(n)e(na)$ for all n.*

(b) *If b is an integer and $h(x) = f(x)e(bx)$ for all x, then $\widehat{h}(n) = \widehat{f}(n-b)$ for all n.*

(c) *If $j(x) = f(-x)$ for all x, then $\widehat{j}(n) = \widehat{f}(-n)$ for all n.*

(d) *If m is a positive integer, and $p(x) = f(mx)$ for all x, then $\widehat{p}(n) = \widehat{f}(n/m)$ if $m|n$, and $\widehat{p}(n) = 0$ otherwise.*

(e) *If $q(x) = \overline{f(x)}$ for all x, then $\widehat{q}(n) = \overline{\widehat{f}(-n)}$ for all n.*

These are the results one would expect if one were to make the corresponding alteration in the Fourier Series of f. For example, in case (d), if we replace x by mx in (3.7), then we have

$$\sum_{n=-\infty}^{\infty} \widehat{f}(n)e(mnx),$$

which is the Fourier Series of a function whose $(mn)^{\text{th}}$ Fourier coefficient is $\widehat{f}(n)$.

Proof. (a) From Definition 3.1 we see that

$$\begin{aligned}\widehat{g}(n) &= \int_0^1 g(x)e(-nx)\,dx = \int_0^1 f(x+a)e(-nx)\,dx \\ &= e(na)\int_0^1 f(x+a)e(-n(x+a))\,dx \\ &= e(na)\int_a^{a+1} f(x)e(-nx)\,dx = e(na)\widehat{f}(n)\end{aligned}$$

by Lemma 3.1.

(b) From Definition 3.1 we see that

$$\widehat{h}(n) = \int_0^1 h(x)e(-nx)\,dx = \int_0^1 f(x)e(-(n-b)x)\,dx = \widehat{f}(n-b)\,.$$

(c) From Definition 3.1 we see that

$$\widehat{j}(n) = \int_0^1 j(x)e(-nx)\,dx = \int_0^1 f(-x)e(-nx)\,dx\,.$$

Put $y = -x$. Then $dy = -dx$, so the above is

$$= -\int_0^{-1} f(y)e(ny)\,dy = \int_{-1}^0 f(y)e(ny)\,dy = \widehat{f}(-n)$$

by Lemma 3.1.

(d) We have

$$\begin{aligned}\widehat{p}(n) &= \int_0^1 f(mx)e(-nx)\,dx = \frac{1}{m}\int_0^m f(x)e(-nx/m)\,dx \\ &= \frac{1}{m}\sum_{k=1}^m \int_{k-1}^k f(x)e(-nm/m)\,dx = \frac{1}{m}\int_0^1 f(x)\Big(\sum_{k=1}^m e(-n(k+x)/m)\Big) \\ &= \frac{1}{m}\Big(\sum_{k=0}^{m-1} e(-nk/m)\Big)\int_0^1 f(x)e(-nx/m)\,dx\,.\end{aligned}$$

By Theorem 2.1, this last sum over k is m if $m|n$, and is 0 otherwise. This gives the result.

(e) We replace n by $-n$ in (3.6), and then take complex conjugates of both sides to see that

$$\overline{\widehat{f}(-n)} = \int_0^1 \overline{f(x)}e(-nx)\,dx = \widehat{q}(n)\,.$$

□

We note some immediate consequences of the above properties.

Corollary 3.3. *Suppose that $f \in L^1(\mathbb{T})$. If f is an even function, then $\widehat{f}(n)$ is an even sequence. If f is an odd function, then $\widehat{f}(n)$ is an odd sequence.*

From (3.9) we see that if f is even, then $b_n = 0$ for all n, so that we have only a cosine series in (3.2). This is reasonable, since it means that we are expressing an even function f as a sum of even terms. Similarly, if f is odd, then $a_n = 0$ for all n in (3.8), and then we have a sine series in (3.2).

Corollary 3.4. *Suppose that $f \in L^1(\mathbb{T})$. If f is real-valued, then $\widehat{f}(-n) = \overline{\widehat{f}(n)}$ for all n.*

In this case, the quantity $\widehat{f}(-n)e(-nx)$ is the complex conjugate of $\widehat{f}(n)e(nx)$, so the sum of these two terms is real. Also, $a_n = 2\operatorname{Re}\widehat{f}(n)$ in (3.8), and $b_n = -2\operatorname{Im}\widehat{f}(n)$ in (3.9), so the coefficients in (3.2) are real.

Example 3.1. Put $s(x) = 0$ when x is an integer, and $s(x) = 1/2 - \{x\}$ otherwise where $\{x\}$ is the fractional part of x. We call this the *sawtooth function*; see Figure 3.1(a). Thus $\widehat{s}(0) = \int_0^1 1/2 - x\,dx = \left[x/2 - x^2/2\right]_0^1 = 0$. For $n \neq 0$ we integrate by parts to see that

$$\widehat{s}(n) = \int_0^1 (1/2-x)e(-nx)\,dx = \left[(1/2-x)\frac{e(-nx)}{-2\pi in}\right]_0^1 - \frac{1}{2\pi in}\int_0^1 e(-nx)\,dx = \frac{1}{2\pi in}.$$

Hence the Fourier Series of $s(x)$ is

$$\sum_{n\neq 0}\frac{e(nx)}{2\pi in} = \sum_{n=1}^{\infty}\frac{\sin 2\pi nx}{\pi n}. \tag{3.16}$$

In §3.4 we show that this Fourier Series converges to $s(x)$ for all x. Here s is real-valued and odd, and we confirm that the sequence $\widehat{s}(n)$ is odd, that $a_n = 0$ for all n, and that all the b_n are real.

Example 3.2. Let f have period 1 and suppose that

$$f(x) = \begin{cases} s(x) & (0 < x \le 1, x \neq 1/4), \\ 10 & (x = 1/4) \end{cases}$$

where $s(x)$ is the sawtooth function of the preceding example. Then $\widehat{f}(n) = \widehat{s}(n)$ for all n, and in particular $\widehat{f}(-n) = -\widehat{f}(n)$ for all n, even though $f(x)$ is not an odd function, because $f(-1/4) = -1/4 \neq -10 = -f(1/4)$. Thus, in contrast to the situation of the DFT where we saw (in Exercise 2.2.2) that $f(n)$ is odd if and only if $\widehat{f}(k)$ is odd, the converse of Corollary 3.3 is false. However, we shall find later (in Exercise 4.1.3) that if the $\widehat{f}(n)$ are even (or odd), then $f(x)$ is almost even (or odd), in a sense that will be made precise.

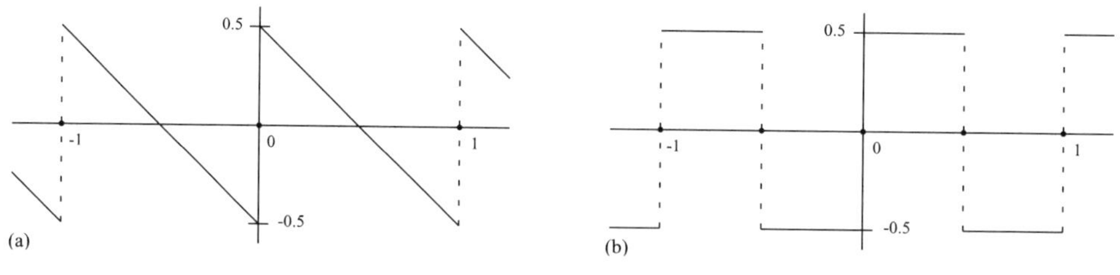

Figure 3.1. Graphs of (a) The sawtooth function $s(x)$; (b) The square wave function $w(x)$.

Example 3.3. Let w be the function with period 1 such that

$$w(x) = \begin{cases} 1/2 & (0 < x < 1/2), \\ -1/2 & (1/2 < x < 1), \\ 0 & (x = 0 \text{ or } 1/2), \end{cases} \tag{3.17}$$

as depicted in Figure 3.1(b). This a *square wave* function. We note that

$$\widehat{w}(n) = \frac{1}{2}\int_0^{1/2} e(-nx)\,dx - \frac{1}{2}\int_{1/2}^{1} e(-nx)\,dx = \frac{1}{2}\Big[\frac{e(-nx)}{-2\pi in}\Big|_0^{1/2} - \frac{1}{2}\Big[\frac{e(-nx)}{-2\pi in}\Big|_{1/2}^{1}$$

(3.18)

$$= -\frac{(-1)^n - 1}{4\pi in} + \frac{1-(-1)^n}{4\pi in} = \frac{1-(-1)^n}{2\pi in} = \begin{cases} \dfrac{1}{\pi in} & (n \text{ odd}), \\ 0 & (n \text{ even}). \end{cases}$$

Thus the Fourier Series of w is

$$\sum_{n \text{ odd}} \frac{e(nx)}{\pi in} = \sum_{n=1}^{\infty} \frac{2}{\pi(2n-1)} \sin 2\pi(2n-1)x\,. \tag{3.19}$$

That this series converges to $w(x)$ for all x will be established in §3.4.

From (3.15) we know that the sequence $|\widehat{f}(n)|$ is bounded. In the examples above, the Fourier coefficients are not just bounded, but tend to 0 as $|n| \to \infty$. We now show that this is always the case. To this end we begin with a useful

Lemma 3.5. *Suppose that $f \in L^1(\mathbb{T})$. Then*

$$\lim_{\delta \to 0} \int_0^1 |f(x+\delta) - f(x)|\,dx = 0\,.$$

Proof. By Theorem 0.22 we know that for any $f \in L^1(\mathbb{T})$ and any $\varepsilon > 0$ there is a continuous function g with period 1 such that $\int_0^1 |f(x) - g(x)|\,dx < \varepsilon$. Since g is continuous, it follows by Theorem 0.5 that g is uniformly continuous. In particular, $|g(x+\delta) - g(x)| < \varepsilon$ for all x, if δ is sufficiently small. Hence $\int_0^1 |g(x+\delta) - g(x)|\,dx < \varepsilon$. Clearly

$$f(x+\delta) - f(x) = \big(g(x+\delta) - g(x)\big) + \big(f(x+\delta) - g(x+\delta)\big) + \big(g(x) - f(x)\big),$$

so by the triangle inequality

$$|f(x+\delta) - f(x)| \le |g(x+\delta) - g(x)| + |f(x+\delta) - g(x+\delta)| + |g(x) - f(x)|\,.$$

We integrate both sides to deduce that

$$\int_0^1 |f(x+\delta) - f(x)|\,dx \le \int_0^1 |g(x+\delta) - g(x)|\,dx + \int_0^1 |f(x+\delta) - g(x+\delta)|\,dx$$
$$+ \int_0^1 |g(x) - f(x)|\,dx < 3\varepsilon\,.$$

Hence the result. □

Theorem 3.6. (The Riemann–Lebesgue Lemma) *Suppose that $f \in L^1(\mathbb{T})$. Then*

$$\lim_{n \to \pm\infty} \widehat{f}(n) = 0\,.$$

Proof. Suppose that $n \ne 0$. By Lemma 3.1 we see that

$$\widehat{f}(n) = \int_{1/(2n)}^{1+1/(2n)} f(x)e(-nx)\,dx = \int_0^1 f\big(x + \tfrac{1}{2n}\big)e\big(-n\big(x + \tfrac{1}{2n}\big)\big)\,dx\,.$$

But $e(-1/2) = -1$, so the above is

$$= -\int_0^1 f(x + \tfrac{1}{2n})e(-nx)\,dx\,.$$

By adding this to the original definition of $\widehat{f}(n)$, we find that

$$2\widehat{f}(n) = -\int_0^1 \big(f(x + \tfrac{1}{2n}) - f(x)\big)e(-nx)\,dx\,.$$

Hence by the triangle inequality,

$$2|\widehat{f}(n)| \le \int_0^1 \big|f(x + \tfrac{1}{2n}) - f(x)\big|\,dx\,. \tag{3.20}$$

The desired result now follows by taking $\delta = 1/(2n)$ in Lemma 3.5. □

Example 3.4. The *Dirac delta function with period* 1 places a unit point mass at every integer. One consequence of this is that $\int_0^1 \delta(x-a)f(x)\,dx = f(a)$ for any continuous function f. In particular, this means that $\widehat{\delta}(n) = 1$ for all n. We note that these coefficients do not decay to 0, which seems to be inconsistent with Theorem 3.6. The simple explanation is that the Dirac delta function is not actually a function, but rather a more general kind of object called a *distribution.* While the Dirac δ is not a function, we can construct families of peak functions that approximate δ. For example, if ε is small and positive, we can define f_ε to be the function with period 1 such that

$$f_\varepsilon(x) = \begin{cases} \dfrac{1}{2\varepsilon} & (-\varepsilon \le x \le \varepsilon), \\ 0 & (\varepsilon < x < 1-\varepsilon). \end{cases}$$

We think of f_ε as approximating δ when ε is small. We compute that

$$\widehat{f_\varepsilon}(n) = \begin{cases} \dfrac{\sin 2\pi n\varepsilon}{2\pi n\varepsilon} & (n \ne 0), \\ 1 & (n = 0), \end{cases}$$

and note that $\lim_{\varepsilon\to 0^+} \widehat{f_\varepsilon}(n) = 1$ for any fixed integer n. Also,

$$\int_0^1 f_\varepsilon(x-a)f(x)\,dx = \frac{1}{2\varepsilon}\int_{a-\varepsilon}^{a+\varepsilon} f(x)\,dx\,.$$

While this may not be exactly $f(a)$, it is approximately $f(a)$ if f is continuous at a and if ε is small.

If we integrate the Fourier Series (3.7) term-by-term, we obtain

$$\widehat{f}(0)x + \sum_{n\ne 0} \frac{\widehat{f}(n)}{2\pi i n} e(nx)\,.$$

We now show that this is indeed what arises when we integrate a function with period 1.

Theorem 3.7. *Suppose that $f \in L^1(\mathbb{T})$, and put*

$$F(x) = \int_0^x f(u)\,du - \widehat{f}(0)x\,.$$

Then $F(x)$ is continuous, $F(x)$ has period 1,

$$\widehat{F}(n) = \frac{\widehat{f}(n)}{2\pi i n}$$

for all $n \neq 0$, and

$$\widehat{F}(0) = \int_0^1 (1/2 - x) f(x)\,dx\,.$$

Proof. That F is continuous follows from principles of integration. To see that F has period 1, we note that

$$F(a+1) - F(a) = \int_a^{a+1} f(x)\,dx - \widehat{f}(0) = 0\,.$$

To compute $\widehat{F}(n)$ we integrate by parts. For $n \neq 0$ we find that

$$\widehat{F}(n) = \int_0^1 F(x)e(-nx)\,dx = \left[F(x)\frac{e(-nx)}{-2\pi i n}\right|_0^1 - \int_0^1 f(n)\frac{e(-nx)}{-2\pi i n}\,dx\,.$$

Here the first term vanishes since the expression has the same value at 0 as it does at 1. The second term gives the stated result. In the case $n = 0$ we again integrate by parts to find that

$$\begin{aligned}\widehat{F}(0) &= \int_0^1 F(x)\,dx = \left[xF(x)\right|_0^1 - \int_0^1 x(f(x) - \widehat{f}(0))\,dx \\ &= F(1) + \frac{1}{2}\widehat{f}(0) - \int_0^1 xf(x)\,dx = \int_0^1 (1/2 - x)f(x)\,dx\end{aligned}$$

since $F(1) = 0$, so we have the stated result. □

Corollary 3.8. *Suppose that f is a continuous function with period 1, and that there is a function $f' \in L^1(\mathbb{T})$ and a constant c such that*

$$f(x) = c + \int_0^x f'(u)\,du \tag{3.21}$$

for all x. Then

$$\widehat{f'}(n) = 2\pi i n \widehat{f}(n)$$

for all n.

Since $f(0) = f(1) = c$ in the above, the hypotheses imply that $\widehat{f'}(0) = 0$. It is essential in the above that f is continuous. For example, if we were to try to apply the above to the sawtooth function $s(x)$, we would find that

$$s(x) = \frac{1}{2} + \int_0^x g(u)\,du$$

for $0 < x < 1$ where $g(u) = -1$ for all u. Hence $g(x) = s'(x)$ for $0 < x < 1$, but $\widehat{g}(0) = -1$, and $\widehat{g}(n) = 0$ when $n \neq 0$. In Example 3.1 we determined the Fourier coefficients of $s(x)$. Thus the corollary, if applied mindlessly, would suggest that $\widehat{g}(0) = 0$ and that $\widehat{g}(n) = 1$ when $n \neq 0$, which, of course, is nonsense.

Example 3.5. Let $h(x) = \|x\|$ as defined in (3.14). Clearly $\widehat{h}(0) = 1/4$. To calculate $\widehat{h}(n)$ for nonzero n, we note that $h(x) = 4\int_0^x w(u)\,du$ where w is the square wave function of Example 3.3. Thus by Theorem 3.7 and (3.18) it follows that

$$\widehat{h}(n) = \begin{cases} \dfrac{-1}{\pi^2 n^2} & (n \text{ odd}), \\ 0 & (n \text{ even}). \end{cases}$$

Hence the Fourier Series of $\|x\|$ is

$$\frac{1}{4} - \sum_{n \text{ odd}} \frac{e(nx)}{\pi^2 n^2} = \frac{1}{4} - \sum_{n=1}^{\infty} \frac{2}{\pi^2(2n-1)^2} \cos 2\pi(2n-1)x\,. \tag{3.22}$$

For $s(x)$ in Example 3.1, the Fourier coefficients are of the order of $1/|n|$. This is typical of a function that is piecewise differentiable but has a finite number of jump discontinuities. In the example above, $\|x\|$ is continuous, piecewise differentiable, its derivative has jump discontinuities, and its Fourier coefficients are of the order of $1/n^2$. This is also typical of such functions. In general, the smoother a function is, the more rapidly its Fourier coefficients decay. On the other hand, some functions are quite jagged, and their Fourier coefficients may decay quite slowly. Instances of this are found in Example 4.2 and in §4.2. We conclude this section with some observations relating to the way in which the shape of a function may imply that its Fourier coefficients are nonnegative.

Theorem 3.9. *Suppose that $f \in L^1(\mathbb{T})$ and that the numbers b_n are defined as in (3.9). If $f(x)$ is weakly decreasing for $0 < x < 1$, then $b_n \ge 0$ for all positive integers n.*

This theorem applies, for example, to the sawtooth function of Example 3.1, and to the square wave function of Example 3.3. If f is weakly increasing, then $-f$ is weakly decreasing, and then $b_n \le 0$.

Proof. We write

$$b_n = 2\int_0^1 f(x)\sin 2\pi nx\,dx = 2\sum_{k=0}^{2n-1} \int_{k/(2n)}^{(k+1)/(2n)} f(x)\sin 2\pi nx\,.$$

Since $\sin 2\pi n(x + 1/(2n)) = -\sin 2\pi nx$, the above is

$$= 2\sum_{k=0}^{n-1} \int_0^{1/(2n)} (f(k/n + x) - f(k/n + 1/(2n) + x))\sin 2\pi nx\,.$$

These integrands are nonnegative since $\sin 2\pi nx > 0$ for $0 < x < 1/(2n)$ and f is decreasing. Hence $b_n \ge 0$, as claimed. □

For our next result we begin with a simple

Lemma 3.10. *Suppose that $f(x)$ is convex in $(0,1)$. Then*

$$\int_0^1 f(x)\cos 2\pi x\,dx \ge 0\,.$$

Proof. Let I denote this integral. Since $\cos 2\pi x = \cos 2\pi(1-x)$, we see that

$$I = \int_0^1 f(1-x)\cos 2\pi x\,dx\,.$$

Let $g(x) = f(x) + f(1-x)$. By adding our two formulas for I, we deduce that

$$2I = \int_0^1 g(x)\cos 2\pi x\,dx\,.$$

The sum of two convex functions is convex, so g is convex. Also, $g(1-x) = g(x)$, so

$$I = \int_0^{1/2} g(x)\cos 2\pi x\,dx\,.$$

Suppose that $0 < x_1 < x_2 \le 1/2$. In order that the point $(x_2, g(x_2))$ should lie below the chord from $(x_1, g(x_1))$ to $(1-x_1, g(1-x_1)) = (1-x_1, g(x_1))$, it is necessary that $g(x_2) \le g(x_1)$. Thus we see that g is decreasing in the interval $(0, 1/2)$. This allows us to argue as in the preceding proof. The function $\cos 2\pi x$ is positive for $0 < x < 1/4$ and negative for $1/4 < x < 1/2$, with $\cos 2\pi(1/2 - x) = -\cos 2\pi x$. Hence

$$I = \int_0^{1/4} (g(x) - g(1/2 - x))\cos 2\pi x\,dx \ge 0,$$

as desired. □

Theorem 3.11. *Suppose that $f \in L^1(\mathbb{T})$ and that the numbers a_n are defined as in (3.8). If f is weakly convex for $0 < x < 1$, then $a_n \ge 0$ for all positive integers n.*

If f is concave, then $-f$ is convex, and then $a_n \le 0$. The function $\|x\|$ is concave in $(0,1)$, and in Example 3.5 we saw that its coefficients (other than the constant term) are ≤ 0 . Another interesting concave function is $\log|2\sin \pi x|$, whose Fourier Series is discussed in Lemma 9.23.

Proof. For $n > 0$ we write

$$a_n = 2\int_0^1 f(x)\cos 2\pi n x\,dx = \sum_{k=0}^{n-1}\int_{k/n}^{(k+1)/n} f(x)\cos 2\pi n x\,dx\,.$$

After making the change of variable $x = (k+u)/n$ we see that the lemma applies, so each of the above integrals is nonnegative, and we have the result. □

Exercises

1. Let f and a sequence of functions f_k be members of $L^1(\mathbb{T})$. Show that if $\lim_{k\to\infty} \int_0^1 |f_k(x) - f(x)|\,dx = 0$, then $\widehat{f_k}(n)$ tends to $\widehat{f}(n)$ uniformly in n.
2. Suppose that f has period 1.
   (a) Show that if $f(x) = f(-x)$ for all x in the interval $-1/2 \le x \le 1/2$, then $f(x) = f(-x)$ for all real x (i.e., f is an even function).
   (b) Show that if $f(x) = -f(-x)$ for all x in the interval $-1/2 \le x \le 1/2$, then $f(x) = -f(-x)$ for all real x (i.e., f is an odd function).
   (c) Show that if $f(x) = f(1-x)$ for $0 \le x \le 1$, then f is even.

(d) Show that if $f(x) = -f(1-x)$ for $0 \le x \le 1$, then f is odd.

3. Suppose that f has period 1 and that $g(x) = f(x+1/2)$.
   (a) Show that if f is even, then g is even.
   (b) Show that if f is odd, then g is odd.
   (c) Suppose that $\int_0^1 |f(x)|\, dx < \infty$. How are the Fourier coefficients of f related to those of g?

4. Suppose that $f \in L^1(\mathbb{T})$. Show that if n is an integer for which $\widehat{f}(-n) = \overline{\widehat{f}(n)}$, then a_n and b_n are real for this same integer n.

5. Suppose that $f \in L^1(\mathbb{T})$. Show that if n is an integer for which a_n and b_n are real, then $\widehat{f}(-n) = \overline{\widehat{f}(n)}$ for this same integer n.

6. Set $W(x) = w(x+1/4)$ where w is the square wave function of Example 3.3.
   (a) Show that W is even.
   (b) Use Theorem 3.2(a) to determine the Fourier coefficients of W.
   (c) Give the Fourier Series of W, both in terms of $e(nx)$ and as a cosine series.

7. Suppose that $f(x)$ and $F(x)$ are defined as in Theorem 3.7, and that the Fourier Series of these functions are

$$\frac{1}{2}a_0 + \sum_{n=1}^{\infty} a_n \cos 2\pi nx + b_n \sin 2\pi nx,$$

$$\frac{1}{2}A_0 + \sum_{n=1}^{\infty} A_n \cos 2\pi nx + B_n \sin 2\pi nx,$$

respectively. Show that

$$A_n = -\frac{b_n}{2\pi n}, \quad \text{and} \quad B_n = \frac{a_n}{2\pi n}$$

for all positive n.

8. Suppose that f has period 1, and that f' is continuous, with $|f'(x)| \le C$ for all x. Show that

$$|\widehat{f}(n)| \le \frac{C}{2\pi |n|}$$

for all $n \ne 0$.

9. Suppose that f has period 1, that f'' is continuous with $|f''(x)| \le C$ for all x. Combine Exercise 8 with Corollary 3.8 to show that

$$|\widehat{f}(n)| \le \frac{C}{(2\pi n)^2}$$

for all $n \ne 0$.

10. Explain the error in the following argument. Let $f(x) = 1/2 - \{x\}$. If $n \ne 0$, then by integration by parts,

$$\begin{aligned}\widehat{f}(n) = \int_0^1 f(x)e(-nx)\, dx &= \left[f(x)\frac{e(-nx)}{-2\pi i n}\right|_0^1 - \int_0^1 f'(x)\frac{e(-nx)}{-2\pi i n}\, dx \\ &= \frac{f(1)}{-2\pi i n} - \frac{f(0)}{-2\pi i n} - \frac{1}{2\pi i n}\int_0^1 e(-nx)\, dx\end{aligned}$$

since $f'(x) = -1$. But f has period 1, so $f(0) = f(1)$, and the last integral vanishes, so $\widehat{f}(n) = 0$ for all $n \neq 0$.

11. Let $f(x) = |\sin \pi x|$. We note that f is continuous and has period 1.
   (a) Show that $\int_0^1 \sin \pi x\, dx = 2/\pi$.
   (b) By writing $\sin \pi x = (e(x/2) - e(-x/2))/(2i)$, show that
$$\widehat{f}(n) = \frac{2}{\pi(1-4n^2)}$$
   for all integers n.
   (c) Deduce that the Fourier Series of f is
$$\frac{2}{\pi} - \frac{4}{\pi}\sum_{n=1}^{\infty} \frac{\cos 2\pi n x}{4n^2-1}\,.$$
   (d) The function f is concave. Are the signs of its Fourier coefficients consistent with Theorem 3.11?
   (e) Suppose that g has period 1 and that $g(x) = \pi \cos \pi x$ for $0 < x < 1$. By writing $\cos \pi x = (e(x/2) + e(-x/2))/2$, show that
$$\widehat{g}(n) = \frac{4ni}{1-4n^2}$$
   for all n.
   (f) Note that $f(x) = \int_0^x g(u)\, du$ for all x. Use Corollary 3.8 to give a second derivation of the $\widehat{g}(n)$.
   (g) Use the identity $g'(x) = -\pi^2 f(x)$ to determine the values of $\widehat{g'}(n)$.
12. Let f have period 1, and suppose that $f(x) = (1-2x)\cos \pi x$ for $0 \le x \le 1$.
   (a) Show that f is continuous, even, that it is differentiable for $0 < x < 1$, and that f' has a jump discontinuity of height 4 at $x = 0$.
   (b) Show that the Fourier Series of f is
$$\frac{4}{\pi^2}\sum_{n=-\infty}^{\infty} \frac{4n^2+1}{(4n^2-1)^2} e(nx) = \frac{4}{\pi^2} + \frac{8}{\pi^2}\sum_{n=1}^{\infty} \frac{4n^2+1}{(4n^2-1)^2}\cos 2\pi n x\,.$$
13. Suppose that f has period 1 and that $|f(x)| \le 1$ for all x. We want to know how large $|\widehat{f}(1) - \widehat{f}(0)|$ can be.
   (a) Explain why
$$\widehat{f}(1) - \widehat{f}(0) = \int_0^1 f(x)(e(-x) - 1)\, dx\,.$$
   (b) Deduce that
$$|\widehat{f}(1) - \widehat{f}(0)| \le \int_0^1 |e(-x) - 1|\, dx\,. \tag{3.23}$$
   (c) Show that $|e(-x) - 1| = 2|\sin \pi x|$.
   (d) Deduce that
$$|\widehat{f}(1) - \widehat{f}(0)| \le \frac{4}{\pi}\,.$$

(e) Show that

$$e(-x) - 1 = -2ie(-x/2)\sin \pi x$$

for all x.

(f) Let f be the function with period 1 such that $f(x) = ie(x/2)$ for $0 \le x < 1$. Show that

$$f(x)(e(-x) - 1) = 2\sin \pi x$$

for $0 \le x \le 1$.

(g) Deduce that for this f, the inequality (3.23) holds with equality.

(h) The chosen f has a jump discontinuity at $x = 0$. How high is this jump?

(i) Show that for the chosen f,

$$\widehat{f}(n) = \frac{2}{\pi(2n-1)}$$

for all n.

(j) Conclude that $\widehat{f}(1) = 2/\pi$ and that $\widehat{f}(0) = -2/\pi$, so that $\widehat{f}(1) - \widehat{f}(0) = 4/\pi$.

14. Suppose that f is convex in $(0,1)$. That is, if $0 < a < x < b < 1$, then

$$f(x) \le \frac{f(b) - f(a)}{b - a}(x - a) + f(a)\,.$$

For $0 < a < b < 1$, let $s(a,b) = (f(b) - f(a))/(b - a)$ denote the slope of the chord from the point $(a, f(a))$ to $(b, f(b))$.

(a) Show that if $0 < a < b < b' < 1$, then $s(a,b) \le s(a,b')$. (Thus $s(a,b)$ is an increasing function of b when $b > a$ and a is fixed.)

(b) Show that if $0 < a < a' < b < 1$, then $s(a,b) \le s(a',b)$. (Thus $s(a,b)$ is an increasing function of a when $a < b$ and b is fixed.)

(c) Suppose that $0 < a < b < 1$, $0 < a' < b' < 1$, $a < a'$, and $b < b'$. Show that $s(a,b) \le s(a',b')$.

(d) Suppose that $0 \le x < 1/4$. Show that $s(x, 1/2 - x) \le s(x + 1/2, 1 - x)$.

(e) Deduce that $f(1/2 - x) - f(x) \le f(1 - x) - f(x + 1/2)$.

(f) Show that $\cos 2\pi x = \cos 2\pi(1-x)$, and that $\cos 2\pi(1/2-x) = \cos 2\pi(1/2+x) = -\cos 2\pi x$.

(g) Show that

$$\int_0^1 f(x)\cos 2\pi x\,dx$$
$$= \int_0^{1/4} \big(f(x) - f(1/2 - x) - f(1/2 + x) + f(1 - x)\big)\cos 2\pi x\,dx\,.$$

(h) Use this to complete a second proof of Lemma 3.10.

15. When computing Fourier coefficients of functions defined by polynomials we integrate by parts repeatedly.

(a) (Kronecker) Show that if $p(x)$ is a polynomial of degree at most n and $f(x)$ is a continuous function, then

$$\int p(x)f(x)\,dx = \sum_{k=0}^{n}(-1)^k p^{(k)}(x)F_{n+1}(x) \ + C \tag{3.24}$$

where $F_1'(x) = f(x)$ and $F_{k+1}'(x) = F_k(x)$ for $1 \le k \le n$.

(b) Deduce that if p is a polynomial of degree at most n, then

$$\int p(x)e(-nx)\,dx = \left(\sum_{r=0}^{n} \frac{p^{(r)}(x)}{(2\pi i n)^r}\right) e(-nx) + C\,.$$

(c) Let p and P be polynomials. Show that the following two assertions are equivalent:
(i) $P(x) = p(x) - p^{(2)}(x) + p^{(4)}(x) - p^{(6)}(x) + \cdots$ for all x;
(ii) $p(x) = P(x) + P''(x)$ for all x.

(d) Suppose that p and P are related as above. Show that

$$\int p(x)\sin x\,dx = P'(x)\sin x - P(x)\cos x + C\,.$$

(e) Let p and P be related as above. Show that

$$\int p(x)\cos x\,dx = P(x)\sin x + P'(x)\cos x + C\,.$$

16. Let f and g be functions that are at least n times differentiable. Show that

$$(fg)^{(n)}(x) = \sum_{k=0}^{n} \binom{n}{k} f^{(k)}(x) g^{(n-k)}(x)\,.$$

17. Let $p(x) = x^n(\pi - x)^n$, let $P(x) = p(x) - p^{(2)}(x) + p^{(4)}(x) - \cdots$, and put

$$I = \int_0^{\pi} p(x)\sin x\,dx\,.$$

(a) Show that $I = P(0) + P(\pi)$.
(b) Show that $p^{(r)}(x) = (-1)^r p^{(r)}(\pi - x)$ for $r = 0, 1, \ldots$.
(c) Show that $P(x) = P(\pi - x)$. Deduce that $I = 2P(0)$.
(d) Show that

$$\left.\frac{d^r}{dx^r}\,x^n\right|_{x=0} = \begin{cases} n! & (r = n), \\ 0 & (\text{otherwise}). \end{cases}$$

(e) Show that

$$p^{(2k)}(0) = \begin{cases} 0 & (2k < n), \\ (-1)^n n(n-1)\cdots(2n-2k+1)\pi^{2n-2k} & (n/2 \le k \le n), \\ 0 & (k > n)\,. \end{cases}$$

(f) Deduce that

$$I = 2\cdot n! \sum_{n/2\le k\le n} (-1)^k n(n-1)\cdots(2n-2k+1)\pi^{2n-2k}\,.$$

(g) Suppose that π^2 is a rational number, say $\pi^2 = a/q$. Show that $q^n I/n!$ is an integer.
(h) Show that $x(\pi - x) \le (\pi/2)^2$ for $0 \le x \le \pi$.
(i) Deduce that $0 < q^n I/n! < \pi(\pi/2)^{2n} q^n/n!$.
(j) Let $A > 0$ be fixed. Show that $A^n/n! \to 0$ as $n \to \infty$.
(k) Deduce that $0 < q^n I/n! < 1$ for all sufficiently large n.
(l) Conclude that π^2 is irrational.
(m) (Lambert, 1761) Deduce that π is also irrational.

18. Let $P(x) = \sum_{k=0}^{n} a_k x^k$ be a polynomial.
    (a) Show that $P(x) = P(-x)$ for all x if and only if $a_{2k+1} = 0$ for $k = 1, 2, \ldots$.
    (b) Show that $P(x) = -P(-x)$ for all x if and only if $a_{2k} = 0$ for $k = 0, 2, \ldots$.
19. Let $f(x) = P(x)\sin x + Q(x)\cos x$ where P and Q are polynomials.
    (a) Show that if f is even, then $Q(2k\pi) = Q(-2k\pi)$ for all integers k. Deduce that $Q(x)$ is an even function.
    (b) Show that if f is even, then $P((2k+1/2)\pi) = P(-(2k+1/2)\pi)$ for $k = 1, 2, \ldots$. Deduce that $P(x)$ is an odd function.
    (c) Suppose that f is an odd function. Show that P is even, and that Q is odd.
20. Let $P(x) = \sum_{k=0}^{n} a_k x^k$ be a polynomial.
    (a) Show that

$$P^{(r)}(x)/r! = \sum_{k=r}^{n} \binom{k}{r} a_k x^{k-r}$$

    for $r = 0, 1, \ldots, n$.
    (b) Deduce that if the coefficients a_k of P are all integers, then the coefficients of $P^{(r)}/r!$ are also all integers.

3.2. Other periods

When we are presented with a problem involving angles in degrees, it is often a good idea to rewrite the problem with angles in radians. The is especially advisable if the problem is going to involve calculus, since a formula such as $(\sin x)' = \cos x$ is valid only when angles are in radians. When the problem has been solved, the answer can be translated back into degrees. We proceed in the same way when presented with a periodic problem in which the period is not 1. We simply translate to an equivalent problem with period 1, solve the problem, and then translate our answer back. It is true that one can do Fourier Analysis with period P, but most of the formulas will involve P. By working with period 1, the constant in various formulas is almost always 1. In Theorem 3.7 and Corollary 3.8 we have rare examples in which the constant is not 1, but in those cases it is easy to see why the $2\pi i n$ arises, and where it belongs.

Suppose that f has period $P > 0$. We put $g(x) = f(Px)$. Thus g has period 1. We note that

$$\widehat{g}(n) = \int_0^1 g(x)e(-nx)\,dx = \int_0^1 f(Px)e(-nx)\,dx = \frac{1}{P}\int_0^P f(x)e(-nx/P)\,dx\,.$$

The Fourier Series of g is $\sum \widehat{g}(n)e(nx)$. We expect this to provide a formula for $g(x) = f(Px)$. On replacing x by x/P, we obtain a formula for $f(x) = g(x/P)$, namely

$$\sum_{n=-\infty}^{\infty} c_n e(nx/P) \qquad (3.25)$$

where

$$c_n = \widehat{g}(n) = \frac{1}{P}\int_0^P f(x)e(-nx/P)\,dx\,. \tag{3.26}$$

We can write

$$\begin{aligned} c_n e(nx/P) + c_{-n}e(-nx/P) &= (c_n + c_{-n})\cos 2\pi nx/P + (ic_n - ic_{-n})\sin 2\pi nx/P \\ &= a_n \cos 2\pi nx/P + b_n \sin 2\pi nx/P \end{aligned}$$

where $a_n = c_n + c_{-n}$ and $b_n = ic_n - ic_{-n}$. Thus (3.3) holds without alteration, for any $P > 0$. However, in view of (3.26) the formulæ (3.8), (3.9) must be replaced by

$$a_n = \frac{2}{P}\int_0^P f(x)\cos 2\pi nx/P\,dx, \qquad b_n = \frac{2}{P}\int_0^P f(x)\sin 2\pi nx/P\,dx\,. \tag{3.27}$$

In this notation, the Fourier Series of f is

$$\frac{1}{2}a_0 + \sum_{n=1}^{\infty}(a_n\cos 2\pi nx/P + b_n \sin 2\pi nx/P)\,. \tag{3.28}$$

Of course, the Fourier Series for f' is

$$\frac{2\pi i}{P}\sum_{n=-\infty}^{\infty} nc_n e(nx/P) \;=\; \frac{2\pi}{P}\sum_{n=1}^{\infty}(nb_n\cos 2\pi nx/P - na_n \sin 2\pi nx/P)\,.$$

3.3. Convolution

Suppose that $f \in L^1(\mathbb{T})$, and that $g \in L^1(\mathbb{T})$. The *convolution* of f and g is the function

$$h(x) = (f * g)(x) = \int_0^1 f(u)g(x-u)\,du\,. \tag{3.29}$$

In the integral on the right, put $v = x - u$, so that $dv = -du$. Then the integral is

$$= -\int_x^{x-1} f(x-v)g(v)\,dv = \int_{x-1}^{x} g(v)f(x-v)\,dv = \int_0^1 g(v)f(x-v)\,dv = (g*f)(x)$$

by Lemma 3.1. Thus

$$f * g = g * f\,. \tag{3.30}$$

These formulas are unfortunately asymmetric, but they are trying to express something symmetric of the sort

$$\iint\limits_{\substack{0\le u\le 1\\ 0\le v\le 1\\ u+v\equiv x \,(\mathrm{mod}\,1)}} f(u)g(v)\,dv\,du\,.$$

The idea is that we are averaging $f(u)g(v)$ over those pairs u, v of arguments that sum to x (mod 1).

Theorem 3.12. *Let f, g, and h be as above. Then $h(x)$ has period* 1,

$$\int_0^1 |h(x)|\,dx \le \Big(\int_0^1 |f(u)|\,du\Big)\Big(\int_0^1 |g(v)|\,dv\Big), \tag{3.31}$$

and

$$\widehat{h}(n) = \widehat{f}(n)\widehat{g}(n) \tag{3.32}$$

for all n.

Recall that in §3.1 we defined

$$\|f\|_1 = \int_0^1 |f(x)|\,dx\,. \tag{3.33}$$

This is the L^1 norm of f. In this notation, the relation (3.31) asserts that $\|f * g\|_1 \le \|f\|_1\|g\|_1$.

Proof. That $h(x)$ has period 1 is clear. By the triangle inequality for integrals we see that

$$\begin{aligned}\int_0^1 |h(x)|\,d\theta &= \int_0^1 \Big|\int_0^1 f(u)g(x-u)\,du\Big|\,dx \\ &\le \int_0^1\int_0^1 |f(u)g(x-u)|\,du\,dx \\ &= \int_0^1 \Big(\int_0^1 |g(x-u)|\,dx\Big)|f(u)|\,du\,. \end{aligned} \tag{3.34}$$

The integral over x is

$$= \int_{-u}^{1-u} |g(x)|\,dx = \int_0^1 |g(x)|\,dx$$

by Lemma 3.1. On inserting this in (3.34), we deduce that

$$\int_0^1 |h(x)|\,dx \;\le\; \Big(\int_0^1 |f(u)|\,du\Big)\Big(\int_0^1 |g(x)|\,dx\Big) \;<\; \infty\,.$$

As for the Fourier coefficients of h, we find that

$$\widehat{h}(n) = \int_0^1 h(x)e(-nx)\,dx = \int_0^1 \Big(\int_0^1 f(u)g(x-u)\,du\Big)\,e(-nx)\,dx\,.$$

We exchange the integrals to see that this is

$$= \int_0^1 f(u)\Big(\int_0^1 g(x-u)e(-nx)\,dx\Big)\,du\,. \tag{3.35}$$

This exchange is justified, since we have already shown that the double integral is absolutely convergent. In the inner integral we make the change of variable $v = x - u$. Thus the inner integral is

$$\begin{aligned}\int_{-u}^{1-u} g(v)e(-n(u+v))\,dv &= e(-nu)\int_{-u}^{1-u} g(v)e(-nv)\,d\phi \\ &= e(-nu)\int_0^1 g(v)e(-nv)\,d\phi = e(-nu)\widehat{g}(n)\,.\end{aligned}$$

On inserting this in (3.35), we obtain (3.32), and the proof is complete. □

From (3.8)–(3.10) we deduce that if

$$\frac{1}{2}a_i(0) + \sum_{n=1}^{\infty}(a_i(n)\cos 2\pi nx + b_i(n)\sin 2\pi nx)$$

for $i = 1, 2, 3$ are the Fourier Series of f_1, f_2, and $f_3 = f_1 * f_2$, then $a_3(0) = \frac{1}{2}a_1(0)a_2(0)$ and

$$\begin{aligned} a_3(n) &= \frac{1}{2}a_1(n)a_2(n) - \frac{1}{2}b_1(n)b_2(n), \\ b_3(n) &= \frac{1}{2}a_1(n)b_2(n) + \frac{1}{2}a_2(n)b_1(n) \end{aligned} \tag{3.36}$$

for $n > 0$.

Example 3.6. Let δ be a fixed small positive real number. Put $f(x) = 1$ if x differs from an integer by no more than δ and set $f(x) = 0$ otherwise. Let $g(x) = f * f$. In the definition of the convolution, $f(u) = 1$ for $-\delta \le u \le \delta$, and $f(x - u) = 1$ for $x - \delta \le u \le x + \delta$. These two intervals are disjoint if $2\delta < x < 1 - 2\delta$. Suppose that $0 \le x \le 2\delta$. Then

$$(f * f)(x) = \int_{x-\delta}^{\delta} 1\, dx = 2\delta - x\,.$$

Similarly, if $-2\delta \le x \le 0$, then

$$(f * f)(x) = \int_{-\delta}^{x+\delta} 1\, dx = x + 2\delta\,.$$

Thus $(f * f)(x) = \max(0, 2\delta - |x|)$ for $-1/2 \le x \le 1/2$. We can confirm (3.32) in this situation by direct calculation of the Fourier coefficients of f and of $f * f$. Clearly $\widehat{f}(0) = 2\delta$, and for $n \ne 0$,

$$\widehat{f}(n) = \int_{-\delta}^{\delta} e(-nx)\, dx = \left[\frac{e(-nx)}{-2\pi i n}\right|_{-\delta}^{\delta} = \frac{\sin 2\pi n\delta}{\pi n}\,.$$

As for $f * f$, we note first that

$$\widehat{f*f}(0) = 2\int_0^{2\delta} 2\delta - x\, dx = \left[4\delta x - x^2\right|_0^{2\delta} = 4\delta^2 = \widehat{f}(0)^2\,.$$

Thus (3.32) holds for $n = 0$. Suppose that $n \ne 0$. Then

$$\widehat{f*f}(n) = \int_{-2\delta}^{2\delta}(2\delta - |x|)e(-nx)\, dx = 2\int_0^{2\delta}(2\delta - x)\cos 2\pi nx\, dx,$$

which by integration by parts is

$$= \left[(2\delta - x)\frac{\sin 2\pi nx}{\pi n}\right|_0^{2\delta} + \int_0^{2\delta}\frac{\sin 2\pi nx}{\pi n}\, dx.$$

Here the first expression vanishes at both endpoints, and the remaining integral is

$$= \left[\frac{-\cos 2\pi nx}{2(\pi n)^2}\right|_0^{2\delta} = \frac{1 - \cos 4\pi n\delta}{2(\pi n)^2} = \Big(\frac{\sin 2\pi n\delta}{\pi n}\Big)^2$$

since $\frac{1}{2}(1-\cos\phi) = (\sin\phi/2)^2$. This is $\widehat{f}(n)^2$, so (3.32) is confirmed also for $n \neq 0$.

A general rule of thumb is that $f * g$ is smoother than either f or g. This is seen in Example 3.6, since f is piecewise continuous with jump discontinuities, while $f * f$ is continuous and piecewise differentiable, and the derivative has jump discontinuities. Of course, since $\widehat{f}(n) \to 0$ and $\widehat{g}(n) \to 0$, it follows that the Fourier coefficients of $f * g$ tend to 0 faster than the coefficients of f, or of g. We now consider a further example in this direction. First we introduce a little notation. For functions with period 1 we set

$$\|f\|_\infty = \sup_{0 \le x \le 1} |f(x)|\,. \tag{3.37}$$

Here "sup" (pronounced "soup") is an abbreviation of *supremum*, which means the least upper bound. So $L^\infty(\mathbb{T})$ is the set of bounded functions with period 1.

Theorem 3.13. *Suppose that f is a bounded function with period 1, and that $g \in L^1(\mathbb{T})$. Then $f * g$ is a continuous function, and* $\max|f * g| \le \|f\|_\infty \|g\|_1$.

The last clause is rather obvious: Since $|f(x)| \le \|f\|_\infty$ for all x,

$$\begin{aligned}|(f*g)(x)| = \Big|\int_0^1 f(x-u)g(u)\,du\Big| &\le \int_0^1 |f(x-u)g(u)|\,du \le \|f\|_\infty \int_0^1 |g(u)|\,du \\ &= \|f\|_\infty \|g\|_1\,.\end{aligned}$$

Thus the important part of the theorem is not that $f * g$ is bounded, but rather that it is continuous. Among bounded functions with period 1, continuous functions form a special (and important) subclass.

Proof. By the definition of convolution we know that

$$(f*g)(x) = \int_0^1 f(u)g(x-u)\,du.$$

Hence

$$(f*g)(x+\delta) - (f*g)(x) = \int_0^1 f(u)\big(g(x+\delta-u) - g(x-u)\big)\,du\,. \tag{3.38}$$

Choose M so that $|f(u)| \le M$ for all u. Then by the triangle inequality,

$$\begin{aligned}|(f*g)(x+\delta) - (f*g)(x)| &\le \int_0^1 \big|f(u)\big(g(x+\delta-u) - g(x-u)\big)\big|\,du \\ &\le M\int_0^1 \big|g(x+\delta-u) - g(x-u)\big|\,du \\ &= M\int_0^1 \big|g(v+\delta) - g(v)\big|\,dv\,.\end{aligned}$$

This last integral tends to 0 as $\delta \to 0$, by Lemma 3.5. Thus $|(f*g)(x+\delta) - (f*g)(x)| < \varepsilon$ for all sufficiently small δ, which gives the result. □

Example 3.7. In Theorem 3.13, the assertion that f is bounded is a very strong hypothesis. We now consider the convolution of two unbounded functions. Let α

and β be fixed, with $0 < \alpha < 1$ and $0 < \beta < 1$. Initially we assume that $\alpha + \beta > 1$. We set

$$f(x) = \|x\|^{-\alpha}, \qquad g(x) = \|x\|^{-\beta}.$$

We show that

$$(f * g)(x) \asymp \|x\|^{1-\alpha-\beta}. \tag{3.39}$$

The implicit constants here may depend on α and β, but this dependence is suppressed, since α and β are fixed. As this is an even function, it suffices to consider $0 \le x \le 1/2$. We write

$$(f * g)(x) = \int_0^1 f(u)g(x-u)\,du = \int_{I_1} + \int_{I_2} + \int_{I_3}$$

where $I_1 = [-x/2, x/2]$, $[x/2, 3x/2]$, $[3x/2, 1-x/2]$. For $u \in I_1$ we have $|x-u| \asymp x$, so $g(x-u) \asymp x^{-\beta}$ for such u. Hence

$$\int_{I_1} \asymp x^{-\beta} \int_0^{x/2} u^{-\alpha}\,du \asymp x^{-\beta} \cdot x^{1-\alpha}.$$

If $u \in I_2$, then $u \asymp x$, so $f(u) \asymp x^{\alpha}$, and so

$$\int_{I_2} \asymp x^{-\alpha} \int_0^{x/2} u^{-\beta}\,du \asymp x^{-\alpha} \cdot x^{1-\beta}.$$

For $u \in I_3$ we have both $\|u\| \asymp x$ and $x - u \asymp x$, so

$$\int_{I_3} \asymp \int_{x/2}^{1-x/2} u^{-\alpha-\beta}\,du \le \int_{x/2}^{\infty} u^{-\alpha-\beta}\,du \asymp x^{1-\alpha-\beta}.$$

These estimates give (3.39). Since $\alpha + \beta > 1$, we see that $f * g$ is an unbounded function. However, the set on which it is large is smaller than for f or g.

As we continue to develop our theory, we increasingly classify functions by the spaces of which they are members. In this context, by "space" we have in mind $L^1(\mathbb{T})$, or $L^\infty(\mathbb{T})$, or $C(\mathbb{T})$, the space of continuous functions with period 1. More generally, for $1 \le p < \infty$

$$\|f\|_p = \Big(\int_0^1 |f(x)|^p\,dx\Big)^{1/p} \tag{3.40}$$

is the L^p norm of f, and $L^p(\mathbb{T})$ is the set of functions with period 1 for which this norm is finite. By Hölder's inequality (Theorem I.11) it follows that if $1 \le p_1 < p_2 < \infty$, then $\|f\|_{p_1} \le \|f\|_{p_2}$. Moreover, if p is chosen so that $p_1 < p < p_2$, and we take $f(x) = \|x\|^{1/p}$, then $f \in L^{p_1}(\mathbb{T})$, but $f \notin L^{p_2}(\mathbb{T})$. Thus $L^{p_2}(\mathbb{T}) \subset L^{p_1}(\mathbb{T})$, and the inclusion is strict.

One way to measure how often a function is large is to consider its p-norms. Let f and g be defined as in Example 3.7. We note that $\int_0^1 f(x)^p\,dx < \infty$ precisely when $p < 1/\alpha$, $\int_0^1 g(x)^q\,dx < \infty$ when $q < 1/\beta$, and $\int_0^1 (f*g)(x)^r\,dx < \infty$ when $r < 1/(\alpha+\beta-1)$. Since $\alpha + \beta - 1 < \alpha$, we see that $r > p$. Similarly, $r > q$. Thus $f * g$ is smoother than either f or g in the sense that its (fractional) pole is of lower order, even though it is still an unbounded function.

If $\alpha+\beta=1$, then the first two integrals are bounded but the third is $\asymp \log 1/x$, so $(f*g)(x) \asymp \log 1/\|x\|$. Thus in this case $f*g$ is still unbounded, but $f*g \in L^p(\mathbb{T})$ for all $p<\infty$. If $\alpha+\beta<1$, then $f*g$ is continuous, by Exercise 5.2.10.

Exercises

1. Suppose that f, g, h are members of $L^1(\mathbb{T})$. Show that $f*(g+h) = f*g+f*h$.
2. Suppose that $f \in L^1(\mathbb{T})$ and $g \in L^1(\mathbb{T})$. Show that $\|f+g\|_1 \le \|f\|_1 + \|g\|_1$.
3. Suppose that $f \in L^1(\mathbb{T})$ and that $g \in L^1(\mathbb{T})$.
   (a) Show that if f and g are even functions, then $f*g$ is even.
   (b) Show that if f is odd and g is even, then $f*g$ is an odd function.
   (c) Show that if f and g are both odd functions, then $f*g$ is an even function.
4. Suppose that f, g, h are members of $L^1(\mathbb{T})$. Show that $f*(g+h) = f*g+f*h$.
5. Suppose that f, g, h are members of $L^1(\mathbb{T})$. Show that $(f*g)*h = f*(g*h)$.
6. Let f be as in Example 3.6, and suppose that $0 < \delta \le 1/6$.
   (a) By convolving f with $f*f$, show that

$$h(\theta) := (f*f*f)(\theta) = \begin{cases} 0 & (-1/2 \le \theta \le -3\delta), \\ \frac{1}{2}(\theta+3\delta)^2 & (-3\delta \le \theta \le -\delta), \\ 3\delta^2 - \theta^2 & (-\delta \le \theta \le \delta), \\ \frac{1}{2}(\theta-3\delta)^2 & (\delta \le \theta \le 3\delta), \\ 0 & (3\delta \le \theta \le 1/2). \end{cases}$$

   (b) For each of the four values $x = -3\delta, -\delta, \delta, 3\delta$, compute $\lim_{\theta\to x^-} h'(\theta)$ and $\lim_{\theta\to x^+} h'(\theta)$. Verify that these one-sided derivatives are equal, and hence prove that h' exists and is continuous everywhere.
   (c) For each of the four values $x = -3\delta, -\delta, \delta, 3\delta$, compute $\lim_{\theta\to x^+} h''(\theta) - \lim_{\theta\to x^-} h''(\theta)$. Thus establish that h'' has jump discontinuities at these four points.
   (d) Use Theorem 3.12 to compute $\widehat{h}(n)$.
7. Let f be as in Example 3.6. By convolving $f*f$ with itself, one can compute that if $0<\delta\le 1/8$, then

$$j(\theta) := (f*f*f*f)(\theta) = \begin{cases} 0 & (-1/2 \le \theta \le -4\delta), \\ \frac{1}{6}(4\delta+\theta)^3 & (-4\delta \le \theta \le -2\delta), \\ \frac{-1}{2}\theta^3 - 2\delta\theta^2 + \frac{16}{3}\delta^3 & (-2\delta \le \theta \le 0), \\ \frac{1}{2}\theta^3 - 2\delta\theta^2 + \frac{16}{3}\delta^3 & (0 \le \theta \le 2\delta), \\ \frac{1}{6}(4\delta-\theta)^3 & (2\delta \le \theta \le 4\delta), \\ 0 & (4\delta \le \theta \le 1/2). \end{cases}$$

   (a) The function f is constant in the intervals $(-1/2,-\delta), (-\delta,\delta), (\delta,1/2)$. What are these constant values?
   (b) Let $g(\theta) = (f*f)(\theta)$. The function g is piecewise linear, so that $g'(\theta)$ is constant in the intervals $(-1/2,-2\delta), (-2\delta,0), (0,2\delta), (2\delta,1/2)$. What are these constant values?

(c) Let $h(\theta) = (f * f * f)(\theta)$. The function h is piecewise quadratic, so that $h''(\theta)$ is constant on the intervals $(-1/2, -3\delta), (-3\delta, -\delta), (-\delta, \delta), (\delta, 3\delta), (3\delta, 1/2)$. What are these constant values?

(d) Let $j(\theta) = (f * f * f * f)(\theta)$. The function $j(\theta)$ is piecewise cubic, so that $j'''(\theta)$ is constant on each of the intervals $(-1/2, -4\delta), (-4\delta, -2\delta), (-2\delta, 0), (0, 2\delta), (2\delta, 4\delta), (4\delta, 1/2)$. What are these constant values?

(e) Formulate a conjecture concerning the n-fold convolution of f with itself.

8. Suppose that f has period 1 and that f has continuous derivatives through the nth order. (The collection of such functions is known as $C^{(n)}(\mathbb{T})$.)

(a) Show that if $k \neq 0$, then

$$\widehat{f}(k) = \frac{1}{(2\pi i k)^n} \widehat{f^{(n)}}(k) .$$

(b) Deduce that

$$|\widehat{f}(k)| \leq \frac{\|f^{(n)}\|_1}{(2\pi |k|)^n}$$

for all $k \neq 0$.

(c) Deduce also that

$$\widehat{f}(k) = o(1/|k|^n)$$

as $|k| \to \infty$.

9. Suppose that $f \in C^{(m)}(\mathbb{T})$ and that $g \in C^{(n)}(\mathbb{T})$.

(a) Show that $f * g$ is n-times differentiable, and that $(f * g)^{(n)}(x) = (f * g^{(n)})(x)$.

(b) Show that $f * g \in C^{(m+n)}(\mathbb{T})$, and that $(f * g)^{(m+n)}(x) = (f^{(m)} * g^{(n)})(x)$.

(c) Deduce that $\widehat{f * g}(k) = o(1/|k|^{m+n})$ as $|k| \to \infty$.

10. Suppose that $f \in L^1(\mathbb{T})$ and that $g \in L^p(\mathbb{T})$ where $1 < p < \infty$.

(a) Explain how one should apply the Hölder inequality for integrals (Theorem I.11) to show that

$$|(f * g)(x)|^p \leq \|f\|_1^{p-1} (|f| * |g|^p)(x) .$$

(b) Use Theorem 3.12 to show that $\big\| |f| * |g|^p \big\|_1 \leq \|f\|_1 \|g\|_p^p$.

(c) Conclude that $f * g \in L^p(\mathbb{T})$, and that $\|f * g\|_p \leq \|f\|_1 \|g\|_p$.

3.4. First Convergence Theorems

In general, a doubly-infinite series $\sum_{n=-\infty}^{\infty} c_n$ is said to converge if

$$\lim_{\substack{M \to -\infty \\ N \to \infty}} \sum_{n=M}^{N} c_n$$

exists. Here M and N act independently. This is equivalent to requiring that both of the two series

$$\sum_{n=0}^{\infty} c_{-n}, \qquad \sum_{n=1}^{\infty} c_n$$

converge. However, in the case of Fourier Series as defined in (3.7), the two terms $\widehat{f}(n)e(nx)$ and $\widehat{f}(-n)e(-nx)$ work together, so we confine ourselves to symmetric partial sums. Thus as far as convergence is concerned, the issue is whether

$$s_N(x) = \sum_{n=-N}^{N} \widehat{f}(n)e(nx) \tag{3.41}$$

tends to a limiting value (hopefully $f(x)$) as $N \to \infty$. In case several functions are under consideration, we identify the partial sum of the function f by writing $s_N(f;x)$. Of course,

$$s_N(x) = \frac{1}{2}a_0 + \sum_{n=1}^{N}(a_n \cos 2\pi nx + b_n \sin 2\pi nx)$$

in the trigonometric notation (3.2).

For any given $f \in L^1(\mathbb{T})$, the partial sum s_N of its Fourier Series is an example of a *trigonometric polynomial*, which is to say a finite sum of the form

$$T(x) = \sum_{n=-N}^{N} t_n e(nx)\,. \tag{3.42}$$

Since $e(nx)$ is continuous and has period 1, it follows that a trigonometric polynomial $T(x)$ is continuous and has period 1, and hence $T \in L^1(\mathbb{T})$. Consequently, T has Fourier coefficients:

$$\widehat{T}(k) = \int_0^1 T(x)e(-kx)\,dx = \sum_{n=-N}^{N} t_n \int_0^1 e((n-k))\,dx = \begin{cases} t_k & (|k| \le N), \\ 0 & (|k| > N) \end{cases} \tag{3.43}$$

by the orthogonality (3.5) that we noted earlier. The *degree* of $T(x)$ is the largest integer k such that at least one of t_k, t_{-k} is non-zero. Thus the collection of all sums of the form (3.42) is the set of all trigonometric polynomials of degree not exceeding N, which is denoted $\mathcal{T}_N$.

Lemma 3.14. *Suppose that $f \in L^1(\mathbb{T})$, and that $T(x)$ is a trigonometric polynomial, as defined in* (3.42). *Then*

$$(T * f)(x) = \sum_{n=-N}^{N} \widehat{f}(n) t_n e(nx)\,.$$

Proof. Let n be a given integer, and put $g(x) = e(nx)$. Then

$$\begin{aligned}(g * f)(x) = (f * g)(x) &= \int_0^1 f(u)e(n(x-u))\,du = e(nx)\int_0^1 f(u)e(-nu)\,du \\ &= \widehat{f}(n)e(nx)\,.\end{aligned}$$

We multiply both sides by t_n, and then sum over n to obtain the stated result. □

Let

$$D_N(x) = \sum_{n=-N}^{N} e(nx)\,. \tag{3.44}$$

This is the *Dirichlet kernel*. By Lemma 3.14 we find that

$$s_N(x) = (D_N * f)(x)\,. \tag{3.45}$$

Our object is to show that if f is suitably well-behaved, then $s_N(x)$ tends to $f(x)$ as N tends to infinity. To do this, we require some information concerning the Dirichlet kernel. First we observe that

$$\int_0^1 D_N(x)\,dx = 1, \tag{3.46}$$

which is merely a special case of (3.43). Next we show that the Dirichlet kernel can be written in closed form,

$$D_N(x) = \begin{cases} \dfrac{\sin(2N+1)\pi x}{\sin \pi x} & (x \notin \mathbb{Z}), \\ 2N+1 & (x \in \mathbb{Z}). \end{cases} \tag{3.47}$$

If $x = 0$, then all summands in the definition (3.44) are 1, so $D_N(0) = 2N+1$. Suppose that x is not an integer. We observe that the ratio of one term to the next in the Dirichlet kernel is

$$\frac{e((n+1)x)}{e(nx)} = e(x),$$

which is independent of n. That is, the Dirichlet kernel is a geometric series. From the formula (0.2) for the sum of a geometric progression we deduce that

$$D_N(x) = \sum_{n=-N}^{N} e(nx) = e(-Nx)\sum_{n=0}^{2N} e(nx) \tag{3.48}$$

$$= e(-Nx)\frac{1 - e((2N+1)x)}{1 - e(x)} = e(-Nx)\frac{e((2N+1)x) - 1}{e(x) - 1}\,. \tag{3.49}$$

In the numerator and denominator we factor out complex exponentials chosen so that the difference takes the form $e(u) - e(-u)$. Thus the above is

$$= e(-Nx)\frac{e((N+1/2)x)\big(e((N+1/2)x) - e(-(N+1/2)x)\big)}{e(x/2)\big(e(x/2) - e(-x/2)\big)} \tag{3.50}$$

$$= \frac{e((N+1/2)x) - e(-(N+1/2)x)}{e(x/2) - e(-x/2)}\,. \tag{3.51}$$

We now divide the numerator and denominator by $2i$ to obtain (3.47).

The function $\sin(2N+1)\pi x$ oscillates between -1 and 1, so $D_N(x)$ oscillates between $-1/\sin \pi x$ and $1/\sin \pi x$. An example of this is displayed in Figure 3.2. From (3.47) we see that $D_N(x) = 0$ precisely when

$$x = 1/(2N+1), 2/(2N+1), \ldots, 2N/(2N+1).$$

Thus $D_N(x)$ has $2N$ roots, which is the maximum possible for a trigonometric polynomial of degree N, in view of

Theorem 3.15. *Let $T(x)$ be a trigonometric polynomial of degree at most N, as in* (3.42). *If T the coefficients of T are not all 0, then T has at most $2N$ roots.*

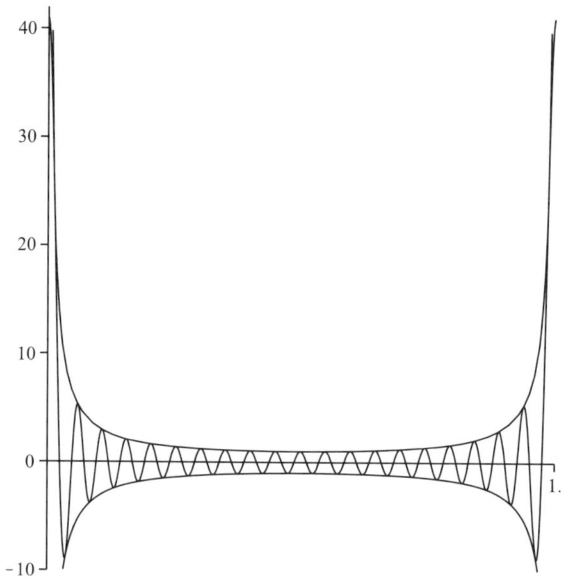

Figure 3.2. Graph of $D_{20}(x)$ for $0 \le x \le 1$, with its envelopes $\pm 1/\sin \pi x$.

One could argue that if $T(x) = 0$, then also $T(x+1) = 0$, $T(x+2) = 0$, and so on, so that T has infinitely many roots. However, when we speak of the number of roots of a function with period 1, we are counting only those in a single period.

Proof. Let $P(z) = \sum_{n=0}^{2N} t_{n-N} z^n$. Thus $T(x) = e(-Nx)P(e(x))$. By the Fundamental Theorem of Algebra (Theorem 1.1), P has exactly $2N$ zeros in the complex plane if $t_N \neq 0$, and fewer zeros otherwise. Thus it has at most $2N$ zeros on the unit circle $|z| = 1$, and it is these zeros that correspond to zeros of T. □

We now apply our knowledge concerning the Dirichlet kernel to show that the Fourier Series of the sawtooth function $s(x)$ converges to $s(x)$ for all x. This would not be so notable if it were an isolated result, but in fact we can parlay this first result into a large family of convergence theorems. Consequently, the sawtooth function $s(x)$ is the most important single function in our entire theory.

Theorem 3.16. *Let*

$$E_N(x) = s(x) - \sum_{n=1}^{N} \frac{\sin 2\pi n x}{\pi n}. \tag{3.52}$$

Then $|E_N(x)| \le \min\big(1/2, 1/(2\pi N \|x\|)\big)$ *for all* x.

An immediate consequence of the above is that

$$s(x) = \sum_{n=1}^{\infty} \frac{\sin 2\pi n x}{\pi n} \tag{3.53}$$

for all x. That is, $\lim_{N\to\infty} s_N(x) = s(x)$ for all x. This is instructive, for in this example we see that a sequence of continuous functions (such as $s_N(x)$) can converge to a discontinuous function (such as $s(x)$). Of course, the convergence is not uniform in the neighborhood of $x = 0$. While the formula (3.53) is satisfying, the quantitative precision of Theorem 3.16 is far more valuable.

Proof. Clearly $E_N(x)$ is an odd function, so it suffices to consider x in the interval $0 \le x \le 1/2$. We note that for $0 < x < 1$,

$$E_N'(x) = -1 - 2\sum_{n=1}^{N} \cos 2\pi n x = -D_N(x)\,.$$

Now $E_N(1/2) = 0$, so if $0 < x < 1/2$, then

$$E_N(x) = -\int_x^{1/2} E_N'(u)\,du = \int_x^{1/2} D_N(u)\,du = \int_x^{1/2} \frac{\sin(2N+1)\pi u}{\sin \pi u}\,du\,.$$

In the integrand on the right, the numerator is oscillating rapidly, while the denominator is monotonic. To estimate such an integral, we integrate by parts. Thus we see that the above is

$$= \frac{\cos(2N+1)\pi x}{(2N+1)\pi \sin \pi x} - \int_x^{1/2} \frac{\cos(2N+1)\pi u \cos \pi u}{(2N+1)(\sin \pi u)^2}\,du\,.$$

This new integral has absolute value not exceeding

$$\frac{1}{2N+1}\int_x^{1/2} \frac{\cos \pi u}{(\sin \pi u)^2}\,du = \frac{1}{(2N+1)\pi}\Big(\frac{1}{\sin \pi x} - 1\Big) < \frac{1}{(2N+1)\pi \sin \pi x}\,.$$

Thus

$$|E_N(x)| \le \frac{2}{(2N+1)\pi \sin \pi x} < \frac{1}{\pi N \sin \pi x} < \frac{1}{2\pi N x}$$

since $\sin \pi x > 2x$ for $0 \le x \le 1/2$. This suffices for $1/(\pi N) \le x \le 1/2$. Now suppose that $0 < x \le 1/(\pi N)$. In this interval $1/2 > s(x) > 1/2 - 1/(\pi N)$. Also, $\sin 2\pi n x \ge 0$ for $1 \le n \le N$, so it is clear that $E_N(x) < 1/2$. To obtain a lower bound for $E_N(x)$ we observe that $\sin 2\pi n x \le 2\pi n x$. Hence

$$\sum_{n=1}^{N} \frac{\sin 2\pi n x}{\pi n} \le 2Nx \le \frac{2}{\pi}\,.$$

Thus $E_N(x) \ge 1/2 - 1/(\pi N) - 2/\pi \ge 1/2 - 3/\pi \ge -1/2$, and so the proof is complete. □

Example 3.8. In Figure 3.3 we see the function $s(x) = 1/2 - x$ and its approximation $s_{20}(x)$. As x increases from 0, the partial sum $s_{20}(x)$ increases rapidly towards its target, overshoots it, then undershoots it, and so on. This is known as the *Gibbs Phenomenon.* To investigate the size of the first overshoot, we note that

$$\begin{aligned} s_N'(x) &= 2\sum_{n=1}^{N} \cos 2\pi n x = D_N(x) - 1 = \frac{\sin(2N+1)\pi x}{\sin \pi x} - 1 \\ &= \frac{\sin(2N+1)\pi x - \sin \pi x}{\sin \pi x} = \frac{2\cos(N+1)\pi x \sin N\pi x}{\sin \pi x} \end{aligned}$$

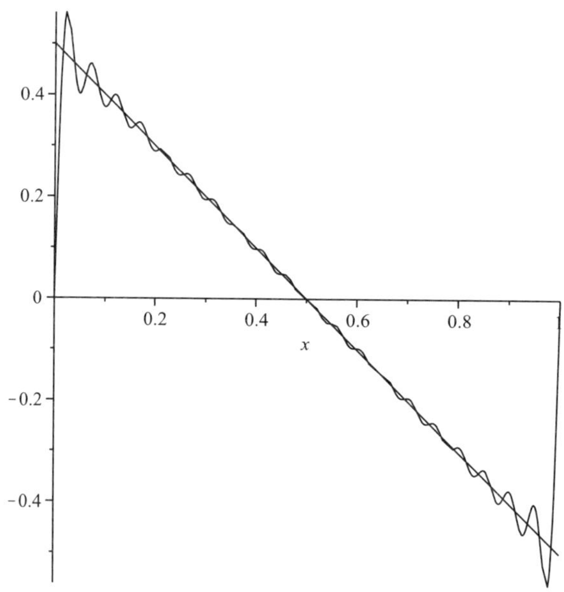

Figure 3.3. Graph of the sawtooth function $s(x) = 1/2 - x$ and its Fourier approximation $s_{20}(x)$.

by (T.27). Thus $s_N'(x) > 0$ for $0 < x < 1/(2(N+1))$ and $1/(2(N+1))$ is a local maximum of $s_N(x)$. We observe that

$$s_N\left(\frac{1}{2(N+1)}\right) = \sum_{n=1}^{N} \frac{\sin\frac{\pi n}{N+1}}{\pi n} = \frac{1}{N+1}\sum_{n=1}^{N} \frac{\sin\frac{\pi n}{N+1}}{\frac{\pi n}{N+1}} = \frac{1}{N+1}\sum_{n=1}^{N+1} \frac{\sin\frac{\pi n}{N+1}}{\frac{\pi n}{N+1}}$$

is a Riemann sum for the integral

$$\int_0^1 \frac{\sin \pi u}{\pi u}\, du = \frac{1}{\pi}\int_0^{\pi} \frac{\sin v}{v}\, dv = 0.5894898721.$$

The function $s(x)$ has a jump discontinuity at $x = 0$ of height 1. The overshoot of $s_N(x)$ is thus approximately 8.9% of the height of the jump. Other functions with jump discontinuity display the same behavior, with an overshoot proportional to the height of the jump.

Let $w(x)$ be the square wave function of Example 3.3. It is easy to see that $w(x) = s(x) - s(x - 1/2)$. By Theorem 3.2(a) it follows that

$$\widehat{w}(n) = \widehat{s}(n)(1 - e(n/2)) = \begin{cases} \dfrac{1-(-1)^n}{2\pi i n} & (n \neq 0), \\ 0 & (n = 0), \end{cases}$$

which is consistent with our calculation in Example 3.3. From (3.53) it now follows that

$$w(x) = \sum_{n=1}^{\infty} \frac{2}{\pi(2n-1)} \sin 2\pi(2n-1)x \tag{3.54}$$

for all x.

The above analysis is not special to the square wave function, but extends to any step function. Let $f(x)$ be a step function with period 1. Suppose that $x_1 < x_2 < \cdots < x_K < x_1 + 1$, that $f(x)$ has a jump discontinuity at x_k of height h_k, and that $f(x)$ is constant in the intervals between the x_k. Let x_0 be slightly smaller than x_1. Then $f(x) = f(x_0)+h_1$ for $x_1 < x < x_2$, $f(x) = f(x_0)+h_1+h_2$ for $x_2 < x < x_3$, and so on, so that $f(x) = f(x_0)+h_1+h_2+\cdots+h_K$ for $x_K < x < x_1$. Since f has period 1, it follows that $\sum_{k=1}^K h_k = 0$. Let

$$g(x) = \sum_{k=1}^{K} h_k s(x - x_k)\,.$$

If x is not one of the x_k, then

$$g'(x) = -\sum_{k=1}^{K} h_k = 0\,.$$

That is, $g(x)$ has the same jump discontinuities, of the same heights and at the same points, and $g(x)$ is constant in intervals between the x_k. Suppose, finally, that $f(x_k) = \big(f(x_k^-) + f(x_k^+)\big)/2$ for all x. The function g also has this property. Thus there is a constant c such that $f(x) = g(x) + c$ for all x. Hence

$$f(x) = c + \sum_{n\neq 0}\Big(\sum_{k=1}^{K} h_k e(nx_k)\Big)\frac{e(nx)}{2\pi i n} \tag{3.55}$$

for all x.

This is all very fulfilling, but we are now in a position to establish a much more general convergence theorem. Let us review a few of the things we know: (1) If $h = f * g$, then $\widehat{h}(n) = \widehat{f}(n)\widehat{g}(n)$; (2) $\widehat{s}(n) = 1/2\pi i n$ for $n \neq 0$; (3) If $F(x) = \int_0^x f(u)\,du - \widehat{f}(0)x$, then $\widehat{F}(n) = \widehat{f}(n)/(2\pi i n)$ for $n \neq 0$; (4) The Fourier Series of $s(x)$ is boundedly convergent to $s(x)$. The first three of these results suggest that it ought to be possible to express $F(x)$, or something similar to it as a convolution of $f(x)$ with $s(x)$. But in Theorem 3.16 we have strong quantitative information concerning the convergence of the Fourier Series of $s(x)$. This allows us to say something useful about the Fourier Series of $F(x)$.

Theorem 3.17. *Suppose that $f \in L^1(\mathbb{T})$, and put*

$$F(x) = \int_0^x f(u)\,du - \widehat{f}(0)x = \int_0^x f(u) - \widehat{f}(0)\,du\,.$$

Then

$$\widehat{F}(0) + \lim_{N\to\infty} \sum_{\substack{n=-N\\ n\neq 0}}^{N} \frac{\widehat{f}(n)}{2\pi i n} e(nx) = F(x) \tag{3.56}$$

uniformly in x.

A formula for $\widehat{F}(0)$ in terms of $f(x)$ was already determined in Theorem 3.7.

Many of the functions of interest to us arise as $F(x)$ for a suitable choice of $f(x)$. For example, in Example 3.5 we determined the Fourier Series of the function $\|x\|$ by observing that $\|x\| = 4\int_0^x w(u)\,du$ where $w(x)$ is the square wave function of Example 3.3. That the series converges, and even that it is uniformly convergent is obvious by comparison with the series $\sum_{n=1}^{\infty} 1/n^2$. What is valuable to us is not that the series converges, but rather that it converges to $\|x\|$.

Proof. We first show that

$$(f * E_N))(x) = F(x) - s_N(F; x) \tag{3.57}$$

where $E_N(x)$ is defined as in Theorem 3.16. This is the crux of the matter, since it is almost trivial to show that the left hand side above tends to 0 as $N \to \infty$.

We start by noting that

$$(f * s)(x) = \int_0^1 f(u)s(x-u)\,du = \int_0^1 \big(f(u) - \widehat{f}(0)\big)s(x-u)\,du$$

since $\widehat{s}(0) = 0$. The integrand has period 1, so by Lemma 3.1 the above is

$$= \int_{x-1}^{x} \big(f(u) - \widehat{f}(0)\big)s(x-u)\,du\,.$$

If $x - 1 < u < x$, then $0 < x - u < 1$, and so $s(x - u) = 1/2 - x + u$. Thus the above is

$$= \int_{x-1}^{x} \big(f(u) - \widehat{f}(0)\big)(1/2 - x + u)\,du,$$

which by integration by parts is

$$\begin{aligned} &= \Big[F(u)(1/2 - x + u)\Big|_{x-1}^{x} - \int_{x-1}^{x} F(u)\,du \\ &= \frac{1}{2}F(x) + \frac{1}{2}F(x-1) - \widehat{F}(0) = F(x) - \widehat{F}(0). \end{aligned}$$

On the other hand, from Lemma 3.14 with $T(x) = s_N(s; x)$, we see that

$$(f * s_N)(x) = \sum_{\substack{n=-N \\ n \neq 0}}^{N} \frac{\widehat{f}(n)}{2\pi i n} e(nx)\,.$$

We note that $s(x) - s_N(s; x) = E_N(x)$ in the notation of Theorem 3.16. Thus when we subtract the above from our former relation we obtain (3.57), in view of Theorem 3.7.

If f is bounded, say $|f(x)| \le M$ for all x, then

$$\Big|\int_0^1 f(u)E_N(x-u)\,du\Big| \le M\int_0^1 |E_N(x-u)|\,du \le \frac{M \log 5N}{N},$$

uniformly in x. Of course a general function $f \in L^1(\mathbb{T})$ is not bounded, but in Theorem 0.22 we recorded the important fact that any $f \in L^1(\mathbb{T})$ is almost continuous, in the sense that for any given $\varepsilon > 0$ there is a continuous function $c(x)$ with period 1 such that $\int_0^1 |f(x) - c(x)|\,dx < \varepsilon$. For such a function $c(x)$, write

$$\int_0^1 f(u)E_N(x-u)\,du = \int_0^1 c(u)E_N(x-u)\,du + \int_0^1 (f(u) - c(u))E_N(x-u)\,du\,.$$

A continuous function, such as $c(x)$, is bounded, so there is an M such that $|c(x)| \le M$. Thus the argument already described applies to $c(x)$, and for the second integral we note that $|E_N(x-u)| \le 1/2$ for all u and x. Hence by the triangle inequality,

$$\begin{aligned}\Big|\int_0^1 f(u)E_N(x-u)\,du\Big| &\le M\int_0^1 |E_N(x-u)|\,du + \frac{1}{2}\int_0^1 |f(u)-c(u)|\,du\\ &< \frac{M\log 5N}{N} + \frac{\varepsilon}{2} < \varepsilon\end{aligned}$$

for all sufficiently large N, and this is uniform in x. □

Example 3.9. Let $B_2(x) = x^2 - x + 1/6$. This is the second Bernoulli polynomial. Let $P_2(x)$ be the function with period 1 such that $P_2(x) = B_2(x)$ for $0 \le x \le 1$. Now $P_2'(x) = -2s(x)$, so by Theorem 3.17 it follows that

$$P_2(x) = \sum_{\substack{n=-\infty\\ n\neq 0}}^{\infty} \frac{e(nx)}{2\pi^2 n^2} = \frac{1}{\pi^2}\sum_{n=1}^{\infty}\frac{\cos 2\pi nx}{n^2} \tag{3.58}$$

uniformly for all x. In particular, since $P_2(0) = 1/6$, we see that

$$\sum_{n=1}^{\infty}\frac{1}{n^2} = \frac{\pi^2}{6}, \tag{3.59}$$

which is a beautiful formula discovered by Euler.

Exercises

1. Figure 3.2 is misleading because of the great discrepancy in the scaling of the x axis as compared to the y axis. On the page, y changes from -41 to 41 in less space than x changes from 0 to 1. The graph looks very flat in the middle, but this is due to the scaling. Demonstrate this effect by computing the upper envelope $1/\sin \pi x$ at $x = 1/2$, $x = 1/4$, $x = 1/8$, $x = 1/16$, $x = 1/32$.
2. Let
$$s_N(x) = \sum_{n=1}^{N}\frac{\sin 2\pi nx}{\pi n}$$
be a partial sum of the Fourier Series of the sawtooth function $s(x)$. In this exercise we show that $s_N(x) \ge 0$ for $0 \le x \le 1/2$.
   (a) Explain why it is enough to show that $s_N(x) \ge 0$ at points $x \in [0, 1/2]$ at which $s_N'(x) = 0$.
   (b) Show that $s_N'(x) = (\sin(2N+1)\pi x)/\sin \pi x - 1$.
   (c) Show that if $s_N'(x) = 0$, then $(1 - \cos 2\pi Nx)\sin \pi x = (\sin 2\pi Nx)\cos \pi x$.
   (d) Deduce that if $s_N'(x) = 0$ and $0 \le x \le 1/2$, then $\sin 2\pi Nx \ge 0$.
   (e) Note that $s_1(x) \ge 0$ for $0 \le x \le 1/2$.
   (f) Suppose that $s_{N-1}(x) \ge 0$ for $0 \le x \le 1/2$. Show that if $s_N'(x) = 0$ and $0 \le x \le 1/2$, then $s_N(x) \ge 0$.
   (g) Conclude that $s_N(x) \ge 0$ for $0 \le x \le 1/2$.
3. (a) Show that $2\cos 2\pi x \cos 2n\pi x = \cos 2(n+1)\pi x + \cos 2(n-1)\pi x$.

(b) Deduce that

$$\begin{aligned}2\cos 2\pi nx\cos 2\pi ny(\cos 2\pi x-\cos 2\pi y)\\ &=\big(\cos 2\pi(n+1)x+\cos 2\pi(n-1)x\big)\cos 2\pi ny\\ &\quad-\big(\cos 2\pi(n+1)y+\cos 2\pi(n-1)y\big)\cos 2\pi x\,.\end{aligned}$$

(c) Sum over n to conclude that

$$1+2\sum_{n=1}^{N}\cos 2\pi nx\cos 2\pi ny=\frac{\cos 2\pi(N+1)x\cos 2\pi Ny-\cos 2\pi Nx\cos 2\pi(N+1)y}{\cos 2\pi x-\cos 2\pi y}$$

Note that this is $D_N(x)$ when $y=0$. This is an example of a Christoffel–Darboux formula.

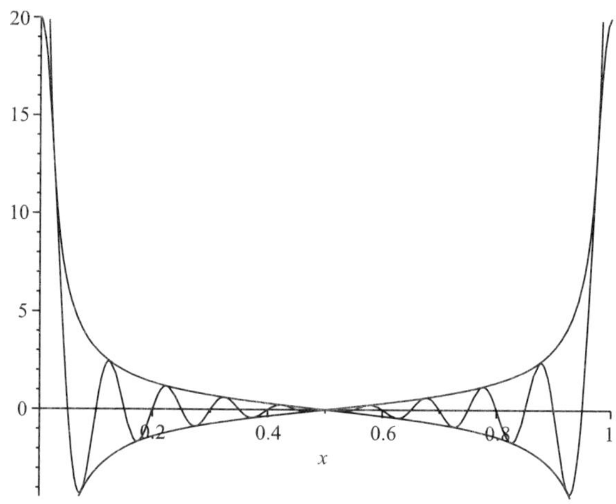

Figure 3.4. The modified Dirichlet kernel $D^*_{10}(x)$ and $\pm\cot\pi x$.

4. We define the *modified Dirichlet kernel* $D^*_N(x)$ to be

$$D^*_N(x)=\cos 2\pi Nx+2\sum_{n=1}^{N-1}\cos 2\pi nx=D_N(x)-\cos 2\pi Nx\,. \tag{3.60}$$

A graph of $D^*_{10}(x)$ is depicted in Figure 3.4.
(a) Show that $D^*_N(x)=\frac{1}{2}D_{N-1}(x)+\frac{1}{2}D_N(x)$.
(b) Show that $D^*_N(x)=\sin 2\pi Nx\cot\pi x$.
(c) Show that $D^*_N(x)$ has $2N$ zeros.

5. We define the *conjugate Dirichlet kernel* $\widetilde{D}_N(x)$ to be

$$\widetilde{D}_N(x)=\sum_{n=1}^{N}(-ie(nx)+ie(-nx))=2\sum_{n=1}^{N}\sin 2\pi nx\,. \tag{3.61}$$

In this situation, the word "conjugate" does not refer to complex conjugate, but rather to harmonic conjugate. A graph of $\widetilde{D}_{10}(x)$ is found in Figure 3.5.

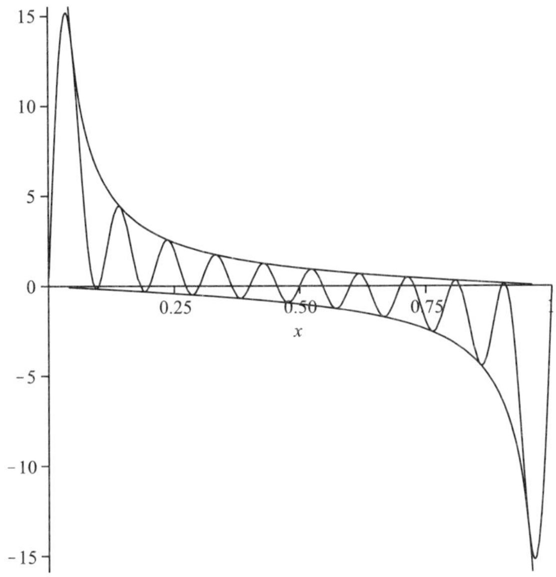

Figure 3.5. Graph of $\widetilde{D}_{10}(x)$, $-\tan\frac{\pi}{2}x$, and $\cot\frac{\pi}{2}x$ for $0 \le x \le 1$.

(a) Show that

$$\sum_{n=1}^{N} e(nx) = e(Nx/2)\frac{\sin(N+1)\pi x}{\sin \pi x}\,.$$

(b) Deduce that

$$\widetilde{D}_N(x) = \frac{2\sin \pi Nx \,\sin \pi(N+1)x}{\sin \pi x}\,.$$

(c) Show that $\widetilde{D}_N(x)$ has $2N$ zeros.

(d) Show that

$$-\tan\frac{\pi}{2}x \le \widetilde{D}_N(x) \le \cot\frac{\pi}{2}x$$

for $0 \le x \le 1$.

(e) Describe the points where the upper bound above is attained. How many such points are there?

(f) Describe the points where the lower bound above is attained. How many such points are there?

(g) Give two proofs that $\widetilde{D}_N(x)$ is an odd function.

6. In the same way that we defined a modified Dirichlet kernel in Exercise 4, we define a *modified conjugate Dirichlet kernel* to be

$$\widetilde{D}^*_N(x) = \sin 2\pi Nx + 2\sum_{n=1}^{N-1}\sin 2\pi nx\,. \tag{3.62}$$

See Figure 3.6 for a plot of $\widetilde{D}^*_{10}(x)$.

(a) In order to derive a closed form formula for this, do the following:

(i) Explain why

$$\sum_{n=0}^{N-1} e(nx) = e\Big(\frac{(N-1)x}{2}\Big)\frac{\sin \pi N x}{\sin \pi x}\,.$$

(ii) Show that $1 + e(x) = 2e(x/2)\cos \pi x$.

(iii) Show that the product of the left hand sides of the above identities is

$$1 + 2\sum_{n=1}^{N-1} e(nx) + e(Nx)\,.$$

(iv) Show that the product of the right hand sides of the same two identities is

$$2e(Nx/2)\frac{\sin \pi N x}{\sin \pi x}\cos \pi x\,.$$

(v) Now take imaginary parts to show that

$$\widetilde{D}_N^*(x) = 2(\sin \pi N x)^2 \cot \pi x\,.$$

(b) Show that $\widetilde{D}_N^*(n/N) = 0$ for all n.

(c) Show that $\frac{d}{dx}\widetilde{D}_N^*(n/N) = 0$ for $0 < n < N$.

(d) Show that $\widetilde{D}_N^*(x)$ has a simple zero at $x = 1/2$ if N is odd, and a triple zero at $x = 1/2$ if N is even.

(e) Show that $\frac{d}{dx}\widetilde{D}_N^*(0) = 2\pi N^2$.

(f) Show that $\widetilde{D}_N^*(x)$ has $2N$ zeros, counting multiplicity.

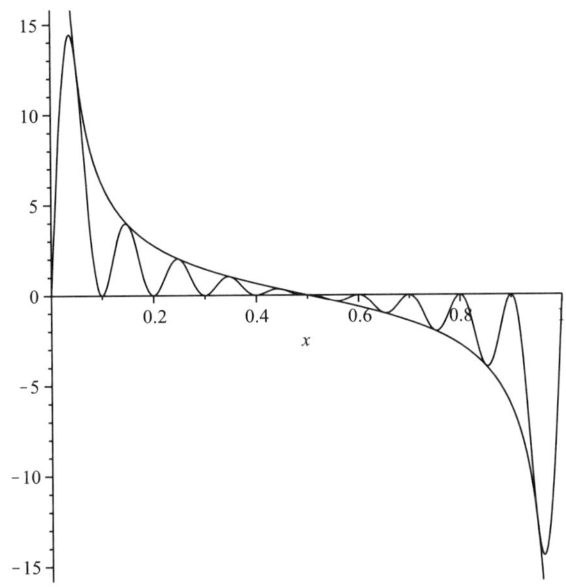

Figure 3.6. The modified conjugate Dirichlet kernel $\widetilde{D}^*_{10}(x)$ and $2\cot \pi x$.

7. Let $s_N(x)$ be defined as in Example 3.8.
   (a) Show that $s'_N(x) < 0$ for $1/(2(N+1)) < x < 1/N$.
   (b) Show that $s_N(1/N)$ is a Riemann sum for
$$\frac{1}{\pi}\int_0^{2\pi} \frac{\sin v}{v}\, dv = 0.4514116666,$$
   which is about 4.86% below the target 1/2.
   (c) Show that $x = k/N$ is a local minimum of $s_N(x)$ for $k < N/2$.
   (d) Note that $s_N(k/N) = s_{N-1}(k/N)$.
   (e) Argue by induction on N that $s_N(x) > 0$ for $0 < x < 1/2$.
8. (a) For nonnegative integers n, put
$$a_n = \int_{n\pi}^{(n+1)\pi} \frac{\sin x}{x}\, dx\,.$$
   Show that the a_n alternate in sign and are decreasing in absolute value.
   (b) Show that the integral
$$\int_0^{\infty} \frac{\sin x}{x}\, dx$$
   converges.
   (c) Suppose that $f \in L^1(\mathbb{T})$. Show that
$$\lim_{t\to\infty} \int_{-1/2}^{1/2} f(x)e(tx)\, dx = 0.$$

(d) Let f have period 1, and for $-1//2 \le x \le 1/2$ set

$$f(x) = \frac{\pi x - \sin \pi x}{\pi x \sin \pi x}$$

for $x \ne 0$ and $f(0) = 0$. Show that f is continuous and bounded for $-1/2 < x < 1/2$, and that f has a jump discontinuity at $1/2$.

(e) Show that

$$\lim_{N\to\infty} \int_{-1/2}^{1/2} f(x) \sin(2N+1)\pi x \, dx = 0 \, .$$

(f) Show that the above integral is

$$\int_{-1/2}^{1/2} D_N(x)\, dx - \frac{1}{\pi} \int_{-(N+1/2)\pi}^{(N+1/2)\pi} \frac{\sin x}{x}\, dx \, .$$

(g) Conclude that

$$\int_0^\infty \frac{\sin x}{x}\, dx = \frac{\pi}{2} \, .$$

Notes

3.1. In Lemma 3.10 it has not been assumed that $f \in L^1(\mathbb{T})$, so in fact the integral in question may be $+\infty$.

3.4. In analysis, an *operator* is a function that maps one function to another. An *integral operator* has the form $F(y) = \int_a^b f(x)K(x,y)\, dx$. Thus f is mapped to F, and K is called the *kernel*. For example, in (3.45) we see that D_N is the kernel that enables us to map $f(x)$ to $s_N(x)$.

Gibbs (1899) noted his phenomenon, without suggesting a proof. Bôcher (1906) gave a proof, and coined the term "Gibbs' phenomenon". It was only in 1925 that it became generally recognized that Wilbraham (1848) had noted the phenomenon and also had given complete proofs much earlier.

Daniel Bernoulli attempted to express a general periodic function as a Fourier Series. His close friend Euler criticized his work for lack of rigor, but nevertheless suggested the integral formulas (3.8) and (3.9) for the coefficients that one should use. Euler did not establish a general theory, but proved the validity of a number of expansions, including (3.53) and (3.54), roughly a half-century before Fourier.

Riemann coined the term "Fourier Series", and gave credit to Dirichlet for having proved the first "Fourier Theorem", i.e., one that establishes that the partial sums $s_N(x)$ converge to $f(x)$ for a general class of functions. Dirichlet (1829) showed that if a function with period 1 has at most finitely many jump discontinuities, and is otherwise continuous and piecewise monotonic, then the partial sums $s_N(x)$ converge to $(f(x^-) + f(x^+))/2$ for all x. Our results of this section are similar, but not exactly the same. Dirichlet's result is contained in Jordan's theorem (our Corollary 8.7), which was proved much later, in 1881.

In Exercise 1.4.10 we established that e is irrational, and in Exercise 3.1.17 we saw that π is irrational. These results should not come as a great surprise, since if e were $2721/1001$ or $\sqrt[3]{21}$, or if π were $355/113$ or $\sqrt{10}$, then we would just know these numbers by such names, and not reserve a special symbol for them. Although it is perhaps a little off our main theme, it turns out that both e and π are not only irrational, but also transcendental. An *algebraic number* is a real or complex number that is the root of a polynomial whose coefficients are integers, not all zero. A number that is not algebraic is called *transcendental.* The first explicit examples of transcendental numbers were given by Liouville (1844), but they were not very natural. Cantor (1874) showed that transcendental numbers must exist because there are only countably many algebraic numbers, while there are uncountably many real numbers. In that sense, almost all real numbers are transcendental. Hermite (1873) showed that e is transcendental, and Lindemann (1882) generalized Hermite's method to show that π is transcendental. The method was further simplified and extended by Weierstrass, Hilbert, Hurwitz, and Gordan. For an introduction to irrational numbers one might read the Carus monograph of Niven (1956), but Siegel (1949) gives a more motivated account. The presentation of Burger and Tubbs (2004) is quite appealing, but the student who aspires to contribute to the field should read Baker (1990).

Chapter 4

Summability of Fourier Series

4.1. Cesàro summability of Fourier Series

When we define a number to be the value of a series, say

$$a = \sum_{n=0}^{\infty} a_n,$$

we think of the series as providing a formula for a. However, all that is being asserted is that the partial sums

$$s_N = \sum_{n=0}^{N} a_n \tag{4.1}$$

form a sequence of approximations to a. If the sequence of partial sums fails to have a limit, or if the limit is approached only very slowly, then sometimes something useful can be salvaged by forming averages of the partial sums,

$$\sigma_N = \frac{1}{N}\sum_{n=0}^{N-1} s_n = \frac{1}{N}\sum_{n=0}^{N-1}\sum_{k=0}^{n} a_k = \frac{1}{N}\sum_{k=0}^{N-1} a_k \sum_{n=k}^{N-1} 1 = \sum_{k=0}^{N-1}\Big(1-\frac{k}{N}\Big)a_k\,. \tag{4.2}$$

We say that the series $\sum_{n=0}^{\infty} a_n$ is *Cesàro summable to* a, and write

$$\sum_{n=0}^{\infty} a_n = a \qquad (C)$$

if

$$\lim_{N\to\infty} \sigma_N = a\,.$$

Example 4.1. Consider the series

$$\sum_{n=0}^{\infty} (-1)^n\,.$$

The series does not converge, and indeed cannot converge, because its terms do not tend to 0. The sequence of its partial sums is $1, 0, 1, 0, \ldots$. Averages of these partial sums are $1, 1/2, 2/3, 1/2, 3/5, 1/2, 4/7, 1/2, \ldots$. In general, $\sigma_{2N} = 1/2$ and $\sigma_{2N+1} = (N+1)/(2N+1)$ and we see that $\sigma_N \to 1/2$ as $N \to \infty$. Accordingly, we write

$$\sum_{n=0}^{\infty} (-1)^n = \frac{1}{2} \qquad (C).$$

The partial sums $s_N(x)$ of a Fourier Series $\sum_{n=-\infty}^{\infty} \widehat{f}(n)e(nx)$ often do not converge very well, but when we form the corresponding Cesàro partial sum

$$\sigma_N(x) = \frac{s_0(x) + s_1(x) + \cdots + s_{N-1}(x)}{N} = \sum_{n=-N}^{N} (1 - |n|/N)\widehat{f}(n)e(nx) \tag{4.3}$$

the behavior is usually much better. By (3.11) and (3.12), $\sigma_N(x)$ can be expressed in trigonometric form,

$$\sigma_N(x) = \frac{1}{2}a_0 + \sum_{n=1}^{N} (1 - n/N)(a_n \cos 2\pi nx + b_n \sin 2\pi nx)\,. \tag{4.4}$$

Let

$$\Delta_N(x) = \sum_{n=-N}^{N} (1 - |n|/N)e(nx) = 1 + 2\sum_{n=1}^{N} (1 - n/N) \cos 2\pi nx\,. \tag{4.5}$$

This trigonometric polynomialis known as the *Fejér kernel.* In the same way that we used Lemma 3.14 to show that the partial sums $s_N(x)$ can be expressed as a convolution of f with the Dirichlet kernel D_N (recall (3.41)), we see by Lemma 3.14 that

$$\sigma_N(x) = (\Delta_N * f)(x) = \int_0^1 \Delta_N(u)f(x-u)\,du\,. \tag{4.6}$$

Our first order of business is to explore the properties of the Fejér kernel. In passing we note that the sums in (4.3) and (4.5) actually run from $-N+1$ to $N-1$, since the coefficient is 0 when $n = \pm N$. Thus $\Delta_N(x)$ is a trigonometric polynomial of degree $N-1$, and $\sigma_N(x)$ is a trigonometric polynomial of degree not exceeding $N-1$.

First we note that from (3.43) and (4.5) it is evident that

$$\int_0^1 \Delta_N(x)\,dx = 1 \tag{4.7}$$

for all N. Next we show that

$$\Delta_N(x) = \begin{cases} \dfrac{1}{N}\Big(\dfrac{\sin \pi N x}{\sin \pi x}\Big)^2 & (x \notin \mathbb{Z}), \\ N & (x \in \mathbb{Z}). \end{cases} \tag{4.8}$$

First assume that x is not an integer. From the formula (0.2) for the sum of a geometric progression we see that

$$\sum_{n=0}^{N-1} e(nx) = \frac{1-e(Nx)}{1-e(x)} = \frac{e(Nx)-1}{e(x)-1}\,.$$

We factor out appropriate exponentials in the numerator and denominator to create expressions of the form $e(u) - e(-u)$. Thus the above is

$$= \frac{e(Nx/2)\big(e(Nx/2)-e(-Nx/2)\big)}{e(x/2)\big(e(x/2)-e(-x/2)\big)}\,.$$

On dividing the numerator and denominator by $2i$, we see that the above is

$$= e((N-1)x/2)\frac{\sin \pi N x}{\sin \pi x}\,. \tag{4.9}$$

Let z denote the common value of the left hand side and right hand side of this identity. Then $|z|^2$ is

$$\Big|\sum_{n=0}^{N-1} e(nx)\Big|^2 = \Big(\frac{\sin \pi N x}{\sin \pi x}\Big)^2\,.$$

We multiply out the left hand side to see that it is

$$\Big(\sum_{m=0}^{N-1} e(mx)\Big)\Big(\sum_{n=0}^{N-1} e(-nx)\Big) = \sum_{m=0}^{N-1}\sum_{n=0}^{N-1} e((m-n)x)\,.$$

On collecting those m and n for which $m - n = k$, we see that the above is

$$= \sum_{k=-N+1}^{N-1} e(kx) \sum_{\substack{0\le m\le N-1\\ 0\le n\le N-1\\ m-n=k}} 1\,.$$

Here the inner sum is $N - |k|$, so on dividing both sides by N we obtain (4.8) when x is not an integer.

Since $\sin u \sim u$ as $u \to 0$, we deduce that

$$\lim_{x\to 0} \Delta_N(x) = \lim_{x\to 0} \frac{1}{N}\Big(\frac{\sin \pi N x}{\sin \pi x}\Big)^2 = N\,.$$

But $\Delta_N(x)$ is a trigonometric polynomial, and hence continuous, so $\Delta_N(0) = N$. Alternatively, from the definition (4.5) we see that

$$\begin{aligned}\Delta_N(0) &= 1 + 2\sum_{n=1}^{N-1}(1-n/N) = 1 + \frac{2}{N}\sum_{n=1}^{N-1} N-n\\ &= 1 + \frac{2}{N}\sum_{n=1}^{N-1} n = 1 + \frac{2}{N}\cdot\frac{N(N-1)}{2} = N\end{aligned}$$

by the formula (0.1) for the sum of an arithmetic progression. Thus (4.8) holds for all x.

The function $\sin N\pi x$ has simple zeros at $x = 0, 1/N, 2/N, \ldots$, and so $\Delta_N(x)$ has double zeros at these points, except for $x = 0$, since $\sin \pi x$ also vanishes there.

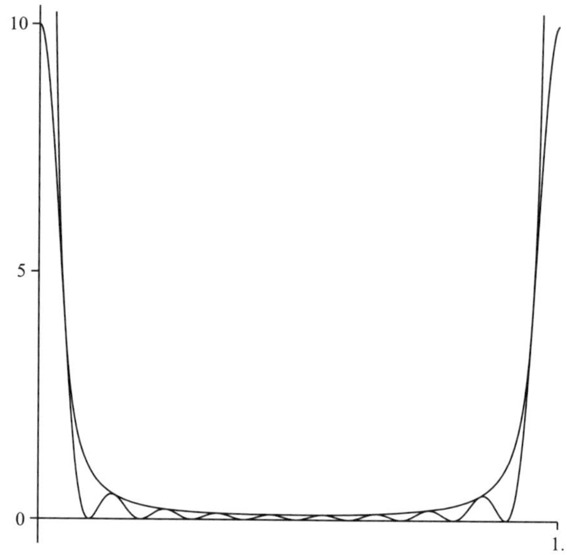

Figure 4.1. Graph of $\Delta_{10}(x)$ showing 9 double zeros and the envelope $1/(10\sin^2 \pi x)$.

Thus $\Delta_N(x)$ has $N-1$ double zeros, making a total of $2N-2$ zeros. Since $\Delta_N(x)$ is a trigonometric polynomial of degree $N-1$, this is the maximum number of zeros that it could have, in view of Theorem 3.15.

As $\Delta_N(x) \ge 0$ for all x and $\int_0^1 \Delta_N(x)\,dx = 1$, it follows that the convolution (4.6) is a weighted average of the values of f. Hence

$$|\sigma_N(x)| \le \max_x |f(x)|, \tag{4.10}$$

and if f is real-valued, then

$$\min_x f(x) \le \sigma_N(x) \le \max_x f(x) \tag{4.11}$$

for all x.

Since $\sin \pi x \ge 2x$ for $0 \le x \le 1/2$, we see that

$$\Delta_N(x) \le \min\left(N, \frac{1}{4Nx^2}\right) \tag{4.12}$$

for $-1/2 \le x \le 1/2$. Thus

$$\int_\delta^{1/2} \Delta_N(x)\,dx \le \int_\delta^{1/2} \frac{1}{4Nx^2}\,dx < \int_\delta^\infty \frac{1}{4Nx^2}\,dx = \frac{1}{4N\delta}\,. \tag{4.13}$$

If $\delta < 1/(2N)$, then we can do better by observing that

$$\int_\delta^{1/2} \Delta_N(x)\,dx \le \int_0^{1/2} \Delta_N(x)\,dx = 1/2.$$

Since $\Delta_N(x)$ is even, it follows that

$$\int_{1/2}^{1-\delta} \Delta_N(x)\,dx = \int_{\delta}^{1/2} \Delta_N(x)\,dx\,.$$

Hence

$$\int_{\delta}^{1-\delta} \Delta_N(x)\,dx \le \min\Big(1, \frac{1}{2N\delta}\Big) \tag{4.14}$$

for $0 < \delta \le 1/2$. We now put the properties of Δ_N to good use.

Theorem 4.1. *If f is a continuous function with period 1, then $\sigma_N(x)$ tends to $f(x)$ uniformly as $N \to \infty$.*

In other words, for any $\varepsilon > 0$, there is an N_0 such that if $N > N_0$, then

$$|f(x) - \sigma_N(x)| < \varepsilon \tag{4.15}$$

for all x.

Let $C(\mathbb{T})$ denote the set of continuous functions with period 1. For such functions we define a norm, called the *uniform norm*:

$$\|f\|_\infty = \max_{x\in\mathbb{T}} |f(x)|\,. \tag{4.16}$$

The distance between two members f, g of $C(\mathbb{T})$ is taken to be $\|f - g\|_\infty$. The import of Theorem 4.1 is that trigonometric polynomials are dense in $C(\mathbb{T})$: For any $f \in C(\mathbb{T})$ and any $\varepsilon > 0$, there is a trigonometric polynomial T such that $\|f - T\|_\infty < \varepsilon$.

Proof. From (4.6) and (4.7) we deduce that

$$f(x) - \sigma_N(x) = \int_0^1 \Delta_N(u)(f(x) - f(x-u))\,du\,.$$

We write this as

$$\int_{-\delta}^{\delta} + \int_{\delta}^{1-\delta} = I_1 + I_2\,.$$

Since f is continuous on $[0, 1]$, it follows (see Theorem 0.5) that f is uniformly continuous. That is, for any $\varepsilon > 0$, there is a $\delta > 0$ such that for all x, $|f(x) - f(x-u)| < \varepsilon$ whenever $|u| \le \delta$. Hence if δ is chosen in this way, then

$$|I_1| \le \varepsilon \int_{-\delta}^{\delta} \Delta_N(u)\,du \le \varepsilon \int_0^1 \Delta_N(u)\,du = \varepsilon\,.$$

On the other hand, since f is continuous it follows that $|f(x)|$ is bounded, say $|f(x)| \le M$ for all x. Hence $|f(x) - f(x-u)| \le |f(x)| + |f(x-u)| \le 2M$, so

$$|I_2| \le 2M \int_{\delta}^{1-\delta} \Delta_N(u)\,du \le \frac{M}{N\delta}$$

by (4.14). Thus $I_2 \to 0$ as $N \to \infty$. In particular, $|I_2| < \varepsilon$ for all sufficiently large N. Thus

$$|f(x) - \sigma_N(x)| < 2\varepsilon$$

for all x and all sufficiently large N, which is the desired result. □

Corollary 4.2. *If $f \in C(\mathbb{T})$ and $\widehat{f}(n) = 0$ for all integers n, then $f(x) = 0$ for all x.*

Proof. We know that for any $|f(x) - \sigma_N(x)| < \varepsilon$ for all x, if N is sufficiently large. But if $\widehat{f}(n) = 0$ for all n, then $\sigma_N(x) = 0$ for all N and all x. Since $|f(x)| < \varepsilon$ for all x and ε is arbitrarily small, it follows that $f(x) = 0$ for all x. □

The following equivalent formulation is sometimes more convenient.

Corollary 4.3. *Suppose that $f \in C(\mathbb{T})$ and that $g \in C(\mathbb{T})$. If $\widehat{f}(n) = \widehat{g}(n)$ for all n, then $f(x) = g(x)$ for all x.*

Proof. Apply Corollary 4.2 to the function $f(x) - g(x)$, whose Fourier coefficients are $\widehat{f}(n) - \widehat{g}(n) = 0$ for all n. □

This has an immediate application to absolutely convergent trigonometric series. The following theorem sets the stage.

Theorem 4.4. *Suppose that*

$$\sum_{n=-\infty}^{\infty} |c_n| < \infty, \tag{4.17}$$

and put

$$g(x) = \sum_{n=-\infty}^{\infty} c_n e(nx)\,. \tag{4.18}$$

Then $g \in C(\mathbb{T})$, and $\widehat{g}(n) = c_n$ for all n.

Proof. The function g has period 1 because each term in its definition has period 1. Suppose that $\varepsilon > 0$ is given. By the convergence of the series (4.17) it follows that if N is a sufficiently large positive integer, then

$$\sum_{|n|>N} |c_n| < \varepsilon\,.$$

Let $s_N(x) = \sum_{n=-N}^{N} c_n e(nx)$. By the triangle inequality it follows that

$$|g(x) - s_N(x)| = \Big| \sum_{|n|>N} c_n e(nx) \Big| \le \sum_{|n|>N} |c_n| < \varepsilon \tag{4.19}$$

for all x. Since the partial sums $s_N(x)$ converge uniformly to $g(x)$, and since each term $c_n e(nx)$ is a continuous function of x, it follows by Theorem 0.5 that g is continuous.

Fix an integer m, and suppose (as we may) that $-N \le m \le N$. From (3.15) we see that

$$|\widehat{g}(m) - \widehat{s}_N(m)| \le \int_0^1 |g(x) - s_N(x)|\, dx < \int_0^1 \varepsilon\, dx = \varepsilon,$$

in view of (4.19). Now $s_N(x)$ is a trigonometric polynomial, and in (3.43) we established that the coefficients of a trigonometric polynomial are its Fourier coefficients.

Thus $\widehat{s}_N(m) = c_m$. Hence $|\widehat{g}(m) - c_m| < \varepsilon$. Since this is true for any $\varepsilon > 0$, it follows that $\widehat{g}(m) = c_m$, and the proof is complete. □

Example 4.2. Let $\delta(1), \delta(2), \ldots$ be a sequence of numbers tending monotonically to 0. Let $n_1, n_2, \ldots$ be a strictly increasing sequence of positive integers such that n_k is so large that $\delta(n_k) < 1/(2k^2)$. Put

$$f(x) = \sum_{k=1}^{\infty} \frac{\cos 2\pi n_k x}{k^2}.$$

By Theorem 4.4 we know that f is a continuous function, and that

$$\widehat{f}(n) = \begin{cases} \frac{1}{2k^2} & (n = \pm n_k \text{ for some } k), \\ 0 & (\text{otherwise}). \end{cases}$$

Thus there exist arbitrarily large n for which $\widehat{f}(n) > \delta(n)$. Here the $\delta(n)$ may tend to 0 slowly, in which case the n_k will increase very rapidly, and then $\widehat{f}(n) = 0$ for all but a very thin set of positive integers.

Suppose that $n_k = k^2$, so that

$$f(x) = \sum_{k=1}^{\infty} \frac{\cos 2\pi k^2 x}{k^2}. \tag{4.20}$$

This formula cannot be obtained from Theorem 3.17, since when the above is differentiated term-by-term we obtain a series whose coefficients do not tend to 0. Thus Theorem 4.4 expands our collection of convergent Fourier Series.

From Corollary 4.3 and Theorem 4.4 we obtain the following useful result.

Corollary 4.5. *Suppose that $f \in C(\mathbb{T})$. If $\sum_{n=-\infty}^{\infty} |\widehat{f}(n)| < \infty$, then*

$$f(x) = \sum_{n=-\infty}^{\infty} \widehat{f}(n) e(nx) \tag{4.21}$$

for all x.

Proof. Let $g(x)$ denote the sum on the right hand side above. By Theorem 4.4 we know that g is a continuous function with period 1, and that $\widehat{g}(n) = \widehat{f}(n)$ for all n. By Corollary 4.3 it follows that $f(x) = g(x)$ for all x. □

From Theorem 4.1 we know that any continuous function with period 1 can by uniformly approximated by a trigonometric polynomial. The corresponding result for algebraic polynomials is a famous result of of Weierstrass:

Corollary 4.6. (Weierstrass) *Suppose that f is continuous on the interval $[a, b]$. For any $\varepsilon > 0$ there is a polynomial $P(x) = \sum_{n=0}^{N} a_n x^n$ such that $|f(x) - P(x)| < \varepsilon$ uniformly for $x \in [a, b]$.*

Proof. Suppose first that $a = 0$ and $b = 1/2$. Put $f_1(x) = f(x)$ for $0 \le x \le 1/2$, and for $1/2 < x \le 1$ let $f_1(x)$ be linear with $f_1(1/2) = f(1/2)$ and $f_1(1) = f(0)$.

Then let $f_1(x)$ have period 1. By Theorem 4.1, there is a trigonometric polynomial $T(x) = \sum_{-N}^{N} t_n e(nx)$ such that $|f_1(x) - T(x)| < \varepsilon_1$ for all x. The power series

$$e^z = \sum_{k=0}^{\infty} \frac{z^k}{k!}$$

converges for all z, and uniformly so for $|z| \le R$, for any given R. Take $R = \pi N$. Then for any $\varepsilon_2 > 0$ there is a K such that

$$\Big| e^z - \sum_{k=0}^{K} \frac{z^k}{k!} \Big| < \varepsilon_2$$

uniformly for $|z| \le \pi N$. In particular, the above holds when $z = 2\pi i n x$, $-N \le n \le N$, and $0 \le x \le 1/2$. Thus

$$\Big| f(x) - \sum_{n=-N}^{N} t_n \sum_{k=0}^{K} \frac{(2\pi i n x)^k}{k!} \Big| < \varepsilon_1 + C\varepsilon_2$$

for $0 \le x \le 1/2$. Here $C = \sum_{-N}^{N} |t_n|$ may depend on ε_1, but we can take $\varepsilon_2 = \varepsilon_1 / C$, and then the above is $< 2\varepsilon_1$. Thus we obtain a good uniform polynomial approximation to $f(x)$, uniformly for $0 \le x \le 1/2$.

Now consider a general interval $[a, b]$. We may assume that $a < b$. Let $\ell(x) = 2(b-a)x + a$. Thus $\ell(x)$ is a linear function with the property that $\ell(0) = a$ and $\ell(1/2) = b$. Put $f_1(x) = f(\ell(x))$ for $0 \le x \le 1/2$. Then f_1 is continuous, so there is a polynomial $P(x)$ such that $|f_1(x) - P(x)| < \varepsilon$ for $0 \le x \le 1/2$. On replacing x by $\ell^{-1}(x)$, we see that $|f_1(\ell^{-1}(x)) - P(\ell^{-1}(x))| < \varepsilon$ for $a \le x \le b$. But $f_1(\ell^{-1}(x)) = f(x)$, and $\ell^{-1}(x) = (x-a)/(2(b-a))$ is linear, so $P(\ell^{-1}(x))$ is a polynomial in x. Thus we have a good polynomial approximation to $f(x)$ for $a \le x \le b$. □

The proof just completed is a bit tortuous and unnatural. The Weierstrass uniform approximation theorem can be proved in many ways, and some of the proofs are very elegant. Our point is not that we have an attractive proof, but rather that Corollary 4.6 is an easy consequence of Theorem 4.1.

The same good properties of $\Delta_N(x)$ that gave rise to Theorem 4.1 can also be applied to an arbitrary function $f \in L^1(\mathbb{T})$.

Theorem 4.7. *Suppose that $f \in L^1(\mathbb{T})$. Then*

$$\lim_{N\to\infty} \int_0^1 |f(x) - \sigma_N(x)|\, dx = 0\,.$$

The relation above is referred to as *convergence in norm.* Since continuous functions are dense in $L^1(\mathbb{T})$ and since we can approximate a continuous function uniformly by a trigonometric polynomial, we already know that trigonometric polynomials are dense in $L^1(\mathbb{T})$. The point of the above is that $\sigma_N(x)$ gives a good approximation.

Proof. Since

$$\sigma_N(x) = \int_0^1 f(x-u)\Delta_N(u)\, du,$$

and since $\int_0^1 \Delta_N(u)\,du = 1$, it follows that

$$f(x) - \sigma_N(x) = \int_0^1 (f(x) - f(x-u))\Delta_N(u)\,du\,.$$

We take absolute values of both sides and integrate to see that

$$\int_0^1 |f(x) - \sigma_N(x)|\,dx = \int_0^1 \Big|\int_0^1 (f(x) - f(x-u))\Delta_N(u)\,du\Big|\,dx\,.$$

By the triangle inequality, the above is

$$\le \int_0^1 \int_0^1 |f(x) - f(x-u)|\Delta_N(u)\,du\,dx\,.$$

On interchanging the order of integration, we see that the above is

$$= \int_0^1 \Delta_N(u)\Big(\int_0^1 |f(x) - f(x-u)|\,dx\Big)\,du$$

$$= \int_{-\delta}^{\delta} + \int_{\delta}^{1-\delta} = I_1 + I_2,$$

say. By Lemma 3.5 we know that for any $\varepsilon > 0$ there is a $\delta > 0$ such that

$$\int_0^1 |f(x) - f(x-u)|\,dx < \varepsilon$$

whenever $|u| < \delta$. With δ chosen in this way, we find that

$$|I_1| \le \varepsilon \int_{-\delta}^{\delta} \Delta_N(u)\,du \le \varepsilon \int_0^1 \Delta_N(u)\,du = \varepsilon\,.$$

As for I_2, we observe that

$$\int_0^1 |f(x) - f(x-u)|\,dx \le \int_0^1 |f(x)|\,dx + \int_0^1 |f(x-u)|\,dx = 2\|f\|_1\,.$$

Hence by (4.14),

$$|I_2| \le 2\|f\|_1 \int_{\delta}^{1-\delta} \Delta_N(u)\,du \le \frac{\|f\|_1}{N\delta}\,.$$

Hence $I_2 \to 0$ as $N \to \infty$. On combining our estimates, we find that

$$\int_0^1 |f(x) - \sigma_N(x)|\,dx < 2\varepsilon$$

for all sufficiently large N. Thus we have the stated result. □

In (3.5) we showed that the functions $e(nx)$ are orthonormal. We now show that this orthonormal system is complete in the sense that if f is orthogonal to all the functions $e(nx)$, then f is 0 almost everywhere.

Corollary 4.8. *If $f \in L^1(\mathbb{T})$, then $\int_0^1 |f(x)|\,dx = 0$ if and only if $\widehat{f}(n) = 0$ for all integers n.*

Proof. If $\int_0^1 |f(x)|\,dx = 0$, then $\widehat{f}(n) = 0$ for all n by inequality (3.15). Suppose conversely that $\widehat{f}(n) = 0$ for all n. Then $\sigma_N(x) = 0$ for all N and all x. From Theorem 4.7 we know that

$$\int_0^1 |f(x) - \sigma_N(x)|\,dx$$

tends to 0 as N tends to infinity. Thus $\int_0^1 |f(x)|\,dx$ tends to 0 as N tends to infinity. Since this quantity is independent of N, in order to tend to 0 it must be 0. □

Suppose that f is a measurable function with period 1. For each integer n, let

$$\mathcal{S}_n = \{x \in [0,1) : 2^{n-1} < |f(x)| \le 2^n\}. \tag{4.22}$$

We recall that the *measure* of a set is its length. For example, if a set $\mathcal{A}$ is a union of some disjoint intervals, then

$$\operatorname{meas} \mathcal{A} = \int_{\mathcal{A}} 1\,dx \tag{4.23}$$

is the sum of the lengths of those intervals. The structure of $\mathcal{A}$ may be more complicated than this, but this is an issue of measure theory that need not concern us, and (4.23) is true for any measurable set. Thus

$$\int_0^1 |f(x)|\,dx = \sum_{n=-\infty}^{\infty} \int_{\mathcal{S}_n} |f(x)|\,dx \le \sum_{n=-\infty}^{\infty} \int_{\mathcal{S}_n} 2^n\,dx = \sum_{n=-\infty}^{\infty} 2^n \operatorname{meas} \mathcal{S}_n = E,$$

say. Similarly,

$$\int_0^1 |f(x)|\,dx = \sum_{n=-\infty}^{\infty} \int_{\mathcal{S}_n} |f(x)|\,dx \ge \sum_{n=-\infty}^{\infty} \int_{\mathcal{S}_n} 2^{n-1}\,dx = \sum_{n=-\infty}^{\infty} 2^{n-1} \operatorname{meas} \mathcal{S}_n = E/2.$$

Thus $E \asymp \|f\|_1$, and in particular $E < \infty$ if and only if $\|f\|_1 < \infty$. Let

$$\mathcal{S} = \bigcup_{n=-\infty}^{\infty} \mathcal{S}_n, \qquad \mathcal{S}^* = \{x \in [0,1) : f(x) = 0\}. \tag{4.24}$$

Here $\mathcal{S}$ is the set of x for which $f(x) \ne 0$, and hence $\mathcal{S} \cup \mathcal{S}^* = [0,1)$. Consequently,

$$1 = \operatorname{meas} \mathcal{S}^* + \operatorname{meas} \mathcal{S} = \operatorname{meas} \mathcal{S}^* + \sum_{n=-\infty}^{\infty} \operatorname{meas} \mathcal{S}_n. \tag{4.25}$$

If $\|f\|_1 = 0$, then $E = 0$, so $\operatorname{meas} \mathcal{S}_n = 0$ for all n, so $\operatorname{meas} \mathcal{S} = 0$. Thus we say that $f(x) = 0$ for almost all x, which simply means that the set of exceptions has Lebesgue measure 0. Conversely, if $\operatorname{meas} \mathcal{S}^* = 1$, then $\operatorname{meas} \mathcal{S} = 0$, so $\operatorname{meas} \mathcal{S}_n = 0$ for all n, so $E = 0$, so $\|f\|_1 = 0$. Thus Corollary 4.8 can be expressed as follows.

Corollary 4.9. *Suppose that $f \in L^1(\mathbb{T})$. Then the following three assertions are equivalent:*

(a) $\displaystyle\int_0^1 |f(x)|\,dx = 0$;

(b) $\widehat{f}(n) = 0$ *for all integers* n;

(c) $f(x) = 0$ *for almost all real* x.

Another way to look at the above is to say that distinct functions have distinct sets of Fourier coefficients. That is, by applying the above to $f(x) - g(x)$ we obtain

Corollary 4.10. *Suppose that f and g are functions in $L^1(\mathbb{T})$. Then the following three assertions are equivalent:*

(a) $\int_0^1 |f(x) - g(x)|\, dx = 0$;

(b) $\widehat{f}(n) = \widehat{g}(n)$ *for all integers* n;

(c) $f(x) = g(x)$ *for almost all real* x.

In the situation of Corollary 4.9, we cannot deduce that $f(x) = 0$ for all x, but we can deduce that $f(x) = 0$ at points of continuity, in view of the following simple observation.

Theorem 4.11. *If $f \in L^1(\mathbb{T})$, if f is continuous at x_0, and if $f(x_0) \neq 0$, then*

$$\int_{x_0-\delta}^{x_0+\delta} |f(x)|\, dx > 0 \tag{4.26}$$

for all $\delta > 0$.

Proof. Let $a = f(x_0)$. Since f is continuous at x_0, for any $\varepsilon > 0$ there is a $\delta > 0$ such that $|f(x) - a| < \varepsilon$ for $x_0 - \delta < x < x_0 + \delta$. Take $\varepsilon = |a|/2$. Then by the triangle inequality, $|f(x)| = |f(x) - a + a| \geq |a| - |f(x) - a| \geq |a|/2$ for $x_0 - \delta < x < x_0 + \delta$. Hence the integral in (4.26) is $\geq |a|\delta$ for all sufficiently small δ. Thus the integral is positive for all $\delta > 0$. □

In particular, if $f \in C(\mathbb{T})$, then $\widehat{f}(n) = 0$ for all n if and only if $f(x) = 0$ for all x. As a second application of the above principle, we derive a useful extension of Corollary 4.5.

Corollary 4.12. *Suppose that $f \in L^1(\mathbb{T})$, and that $\sum_{n=-\infty}^{\infty} |\widehat{f}(n)| < \infty$. Then*

$$f(x) = \sum_{-\infty}^{\infty} \widehat{f}(n)e(nx) \tag{4.27}$$

for any x at which f is continuous.

Proof. We let $g(x)$ be the sum on the right hand side above. By Theorem 4.4 we know that g is a continuous function with period 1, and that $\widehat{g}(n) = \widehat{f}(n)$ for all n. By Corollary 4.10 it follows that $\int_0^1 |f(x) - g(x)|\, dx = 0$. Hence by Theorem 4.11, $f(x) = g(x)$ at any point where f is continuous. □

Theorem 4.7 is fundamental, as it allows one to show—in some situations—that if something is true for trigonometric polynomials, then it is also true for general functions. The following is an example of this line of reasoning.

Theorem 4.13. (Fejér's Lemma) *Suppose that $f \in L^1(\mathbb{T})$, that g is a bounded function with period 1. Then*

$$\lim_{m\to\infty} \int_0^1 f(x)g(mx)\, dx = \widehat{f}(0)\widehat{g}(0)\,.$$

We note that the Riemann–Lebesgue Lemma (Theorem 3.6) is the special case $g(x) = e(-x)$ of the above.

Proof. Suppose that N is so large that $\int_0^1 |f(x) - \sigma_N(x)|\,dx < \varepsilon$, and let C be a constant chosen so that $|g(x)| \le C$ for all x. Then by the triangle inequality

$$\left|\int_0^1 f(x)g(mx)\,dx - \widehat{f}(0)\widehat{g}(0)\right| \tag{4.28}$$

$$= \left|\int_0^1 \sigma_N(x)g(mx)\,dx - \widehat{f}(0)\widehat{g}(0) + \int_0^1 (f(x) - \sigma_N(x))g(mx)\,dx\right| \tag{4.29}$$

$$\le \left|\int_0^1 \sigma_N(x)g(mx)\,dx - \widehat{f}(0)\widehat{g}(0)\right| + \left|\int_0^1 (f(x) - \sigma_N(x))g(mx)\,dx\right|.$$

Again by the triangle inequality, the last term is

$$\le \int_0^1 |f(x) - \sigma_N(x)||g(mx)|\,dx \le C\int_0^1 |f(x) - \sigma_N(x)|\,dx < C\varepsilon\,. \tag{4.30}$$

Now

$$\int_0^1 \sigma_N(x)g(mx)\,dx = \sum_{n=-N}^{N} (1 - |n|/N)\widehat{f}(n)\int_0^1 e(nx)g(mx)\,dx,$$

which by Theorem 3.2 (d) is

$$= \sum_{\substack{-N<n<N\\ m|n}} (1 - |n|/N)\widehat{f}(n)\widehat{g}(-n/m)\,.$$

If $m \ge N$, then the only multiple of m in the interval $(-N, N)$ is $n = 0$. Hence the above is exactly $\widehat{f}(0)\widehat{g}(0)$ for all $m \ge N$. The stated result now follows by combining this with (4.28) and (4.30). □

Before continuing, we make a comment about notation: We let $f(a^-)$ denote the limit of $f(x)$ as x approaches a from below, and $f(a^+)$ denote the limit of $f(x)$ as x approaches a from above:

$$f(a^-) = \lim_{x\to a^-} f(x), \qquad f(a^+) = \lim_{x\to a^+} f(x)\,. \tag{4.31}$$

Theorem 4.14. (Fejér) *Suppose that $f \in L^1(\mathbb{T})$. If $f(x^-)$ and $f(x^+)$ both exist and are finite, then*

$$\lim_{N\to\infty} \sigma_N(x) = \frac{f(x^-) + f(x^+)}{2}\,.$$

Note that the value of $f(x)$ is irrelevant. However, to say that f is continuous at x is equivalent to saying that $f(x^-) = f(x^+) = f(x)$, and in that case the above asserts that $\sigma_N(x)$ tends to $f(x)$.

Proof. We write

$$\sigma_N(x) = \int_0^1 f(x-u)\Delta_N(u)\,du = \int_0^\delta + \int_\delta^{1-\delta} + \int_{1-\delta}^1 = I_1 + I_2 + I_3,$$

say. Here we suppose that δ has been chosen small enough to ensure that $|f(x-u)-f(x^-)|<\varepsilon$ uniformly for $0\le u\le\delta$ and $|f(x-u)-f(x^+)|<\varepsilon$ uniformly for $1-\delta\le u\le 1$. We write

$$I_1=\int_0^\delta (f(x-u)-f(x^-))\Delta_N(u)\,du+f(x^-)\int_0^\delta \Delta_N(u)\,du=T_1+T_1',$$

say. Now

$$|T_1|\le\int_0^\delta |f(x-u)-f(x^-)|\Delta_N(u)\,du<\varepsilon\int_0^\delta \Delta_N(u)\,du\le\varepsilon\int_{-1/2}^{1/2}\Delta_N(u)\,du=\varepsilon$$

by (4.7). From (4.7) and (4.13) we deduce that

$$\frac{1}{2}-\frac{1}{4N\delta}\le\int_0^\delta \Delta_N(u)\,du\le\frac{1}{2}\,.$$

Hence

$$\left|T_1'-\frac{f(x^-)}{2}\right|\le\frac{|f(x^-)|}{4N\delta}\,.$$

By (4.12) we see that

$$|I_2|\le\frac{1}{4N\delta^2}\int_\delta^{1-\delta}|f(x-u)|\,du\le\frac{1}{4N\delta^2}\int_0^1|f(u)|\,du\,.$$

We treat I_3 in the same way as I_1, with terms T_3 and T_3'. Thus

$$\sigma_N(x)=T_1+T_1'+I_2+T_3+T_3'\,.$$

Here $|T_1|<\varepsilon$ and $|T_3|<\varepsilon$ for all N, while $T_1'\to f(x^-)/2$, $I_2\to 0$, and $T_3'\to f(x^+)/2$ as $N\to\infty$. Hence $|\sigma_N(x)-(f(x^-)+f(x^+)/2|<3\varepsilon$ for all sufficiently large N so the proof is complete. □

As was discussed in §0.3, if f is a Lebesgue-integrable function, then x is said to be a *Lebesgue point* of f if

$$\lim_{h\to0+}\frac{1}{h}\int_{x-h}^{x+h}|f(u)-f(x)|\,du=0\,.$$

Theorem 4.15. *If $f\in L^1(\mathbb{T})$, and if x is a Lebesgue point of f, then $\sigma_N(x)\to f(x)$ as $N\to\infty$.*

Proof. We know that

$$f(x)-\sigma_N(x)=\int_0^1(f(x)-f(x-u))\Delta_N(u)\,du\,.$$

Let $R_N(x) = \min(N, 1/(N\|x\|^2))$, so that $\Delta_N(x) \le R_N(x)$ for all x by (4.12). Thus by the triangle inequality,

$$|f(x) - \sigma_N(x)| \le \int_0^1 |f(x) - f(x-u)| R_N(u)\, du$$

$$= \int_0^{1/2} \big(|f(x) - f(x-u)| + |f(x) - f(x+u)|\big) R_N(u)\, du$$

$$= N \int_0^{1/N} \big(|f(x) - f(x-u)| + |f(x) - f(x+u)|\big)\, du \tag{4.32}$$

$$+ \frac{1}{N} \int_{1/N}^{1/2} \big(|f(x) - f(x-u)| + |f(x) - f(x+u)|\big) \frac{du}{u^2}\,. \tag{4.33}$$

Let

$$I(h) = \int_0^h |f(x) - f(x-u)| + |f(x) - f(x+u)|\, du\,.$$

By integrating by parts we see that the term (4.33) is

$$= \frac{4}{N} I(1/2) - N I(1/N) + \frac{2}{N} \int_{1/N}^{1/2} I(u)\, \frac{du}{u^3}\,.$$

The term (4.32) is $NI(1/N)$, so on adding these two terms we deduce that

$$|f(x) - \sigma_N(x)| \le \frac{4}{N} I(1/2) + \frac{2}{N} \int_{1/N}^{1/2} I(u)\, \frac{du}{u^3}\,. \tag{4.34}$$

Suppose that x is a Lebesgue point of f. Then for any $\varepsilon > 0$ there is a $\delta > 0$ such that $I(h) < \varepsilon h$ when $0 \le h \le \delta$. Clearly $I(1/2) \le 2|f(x)| + \|f\|_1$. Thus the first quantity on the right above is small if N is large. We write the remaining expression above as

$$\frac{2}{N} \int_{1/N}^{\delta} I(u)\, \frac{du}{u^3} + \frac{2}{N} \int_{\delta}^{1/2} I(u)\, \frac{du}{u^3} = T_1 + T_2,$$

say. Since $I(u) < \varepsilon u$ for $u < \delta$, we see that

$$T_1 < \frac{2\varepsilon}{N} \int_{1/N}^{\delta} u^{-2}\, du < \frac{2\varepsilon}{N} \int_{1/N}^{\infty} u^{-2}\, du = 2\varepsilon\,.$$

Since $I(u) \le I(1/2)$ for $u \le 1/2$, we see that

$$T_2 \le \frac{2I(1/2)}{N} \int_{\delta}^{1/2} u^{-3}\, du < \frac{2I(1/2)}{N} \int_{\delta}^{\infty} u^{-3}\, du = \frac{I(1/2)}{N\delta^2}\,,$$

which is small if N is large. □

Corollary 4.16. *If $f \in L^1(\mathbb{T})$, then for almost all x, $\sigma_N(x) \to f(x)$ as $N \to \infty$.*

Proof. As was noted in §0.3, almost all x are Lebesgue points of f. □

Exercises

1. Suppose that $f \in L^1(\mathbb{T})$.
   (a) Show that
   $$\int_0^1 (1 \pm \cos 2\pi nx) f(x)\, dx = \frac{1}{2}a_0 \pm \frac{1}{2}a_n$$
   in the usual trigonometric notation.
   (b) Assuming that $f(x) \geq 0$ for all x, explain why the integrand above is everywhere nonnegative.
   (c) Deduce that if $f(x) \geq 0$ for all x, then $|a_n| \leq a_0$ for all positive n.
   (d) Show that a_n can be nearly as big as a_0 by considering $f(x) = \Delta_N(x)$ with N large compared with n.

2. Suppose that f is a bounded function with period 1 and that $g \in L^1(\mathbb{T})$. Explain how you know that
   $$\sum_{n=-N}^{N} \left(1 - \frac{|n|}{N}\right) \widehat{f}(n)\widehat{g}(n)e(n\theta) \longrightarrow (f * g)(\theta)$$
   uniformly in θ. (Hint: Recall Theorem 3.13.)

3. Suppose that f is defined on the real line, and put
   $$f_e(x) = \frac{f(x) + f(-x)}{2}, \qquad f_o(x) = \frac{f(x) - f(-x)}{2}.$$
   (a) Show that f_e is an even function and that f_o is an odd function.
   (b) Show that $f(x) = f_e(x) + f_o(x)$ for all x.
   (c) Suppose that $f(x) = g(x) + h(x)$ for all x, that g is an even function, and that h is an odd function. Show that $g(x) = f_e(x)$ and that $h(x) = f_o(x)$ for all x. We call f_e the *even part* of f, and f_o the *odd part* of f.
   (d) Show that if $f \in L^1(\mathbb{T})$, then
   $$\widehat{f_e}(n) = \frac{\widehat{f}(n) + \widehat{f}(-n)}{2}, \qquad \widehat{f_o}(n) = \frac{\widehat{f}(n) - \widehat{f}(-n)}{2}.$$
   (e) Suppose that $f \in L^1(\mathbb{T})$. Show that if $\widehat{f}(n)$ is an even sequence, then $\widehat{f_o}(n) = 0$ for all n. Explain why it follows that there is an even function $f_e \in L^1(\mathbb{T})$ such that $f(x) = f_e(x)$ for almost all x.
   (f) Suppose that $f \in L^1(\mathbb{T})$. Show that if $\widehat{f}(n)$ is an odd sequence, then $\widehat{f_e}(n) = 0$ for all n. Explain why it follows that there is an odd function $f_o \in L^1(\mathbb{T})$ such that $f(x) = f_o(x)$ for almost all x. Note that these last two results give a partial converse of Corollary 3.3.

4. (a) For real x let $f(x) = 1/(x-1)$ when $x \neq 1$, and put $f(1) = 0$. Express f as the sum of an even function and an odd function.
   (b) Let f have period 1, $f(x) = x^2$ for $0 < x < 1$, and $f(0) = 1/2$. Express f as the sum of an even function and an odd function.

5. Suppose that $f \in L^1(\mathbb{T})$, and put

$$f_r(x) = \frac{f(x) + \overline{f(x)}}{2} = \operatorname{Re} f(x), \qquad f_i(x) = \frac{f(x) - \overline{f(x)}}{2i} = \operatorname{Im} f(x)\,.$$

   (a) Show that $f(x) = f_r(x) + i f_i(x)$ for all x.
   (b) Show that

$$\widehat{f_r}(n) = \frac{\widehat{f}(n) + \overline{\widehat{f}(-n)}}{2}, \qquad \widehat{f_i}(n) = \frac{\widehat{f}(n) - \overline{\widehat{f}(-n)}}{2i}\,.$$

   (c) Conclude that if $\widehat{f}(-n) = \overline{\widehat{f}(n)}$ for all n, then f is almost a real-valued function in the sense that there is a real-valued function, namely f_r, such that $f(x) = f_r(x)$ for almost all x. Note that this gives a partial converse of Corollary 3.4.

6. Suppose that $f \in L^1(\mathbb{T})$. Show that if $\widehat{f}(n) = 0$ for all but finitely many n, then there is a trigonometric polynomial T such that $f(x) = T(x)$ for almost all x.

7. (a) Show that if $f \in C(\mathbb{T})$ and $\widehat{f}(-n) = \widehat{f}(n)$ for all n, then f is an even function.
   (b) Show that if $f \in C(\mathbb{T})$ and $\widehat{f}(-n) = -\widehat{f}(n)$ for all n, then f is an odd function.
   (c) Show that if $f \in C(\mathbb{T})$ and $\widehat{f}(-n) = \overline{\widehat{f}(n)}$ for all n, then f is a real-valued function.

8. Suppose that $f \in L^1(\mathbb{T})$, that n is an integer for which $|n|$ is large, set $N = |n| - 1$, and put $g(x) = f(x) - \sigma_N(f; x)$.
   (a) Explain why $\|g\|_1$ is small.
   (b) Explain why $|\widehat{g}(n)|$ is small.
   (c) Show that $\widehat{f}(n) = \widehat{g}(n)$.
   (d) Thus give a second proof of the Riemann–Lebesgue Lemma (Theorem 3.6). Of course, both proofs depend on Lemma 3.5.

9. Suppose that $p > 0$ and that $M_p = \int_a^b |f(x)|^p\,dx < \infty$.
   (a) Show that for $V > 0$,

$$\operatorname{meas}\{x \in [a,b] \,:\, |f(x)| \ge V\} \le \frac{M_p}{V^p}\,.$$

   (b) Show that $\operatorname{meas}\{x \in [a,b] \,:\, |f(x)| \ge V\} = o(1/V^p)$ as $V \to \infty$. (For a discussion of the "little oh" notation, see Appendix O.)

   If there is a constant C such that

$$\operatorname{meas}\{x \in [a,b] \,:\, |f(x)| \ge V\} \le C/V^p \tag{4.35}$$

   for all $V > 0$, then we say that f is of *weak type* L^p.

   (c) Suppose that $a < b$ and $p > 0$ are given. Give an example of a function f such that f is of weak type L^p on $[a,b]$, but $\int_a^b |f(x)|^p\,dx = \infty$.
   (d) Suppose that $a < b$, that $0 < p_2 < p_1$, and that f is of weak type L^{p_1} on $[a,b]$. Show that $\int_a^b |f(x)|^{p_2}\,dx < \infty$.

10. Suppose that $f \in L^1(\mathbb{T})$. Show that if f is continuous at x, then

$$|f(x)| \le \sum_{n=-\infty}^{\infty} |\widehat{f}(n)|.$$

(The sum on the right may diverge, but this is analogous to (2.15) for the Discrete Fourier Transform.)

11. (a) Show that $e(\alpha) - e(\beta) = 2ie((\alpha+\beta)/2)\sin \pi(\alpha-\beta)$.
    (b) Show that $e(\alpha) + e(\beta) = 2e((\alpha+\beta)/2)\cos \pi(\alpha-\beta)$.

12. Show that

$$\frac{1}{1-e(x)} = \frac{1}{2} + \frac{i}{2}\cot \pi x\,.$$

13. We define the *conjugate Fejér kernel* (plotted in Figure 4.2) to be

$$\widetilde{\Delta}_N(x) = \sum_{n=1}^{N-1}\Big(1-\frac{n}{N}\Big)\big(-ie(nx)+ie(-nx)\big) = 2\sum_{n=1}^{N-1}\Big(1-\frac{n}{N}\Big)\sin 2\pi n x\,. \tag{4.36}$$

(a) Let $s_N(x) = 1 + e(x) + e(2x) + \cdots + e((N-1)x)$. Show that

$$s_N(x) = \frac{1-e(Nx)}{1-e(x)}\,.$$

(b) Put

$$\sigma_N(x) = \sum_{n=0}^{N-1}(N-n)e(nx)\,.$$

Note that $\sigma_N = s_1(x) + \cdots + s_N(x)$. Show that

$$\sigma_N(x) = \frac{N}{1-e(x)} - e(x)\frac{1-e(Nx)}{(1-e(x))^2}\,.$$

(c) Show that the above is

$$= N\Big(\frac{1}{2}+\frac{i}{2}\cot\pi x\Big) - \frac{i}{2}e(Nx/2)\frac{\sin \pi x}{(\sin\pi x)^2}\,.$$

(d) Take imaginary parts and multiply by $2/N$ to deduce that

$$\widetilde{\Delta}_N(x) = \cot \pi x - \frac{\cos \pi N x \,\sin \pi N x}{N(\sin \pi x)^2}\,.$$

(e) Conclude that

$$\widetilde{\Delta}_N(x) = \cot \pi x - \frac{\sin 2\pi N x}{2N(\sin\pi x)^2}\,. \tag{4.37}$$

14. (a) Show that if n is an integer, $0 < n < 2N$, then $\widetilde{\Delta}_N(n/(2N)) = \cot \pi n/(2N)$.
    (b) Show that if n is an integer, $0 < n < 2N$, then

$$\widetilde{\Delta}'_N(n/(2N)) = \begin{cases} 0 & (n \text{ odd}),\\ -2\pi \csc^2 \pi n/(2N) & (n \text{ even}).\end{cases}$$

(c) Show that if $0 < x < 1$, then

$$\lim_{N\to\infty} \widetilde{\Delta}_N(x) = \cot \pi x\,.$$

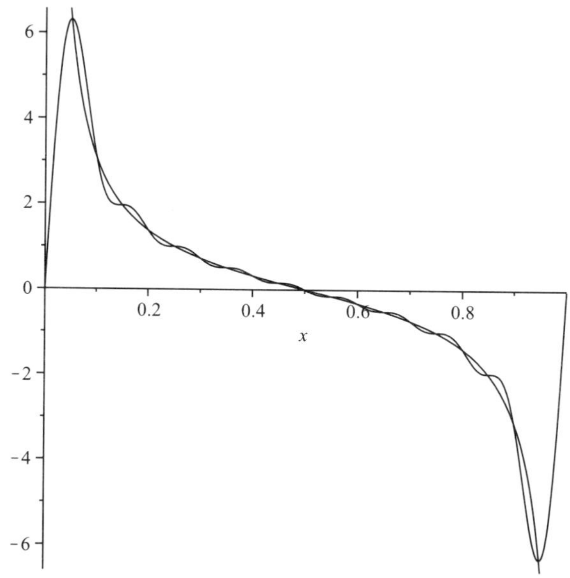

Figure 4.2. The conjugate Fejér kernel $\widetilde{\Delta}_{10}(x)$ and $\cot \pi x$.

(d) Show that for any integer n,

$$\lim_{N\to\infty} \widehat{\widetilde{\Delta}}_N(n) = \begin{cases} -i & (n>0), \\ 0 & (n=0), \\ i & (n<0). \end{cases}$$

Let $K(x) = \cot \pi x$. In some sense,

$$\widehat{K}(n) = \begin{cases} -i & (n>0), \\ 0 & (n=0), \\ i & (n<0). \end{cases} \tag{4.38}$$

Since these Fourier coefficients do not tend to 0 it is clear at once (if it were not already clear) that $K \notin L^1(\mathbb{T})$. Nevertheless, one might hope that if $f \in L^1(\mathbb{T})$, then

$$\int_0^1 K(u)f(x-u)\,du = \widetilde{f}(x), \tag{4.39}$$

the conjugate function of f, whose Fourier coefficients are

$$\widehat{\widetilde{f}}(n) = \begin{cases} -i\widehat{f}(n) & (n>0), \\ 0 & (n=0), \\ i\widehat{f}(n) & (n<0). \end{cases}$$

There are problems here, since the integral (4.39) is unlikely to converge, although

$$\lim_{\varepsilon\to 0^+} \int_\varepsilon^{1-\varepsilon} K(u)f(x-u)\,du$$

will exist if f is smooth enough at x. Still, it is known that there exist $f \in L^1(\mathbb{T})$ such that $\widehat{f} \notin L^1(\mathbb{T})$. In the next several exercises, we show that (4.38) is in some sense true.

15. Explain why

$$\int_{\varepsilon}^{1-\varepsilon} (\cot \pi u) \cos(2\pi n u \, du = 0$$

for all integers n and all $\varepsilon > 0$.

16. Put

$$g_1(x) = \frac{1}{\pi} \sum_{k=-K}^{K} \frac{1}{x-k},$$

$$g_2(x) = \frac{1}{\pi} \sum_{k>K} \Big(\frac{1}{x-k} + \frac{1}{x+k} \Big).$$

Then by (T.88) we know that $g_1(x) + g_2(x) = \cot \pi x$.

(a) Show that

$$\int_{-1/2}^{1/2} g_1(u) \sin 2\pi n u \, du = \frac{1}{\pi} \int_{-K-1/2}^{K+1/2} \frac{2\pi n u}{u} \, du .$$

(b) Suppose that $n > 0$. Show that

$$\int_{-K-1/2}^{K+1/2} \frac{\sin 2\pi n u}{u} \, du = \int_{-\pi n(2K+1)}^{\pi n(2K+1)} \frac{\sin v}{v} \, dv .$$

(c) Show that if $0 < a < b$, then

$$\int_a^b \frac{\sin v}{v} \, dv = \frac{\cos a}{a} - \frac{\cos b}{b} - \int_a^b \frac{\cos v}{v^2} \, dv .$$

(d) Deduce that if $1 \le a < b$, then

$$\int_a^b \frac{\sin v}{v} \, dv = O(1/a) .$$

(e) Deduce that

$$\lim_{V \to \infty} \int_{-V}^{V} \frac{\sin v}{v} \, dv$$

exists.

(f) Show that

$$\frac{1}{x-k} + \frac{1}{x+k} = O(1/k^2)$$

if $|k| \ge 1$ and $1/2 \le x \le 1/2$.

(g) Deduce that $g_2(x) = O(1/K)$ for $-1/2 \le x \le 1/2$.

(h) Deduce that $\int_{-1/2}^{1/2} g_2(u) \sin 2\pi n u \, du = O(1/K)$.

(i) Conclude that if n is a positive integer, then

$$\int_0^1 (\cot \pi u) \sin 2\pi n u \, du = \frac{1}{\pi} \int_{-\infty}^{\infty} \frac{\sin v}{v} \, dv .$$

17. Let $f(x) = \csc x - 1/(\pi x)$. Note that by (T.89) this is a bounded function in the interval $-1/2 \le x \le 1/2$. Let

$$I = \int_{-1/2}^{1/2} f(x) \sin(2N+1)\pi x \, dx \,.$$

(a) Show that

$$I = \int_{-1/2}^{1/2} f'(x) \frac{\cos(2N+1)\pi x}{(2N+1)\pi} \, dx \,.$$

(b) By term-by-term differentiation of (T.89) show that $f'(x)$ is bounded for $-1/2 \le x \le 1/2$. Deduce that $I = O(1/N)$.

(c) Show that

$$I = \int_{-1/2}^{1/2} D_N(x) \, dx - \frac{1}{pi} \int_{-1/2}^{1/2} \frac{\sin(2N+1)\pi x}{x} \, dx \,.$$

(d) Show that

$$\int_{-1/2}^{1/2} \frac{\sin(2N+1)\pi x}{x} \, dx = \int_{(-N+1/2)\pi}^{(N+1/2)\pi} \frac{\sin v}{v} \, dv \,.$$

(e) Deduce that

$$\int_{-\infty}^{\infty} \frac{\sin v}{v} \, dv = \pi \,.$$

(f) Deduce that

$$\int_0^1 (\cot \pi x) \sin 2\pi n x \, dx = 1$$

for all positive integers n.

(g) Show that

$$\lim_{\varepsilon \to 0+} \int_\varepsilon^{1-\varepsilon} \cot \pi x e(-nx) \, dx = -i \int_0^1 (\cot \pi x) \sin 2\pi x \, dx \,.$$

(h) Conclude that

$$\lim_{\varepsilon \to 0+} \int_\varepsilon^{1-\varepsilon} (\cot \pi x) e(-nx) \, dx = \begin{cases} -i & (n > 0), \\ 0 & (n = 0), \\ i & (n > 1). \end{cases}$$

18. Suppose that $f \in L^1(\mathbb{T})$, and recall the old-fashioned notation (3.2).

(a) Show that $(f * D_N)(x) = \frac{1}{2}a_0 + \sum_{n=1}^N (a_n \cos 2\pi n x + b_n \sin 2\pi n x)$.

(b) Show that $(f * D_N^*)(x) = \frac{1}{2}a_0 + \sum_{n=1}^{N-1} (a_n \cos 2\pi n x + b_n \sin 2\pi n x)$ $+\frac{1}{2}(a_N \cos 2\pi N x + b_N \sin 2\pi N x)$.

(c) Show that $(f * \widetilde{D}_N)(x) = \sum_{n=1}^N (-b_n \cos 2\pi n x + a_n \sin 2\pi n x)$.

(d) Show that $(f * \Delta_N)(x) = \frac{1}{2}a_0 + \sum_{n=1}^N (1 - n/N)(a_n \cos 2\pi n x + b_n \sin 2\pi n x)$.

(e) Show that $(f * \widetilde{\Delta}_N)(x) = \sum_{n=1}^{N-1} (1 - n/N)(-b_n \cos 2\pi n x + a_n \sin 2\pi n x)$.

4.2. Special coefficients

We begin with a short review of summation by parts (aka partial summation), which is a discrete analogue of integration by parts.

Lemma 4.17. *For $1 \le n \le N$, let c_n and e_n be real or complex numbers, and let $S_n = e_1 + e_2 + \cdots + e_n$. Then*

$$\sum_{n=1}^{N} c_n e_n = \sum_{n=1}^{N-1} (c_n - c_{n+1}) S_n \; + c_N S_N \, . \tag{4.40}$$

Proof. Put $S_0 = 0$, so that $e_n = S_n - S_{n-1}$ for $1 \le n \le N$. Thus

$$\begin{aligned} \sum_{n=1}^{N} c_n e_n &= \sum_{n=1}^{N} c_n (S_n - S_{n-1}) \\ &= \sum_{n=1}^{N} c_n S_n - \sum_{n=1}^{N} c_n S_{n-1} \, . \end{aligned}$$

We reindex the second sum with $m = n - 1$ to see that the above is

$$= \sum_{n=1}^{N} c_n S_n - \sum_{m=0}^{N-1} c_{m+1} S_m \, .$$

For the indices running from 1 to $N-1$ we combine the terms in the two sums. The term $c_N S_N$ for $n = N$ in the first sum is reported in (4.40). The term $c_1 S_0$ in the second sum makes no contribution since $S_0 = 0$. Thus we have the stated identity. □

If $0 < M < N$ and we apply the partial summation identity (4.40) first for M and then for N and subtract, we find that

$$\sum_{n=M+1}^{N} c_n e_n = \sum_{n=M}^{N-1} (c_n - c_{n+1}) S_n - c_M S_M + c_N S_N \, . \tag{4.41}$$

The two most familiar applications of partial summation are as follows.

Theorem 4.18. (Abel's Test) *If $\sum_{n=1}^{\infty} e_n$ is a convergent series, and if $\{c_n\}$ is a bounded monotonically decreasing sequence, then the series $\sum_{n=1}^{\infty} c_n e_n$ is convergent.*

Theorem 4.19. (Dirichlet's Test) *Let $S_n = e_1 + e_2 + \cdots + e_n$. If the sequence $\{S_n\}$ is bounded, and the sequence $\{c_n\}$ is monotonically decreasing to 0, then the series $\sum_{n=1}^{\infty} c_n e_n$ converges.*

Combined proof of both Tests. By the Cauchy principle for convergence, it suffices to show that the left hand side of (4.41) tends to 0 as $M, N \to \infty$ with $M < N$. In both tests the sequence S_n of partial sums is bounded; let S be chosen so that $|S_n| \le S$ for all n. By the triangle inequality, the sum on the right hand side of (4.41) has absolute value not exceeding

$$\sum_{n=M}^{N-1} (c_n - c_{n+1})S = S \sum_{n=M}^{N-1} (c_n - c_{n+1}) = S(c_M - c_N)\,.$$

By Theorem 0.3, a bounded monotonic sequence has a limit; let c denote the limit of the numbers c_n. Since $c_M \to c$ and $c_N \to c$, it follows that $c_M - c_N \to 0$ as $M, N \to \infty$. We note that

$$\begin{aligned}|c_N S_N - c_M S_M| &= |(c_N - c)S_N - (c_M - c)S_M + c(S_N - S_M)| \\ &\le |c_N - c|S + |c_M - c|S + |c||S_N - S_M|\,.\end{aligned}$$

The first two terms on the right hand side tend to 0 since c_n is tending to c and S is finite. It is only in treating the final term that we argue differently for the two tests. In the case of Abel's Test, the S_n tend to a limit, say $S_n \to s$. Then $S_N - S_M = (S_N - s) - (S_M - s) \to 0$, and $|c|$ is finite, so the term tends to 0. In Dirichlet's Test, $c = 0$, so the term is 0 for all M and N. □

Lemma 4.20. *Suppose that $c_0, c_1, \ldots$ are positive real numbers tending monotonically to 0. Then the series*

$$f(x) = \sum_{n=0}^{\infty} c_n e(nx) \tag{4.42}$$

converges for $0 < x < 1$, and converges uniformly for $\delta \le x \le 1 - \delta$ if $0 < \delta \le 1/2$.

Since each summand is continuous, the uniform convergence implies that $f(x)$ is continuous for $0 < x < 1$ (recall Theorem 0.5). However, $f(x)$ is not necessarily in $L^1(\mathbb{T})$, because it may be very large when x is near 0.

Proof. Let $S_0(x) = 0$, and for positive integers N put $S_N(x) = 1 + e(x) + \cdots + e((N-1)x)$. Then by (4.9) we know that $|S_N(x)| \le 1/\sin \pi x$ for $0 < x < 1$, and that $|S_N(x)| \le 1/\sin \pi\delta$ uniformly for $\delta \le x \le 1 - \delta$. By summing by parts we see that

$$\sum_{n=0}^{N} c_n e(nx) = \sum_{n=1}^{N-1} (c_{n-1} - c_n) S_n(x) + c_N S_N(x)\,. \tag{4.43}$$

Here the last term tends to 0 uniformly for $\delta \le x \le 1-\delta$, since $|S_{N+1}(x)| \le 1/\sin \pi\delta$ and $c_N \searrow 0$. The sum is uniformly convergent, since

$$\Big| \sum_{n=N+1}^{\infty} (c_{n-1} - c_n) S_n(x) \Big| \le \frac{1}{\sin \pi\delta} \sum_{n=N+1}^{\infty} (c_{n-1} - c_n) = \frac{c_N}{\sin \pi\delta} \to 0\,.$$

□

Although f is continuous in the open interval $(0, 1)$, it may be that f tends to infinity so rapidly as $x \to 0$ that $f \notin L^1(\mathbb{T})$. Before considering conditions that ensure that $f \in L^1(\mathbb{T})$, we show that if $f \in L^1(\mathbb{T})$, then the series (4.42) that defines f is indeed the Fourier Series of f.

Theorem 4.21. *Let f be defined as in (4.42) where $c_n \searrow 0$. If $f \in L^1(\mathbb{T})$, then $\widehat{f}(n) = c_n$ if $n \ge 0$, and $\widehat{f}(n) = 0$ if $n < 0$.*

Proof. Let k be an integer. We suppose first that x is not an integer. Then by partial summation we see that

$$(4.44)\qquad \sum_{n=M+1}^{N} c_n e(nx)(1-e(-kx)) = (1-e(-kx))\sum_{M}^{N-1}(c_n - c_{n+1})S_n(x) + (1-e(-kx))\big(c_N S_N(x) - c_M S_M(x)\big).$$

We know that $|S_n(x)| \le 1/|\sin \pi x|$. Also, $|\sin \pi x| \ge 2\|x\|$, so $|S_n(x)| \le 1/(2\|x\|)$. Now $|e(\delta)-1| = |2\sin \pi\delta| \le 2\pi|\delta|$ for any real δ. Hence $|1-e(-kx)| \le 2\pi|k||x|$. Since $|1-e(-kx)|$ has period 1, we may suppose that $-1/2 \le x \le 1/2$. Thus we conclude that $|1-e(-kx)| \le 2\pi|k|\|x\|$. By these estimates we see that the absolute value of the right hand side above is

$$\le \frac{|1-e(-kx)|}{|\sin \pi x|}(c_M - c_N) + \frac{|1-e(-kx)|}{|\sin \pi x|}(c_M + c_N)$$

$$(4.45)\qquad = \frac{2|1-e(-kx)|}{|\sin \pi x|}c_M \le \frac{2\pi|k|\|x\|}{\|x\|}c_M = 2\pi|k|c_M\,.$$

Since $c_M \to 0$ as $M \to \infty$, this tends to 0 uniformly in x. If x is an integer, then both sides of (4.44) are 0, so the upper bound (4.45) holds in this case also. Since the series

$$\sum_{n=1}^{\infty} c_n e(nx)(1-e(-kx))$$

converges uniformly, we may integrate it term-by-term. Thus we find that

$$\widehat{f}(0) - \widehat{f}(k) = \int_0^1 f(x)(1-e(-kx))\,dx = \sum_{n=1}^{\infty} c_n \int_0^1 e(nx)(1-e(-kx))\,dx$$

$$= \begin{cases} c_0 - c_k & (k \ge 0), \\ c_0 & (k < 0). \end{cases}$$

Now $c_k \to 0$ as $k \to \infty$ by hypothesis, and $\widehat{f}(k) \to 0$ as $k \to \infty$ by the Riemann–Lebesgue Lemma. Hence $\widehat{f}(0) = c_0$. On cancelling this from both sides we deduce also that $\widehat{f}(k) = c_k$ if $k > 0$, and that $\widehat{f}(k) = 0$ if $k < 0$. □

In the sum that defines $f(x)$ there is little to no cancelation among the first terms if x is near an integer. By partial summation we know that

$$\Big|\sum_{n=N+1}^{\infty} c_n e(nx)\Big| \le \frac{c_N}{\|x\|}\,.$$

Thus

$$|f(x)| \le c_0 + c_1 + \cdots + c_N + \frac{c_N}{\|x\|}\,.$$

If $\|x\| \ge 1/(N+1)$, then the above is

$$\le 2(c_0 + c_1 + \cdots + c_N) \tag{4.46}$$

since $c_n \ge c_N$ for $0 \le n \le N$. We use this bound for $1/(N+1) \le \|x\| \le 1/N$.

Theorem 4.22. *Suppose that f is defined by (4.42) where $c_n \searrow 0$. Then*

$$\|f\|_1 \le 2c_0 + 4\sum_{n=1}^{\infty} \frac{c_n}{n}.$$

Proof. Since $f(-x) = \overline{f(x)}$, it follows that $|f(-x)| = |f(x)|$. Hence by (4.46) we see that

$$\int_0^1 |f(x)|\,dx = 2\int_0^{1/2} |f(x)| = 2\sum_{n=2}^{\infty} \int_{1/(n+1)}^{1/n} |f(x)|\,dx$$

$$\le 4\sum_{N=2}^{\infty} \frac{(c_0 + c_1 + \cdots + c_N)}{N(N+1)}. \tag{4.47}$$

Now

$$\sum_{N=n}^{\infty} \frac{1}{N(N+1)} = \sum_{N=n}^{\infty} \Big(\frac{1}{N} - \frac{1}{N+1}\Big) = \frac{1}{n},$$

so the quantity (4.47) is

$$= 2c_0 + 2c_1 + 4\sum_{k=2}^{\infty} \frac{c_n}{n},$$

which gives the stated bound. □

For f defined as in (4.42), we can take the real part of f to obtain a cosine series and the imaginary part to obtain a sine series. Thus far we have treated the two equally, and we can write

$$f(x) = \sum_{n=0}^{\infty} (c_n - c_{n+1}) S_n(x),$$

but when we take a closer look at the sums $S_n(x)$ we notice a difference in the distribution of the real part and imaginary parts. With a little calculation one finds that

$$S_n(x) = \frac{ie(-x/2)}{2\sin \pi x} - \frac{ie((n+1/2)x)}{2\sin \pi x}.$$

Thus for fixed x, the $S_n(x)$ lie on a circle with center $ie(-x/2)/(2\sin \pi x)$ and radius $1/(2|\sin \pi x|)$. Thus if x is small and positive, then

$$\operatorname{Re} S_n \doteqdot \frac{\sin 2\pi n x}{2\pi x}, \tag{4.48}$$

$$\operatorname{Im} S_n \doteqdot \frac{1 - \cos 2\pi n x}{2\pi x}. \tag{4.49}$$

Since $\operatorname{Re} S_n(x)$ is oscillating sinusoidally with a mean near 0, this suggests that if the differences $c_n - c_{n+1}$ are monotonically decreasing, then we could obtain a better estimate by summing by parts a second time. On the other hand, $\operatorname{Im} S_n(x)$, though oscillating, has a large mean value, which suggests that the estimates that we have already derived are sharp. We now show that both of these first impressions are correct.

Theorem 4.23. *If*

$$f(x)=\sum_{n=1}^{\infty} b_n \sin 2\pi n x \tag{4.50}$$

where $b_n \searrow 0$, *then*

$$\frac{1}{\pi}\sum_{n=1}^{\infty}\frac{b_n}{n} \le \|f\|_1 \le 4\sum_{n=1}^{\infty}\frac{b_n}{n}. \tag{4.51}$$

It should be understood that in the above situation, $f \in L^1(\mathbb{T})$ if and only if $\sum_{n=1}^{\infty} b_n/n < \infty$.

Proof. For the lower bound, suppose that f is an odd function in $L^1(\mathbb{T})$. Then $\widehat{f}(0)=0$, and from Theorem 3.7 we know that if $F(x)=\int_0^x f(u)\,du$, then $\widehat{F}(0)=\int_0^1(1/2-u)f(u)\,du$. Now $F(0)=0$, so from Theorem 3.17 we see that

$$\widehat{F}(0) = -\lim_{N\to\infty}\sum_{\substack{n=-N\\ n\ne 0}}^{N}\frac{\widehat{f}(n)}{2\pi i n} = \frac{1}{2\pi}\sum_{n=1}^{\infty}\frac{b_n}{n}.$$

That is,

$$\sum_{n=1}^{\infty}\frac{b_n}{n} = \pi\int_0^1 (1-2u)f(u)\,du. \tag{4.52}$$

Note that this applies to any odd function $f\in L^1(\mathbb{T})$, without any condition on the b_n. Now $-1\le 1-2u\le 1$ for $0\le u\le 1$, so

$$\Big|\int_0^1 (1-2u)f(u)\,du\Big| \le \int_0^1 |(1-2u)f(u)|\,du \le \int_0^1 |f(u)|\,du = \|f\|_1.$$

Thus we have the lower bound in (4.51). The upper bound was already established in Theorem 4.22. □

For odd functions whose sine coefficients tend monotonically to 0, we can also characterize when the function is bounded and when it is continuous.

Theorem 4.24. *Suppose that* $b_n \searrow 0$, *and that* $f(x)$ *is defined as in* (4.50). *Then the following are equivalent:*

(a) $|f(x)|$ *is bounded;*

(b) nb_n *is a bounded sequence;*

(c) *The series* (4.50) *is boundedly convergent.*

Proof. (a) $\Longrightarrow$ (b). Let M be chosen so that $|f(x)|\le M$ for all x. As noted in (4.10), it follows that $|\sigma_N(x)|\le M$ for all x and all N. To bound b_n, we take $x=1/(4n)$ and $N=2n$. Thus

$$\sum_{k=1}^{2n}\Big(1-\frac{k}{2n}\Big)b_k\sin\frac{\pi k}{2n} \le M.$$

Now $\sin \pi x \ge 0$ for $0 \le x \le 1$, so the sine factor is nonnegative in every term. Thus the above sum is

$$\ge \sum_{k=1}^{n} \Big(1 - \frac{k}{2n}\Big) b_k \sin \frac{\pi k}{2n} .$$

Now $1 - k/(2n) \ge 1/2$ for $1 \le k \le n$, $b_k \ge b_n$, and $\sin \pi x \ge 2x$ for $0 \le x \le 1/2$, so the above is

$$\ge \frac{1}{2} b_n \sum_{k=1}^{n} \frac{k}{n} = (n+1)b_n/4 .$$

Thus $nb_n \le 4M$ for all positive n.

(b) $\implies$ (c). Suppose that B is chosen so that $nb_n \le B$ for all positive n. That the series (4.50) converges for all x was established already in Lemma 4.20. In order to show that the series is boundedly convergent we need to establish a bound for

$$\sum_{n=M+1}^{N} b_n \sin 2\pi n x$$

that holds uniformly for all M, N, and x. Since f is an odd function with period 1, we may assume that $0 \le x \le 1/2$. We consider three cases.

Case 1. $Nx \le 2$. Since $\sin u \le u$ for all u, we see that

$$\sum_{n=M+1}^{N} b_n \sin 2\pi n x \le 2\pi x \sum_{n=M+1}^{N} n b_n \le 2\pi N x B \le 4\pi B .$$

Case 2. $Mx \ge 1$. Let $S_n(x) = \sum_{k=0}^{n-1} e(kx)$. By the summation by parts formula (4.41) we see that

$$\sum_{n=M+1}^{N} b_n e(nx) = \sum_{n=M}^{N-1} (b_n - b_{n+1}) S_n(x) \; - b_M S_M(x) + b_N S_N(x) .$$

From (4.9) we know that $|S_n(x)| \le 1/\sin \pi x$ for all n. Thus the above has absolute value not exceeding

$$\frac{1}{\sin \pi x} \sum_{n=M}^{N-1} (b_n - b_{n+1}) + \frac{b_M}{\sin \pi x} + \frac{b_N}{\sin \pi x} = \frac{2b_M}{\sin \pi x} .$$

But $\sin \pi x \ge 2x$, so the above is

$$\le \frac{b_M}{x} = \frac{M b_M}{M x} \le B.$$

Case 3. $Mx < 1$, $Nx > 2$. Choose K so that $1 \le Kx \le 2$, and write

$$\sum_{n=M+1}^{N} b_n \sin 2\pi n x = \sum_{n=M+1}^{K} b_n \sin 2\pi n x + \sum_{n=K+1}^{N} b_n \sin 2\pi n x .$$

Then apply Case 1 to the first sum and Case 2 to the second sum. Thus

$$\Big| \sum_{n=M+1}^{N} b_n \sin 2\pi n x \Big| \le (6\pi + 1) B$$

in all cases.

(c) $\implies$ (a). This is trivial. We know that the series converges to $f(x)$ for all x. Since the partial sums are uniformly bounded, it follows that the limit $f(x)$ that they are tending to is also bounded. □

The same line of reasoning yields the following similar result.

Theorem 4.25. *Suppose that $b_n \searrow 0$, and that $f(x)$ is defined as in (4.50). Then the following are equivalent:*

(a) *$|f(x)|$ is continuous;*

(b) *$nb_n \to 0$ as $n \to \infty$;*

(c) *The series (4.50) is uniformly convergent.*

We now consider cosine series with convex coefficients.

Lemma 4.26. *Let $a_0, a_1, a_2, \ldots$ be a sequence of real numbers such that a_n tends to 0 as n tends to infinity, and $a_{n-1} - 2a_n + a_{n+1} \ge 0$ for all $n \ge 1$. Then*

(a) *The sequence a_n is monotonically decreasing;*

(b) *$(a_n - a_{n+1})n \to 0$ as $n \to \infty$;*

and

(c) $$\sum_{n=k}^{\infty}(a_{n-1} - 2a_n + a_{n+1})n = (a_{k-1} - a_k)k + a_k = ka_{k-1} - (k-1)a_k$$

for any positive integer k.

Proof. (a) Our second hypothesis concerning the a_n asserts that $a_{n-1} - a_n \ge a_n - a_{n+1}$. Thus $a_0 - a_1 \ge a_1 - a_2 \ge a_2 - a_3 \ge \cdots$, and these differences tend to 0 because the a_n tend to 0. Hence these differences are nonnegative, which is to say that $a_n \ge a_{n+1}$ for all $n \ge 0$.
(b) Suppose that $0 < M < N$. Then

$$a_M \ge a_M - a_N = \sum_{n=M+1}^{N} (a_{n-1} - a_n)\,.$$

The summands $a_{n-1} - a_n$ are decreasing, so they are all $\ge a_{N-1} - a_N$. Moreover, there are $N - M$ summands, so the sum above is

$$\ge (N - M)(a_{N-1} - a_N)\,.$$

Now a_M is small if M is large, and $N - M \ge N/2$ if $N \ge 2M$, so $N(a_{N-1} - a_N) \to 0$.
(c) Clearly

$$\sum_{n=k}^{N}(a_{n-1} - 2a_n + a_{n+1})n = \sum_{n=k}^{N}(a_{n-1} - a_n)n - \sum_{n=k}^{N}(a_n - a_{n+1})n\,.$$

After reindexing the second sum, we find that the above is

$$\begin{aligned} &= \sum_{n=k}^{N}(a_{n-1}-a_n)n - \sum_{n=k+1}^{N+1}(a_{n-1}-a_n)(n-1) \\ &= (a_{k-1}-a_k)k - (a_N - a_{N+1})N + \sum_{n=k+1}^{N}(a_{n-1}-a_n) \\ &= (a_{k-1}-a_k)k - (a_N - a_{N+1})N + a_k - a_N\,. \end{aligned}$$

By parts (a) and (b) this tends to $(a_{k-1}-a_k)k + a_k$ as $N \to \infty$. □

Theorem 4.27. *If $a_0, a_1, a_2, \ldots$ are real numbers such that $a_n \to 0$ as $n \to \infty$, and $a_{n-1} - 2a_n + a_{n+1} \ge 0$ for all $n \ge 1$, then*

(a)
$$\frac{1}{2}a_0 + \sum_{n=1}^{N} a_n \cos 2\pi x = \frac{1}{2}\sum_{n=1}^{N-1}(a_{n-1}-2a_n+a_{n+1})n\Delta_n(x) + \frac{1}{2}(a_{N-1}-a_N)N\Delta_N(x) + \frac{1}{2}a_N D_N(x)$$

for all x;
(b) *the series*

$$f(x) = \frac{1}{2}a_0 + \sum_{n=1}^{\infty} a_n \cos 2\pi n x$$

converges for $0 < x < 1$;

(c)
$$f(x) = \frac{1}{2}\sum_{n=1}^{\infty}(a_{n-1}-2a_n+a_{n+1})n\Delta_n(x)$$

for $0 < x < 1$;
(d) *$f(x) \ge 0$ for $0 < x < 1$;*

(e)
$$\int_0^1 f(x)\,dx = \frac{1}{2}a_0;$$

(f) *$\widehat{f}(\pm n) = \frac{1}{2}a_n$ for $n \ge 0$.*

The coefficients a_n may tend to 0 arbitrarily slowly. In Example 4.2 we saw that Fourier coefficients may tend to 0 arbitrarily slowly, but in that construction the Fourier Series was absolutely convergent, but most Fourier coefficients were 0. In the above we see that Fourier coefficients can tend to 0 arbitrarily slowly even when they are monotonic.

Consider the two series

$$f(x) = \sum_{n=1}^{\infty}\frac{\sin 2\pi n x}{\log(n+2)}, \qquad g(x) = \sum_{n=1}^{\infty}\frac{\cos 2\pi n x}{\log(n+2)}. \tag{4.53}$$

The coefficients here are monotonically decreasing to 0, so by taking real and imaginary parts in Lemma 4.20 we know that these series are convergent for $0 < x < 1$, and that f and g are continuous in this interval. Since

$$\sum_{n=1}^{\infty}\frac{1}{n\log(n+2)} = \infty,$$

we know by Theorem 4.23 that $f \notin L^1(\mathbb{T})$. By applying Theorem 4.27 to $\frac{1}{2\log 2}+g(x)$ we find that $g \in L^1(\mathbb{T})$. From the Riemann–Lebesgue Lemma we know that a sequence must tend to 0 at $\pm\infty$ in order to be a set of Fourier coefficients, but by this example we see that this condition by itself is not sufficient.

Put

$$G(x) = \sum_{n=2}^{\infty} \frac{\sin 2\pi nx}{n \log n}\,.$$

Since $g \in L^1(\mathbb{T})$, it follows by Theorem 3.17 that the Fourier Series of G converges to G. However, $G \notin A(\mathbb{T})$, so we see that Corollary 4.5 does not include Theorem 3.17, even when the coefficients are monotonic.

Proof. (a) We observe that

$$\begin{aligned}\frac{1}{2}a_0 + \sum_{n=1}^{N} a_n \cos 2\pi nx &= \frac{1}{2}a_0 + \frac{1}{2}\sum_{n=1}^{N} a_n(D_n(x) - D_{n-1}(x)) \\ &= \frac{1}{2}a_0 + \frac{1}{2}\sum_{n=1}^{N} a_n D_n(x) - \frac{1}{2}\sum_{n=0}^{N-1} a_{n+1}D_n(x) \\ &= \frac{1}{2}\sum_{n=0}^{N-1}(a_n - a_{n-1})D_n(x) \ + \frac{1}{2}a_N D_N(x)\,. \qquad (4.54)\end{aligned}$$

Here the sum over n is

$$\begin{aligned}&\sum_{n=0}^{N-1}(a_n - a_{n+1})\big((n+1)\Delta_{n+1}(x) - n\Delta_n(x)\big) \\ &\quad= \sum_{n=0}^{N-1}(a_n - a_{n+1})(n+1)\Delta_{n+1}(x) - \sum_{n=0}^{N-1}(a_n - a_{n+1})n\Delta_n(x) \\ &\quad= \sum_{n=1}^{N}(a_{n-1} - a_n)n\Delta_n(x) - \sum_{n=0}^{N-1}(a_n - a_{n+1})n\Delta_n(x) \\ &\quad= \sum_{n=1}^{N-1}(a_{n-1} - 2a_n + a_{n+1})n\Delta_n(x) + (a_{N-1} - a_N)N\Delta_N(x)\,.\end{aligned}$$

On inserting this in (4.54), obtain the identity (a).
(b) By Lemma 4.26 (a) we know that the a_n are monotonically decreasing. Thus our series is the real part of the convergent series in Lemma 4.20.
(c) Suppose that $0 < x < 1$. Then the quantities $N\Delta_N(s)$ and $D_N(x)$ in (a) are bounded functions of N. Hence by Lemma 4.26 (a),(b) the last two terms in (a) tend to 0 as $N \to \infty$, and so we have the stated formula for $f(x)$.
(d) This follows from (c), since all terms are nonnegative.
(e) Integration and summation may be exchanged if the functions being summed are nonnegative, and $\int_0^1 \Delta_N(x)\,dx = 1$, so the stated result follows by taking $k = 1$ in Lemma 4.26 (c).
(f) Since $f(x) \geq 0$ and $\int_0^1 f(x)\,dx < \infty$, it follows by the principle of dominated

convergence (see Theorem 0.19) that for $k > 0$,

$$\widehat{f}(k) = \frac{1}{2}\sum_{n=1}^{\infty}(a_{n-1} - 2a_n + a_{n+1})n\widehat{\Delta_n}(k) = \frac{1}{2}\sum_{n=k+1}^{\infty}(a_{n-1} - 2a_n + a_{n+1})(n-k).$$

This is the limit as $N \to \infty$ of

$$\begin{aligned}
\frac{1}{2}&\sum_{n=k+1}^{N}(a_{n-1} - 2a_n + a_{n+1})(n-k)\\
&= \frac{1}{2}\sum_{n=k+1}^{N}(a_{n-1} - a_n)(n-k) - \frac{1}{2}\sum_{n=k+1}^{N}(a_n - a_{n+1})(n-k)\\
&= \frac{1}{2}\sum_{n=k}^{N-1}(a_n - a_{n+1})(n+1-k) - \frac{1}{2}\sum_{n=k+1}^{N}(a_n - a_{n+1})(n-k)\\
&= \frac{1}{2}(a_k - a_{k+1}) - (a_N - a_{N+1})(N-k) + \frac{1}{2}\sum_{n=k+1}^{N-1}(a_n - a_{n+1})\\
&= \frac{1}{2}a_k - (a_N - a_{N+1})(N-k) - a_N.
\end{aligned}$$

By Lemma 4.26 (a),(b), the last two terms above tend to 0 as $N \to \infty$. Thus $\widehat{f}(k) = \frac{1}{2}a_k$ when $k > 0$. Since f is even, $\widehat{f}(-k) = \widehat{f}(k)$, so the proof is complete. □

4.3. Summability

Sometimes the existence of one limit implies the existence of another. We begin with a simple example of this phenomenon.

Theorem 4.28. *Suppose that $u_1, u_2, u_3, \ldots$ is an infinite sequence such that $\lim u_n = a$. Put $v_1 = u_1$, $v_2 = (u_1 + u_2)/2$, $v_3 = (u_1 + u_2 + u_3)/3$, and in general*

$$v_n = \frac{1}{n}\sum_{k=1}^{n} u_k .$$

Then $\lim_{n\to\infty} v_n = a$.

Proof. Put $U_n = u_n - a$, $V_n = v_n - a$. Then $\lim_{n\to\infty} U_n = 0$,

$$V_n = \frac{1}{n}\sum_{k=1} U_k,$$

and we wish to prove that $\lim_{n\to\infty} V_n = 0$. That is, it suffices to prove the theorem in the case $a = 0$.

Since $\lim_{n\to\infty} U_n = 0$, for any given $\varepsilon > 0$ there is an N such that $|U_n| < \varepsilon$ whenever $n > N$. By the triangle inequality,

$$|V_n| \le \frac{1}{n}\sum_{k=1}^{n}|U_k| = \frac{1}{n}\sum_{k=1}^{N}|U_k| + \frac{1}{n}\sum_{k=N+1}^{n}|U_k| .$$

Here the first sum is independent of n, and each summand in the second sum is $< \varepsilon$. Hence the above is

$$\le \frac{1}{n}\sum_{k=1}^{N} |U_k| + \frac{n-N}{n}\varepsilon .$$

Here the second term is $< \varepsilon$, and the first term tends to 0 as $n \to \infty$. Thus $|V_n| < 2\varepsilon$ for all sufficiently large n. Since ε can be taken to be arbitrarily small, it follows that $\lim_{n\to\infty} V_n = 0$. □

Whether it is a good idea to average a sequence depends on the situation. If the sequence u_n is oscillatory and tends to a slowly, then the v_n may tend to a much more quickly. On the other hand, if u_n tends to a very quickly, then the sequence v_n of averages tends to a much more slowly. For example, if $u_n = 1/2^n$, then $v_n \asymp 1/n$.

Let s_n denote a partial sum of an infinite series, as in (4.1), and let σ_N be the average of the first N partial sums, as in (4.2). On taking $u_n = s_{n-1}$ and $v_n = \sigma_{n-1}$ we see from Theorem 4.28 that if the s_n tend to a limit, then the σ_N tend to the same limit. In other words, if a series converges to a, then it is also Cesàro summable to a. This has an immediate application to Fourier series, in view of Fejér's Theorem (Theorem 4.14):

Corollary 4.29. *Suppose that $f \in L^1(\mathbb{T})$. If f is continuous at x, and if the partial sums $s_N(x)$ tend to a limit, then that limit is $f(x)$.*

For power series we have the following fundamental result.

Theorem 4.30. (Abel) *If $\sum_{n=0}^{\infty} a_n$ converges, then*

$$\lim_{x\to 1^-} \sum_{n=0}^{\infty} a_n x^n = \sum_{n=0}^{\infty} a_n .$$

Proof. Let $s_n = \sum_{k=0}^{n} a_k$. We begin by observing that

$$(1-x)\sum_{n=0}^{\infty} s_n x^n = \sum_{n=0}^{\infty} a_n x^n \tag{4.55}$$

when $|x| < 1$. To see why this is so, observe that the left hand side is

$$\sum_{n=0}^{\infty} s_n x^n - \sum_{n=0}^{\infty} s_n x^{n+1} = \sum_{n=0}^{\infty} s_n x^n - \sum_{n=1}^{\infty} s_{n-1} x^n = s_0 + \sum_{n=1}^{\infty} (s_n - s_{n-1}) x^n .$$

But $s_n - s_{n-1} = a_n$ for $n > 0$, so we have (4.55). Let $S = \sum_{n=0}^{\infty} a_n$, and put $S_n = s_n - S$. Since

$$\sum_{n=0}^{\infty} x^n = \frac{1}{1-x} \tag{4.56}$$

for $|x| < 1$, it follows that

$$(1-x)\Big(\sum_{n=0}^{\infty} s_n x^n\Big) - S = (1-x)\Big(\sum_{n=0}^{\infty} s_n x^n\Big) - (1-x)\sum_{n=0}^{\infty} S x^n = (1-x)\sum_{n=0}^{\infty} S_n x^n .$$

Our task now is to show that if the S_n tend to 0, then the right hand side above also tends to 0. That is, we have reduced our problem to the special case $S = 0$. Let $\varepsilon > 0$ be given, and let N be so large that $|S_n| < \varepsilon$ for all $n \geq N$. Write the right hand side above as

$$(1-x)\sum_{n=0}^{N-1} S_n x^n + (1-x)\sum_{n=N}^{\infty} S_n x^n = T_1 + T_2,$$

say. Now

$$|T_1| \leq (1-x)\sum_{n=0}^{N-1} |S_n| \tag{4.57}$$

for $0 \leq x \leq 1$, and

$$|T_2| \leq (1-x)\sum_{n=N}^{\infty} \varepsilon x^n \leq \varepsilon(1-x)\sum_{n=0}^{\infty} x^n = \varepsilon$$

by (4.56). From (4.57) we see that $T_1 \to 0$ as $x \to 1^-$. Thus there is a $\delta > 0$ such that $|T_1| < \varepsilon$ when $1 - \delta < x < 1$. Hence

$$\left|(1-x)\sum_{n=0}^{\infty} S_n x^n\right| < 2\varepsilon$$

for $1 - \delta < x < 1$. Since ε can be taken arbitrarily small, it follows that the limit is 0 as $x \to 1^-$, and the proof is complete. □

Although it was a simple matter to verify (4.55), there is a different approach that is also instructive. By dividing both sides by $1 - x$ and using (4.56), we see that (4.55) is equivalent to the identity

$$\left(\sum_{j=0}^{\infty} x^j\right)\left(\sum_{k=0}^{\infty} a_k x^k\right) = \sum_{n=0}^{\infty} s_n x^n .$$

Here the sums on the left hand side are absolutely convergent for $|x| < 1$, and hence can be arbitrarily rearranged. The general term is $a_k x^{j+k}$. We choose an n, and consider those pairs j, k for which $j + k = n$. Thus we see that the coefficient of x^n is $a_0 + a_1 + \cdots + a_n = s_n$. Thus we have the identity above, and hence (4.55).

Corollary 4.31. *If* $\sum_{n=0}^{\infty} a_n e(n\theta)$ *converges, then*

$$\lim_{r\to 1^-} \sum_{n=0}^{\infty} a_n r^n e(n\theta) = \sum_{n=0}^{\infty} a_n e(n\theta) .$$

If a power series $f(z)$ has radius of convergence $R > 1$, then f is continuous in the closed disk $|z| \leq 1$, which means that the above is trivial in this case. Thus the main interest in Corollary 4.31 is when f has radius of convergence 1. One should note that in Corollary 4.31 we take only a radial limit, which is less than asserting that $f(z)$ is continuous at $e(\theta)$. Our method of proof could be developed to show that $f(z)$ tends to $f(e(\theta))$ as z tends to $e(\theta)$ within a sector, as depicted in Figure 4.3 (a). However, examples can be constructed to show that the limit need not exist if z approaches $e(\theta)$ along a path that approaches $|z| = 1$ tangentially, as in Figure 4.3 (b).

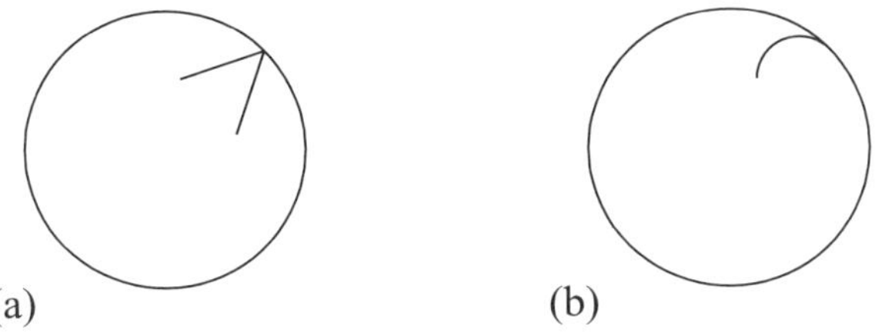

Figure 4.3. (a) A sector that bounds the approach; (b) A tangential approach.

One may note that the proof of Theorem 4.30 bears a striking resemblance to the proof of Theorem 4.28. The similarity is more than just superficial. In Theorem 4.28 we consider averages of the u_n. Put $w_n = (1-x)x^n$. Then the left hand side of (4.55) is $\sum_{n=0}^{\infty} w_n s_n$, which is a weighted average of the s_n, in the sense that $w_n \ge 0$ for all n, and $\sum_{n=0}^{\infty} w_n = 1$ by (0.3). Moreover, when x is just slightly less than 1, the initial weights $w_0, w_1, \ldots$ are very small but approximately equal. Thus having x near 1 in the proof of Theorem 4.30 is analogous to having n large in the proof of Theorem 4.28.

If $\lim_{x\to 1^-} \sum_{n=0}^{\infty} a_n x^n$ exists and has the value a, then we say that the series $\sum a_n$ is *Abel-summable* to a, and we write

$$\sum_{n=0}^{\infty} a_n = a \quad (A).$$

Example 4.3. Let $a_n = (-1)^n$. Then

$$\sum_{n=0}^{\infty} a_n x^n = \sum_{n=0}^{\infty} (-1)^n x^n = \sum_{n=0}^{\infty} (-x)^n = \frac{1}{1+x}$$

for $|x| < 1$, by (4.56). On letting x tend to 1 from below, we find that

$$\sum_{n=0}^{\infty} (-1)^n = 1/2 \quad (A).$$

Theorems 4.28 and 4.30 are called *abelian* in honor of Abel, who proved Theorem 4.30. In general, an abelian theorem is one in which one limit implies another. Usually, the second limit is some sort of weighted average of the first. From Theorem 4.28 we saw that if a series is convergent, then it is Cesàro-summable to the same value. In Example 4.1 we found that the converse is false, and in Example 4.3 we found that the converse of Abel's Theorem (Theorem 4.30) is also false. While converses of abelian theorems are usually false, often one can prove a partial converse by imposing an extra hypothesis. Such partial converses are called *tauberian*, after Tauber, who proved a partial converse of Abel's Theorem. The following is a good first example of a tauberian theorem.

Theorem 4.32. (Hardy) *Suppose that the series* $\sum_{n=1}^{\infty} a_n$ *is Cesàro-summable to* a *and that there is a constant* $C > 0$ *such that*

$$|a_n| \le \frac{C}{n} \tag{4.58}$$

for all n. *Then* $\sum_{n=1}^{\infty} a_n$ *converges to* a.

Here (4.58) is the new hypothesis, called the *tauberian hypothesis*, which permits the converse. Suppose the $f(n)$ is positive and increasing to $+\infty$. No matter how slowly f increases, one can construct examples to show that if (4.58) were replaced by the weaker hypothesis

$$|a_n| \le \frac{f(n)}{n},$$

then Theorem 4.32 would become false. That is, the hypothesis (4.58) is the weakest hypothesis of its kind that enables the converse implication.

Proof. Let $s_n = \sum_{k=1}^{n} a_k$ and $\sigma_n = \sum_{j=1}^{n} \left(1 - \frac{j-1}{n}\right) a_j$. We have to show that if $\sigma_n \to a$ as $n \to \infty$, then also $s_n \to a$. We note that

$$(n+h)\sigma_{n+h} - n\sigma_n = \sum_{j=1}^{n+h} (n+h+1-j)a_j \;-\; \sum_{j=1}^{n} (n+1-j)a_j\,.$$

For $1 \le j \le n$, the difference between the coefficients of a_j is simply h. Thus the above is

$$= hs_n + \sum_{j=n+1}^{n+h} (n+h+1-j)a_j$$

We subtract ha from both sides, and divide through by h. Thus the above can be written

$$s_n - a = \frac{n+h}{h}(\sigma_{n+h} - a) - \frac{n}{h}(\sigma_n - a) - \sum_{j=n+1}^{n+h} \frac{n+h+1-j}{h} a_j\,.$$

In the remaining sum, $|a_j| \le C/j \le C/n$, the coefficient of a_j is at most 1, and there are h terms. Hence by the triangle inequality,

$$|s_n - a| \le \frac{n+h}{h}|\sigma_{n+h} - a| + \frac{n}{h}|\sigma_n - a| + \frac{Ch}{n}\,. \tag{4.59}$$

Choose an $\varepsilon > 0$, take $h = [\varepsilon n]$ and let n tend to infinity. The ratio $(n+h)h$ is bounded, and $\sigma_{n+h} - a$ tends to 0, so the first term on the right hand side tends to 0. The ratio n/h is also bounded, and $\sigma_n - a$ tends to 0, so the second term also tends to 0. Finally, the last term is $\le C\varepsilon$. That is,

$$|s_n - a| \le (C+1)\varepsilon$$

for all sufficiently large n. But ε can be arbitrarily small, so we conclude that $s_n \to a$ as $n \to \infty$. □

We might say that the σ_n is obtained from the s_n by smoothing. In the proof just completed, we recovered the s_n from the σ_n by an "unsmoothing" that hinges on the identity

$$\max(n+h-j, 0) - \max(n-j, 0) = \begin{cases} h & (0 \le j \le n), \\ n+h-j & (n \le j \le n+h), \\ 0 & (j \ge n+h). \end{cases} \tag{4.60}$$

A graphical presentation of this identity is given in Figure 4.4.

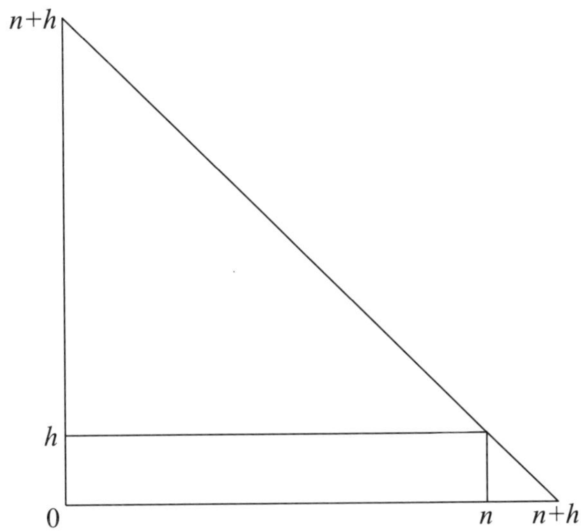

Figure 4.4. Graphical depiction of the unsmoothing identity (4.60).

On combining Fejér's theorem (Theorem 4.14) with Hardy's tauberian theorem (Theorem 4.32) we obtain

Corollary 4.33. *Suppose that $f \in L^1(\mathbb{T})$ and that $|n\widehat{f}(n)|$ is bounded. Then the partial sums $s_N(x)$ of the Fourier series of f converge to $\frac{1}{2}(f(x^-)+f(x^+))$ whenever the one-sided limits $f(x^-)$ and $f(x^+)$ both exist and are finite.*

In the context of Hardy's tauberian theorem, suppose that a sequence $\varepsilon_n \searrow 0$ has the property that $|\sigma_n - a| \le \varepsilon_n$ for all n. Then from the key inequality (4.59) we find that

$$|s_n - a| \le \frac{3n}{n}\varepsilon_n + \frac{Ch}{n}.$$

On taking $h = [\sqrt{\varepsilon_n} n]$ we deduce that $|s_n - a| \le C_1\sqrt{\varepsilon_n}$. Thus a quantitative form of the hypothesis that $\sigma_n \to a$ yields a corresponding quantitative form of the conclusion that $s_n \to a$. This observation is particularly valuable when we apply Hardy's theorem to a Fourier Series. If $|n\widehat{f}(n)| \le C$ for all n, then with $a_n = \widehat{f}(n)e(nx) + \widehat{f}(-n)e(-nx)$ we have $|na_n| \le 2C$ *uniformly* in x. Thus if $\sigma_N(x) \to f(x)$ uniformly for x in a certain set, then $s_N \to f(x)$ uniformly for all x in that same set. Of course this applies only to those Fourier Series for which $|n\widehat{f}(n)|$ is a bounded sequence.

Abel summability involves considerably more smoothing than Cesàro summability, but we can still unsmooth, as follows.

Theorem 4.34. (Littlewood) *If $\sum_{n=0}^\infty a_n = a$ (A), and (4.58) holds for all n, then the series $\sum_{n=0}^\infty a_n$ converges to a.*

It is remarkable that when unsmoothing the abelian weights, the tauberian hypothesis needed is still the same one as in the Cesàro case.

In the argument below we employ a useful notational device: If S is a set, then the *characteristic function* of S, denoted χ_S, is

$$\chi_S(x) = \begin{cases} 1 & (\text{if } x \in S), \\ 0 & (\text{otherwise}). \end{cases} \tag{4.61}$$

This convention is observed throughout mathematics, but in probability theory this function is called the *indicator function*, denoted I_S, and what a probabilist calls a characteristic function is what we call a Fourier transform.

Proof. Suppose that $\varepsilon > 0$ is given. Let $\mathcal{J} = [1/e, 1]$ and $\mathcal{K} = [e^{-1-\varepsilon}, e^{-1+\varepsilon}]$. Assume for the moment that there exists a polynomial $P(x)$ such that

$$|P(x) - \chi_{\mathcal{J}}(x)| \le \varepsilon x(1-x) + 5\chi_{\mathcal{K}}(x) \tag{4.62}$$

for $0 \le x \le 1$. We note that

$$s_N = \sum_{n=0}^{N} a_n = \sum_{n=0}^{\infty} a_n \chi_{\mathcal{J}}\big(e^{-n/N}\big) . \tag{4.63}$$

From (4.62) we see that $P(0) = 0$ and that $P(1) = 1$. Thus P is of the form $P(x) = \sum_{r=1}^{R} c_r x^r$ with $\sum_{r=1}^{R} c_r = 1$. If on the right hand side above we replace $\chi_{\mathcal{J}}$ by P, the expression becomes

$$\sum_{n=0}^{\infty} a_n P(e^{-n/N}) = \sum_{n=0}^{\infty} a_n \sum_{r=1}^{R} c_r e^{-rn/N} = \sum_{r=1}^{R} c_r \sum_{n=0}^{\infty} a_n e^{-rn/N} = \sum_{r=1}^{R} c_r f\big(e^{-r/N}\big) \tag{4.64}$$

where $f(x) = \sum_{n=0}^{\infty} a_n x^n$. But $f(e^{-r/N}) \to a$ as $N \to \infty$, and $\sum_{r=1}^{R} c_r = 1$, so

$$\lim_{N\to\infty} \sum_{r=1}^{R} c_r f\big(e^{-r/N}\big) = a . \tag{4.65}$$

By subtracting (4.64) from (4.63) and applying the triangle inequality we see that

$$\Big| s_N - \sum_{r=1}^{R} c_r f\big(e^{-r/N}\big) \Big| \le \sum_{n=0}^{\infty} |a_n| |P(e^{-n/N}) - \chi_{\mathcal{J}}(e^{-n/N})| \tag{4.66}$$

which by (4.62) is

$$\le \varepsilon \sum_{n=0}^{\infty} |a_n| e^{-n/N} \big(1 - e^{-n/N}\big) \; + \; 5 \sum_{(1-\varepsilon)N \le n \le (1+\varepsilon)N} |a_n| \tag{4.67}$$

$$= T_1 + T_2, \tag{4.68}$$

say. If $0 \le u \le 1$, then $1 - e^{-u} \asymp u$. Thus

$$\sum_{n=0}^{N} |a_n| e^{-n/N} \big(1 - e^{-n/N}\big) \; \le \; \sum_{n=0}^{N} |a_n| \big(1 - e^{-n/N}\big) \; \le \; \frac{C_1}{N} \sum_{n=0}^{N} n|a_n| .$$

Here the term $n = 0$ makes no contribution, and $n|a_n|$ is bounded for $n > 0$ by (4.58), so the above is $\le C_2$. To estimate $\sum_{n>N} |a_n| e^{-n/N}$ we consider dyadic blocks. Suppose that $2^k N < n \le 2^{k+1} N$. There are $2^k N$ terms in this block, and

$|a_n| \le C/(2^k N)$ for all n in this range, so the contribution is $\le Ce^{-2^k}$. The sum of this over k converges, so $T_1 < C_3\varepsilon$. As for T_2, we note that there are $\asymp \varepsilon N$ summands, each of which is $< C_4/N$, so $T_2 < C_5\varepsilon$. On combining this with (4.65), we deduce that

$$|s_N - a| < C_6\varepsilon$$

for all sufficiently large N. Since ε may be taken arbitrarily small, it follows that $s_N \to a$ as $N \to \infty$.

It remains to show that there is a polynomial P with the property (4.62). Let $g(x) = -1/(1-x)$ for $0 \le x \le e^{-1-\varepsilon}$, let $g(x) = 1/x$ for $e^{-1+\varepsilon} \le x \le 1$, and in the interval $(e^{-1-\varepsilon}, e^{-1+\varepsilon})$ let g be linear, interpolating between $g(e^{-1-\varepsilon})$ and $g(e^{-1+\varepsilon})$, so that g is continuous on $[0,1]$. By the Weierstrass approximation theorem (Corollary 4.6) there is a polynomial $Q(x)$ such that $|Q(x) - g(x)| < \varepsilon$ for all $x \in [0,1]$. Now put

$$h(x) = \frac{\chi_{\mathcal{J}}(x) - x}{x(1-x)} = \begin{cases} -1/(1-x) & (0 \le x < 1/e), \\ 1/x & 1/e \le x \le 1). \end{cases} \tag{4.69}$$

Thus h is the same as g except that in the interval $\mathcal{K}$, h has a jump discontinuity while g is continuous. Thus $|Q(x) - h(x)| < 5$ for $x \in \mathcal{K}$ (assuming that ε is small), and $|Q(x) - h(x)| < \varepsilon$ otherwise. On rewriting (4.69), we see that

$$\chi_{\mathcal{J}}(x) = x + x(1-x)h(x)\,.$$

If we replace h by Q, then we obtain a polynomial $P(x) = x + x(1-x)Q(x)$ that has the property (4.62). □

Exercises

1. Put $s_N = \sum_{n=1}^N a_n$, and suppose that

$$\lim_{N\to\infty} \frac{s_N}{N} = a\,. \tag{4.70}$$

(a) Put $S_N = s_N - aN = \sum_{n=1}^N (a_n - a)$. Show that (4.70) is equivalent to

$$\lim_{N\to\infty} \frac{S_N}{N} = 0\,. \tag{4.71}$$

(b) Let $\sigma_N = \frac{1}{N}\sum_{n=1}^N s_n$. Show that

$$\sigma_N = \frac{1}{N}\sum_{k=1}^N (N+1-k)a_k\,.$$

Explain why

$$\frac{1}{N}\sum_{n=1}^N S_n = \sigma_N - \frac{1}{2}(N+1)a\,.$$

(c) Show that

$$\lim_{N\to\infty} \frac{2}{N+1}\sum_{n=1}^N S_n = 0\,.$$

(d) Conclude that (4.70) implies that

$$\lim_{N\to\infty} \frac{2}{N(N+1)} \sum_{n=1}^{N} (N+1-n)a_n = a\,. \tag{4.72}$$

2. (Frobenius) The object of this exercise is to prove that

$$\sum_{n=0}^{\infty} a_n = a \quad \text{(C)} \qquad \Longrightarrow \qquad \sum_{n=0}^{\infty} a_n \quad \text{(A)}\,.$$

Thus it is no coincidence that Examples 4.1 and 4.3 produced the same values.

(a) Let

$$s_n = \sum_{k=0}^{n} a_k, \qquad \sigma_n = \frac{1}{n} \sum_{k=0}^{n-1} s_k\,.$$

Show that

$$\sum_{n=0}^{\infty} s_n x^n = (1-x) \sum_{n=0}^{\infty} (n+1)\sigma_{n+1} x^n\,.$$

(b) Deduce that

$$\sum_{k=0}^{\infty} a_k x^k = (1-x)^2 \sum_{n=0}^{\infty} (n+1)\sigma_{n+1} x^n$$

for $|x| < 1$.

(c) Show that

$$\sum_{n=0}^{\infty} (n+1)x^n = \frac{1}{(1-x)^2}$$

for $|x| < 1$.

(d) Deduce that

$$\sum_{k=0}^{\infty} a_k x^k \ - a = (1-x)^2 \sum_{n=0}^{\infty} (n+1)(\sigma_{n+1} - a)x^n\,.$$

(e) Suppose that $|\sigma_n - a| < \varepsilon$ for all $n > N$, and write the right hand side above as

$$(1-x)^2 \sum_{n=0}^{N-1} (n+1)(\sigma_{n+1} - a)x^n + (1-x)^2 \sum_{n=N}^{\infty} (n+1)(\sigma_{n+1} - a)x^n = T_1 + T_2\,.$$

(f) Explain why

$$|T_1| \le (1-x)^2 \sum_{n=0}^{N-1} (n+1)|\sigma_{n+1} - a|$$

uniformly for $0 \le x \le 1$.

(g) Show that

$$|T_2| \le \varepsilon$$

for all x, $0 \le x < 1$.

(h) Conclude that $\sum a_n = a$ (A).

3. In this Exercise, we derive a variant of Theorem 4.32. We assume that $\sum_{n=1}^{\infty} a_n = a$ (C). Instead of the tauberian hypothesis (4.58), we assume that

$$a_n \ge 0 \tag{4.73}$$

for all n. Let s_n and σ_n be defined as in the proof of Theorem 4.32.

(a) Explain why

$$n\sigma_n - (n-h)\sigma_{n-h} \le hs_n \le (n+h)\sigma_{n+h} - n\sigma_n\,.$$

(b) Deduce that

$$\frac{n(\sigma_n - a) - (n-h)(\sigma_{n-h} - a)}{h} \le s_n - a \le \frac{(n+h)(\sigma_{n+h} - a) - n(\sigma_n - a)}{n}\,.$$

(c) Take $h = [\varepsilon n]$, and let n tend to infinity. Conclude that $s_n \to a$ as $n \to \infty$. That is, the series $\sum_{n=1}^{\infty} a_n$ converges to a.

4. In this exercise, we derive a further variant of Theorem 4.32. We assume that $\sum_{n=1}^{\infty} a_n = a$ (C). In place of the tauberian hypothesis (4.73) we have a weaker hypothesis, namely that there is a constant $C > 0$ such that

$$a_n \ge -C/n \tag{4.74}$$

for all n.

(a) Explain why

$$n\sigma_n - (n-h)\sigma_{n-h} - \frac{Ch^2}{n} - h \le hs_n \le (n+h)\sigma_{n+h} - n\sigma_n + \frac{Ch^2}{n}\,.$$

(b) Deduce that

$$\frac{n(\sigma_n - a) - (n-h)(\sigma_{n-h} - a)}{h} - \frac{Ch}{n-h} \le s_n - a \le \frac{(n+h)(\sigma_{n+h} - a) - n(\sigma_n - a)}{h} + \frac{Ch}{n}\,.$$

(c) Take $h = [\varepsilon n]$, and let n tend to infinity. Deduce that $\sum_{n=1}^{\infty} a_n$ converges to a.

(d) Explain why this result implies the result of the preceding exercise.

(e) Explain why this result implies Theorem 4.32. (Caution: The a_n in Theorem 21 may be complex, so it is necessary to separate the real and imaginary parts, and treat them separately.)

5. We now establish a tauberian companion to the abelian result of Exercise 1. We suppose that (4.72) holds, and our tauberian hypothesis is that there is a constant $C > 0$ such that

$$|a_n| \le C \tag{4.75}$$

for all n.

(a) Explain why

$$\sum_{n=1}^{N+H}(N+H+1-n)(a_n-a)-\sum_{n=1}^{N}(N+1-n)(a_n-a)$$
$$=H(s_N-aN)+\sum_{n=N+1}^{N+H}(N+H+1-n)(a_n-a).$$

(b) Deduce that

$$\begin{aligned}\frac{s_N}{N}-a=\frac{1}{NH}\sum_{n=1}^{N+H}(N+H+1-n)(a_n-a)-\frac{1}{NH}\sum_{n=1}^{N}(N+1-n)(a_n-a)\\ -\frac{1}{NH}\sum_{n=N+1}^{N+H}(N+H+1-n)(a_n-a)\\ =T_1+T_2+T_3.\end{aligned}$$

(c) Show that $|a|\le C$.
(d) Deduce that $|a_n-a|\le 2C$ for all n.
(e) Take $H=[\varepsilon N]$ and let $N\to\infty$. Show that $T_1\to 0$ and that $T_2\to 0$.
(f) Show that $|T_3|\le 2CH/N$.
(g) Deduce that $s_N/N\to a$.

6. Let f be a function with a continuous second derivative.
(a) Explain why

$$f(x+h)=f(x)+hf'(x)+\int_x^{x+h}(x+h-u)f''(u)\,du.$$

(b) Deduce that

$$f'(x)=\frac{f(x+h)}{h}-\frac{f(x)}{h}-\int_0^h(1-v/h)f''(x+v)\,dv=T_1+T_2+T_3.$$

(c) Suppose that $|f(x)|\le C_0$ for all $x\ge X$, and that $|f''(x)|\le C_2$ for all $x\ge X$. Show that $|T_1|\le C_0/h$, that $|T_2|\le C_0/h$, and that $|T_3|\le C_2h/2$.
(d) By making a suitable choice of h, conclude that $|f'(x)|\le 2\sqrt{C_0C_2}$ for all $x\ge X$.
(e) Suppose that $f\to 0$ as $x\to\infty$, and that f'' is bounded. Deduce that $f'\to 0$ as $x\to\infty$.

7. Suppose that $\int_0^\infty g(x)\,dx$ converges, and that $g'(x)$ is bounded. Deduce that $g(x)\to 0$ as $x\to\infty$. (Hint: Take $f(x)=\int_x^\infty g(u)\,du$ in the preceding exercise.)

4.4. Summability kernels

We say that a sequence $K_N(x)$ of functions in $L^1(\mathbb{T})$ constitute a *summability kernel* if

$$\int_0^1 K_N(x)\,dx=1 \tag{4.76}$$

for all N, if there is a constant $C > 0$ such that

$$\int_0^1 |K_N(x)|\, dx \le C \tag{4.77}$$

for all N, and if

$$\lim_{N\to\infty} \int_\delta^{1-\delta} |K_N(x)|\, dx = 0\,. \tag{4.78}$$

In the case of the Fejér kernel, (4.76) is given by (4.7), (4.77) holds with $C = 1$ because by (4.8) the Fejér kernel is nonnegative, and (4.78) follows from (4.14). The Fejér kernel has additional helpful properties, but it is the three properties (4.76)–(4.78) that enable us to prove the important uniform approximation theorem of continuous functions (Theorem 4.1), and the convergence in norm for arbitrary $f \in L^1(\mathbb{T})$ (Theorem 4.7). Thus these theorems can be generalized to other summability kernels. The Dirichlet kernel satisfies (4.76), in view of (3.46), but

$$\int_0^1 |D_N(x)|\, dx \asymp \log N,$$

and

$$\int_\delta^{1-\delta} |D_N(x)|\, dx \asymp \log \frac{1}{\delta},$$

so both (4.77) and (4.78) fail for the Dirichlet kernel. Theorems 4.1, 4.7, 4.14, 4.15 all fail when the Cesàro partial sum $\sigma_N(x)$ is replaced by the unweighted partial sum $s_N(x)$. In each case, the reason for the failure can be traced back to the fact that $\int_0^1 |D_N(x)|\, dx$ is unbounded.

A summability kernel is not necessarily nonnegative. The inequalities (4.10) and (4.11) follow from from the nonnegativity of the Fejér kernel, and similar inequalities would apply to $(K_N * f)(x)$ if $K_N(x) \ge 0$ for all x and (4.76) holds. Fejér's pointwise convergence theorem (Theorem 4.14) also does not extend to arbitrary summability kernels, as its proof depends not only on (4.77) and (4.78) but also on a uniform decay property:

$$\lim_{N\to\infty} \max_{\delta \le x \le 1-\delta} |K_N(x)| = 0,$$

which for the Fejér kernel is given by (4.12).

Exercises

1. Using the proof of Theorem 4.1 as a model, give a detailed proof that if $f \in C(\mathbb{T})$ and K_N is a summability kernel,
$$|f(x) - (f * K_N)(x)| < \varepsilon$$
holds uniformly in x, if $N > N_0(\varepsilon)$.
2. Using the proof of Theorem 4.7 as a model, give a detailed proof that if $f \in L^1(\mathbb{T})$ and K_N is a summability kernel, then
$$\lim_{N\to\infty} \int_0^1 |f(x) - (f * K_N)(x)|\, dx = 0\,.$$

3. For $0 \le r < 1$, the *Poisson kernel* is

$$P_r(x) = \sum_{n=-\infty}^{\infty} r^{|n|} e(nx) = 1 + 2\sum_{n=1}^{\infty} r^n \cos 2\pi nx \,. \tag{4.79}$$

In this context, $r \to 1^-$ corresponds to $N \to \infty$ for discretely indexed kernels.
   (a) Let r be fixed, $0 \le r < 1$. Show the series defining $P_r(x)$ is absolutely convergent, that $P_r(x)$ is a continuous function of x, and that $\widehat{P_r}(n) = r^{|n|}$ for all n.
   (b) By two applications of the formula for the sum of a geometric series, show that
$$P_r(x) = \frac{re(x)}{1 - re(x)} + \frac{1}{1 - re(-x)} \,.$$
   (c) Deduce that
$$P_r(x) = \frac{1 - r^2}{1 - 2r\cos 2\pi x + r^2} \,.$$
   (d) Show that
$$P_r(x) = \frac{1 - r^2}{(1-r)^2 + 4r\sin^2 \pi x} \,.$$
   (e) Show that the denominator above is always $\ge (1-r)^2$.
   (f) Show that for fixed r, $P_r(x)$ is a decreasing function of x for $0 \le x \le 1/2$.
   (g) Show that $\int_0^1 P_r(x)\,dx = 1$.
   (h) Show that $P_r(x) \ge 0$ for all x.
   (i) Show that if $1/2 \le r < 1$, then
$$P_r(x) \le \frac{1-r}{\sin^2 \pi x}$$
   uniformly in x.
   (j) Deduce that P_r is a summability kernel. Show that if $f \in L^1(\mathbb{T})$, then
$$(f * P_r)(x) = \sum_{n=-\infty}^{\infty} r^{|n|} \widehat{f}(n) e(nx) \,.$$
   (k) Show that $P_r(0) \sim 2/(1-r)$ as $r \to 1^-$.
   (l) $P_r(x)$ has a peak at $x = 0$. About how wide is this peak?
4. Suppose that $f \in L^1(\mathbb{T})$, that f is odd, and that $f(x) \ge 0$ for $0 \le x \le 1/2$.
   (a) Show that
$$(f * P_r)(x) = \int_0^{1/2} f(u)\big(P_r(x-u) - P_r(x+u)\big)\,du \,.$$
   (b) Show that if $0 \le x \le 1/2$ and $0 \le u \le 1/2$, then $P_r(x-u) \ge P_r(x+u)$.
   (c) Conclude that $(f * P_r)(x) \ge 0$ for $0 \le x \le 1/2$.
5. (a) Show that $x^n - y^n = (x-y)(x^{n-1} + x^{n-2}y + \cdots + y^{n-1})$.
   (b) Deduce that
$$\sin 2\pi nx = (\sin 2\pi x)(e((n-1)x) + e((n-3)x) + \cdots + e(-(n-1)x)) \,.$$
   (c) Deduce that $|\sin 2\pi nx| \le n|\sin 2\pi x|$ for all x.

(d) From now on, assume that f is an odd function in $L^1(\mathbb{T})$ such that $f(x) \geq 0$ for $0 \leq x \leq 1/2$. In the formula

$$b_1 = 2\int_0^1 f(x)\sin 2\pi x\,dx,$$

explain why the integrand is everywhere nonnegative.

(e) Deduce that $b_1 \geq 0$.

(f) Show that

$$\int_0^1 (n\sin 2\pi x \pm \sin 2\pi nx)f(x)\,dx = \frac{n}{2}b_0 \pm \frac{1}{2}b_n\,.$$

(g) Explain why the integrand above is everywhere nonnegative.

(h) Deduce that $|b_n| \leq nb_1$.

(i) Show that if $f(x) = \frac{\partial}{\partial x}P_r(x)$, then f has the required properties, and that b_n is nearly as large as nb_1 if r is near 1 and n is fixed.

6. Let $V_N(x) = 2\Delta_{2N}(x) - \Delta_N(x)$. We call this the *de la Vallée Poussin kernel.*

(a) Show that $\int_0^1 V_N(x)\,dx = 1$.

(b) Define a continuous function $v(u)$ in such a way that

$$V_N(x) = \sum_n v(n/N)e(nx)\,.$$

(c) Show that $|V_N(x)| \leq 2\Delta_{2N}(x) + \Delta_N(x)$.

(d) Show that $\int_0^1 |V_N(x)|\,dx \leq 3$.

(e) Derive an upper bound for $\int_\delta^{1-\delta} |V_N(x)|\,dx$, one that tends to 0 as N tends to infinity, for any fixed $\delta \in (0, 1/2]$.

(f) Conclude that V_N is a summability kernel.

(g) Show that the zeros of V_N are in arithmetic progression, and that every third one is a double zero.

(h) Show that $\deg V_N = 2N - 1$. (Thus by Theorem 3.15, V_N can have at most $4N - 2$ zeros.)

(i) Show that V_N has exactly $4N - 2$ zeros.

(j) Show that $-1 \leq (\sin 2\theta)^2 - (\sin\theta)^2 \leq 9/16$ for all θ. Deduce that

$$\frac{-1}{N(\sin \pi x)^2} \leq V_N(x) \leq \frac{9/16}{N(\sin \pi x)^2}$$

for all x.

7. We define the *de la Vallée Poussin power kernel* to be

$$p_n(x) = \frac{(2\cos \pi x)^{2n}}{\binom{2n}{n}}\,.$$

(a) Express $\cos \pi x$ in terms of complex exponentials, and apply the binomial theorem, to show that

$$p_n(x) = \sum_{k=-n}^{n} \frac{\binom{2n}{k+n}}{\binom{2n}{n}} e(kx).$$

(b) Show that $\int_0^1 p_n(x)\,dx = 1$.
(c) Explain why $p_n(x) \ge 0$ for all x.
(d) Deduce that $\int_0^1 |p_n(x)|\,dx = 1$ for all n.
(e) Explain why $\sum_{k=0}^{2n} \binom{2n}{k} = 2^{2n}$.
(f) Deduce that $\binom{2n}{n} \ge 2^{2n}/(2n+1)$.
(g) Deduce that $p_n(x) \le (2n+1)(\cos \pi x)^{2n}$.
(h) Deduce that p_n is a summability kernel.
(i) What is the order of magnitude of $p_n(0)$, as a function of n? (Hint: Recall Stirling's formula, which asserts that $n! \sim \sqrt{2\pi n}(n/e)^n$.)
(j) $p_n(x)$ has a peak at $x = 0$. About how wide is this peak?

Notes

Although convergence criteria for Fourier Series were established by Dirichlet, Riemann, Dini, and Jordan, by the end of the nineteenth century it was clear convergence is problematic, and prospects for the future were guarded. Then in 1900 a twenty year old graduate student named Fejér submitted for publication a paper that revolutionized the subject. He showed that if one works with Cesàro partial sums instead of unweighted partial sums, the difficulties disappear.

For a collection of attractive proofs of the Weierstrass approximation theorem (Corollary 4.6), see Duren (2012, Chapter 6).

Recall the Dirac delta function with period 1 considered in Example 3.4. One could consider the Fejér kernel to be a Cesàro partial sum of the Fourier Series of $\delta(x)$. The lesson of Exercise 4.1.17 is that $\cot \pi x$ is the harmonic conjugate of δ. Thus the conjugate Fejér kernel is a Cesàro partial sum of the Fourier Series of $\cot \pi x$. Although $\cot \pi x \notin L^1(\mathbb{T})$, at least it is a function, whereas its conjugate, $\delta(x)$, is a distribution, but not a function. Of course in (4.53) we encountered a function $g \in L^1(\mathbb{T})$ whose conjugate $f = \widetilde{g} \notin L^1(\mathbb{T})$. M. Riesz (1928) showed that if $1 < p < \infty$ and $f \in L^p(\mathbb{T})$, then the conjugate function $\widetilde{f} \in L^p(\mathbb{T})$.

An undergraduate, Harsh Mehta, who studied from a preliminary draft of this book, asked for an REU project, and was given the task of determining the asymptotics of $\|V_N\|_1$, where V_N is the de la Vallée Poussin kernel, as defined in Exercise 4.4.6. Clearly $1 \le \|V_N\|_1 \le 3$, and one would expect that these numbers tend to a limit. In fact, Mehta (2014) established the surprising fact that

$$\int_0^1 |V_N(x)|\,dx = \frac{1}{3} + \frac{2\sqrt{3}}{\pi}$$

for all N.

Chapter 5

Fourier Series in Mean Square

5.1. Vector spaces of functions

We now consider functions with period 1, and think of them as vectors with infinitely many coordinates. The function that is identically 0 is our **0** vector. To obtain an inner product, we restrict our attention to functions f with period 1 such that $\int_0^1 |f(x)|^2\,dx < \infty$, which is to say $f \in L^2(\mathbb{T})$. For such functions we define the inner product

$$\langle f, g\rangle = \int_0^1 f(x)\overline{g(x)}\,dx\,. \tag{5.1}$$

We note that

$$\langle \alpha f + \beta g, h\rangle = \alpha\langle f, h\rangle + \beta\langle g, h\rangle \tag{5.2}$$

and that

$$\langle g, f\rangle = \int_0^1 \overline{g(x)}f(x)\,dx = \overline{\langle f, g\rangle}\,. \tag{5.3}$$

On combining these observations we deduce that

$$\langle f, \alpha g + \beta h\rangle = \overline{\alpha}\langle f, g\rangle + \overline{\beta}\langle f, h\rangle\,. \tag{5.4}$$

For functions $f \in L^2(\mathbb{T})$ we use the inner product to define a norm, namely

$$\sqrt{\langle f, f\rangle} = \Big(\int_0^1 |f(x)|^2\,dx\Big)^{1/2},$$

which is more briefly referred to as $\|f\|_2$. In case of ambiguity among several norms, one could write $\|f\|_{L^2(\mathbb{T})}$.

We now establish three fundamental inequalities.

Theorem 5.1. (The Cauchy–Schwarz Inequality) *If f and g are in $L^2([a,b])$, then*

$$\Big|\int_a^b f(x)\overline{g(x)}\,dx\Big| \le \Big(\int_a^b |f(x)|^2\,dx\Big)^{1/2}\Big(\int_a^b |g(x)|^2\,dx\Big)^{1/2}. \tag{5.5}$$

Proof. It suffices to observe that

$$\int_a^b |f(x)|^2\,dx \int_a^b |g(x)|^2\,dx - \left|\int_a^b f(x)\overline{g(x)}\,dx\right|^2$$
$$= \frac{1}{2}\int_a^b \int_a^b |f(x)g(y) - f(y)g(x)|^2\,dx\,dy\,.$$

□

By the Cauchy–Schwarz inequality (5.5) we see that if $f \in L^2(\mathbb{T})$, then

$$\|f\|_1 = \int_0^1 |f(x)|\,dx = \int_0^1 |f(x)|\cdot 1\,dx \le \Big(\int_0^1 |f(x)|^2\,dx\Big)^{1/2}\Big(\int_0^1 1^2\,dx\Big)^{1/2} = \|f\|_2.$$

In particular, if $\|f\|_2 < \infty$, then $\|f\|_1 < \infty$. Thus the space $L^2(\mathbb{T})$ is a subspace of $L^1(\mathbb{T})$. In symbols, $L^2(\mathbb{T}) \subseteq L^1(\mathbb{T})$. The inclusion is strict, for if $f(x) = \|x\|^{-3/4}$, then $f \in L^1(\mathbb{T})$, but $\int_0^1 |f(x)|^2\,dx = +\infty$, so $f \notin L^2(\mathbb{T})$.

Theorem 5.2. (The Triangle Inequality) *If $f \in L^2(\mathbb{T})$ and $g \in L^2(\mathbb{T})$, then*

$$\|f+g\|_2 \le \|f\|_2 + \|g\|_2\,.$$

Proof. We observe that

$$\|f+g\|_2^2 = \langle f+g, f+g\rangle = \langle f,f\rangle + \langle f,g\rangle + \langle g,f\rangle + \langle g,g\rangle = \|f\|_2^2 + 2\,\mathrm{Re}\langle f,g\rangle + \|g\|_2^2\,.$$

Now $|\langle f,g\rangle| \le \|f\|_2\|g\|_2$ by the Cauchy–Schwarz inequality, so the right hand side above is

$$\le \|f\|_2^2 + 2\|f\|_2\|g\|_2 + \|g\|_2^2 = \big(\|f\|_2 + \|g\|_2\big)^2,$$

which gives the desired inequality. □

Theorem 5.3. (Bessel's Inequality for Fourier Series) *If $f \in L^2(\mathbb{T})$, then*

$$\sum_{n=-\infty}^{\infty} |\widehat{f}(n)|^2 \le \int_0^1 |f(x)|^2\,dx\,.$$

Proof. It suffices to show that

$$\int_0^1 \Big|f(x) - \sum_{n=-N}^{N} \widehat{f}(n)e(nx)\Big|^2\,dx = \sum_{n=-N}^{N} |\widehat{f}(n)|^2, \tag{5.6}$$

and then let $N \to \infty$. To establish this, we observe that the left hand side is

$$\int_0^1 \Big(f(x) - \sum_{m=-N}^{N} \widehat{f}(m)e(mx)\Big)\Big(\overline{f(x)} - \sum_{n=-N}^{N} \overline{\widehat{f}(n)}e(-nx)\Big)\,dx$$
$$= \int_0^1 |f(x)|^2\,dx - \sum_{n=-N}^{N} \overline{\widehat{f}(n)}\int_0^1 f(x)e(-nx)\,dx$$
$$- \sum_{m=-N}^{N} \widehat{f}(m)\int_0^1 \overline{f(x)}e(mx)\,dx + \sum_{m=-N}^{N} \widehat{f}(m)\sum_{n=-N}^{N} \overline{\widehat{f}(n)}\int_0^1 e((m-n)x)\,dx\,.$$

In the second term above, the integral is $\widehat{f}(n)$. In the third term, the integral is $\overline{\widehat{f}(m)}$. In the fourth term, the integral is 1 if $m = n$, and is otherwise 0. Thus these terms combine to give (5.6). □

Lemma 5.4. *If $f \in L^2(\mathbb{T})$, then*

$$\lim_{\delta\to 0}\int_0^1 |f(x+\delta) - f(x)|^2\,dx = 0\,.$$

This is analogous to Lemma 3.5 for functions in $L^1(\mathbb{T})$.

Proof. As we noted in remarks following Theorem 0.22, for any $\varepsilon > 0$ there is a continuous function $c(x)$ with period 1 such that $\|f - c\|_2 < \varepsilon$. We write

$$f(x+\delta) - f(x) = \big(f(x+\delta) - c(x+\delta)\big) - \big(f(x) - c(x)\big) + \big(c(x+\delta) - c(x)\big)\,.$$

Hence by the Triangle Inequality,

$$\|f(x+\delta) - f(x)\|_2 \le \|f(x+\delta) - c(x+\delta)\|_2 + \|f(x) - c(x)\|_2 + \|c(x+\delta) - c(x)\|_2\,.$$

Here the first two terms on the right are $< \varepsilon$. Since $c(x)$ is continuous, it is uniformly continuous, so $|c(x+\delta) - c(x)| < \varepsilon$ for all x, if δ is sufficiently small. Thus we have the result. □

Theorem 5.5. *If $f \in L^2(\mathbb{T})$ and $g \in L^2(\mathbb{T})$, then $f * g$ is a continuous function, and $\|f * g\|_\infty \le \|f\|_2\|g\|_2$.*

This is analogous to Theorem 3.13.

Proof. Clearly

$$(f*g)(x+\delta) - (f*g)(x) = \int_0^1 f(u)\big(g(x+\delta-u) - g(x-u)\big)\,du\,.$$

By the Cauchy–Schwarz inequality it follows that

$$|(f*g)(x+\delta) - (f*g)(x)| \le \Big(\int_0^1 |f(u)|^2\,du\Big)^{1/2}\Big(\int_0^1 \big|g(x+\delta-u) - g(x-u)\big|^2\,du\Big)^{1/2}.$$

Here the first factor on the right is $\|f\|_2$, and the second factor tends to 0 as $\delta \to 0$ by Lemma 5.4, so $f * g$ is continuous. As for the last clause, we note that

$$\begin{aligned}|(f*g)(x)| &= \Big|\int_0^1 f(u)g(x-u)\,du\Big| \le \int_0^1 |f(u)g(x-u)|\,du\\ &\le \Big(\int_0^1 |f(u)|^2\,du\Big)^{1/2}\Big(\int_0^1 |g(x-u)|^2\,du\Big)^{1/2} = \|f\|_2\|g\|_2\end{aligned}$$

for any x. □

Since the series in Bessel's Inequality converges, it follows that its terms tend to 0, which is to say that $\widehat{f}(n) \to 0$ as $n \to \pm\infty$. We proved this earlier, under the weaker hypothesis that $\int_0^1 |f(x)|\,dx < \infty$.

5.2. Parseval's Identity

Suppose that

$$T(x) = \sum_{n=-N}^{N} t_n e(nx)$$

is a trigonometric polynomial. Then

$$\begin{aligned}\int_0^1 |T(x)|^2\,dx &= \int_0^1 \Big(\sum_{m=-N}^{N} t_m e(mx)\Big)\Big(\sum_{n=-N}^{N} \overline{t_n} e(-nx)\Big)\,dx \\ &= \sum_{m=-N}^{N} \sum_{n=-N}^{N} t_m \overline{t_n} \int_0^1 e((m-n)x)\,dx\,.\end{aligned}$$

Here the last integral on the right is 1 if $m = n$, and is 0 otherwise. Hence the above is

$$= \sum_{n=-N}^{N} |t_n|^2\,.$$

But $t_n = \widehat{T}(n)$ for all n, so we see that equality holds in Bessel's inequality when the function in question is a trigonometric polynomial. In order to show that Bessel's inequality is always an equality for Fourier Series, we need to show that any function $f \in L^2(\mathbb{T})$ can be well approximated by trigonometric polynomials.

Theorem 5.6. *Suppose that $f \in L^2(\mathbb{T})$, and that $\sigma_N(x)$ is the Nth Cesàro partial sum of the Fourier Series of f. Then*

$$\lim_{N\to\infty} \int_0^1 |f(x) - \sigma_N(x)|^2\,dx = 0\,.$$

This is analogous to the uniform approximation theorem for $C(\mathbb{T})$ (Theorem 4.1) and the convergence in the L^1-norm (Theorem 4.7).

From the above we see that trigonometric polynomials are dense in $L^2(\mathbb{T})$.

Proof. By (4.6) we know that

$$\sigma_N(x) = \int_0^1 \Delta_N(u) f(x-u)\,du$$

where $\Delta_N(u)$ is the Fejér kernel. Since $\int_0^1 \Delta_N(u)\,du = 1$, (recall (4.7)) it follows that

$$f(x) - \sigma_N(x) = \int_0^1 (f(x) - f(x-u))\Delta_N(u)\,du\,.$$

By the Cauchy–Schwarz inequality (5.5) applied to the two functions

$$(f(x) - f(x-u))\sqrt{\Delta_N(u)}, \qquad \sqrt{\Delta_N(u)},$$

we deduce that

$$|f(x) - \sigma_N(x)|^2 = \Big|\int_0^1 (f(x) - f(x-u))\Delta_N(u)\,du\Big|^2$$
$$\le \Big(\int_0^1 |f(x) - f(x-u)|^2\Delta_N(u)\,du\Big)\Big(\int_0^1 \Delta_N(u)\,du\Big).$$

Here the last integral is 1, so the above is

$$= \int_0^1 |f(x) - f(x-u)|^2\Delta_N(u)\,du\,.$$

Next we integrate this bound with respect to x to see that

$$\int_0^1 |f(x) - \sigma_N(x)|^2\,dx \le \int_0^1\int_0^1 |f(x) - f(x-u)|^2\Delta_N(u)\,du\,dx\,.$$

On interchanging the order of integration, we see that this is

$$= \int_0^1 \Delta_N(u)\Big(\int_0^1 |f(x) - f(x-u)|^2\,dx\Big)\,du\,.$$

We write the integral with respect to u as

$$\int_{-\delta}^{\delta} + \int_{\delta}^{1-\delta} = I_1 + I_2,$$

say. If δ is sufficiently small, then

$$\int_0^1 |f(x) - f(x-u)|^2\,dx < \varepsilon$$

whenever $|x| < \delta$, by Lemma 5.4. Thus

$$I_1 \le \varepsilon\int_{-\delta}^{\delta}\Delta_N(x)\,dx \le \varepsilon\int_0^1 \Delta_N(x)\,dx = \varepsilon\,.$$

As for I_2, we first observe that if a and b are arbitrary complex numbers, then by Cauchy's inequality (0.4)

$$|a+b|^2 = |1\cdot a + 1\cdot b|^2 \le (1^2+1^2)(|a|^2+|b|^2) = 2|a|^2 + 2|b|^2\,.$$

On applying this with $a = f(x)$ and $b = -f(x-u)$, we deduce that

$$|f(x) - f(x-u)|^2 \le 2|f(x)|^2 + 2|f(x-u)|^2\,.$$

Let $C = \int_0^1 |f(x)|^2\,dx$. Then

$$\int_0^1 |f(x) - f(x-u)|^2\,dx \le 4C$$

for all u. Hence

$$I_2 \le 4C\int_{\delta}^{1-\delta}\Delta_N(x)\,dx \le \frac{2C}{N\delta}$$

by (4.14). This is $< \varepsilon$ for all sufficiently large N. This completes the proof. □

We are now in a position to prove

Theorem 5.7. (Parseval's Identity, First Form) *If* $f \in L^2(\mathbb{T})$, *then*

$$\int_0^1 |f(x)|^2\,dx = \sum_{n=-\infty}^{\infty} |\widehat{f}(n)|^2 .$$

For $p \ge 1$ we can define an L^p norm,

$$\|f\|_p = \left(\int_0^1 |f(x)|^p\,dx \right)^{1/p} .$$

Thus $L^p(\mathbb{T})$ is the set of functions with period 1 for which this norm is finite. For $q \ge 1$ we can also define a norm on sequences $\{a_n\}$,

$$\|a\|_q = \left(\sum_{n=-\infty}^{\infty} |a_n|^q \right)^{1/q} .$$

To distinguish it from the L^p norm obtained by integrating, this is called the ℓ^q or "little ell q" norm. It is natural to ask for relationships between the L^p norms of f for various p and the ℓ^q norms of $\widehat{f}$ for various q. While there are some results, these are generally only inequalities, and indeed such inequalities are often very weak. Parseval's Identity is of great importance because it stands alone as a situation in which we have a precise relationship: $\|f\|_2 = \|\widehat{f}\|_2$. An example of an inequality between two norms is seen in (3.15), which may be thought of as asserting that $\|\widehat{f}\|_\infty \le \|f\|_1$. This is sometimes a very useful inequality, and it holds with equality if $f(x) = e(nx)$ for some n. However, it is often very weak. In §6.4 we shall construct trigonometric polynomials of degree N for which $\|\widehat{f}\|_\infty = 1$ but $\|f\|_1 \asymp \sqrt{N}$. For general $f \in L^1(\mathbb{T})$ the ratio $\|f\|_1/\|\widehat{f}\|_\infty$ can be arbitrarily large.

Proof. By the Triangle Inequality of Theorem 5.2 we see that

$$\|f\|_2 = \|(f - \sigma_N) + \sigma_N\|_2 \le \|f - \sigma_N\|_2 + \|\sigma_N\|_2$$

and that

$$\|\sigma_N\|_2 = \|(\sigma_N - f) + f\|_2 \le \|f - \sigma_N\|_2 + \|f\|_2,$$

so

$$-\|f - \sigma_N\|_2 \le \|f\|_2 - \|\sigma_N\|_2 \le \|f - \sigma_N\|_2 .$$

But $\|f - \sigma_N\|_2$ tends to 0 as N tends to infinity by Theorem 5.6, so

$$\|\sigma_N\|_2 \to \|f\|_2 \tag{5.7}$$

as $N \to \infty$. As σ_N is a trigonometric polynomial, we know that

$$\|\sigma_N\|_2^2 = \sum_{n=-N}^{N} \left(1 - \frac{|n|}{N}\right)^2 |\widehat{f}(n)|^2.$$

By Bessel's inequality we know that $\sum_n |\widehat{f}(n)|^2 < \infty$. Clearly

$$(1-\varepsilon)^2 \sum_{n=-\varepsilon N}^{\varepsilon N} |\widehat{f}(n)|^2 \le \|\sigma_N\|_2^2 \le \sum_{n=-\infty}^{\infty} |\widehat{f}(n)|^2$$

for all N. Here the left hand side tends to $(1-\varepsilon)^2 \sum_{n-\infty}^{\infty} |\widehat{f}(n)|^2$, so by the squeeze theorem (0.6) it follows that $\|\sigma_N\|_2^2 \to \sum_{-\infty}^{\infty} |\widehat{f}(n)|^2$ as $N \to \infty$. When combined with (5.7) we obtain the desired result. □

Suppose that $f \in L^2(\mathbb{T})$ and that $T(x)$ is a trigonometric polynomial. By applying Parseval's Identity to $f - T$ we find that

$$\int_0^1 |f(x) - T(x)|^2 \, dx = \sum_{n=-N}^{N} |\widehat{f}(n) - t_n|^2 + \sum_{|n|>N} |\widehat{f}(n)|^2 .$$

Here the right hand side is minimized by taking $t_n = \widehat{f}(n)$ for all n, $-N \le n \le N$. With this choice of the t_n, $T(x)$ is the unweighted partial sum $s_N(x)$ of the Fourier Series of f. Thus

$$\int_0^1 |f(x) - s_N(x)|^2 \, dx = \sum_{|n|>N} |\widehat{f}(n)|^2 . \tag{5.8}$$

Thus Theorem 5.6 is valid with $\sigma_N(x)$ replaced by $s_N(x)$, and indeed $s_N(x)$ gives the better approximation, although Theorem 5.6 is still part of our chain of reasoning.

Example 5.1. We revisit Example 2.6, now for continuous rather than discrete random variables. Suppose that we have a random number generator that is supposed to provide values that are uniformly distributed in the interval $[0, 1]$, but actually it produces values with a density $f(x)$ that is not quite uniform, say $f(x) = 1 + \delta(x)$ where $\delta(x)$ is generally small. The following procedure is proposed: Let X and Y be independent random variables with the density $f(x)$, and put

$$Z = \begin{cases} X + Y & (0 \le X + Y \le 1), \\ X + Y - 1 & (1 < X + Y \le 2). \end{cases}$$

We want to determine the extent to which Z is closer to being uniform. Let $f_Z(x)$ denote the density of Z. Then $f_Z(x) = 0$ if $x > 1$ or $x < 0$, and for $0 \le x \le 1$,

$$f_Z(x) = (f*f)(x) + (f*f)(x+1) = \int_0^x f(u)f(x-u)\,du + \int_x^1 f(u)f(x+1-u)\,du .$$

Let us now extend the definitions of f and δ so that they have period 1. Thus

$$f_Z(x) = \int_0^1 f(u)g(x-u)\,du = \int_0^1 (1+\delta(u))(1+\delta(x-u))\,du .$$

Now $\int_0^1 f(u)\,du = 1$ since f was originally the density function of a random variable taking values in $[0, 1]$, so $\int_0^1 \delta(u)\,du = 0$. Hence

$$f_Z(x) = 1 + \int_0^1 \delta(u)\delta(x-u)\,du .$$

The Fourier Series of $\delta * \delta$ is

$$\sum_{n=-\infty}^{\infty} \widehat{\delta}(n)^2 e(nx) . \tag{5.9}$$

By Parseval's Identity,

$$\int_0^1 |\delta(x)|^2\,dx = \sum_{n=-\infty}^{\infty} |\widehat{\delta}(n)|^2,$$

which is finite since we are assuming that δ is small, at least in mean square. Thus the Fourier Series (5.9) is absolutely convergent. By Theorem 3.13 we know that $\delta * \delta$ is continuous. Thus $\delta * \delta \in A(\mathbb{T})$, and by Corollary 4.5 it follows that

$$(\delta * \delta)(x) = \sum_{n=-\infty}^{\infty} \widehat{\delta}(n)^2 e(nx)$$

for all x. By the triangle inequality it follows that

$$|(\delta * \delta)(x)| \le \sum_{n=-\infty}^{\infty} |\widehat{\delta}(n)|^2 = \int_0^1 |\delta(u)|^2\,du\,.$$

Thus $f_Z(x) = 1 + \varepsilon(x)$ where

$$|\varepsilon(x)| \le \int_0^1 |\delta(u)|^2\,du$$

for all x. In particular, if $|\delta(x)| \le \delta$ for all x, then $|\varepsilon(x)| \le \delta^2$ for all x. The actual size of $\varepsilon(x)$ may be smaller, since in applying the triangle inequality we may have thrown away some cancellation.

Like $L^2(\mathbb{T})$, the space ℓ^2 not only has a norm, but also an inner product,

$$\langle a, b\rangle = \sum_{n=-\infty}^{\infty} a_n \overline{b_n}\,.$$

We now derive a Second Form of Parseval's Identity, which asserts that $\langle f, g\rangle = \langle \widehat{f}, \widehat{g}\rangle$.

Theorem 5.8. (Parseval's Identity, Second Form) *Suppose that $f \in L^2(\mathbb{T})$, and that $g \in L^2(\mathbb{T})$. Then*

$$\int_0^1 f(x)\overline{g(x)}\,dx = \sum_{n=-\infty}^{\infty} \widehat{f}(n)\overline{\widehat{g}(n)}\,.$$

When $g = f$ we have the original First Form of Parseval's Identity. We give two proofs.

First Proof. Let z and w be complex numbers. It is easy to verify that

$$z\overline{w} = \frac{1}{4}\big(|z+w|^2 + i|z+iw|^2 - |z-w|^2 - i|z-iw|^2\big)\,. \tag{5.10}$$

(Alternatively, in Exercise 2.2.18 take $q = 4$, $h = -1$, $f(0) = z$, $f(1) = w$, $f(2) = f(3) = 0$.) Thus

$$\begin{aligned}\int_0^1 f(x)\overline{g(x)}\,dx &= \frac{1}{4}\int_0^1 |f(x)+g(x)|^2\,dx + \frac{i}{4}\int_0^1 |f(x)+ig(x)|^2\,dx \\ &\quad - \frac{1}{4}\int_0^1 |f(x)-g(x)|^2\,dx - \frac{i}{4}\int_0^1 |f(x)-ig(x)|^2\,dx\,.\end{aligned}$$

By four applications of Parseval's Identity this is

$$= \frac{1}{4}\sum_{n=-\infty}^{\infty} |\widehat{f}(n)+\widehat{g}(n)|^2 + \frac{i}{4}\sum_{n=-\infty}^{\infty} |\widehat{f}(n)+i\widehat{g}(n)|^2$$
$$-\frac{1}{4}\sum_{n=-\infty}^{\infty} |\widehat{f}(n)-\widehat{g}(n)|^2 - \frac{i}{4}\sum_{n=-\infty}^{\infty} |\widehat{f}(n)-i\widehat{g}(n)|^2 .$$

By (5.10) this is

$$= \sum_{n=-\infty}^{\infty} \widehat{f}(n)\overline{\widehat{g}(n)} .$$

This completes the proof. □

Second Proof. Let $h(x) = \overline{g(-x)}$. By Theorem 5.5 we know that $f * h$ is continuous. By Theorem 3.12 we know that the Fourier Series of $f * h$ is

$$\sum_{n=-\infty}^{\infty} \widehat{f}(n)\widehat{h}(n)e(nx) .$$

By Cauchy's inequality and Bessel's inequality we see that this series is absolutely convergent:

$$\sum_{n=-\infty}^{\infty} |\widehat{f}(n)\widehat{h}(n)| \le \Big(\sum_{n=-\infty}^{\infty} |\widehat{f}(n)|^2\Big)^{1/2}\Big(\sum_{n=-\infty}^{\infty} |\widehat{h}(n)|^2\Big)^{1/2}$$
$$\le \|f\|_2\|h\|_2 = \|f\|_2\|g\|_2 < \infty .$$

In Corollary 4.5 we saw that if the Fourier Series of a continuous function is absolutely convergent, then it converges to the function for all x. Thus

$$\int_0^1 f(u)h(x-u)\,du = \sum_{n=-\infty}^{\infty} \widehat{f}(n)\widehat{h}(n)e(nx)$$

for all x. From Theorem 3.2 (c) and (e) we know that $\widehat{h}(n) = \overline{\widehat{g}(n)}$. Hence

$$\int_0^1 f(u)\overline{g(u-x)}\,du = \sum_{n=-\infty}^{\infty} \widehat{f}(n)\overline{\widehat{g}(n)}e(nx)$$

for all x. Now take $x = 0$ to obtain the stated formula. □

Suppose that $f \in L^2(\mathbb{T})$, that $h \in L^2(\mathbb{T})$, and put $j(x) = f(x)h(x)$. Then j has period 1 and by the Cauchy–Schwarz inequality (5.5) we find that $\|j\|_1 \le \|f\|_2\|h\|_2 < \infty$. Since $j \in L^1(\mathbb{T})$, it is therefore reasonable to wonder how the Fourier coefficients of j might be expressed in terms of those of f and h. This natural question is answered as follows

Theorem 5.9. *If $f \in L^2(\mathbb{T})$ and $h \in L^2(\mathbb{T})$, then $fh \in L^1(\mathbb{T})$, and*

$$\widehat{fh}(k) = \sum_{n=-\infty}^{\infty} \widehat{f}(n)\widehat{h}(k-n) \tag{5.11}$$

for every integer k. Moreover, the above series is absolutely convergent.

While the above formula for $\widehat{fh}(k)$ involves an absolutely convergent series, the Fourier Series of $f(x)h(x)$ is not necessarily absolutely convergent, as one can see from any number of examples; for instance, if $f(x)$ is the sawtooth function of Example 3.1, and $h(x) \equiv 1$. In any case, in the above we encounter another situation in which a pointwise product on one side corresponds to a convolution on the other.

Formally (i.e., unrigorously) one would expect that

$$\int_0^1 f(x)h(x)e(-kx)\,dx = \int_0^1 \Big(\sum_{n=-\infty}^{\infty} \widehat{f}(n)e(nx)\Big)\Big(\sum_{m=-\infty}^{\infty} \widehat{h}(m)e(mx)\Big)e(-kx)\,dx$$
$$= \sum_{m,n} \widehat{f}(n)\widehat{h}(m)\int_0^1 e((m+n-k)x)\,dx = \sum_n \widehat{f}(n)\widehat{h}(k-n)$$

by orthogonality. Thus this theorem confirms what one would expect.

Proof. By the Cauchy–Schwarz inequality

$$\int_0^1 |f(x)h(x)|\,dx \le \Big(\int_0^1 |f(x)|^2\,dx\Big)^{1/2}\Big(\int_0^1 |h(x)|^2\,dx\Big)^{1/2} = \|f\|_2\|h\|_2 < \infty\,.$$

Thus $fh \in L^1(\mathbb{T})$. Let $g(x) = \overline{h(x)}e(kx)$ in Theorem 5.8. Then $h(x)e(-kx) = \overline{g(x)}$, so

$$\widehat{fh}(k) = \int_0^1 f(x)h(x)e(-kx)\,dx = \int_0^1 f(x)\overline{g(x)}\,dx = \sum_{-\infty}^{\infty} \widehat{f}(n)\overline{\widehat{g}(n)}\,.$$

Now

$$\overline{\widehat{g}(n)} = \int_0^1 \overline{g(x)}e(nx)\,dx = \int_0^1 h(x)e((n-k)x)\,dx = \widehat{h}(k-n),$$

which gives the stated identity. To show that the series (5.11) is absolutely convergent, we use Cauchy's inequality, from which we see that

$$\sum_{n=-\infty}^{\infty} |\widehat{f}(n)\widehat{h}(k-n)| \le \Big(\sum_{n=-\infty}^{\infty} |\widehat{f}(n)|^2\Big)^{1/2}\Big(\sum_{n=-\infty}^{\infty} |\widehat{h}(k-n)|^2\Big)^{1/2} = \|f\|_2\|h\|_2 < \infty\,.$$

□

As discussed in §0.2, the real numbers are complete in the sense that any Cauchy sequence converges to a real number. Similarly, as remarked in §0.3, the spaces $L^p(\mathbb{T})$ are complete, for all $p \ge 1$. This enables a converse of Parseval's Identity.

Theorem 5.10. (The Riesz–Fischer Theorem) *Suppose that $\{c_n\}$ is a sequence of real or complex numbers such that*

$$\sum_{n=-\infty}^{\infty} |c_n|^2 < \infty\,. \tag{5.12}$$

Then there is an $f \in L^2(\mathbb{T})$ such that $\widehat{f}(n) = c_n$ for all n.

Proof. Let

$$s_N(x) = \sum_{n=-N}^{N} c_n e(nx) .$$

Suppose that $0 < M < N$. Since $s_N - s_M$ is a trigonometric polynomial, it is trivial that

$$\int_0^1 |s_M(x) - s_N(x)|^2 \, dx = \sum_{M<|n|\le N} |c_n|^2 .$$

By the hypothesis (5.12) we see that this is small if M and N are large. That is, $\{s_N\}$ is a Cauchy sequence in $L^2(\mathbb{T})$. Hence there is an $f \in L^2(\mathbb{T})$ such that

$$\lim_{N\to\infty} \|f - s_N\|_2 = 0 .$$

By (3.15) we know that $|\widehat{f}(n) - \widehat{s_N}(n)| \le \|f - s_N\|_1$. By the Cauchy–Schwarz inequality we know that $\|f - s_N\|_1 \le \|f - s_N\|_2$. From the above we know that this is $< \varepsilon$ if N is large. But $\widehat{s_N}(n) = c_n$ if N is large, so $|\widehat{f}(n) - c_n| < \varepsilon$. But ε can be arbitrarily small, so we conclude that $\widehat{f}(n) = c_n$ for all n. □

Corollary 5.11. *If $f \in L^1(\mathbb{T})$, and if*

$$\sum_{n=-\infty}^{\infty} |\widehat{f}(n)|^2 < \infty,$$

then $f \in L^2(\mathbb{T})$ (and so Parseval's Identity applies to f).

Equivalently, if $f \in L^1(\mathbb{T})$ but $\int_0^1 |f(x)|^2 \, dx = \infty$, then $\sum_{n=-\infty}^{\infty} |\widehat{f}(n)|^2 = \infty$.

Proof. By taking $c_n = \widehat{f}(n)$ in Theorem 5.10, we see that there is a function $g \in L^2(\mathbb{T})$ such that $\widehat{g}(n) = \widehat{f}(n)$ for all n. By Corollary 4.8 it follows that $\int_0^1 |f(x) - g(x)| \, dx = 0$. This implies that $f(x) = g(x)$ for almost all x, which in turn implies that $\int_0^1 |f(x) - g(x)|^p \, dx = 0$ for all $p > 0$, and in particular for $p = 2$. By the triangle inequality (5.2) we conclude that $\|f\|_2 = \|(f-g)+g\|_2 \le \|f-g\|_2 + \|g\|_2 < \infty$, as desired. □

Exercises

1. Suppose that numbers a_n and b_n are defined in terms of the $\widehat{f}(n)$ as in (3.8) and (3.9).
   (a) Show that if $n > 0$, then $|a_n|^2 + |b_n|^2 = 2|\widehat{f}(n)|^2 + 2|\widehat{f}(-n)|^2$.
   (b) Show that $a_0 = 2\widehat{f}(0)$.
   (c) Conclude that if $f \in L^2(\mathbb{T})$, then

$$\int_0^1 |f(x)|^2 \, dx = \frac{1}{4}|a_0|^2 + \frac{1}{2}\sum_{n=1}^{\infty}(|a_n|^2 + |b_n|^2) .$$

2. Show that if $f \in L^2(\mathbb{T})$ and $g \in L^2(\mathbb{T})$, then

$$\int_0^1 f(x)g(x)\,dx = \sum_{n=-\infty}^{\infty} \widehat{f}(n)\widehat{g}(-n)\,. \tag{5.13}$$

3. Let $P_2(x)$ be defined as in Example 3.9. By applying Parseval's Identity, show that

$$\sum_{n=1}^{\infty} \frac{1}{n^4} = \frac{\pi^4}{90}\,.$$

4. Suppose that $f \in L^2(\mathbb{T})$.
   (a) Use Parseval's Identity to show that

$$\int_0^1 |f(x) - \sigma_N(x)|^2\,dx = \frac{1}{N^2}\sum_{n=-N}^{N} n^2|\widehat{f}(n)|^2 \; + \sum_{|n|>N} |\widehat{f}(n)|^2\,.$$

   (b) We used Theorem 5.6 to prove Parseval's Identity. Now use the above to prove Theorem 5.6.
   (c) Suppose that $\widehat{f}(n) = 1/n^2$ for $n \ne 0$. Show that

$$\int_0^1 |f(x) - \sigma_N(x)|^2\,dx \asymp 1/N^2,$$

   and that

$$\int_0^1 |f(x) - s_N(x)|^2\,dx = O(1/N^3)\,.$$

5. Let f be given, $f \in L^1(\mathbb{T})$, and put $g_\delta(x) = f(x+\delta) - f(x)$.
   (a) Show that $\widehat{g_\delta}(n) = (e(n\delta) - 1)\widehat{f}(n)$.
   (b) Now suppose that $f \in L^2(\mathbb{T})$. Show that

$$\int_0^1 |g_\delta(x)|^2\,dx = 4\sum_{n=-\infty}^{\infty} |\widehat{f}(n)|^2(\sin \pi n\delta)^2\,.$$

   (c) We used Lemma 5.4 to prove Parseval's Identity. Use the identity above to prove Lemma 5.4.
   (d) For a specific f with known Fourier coefficients, the identity of part (b) allows us to derive a quantitative measure of the rate at which the limit in Lemma 5.4 is attained. For example, suppose that $\widehat{f}(n) = O(1/|n|)$ for $n \ne 0$. Show that $f \in L^2(\mathbb{T})$ and that

$$\int_0^1 |f(x+\delta) - f(x)|^2\,dx = O(\delta)$$

   for $0 \le \delta \le 1/2$.

6. Show that

$$\int_0^1 \Delta_N(x)^2\,dx = \frac{2}{3}N + \frac{1}{3N}\,.$$

(Hint: See page 225.)

7. Suppose that $f \in L^2(\mathbb{T})$. Show that there exists a $\widetilde{f} \in L^2(\mathbb{T})$ such that

$$\widehat{\widetilde{f}}(n) = \begin{cases} -i\widehat{f}(n) & (n > 0), \\ 0 & (n = 0), \\ i\widehat{f}(n) & (n < 0). \end{cases}$$

8. Suppose that $f \in L^1(\mathbb{T})$. Put $g(x) = \sqrt{|f(x)|}$ and $h(x) = f(x)/g(x)$.
   (a) Explain why $g \in L^2(\mathbb{T})$ and $h \in L^2(\mathbb{T})$.
   (b) Explain why $\widehat{g} \in \ell^2$ and $\widehat{h} \in \ell^2$.
   (c) Show that $\widehat{f}(n) = (\widehat{g} * \widehat{h})(n)$.
   (d) Suppose that $a \in \ell^2$ and that $b \in \ell^2$. Put $c = a * b$. Show that there is an $f \in L^1(\mathbb{T})$ such that $\widehat{f}(n) = c_n$ for all n.

   (Thus a sequence c_n is a sequence of Fourier coefficients if and only if it can be written as a convolution of two sequences in ℓ^2. Unfortunately, this seems to be more amusing than useful.)
9. Suppose that $1 < p < \infty$. By arguing as in the proof of Lemma 5.4, and using Minkowski's inequality for integrals (Theorem I.12), show that if $f \in L^p(\mathbb{T})$, then

$$\lim_{\delta \to 0} \int_0^1 |f(x+\delta) - f(x)|\, dx = 0\,.$$

10. Suppose that $1 < p < \infty$, that $1/p + 1/q = 1$, that $f \in L^p(\mathbb{T})$, and that $g \in L^q(\mathbb{T})$.
    (a) By arguing as in the proof of Theorem 5.5, and using Hölder's inequality for integrals (Theorem I.11), show that $f * g$ is continuous.
    (b) Show that $\|f * g\|_\infty \le \|f\|_p \|g\|_q$.
11. Suppose that $1 < p < \infty$. Show that if $f \in L^p(\mathbb{T})$, then

$$\lim_{N \to \infty} \int_0^1 |f(x) - \sigma_N(x)|^p\, dx = 0\,.$$

12. Suppose that $f \in L^1(\mathbb{T})$ and that $F \in L^1(\mathbb{T})$. We say that F *majorizes* f, and write $f \prec F$, if $|\widehat{f}(n)| \le \widehat{F}(n)$ for all n. The Hardy–Littlewood Majorant Conjecture proposes that for each p, $2 \le p < \infty$, there is a constant $c(p)$ such that

$$\int_0^1 |f(x)|^p\, dx \le c(p) \int_0^1 |F(x)|^p\, dx \tag{5.14}$$

    if $f \prec F$. In this exercise we determine for which p the conjecture is true.
    (a) Show that if $f \prec F$, then (5.14) holds for $p = 2$ with $c(2) = 1$.
    (b) Let $p \ge 2$ be fixed, and suppose that it is known that (5.14) holds for some particular $c(p)$ when f and F are trigonometric polynomials with $f \prec F$. Show that (5.14) then follows with the same constant for arbitrary f and $F \in L^p(\mathbb{T})$, with $f \prec F$. Hint: If $f \prec F$, then $\sigma_N(f; x) \prec \sigma_N(F; x)$.
    (c) Suppose that f and F are trigonometric polynomials with $f \prec F$. Show that $f^2 \prec F^2$.
    (d) Show that if $F \in L^4(\mathbb{T})$ and $f \prec F$, then (5.14) holds for $p = 4$ with $c(4) = 1$. (This can be extended to $p = 6, 8, 10, \ldots$, all with $c(p) = 1$.)

(e) Fix a real number $p > 2$, and suppose that there exist trigonometric polynomials f and F with $f \prec F$ for which

$$\int_0^1 |f(x)|^p\,dx = c\int_0^1 |F(x)|^p\,dx$$

with $c > 1$. Assume that among constants c for which there exist such f and F, that c is nearly as large as possible. Put $g(x) = f(x)f(Mx)$, and $G(x) = F(x)F(Mx)$. Show that $g \prec G$ if M is sufficiently large.

(f) Use Fejér's Lemma (Theorem 4.13) to show that if M is sufficiently large, then

$$\int_0^1 |g(x)|^p\,dx \doteqdot \Big(\int_0^1 |f(x)|^p\,dx\Big)^2, \quad \int_0^1 |G(x)|^p\,dx \doteqdot \Big(\int_0^1 |F(x)|^p\,dx\Big)^2.$$

Thus $\int_0^1 |g(x)|^p\,dx > (c^2 - \varepsilon)\int_0^1 |G(x)|^p\,dx$.

(g) Conclude that the Majorant Conjecture is false for the exponent p.

(h) Suppose that $p \ge 2$ and that p is not an even integer. Choose k so that $2k - 4 < p < 2k - 2$. Set $f(x) = 1 + re(x) - r^k e(kx)$ and $F(x) = 1 + re(x) + r^k e(kx)$ where r is small. The binomial theorem asserts that $(1+w)^{p/2} = \sum_{m=0}^\infty \binom{p/2}{m} w^m$. By taking $w = z + z^k$, show that

$$(1 + z \pm z^k)^{p/2} = \sum_{m=0}^{k-1}\binom{p/2}{m} z^m + \left(\binom{p/2}{k} \pm \frac{p}{2}\right) z^k + \text{higher order terms}.$$

(i) Show that $\binom{p/2}{j} \ge 0$ for $j < k$ and that $\binom{p/2}{k} < 0$. By applying the Parseval Identity to $f(x)^{p/2}$ and to $F(x)^{3/2}$ show that

$$\int_0^1 |f(x)|^p\,dx = \sum_{m=0}^{k-1}\binom{p/2}{m} 2r^{2m} + \left(\binom{p/2}{k} - \frac{p}{2}\right)^2 r^{2k} + O\big(r^{2k+2}\big),$$

$$\int_0^1 |F(x)|^p\,dx = \sum_{m=0}^{k-1}\binom{p/2}{m}^2 r^{2m} + \left(\binom{p/2}{k} + \frac{p}{2}\right)^2 r^{2k} + O\big(r^{2k+2}\big).$$

Here $\left|\binom{p/2}{k} - p/2\right| > \left|\binom{p/2}{k} + p/2\right|$ because $\binom{p/2}{k} < 0$. The error term is negligible if r is small enough. Thus we have trigonometric polynomials f, F with $f \prec F$ such that $\int_0^1 |f(x)|^p\,dx > \int_0^1 |F(x)|^p\,dx$.

(j) Conclude that the Hardy–Littlewood Majorant Conjecture is true precisely when p is a positive even integer.

Notes

Parseval (1799) noted the identity that bears his name, but did not prove it, saying only that it is self-evident. It was first proved rigorously by Fatou (1906). Riesz (1907) and Fischer (1907) found almost identical results independently and simultaneously. For an interesting analysis of what they proved, see Horváth (2004).

Chapter 6

Trigonometric Polynomials

As we have already said, a *trigonometric polynomial* is a sum of the form

$$T(x) = \sum_{n=-N}^{N} t_n e(nx) \tag{6.1}$$

for some N. For any given N, the set of such trigonometric polynomials is denoted $\mathcal{T}_N$. Thus $\mathcal{T}_N$ consists of those trigonometric polynomials of degree not exceeding N, and

$$\widehat{T}(n) = \begin{cases} t_n & (|n| \le N), \\ 0 & (|n| > N). \end{cases}$$

We have already seen (in Theorem 3.15) that a trigonometric polynomial of degree N can have at most $2N$ roots, unless it is identically 0. We now address additional properties of trigonometric polynomials.

6.1. Sampling and interpolation

We begin with a simple observation concerning averages at equally spaced points.

Theorem 6.1. *Let $T(x)$ be defined as in (6.1), and suppose that q is a positive integer. Then*

$$\frac{1}{q}\sum_{a=1}^{q} T(a/q) = \sum_{\substack{-N\le n\le N \\ q\mid n}} t_n \,.$$

Proof. The left hand side above is

$$= \frac{1}{q}\sum_{n=-N}^{N} t_n \sum_{a=1}^{q} e(an/q),$$

which is equal to the right hand side above, in view of Theorem 2.1. □

Corollary 6.2. *Let $T(x)$ be defined as in (6.1), and suppose that q is an integer with $q > N$. Then*

$$\frac{1}{q}\sum_{a=1}^{q} T(a/q) = t_0\,.$$

We can similarly capture other Fourier coefficients of a trigonometric polynomial.

Theorem 6.3. *Let $T(x)$ be defined as in (6.1), and suppose that q is a positive integer. Then*

$$\frac{1}{q}\sum_{a=1}^{q} e(-ak/q)T(a/q) = \sum_{\substack{-N\le n\le N\\ n\equiv k \pmod q}} t_n$$

for any integer k.

Proof. The left hand side above is

$$= \frac{1}{q}\sum_{n=-N}^{N} t_n \sum_{a=1}^{q} e((n-k)a/q),$$

which is equal to the right hand side above, in view of Theorem 2.1. □

If $q > 2N$, then there can be at most one term in the sum, and so we have

Corollary 6.4. *Let $T(x)$ be defined as in (6.1), and suppose that q is an integer with $q > 2N$. If $-N \le k \le N$, then*

$$\frac{1}{q}\sum_{a=1}^{q} e(-ak/q)T(a/q) = t_k\,.$$

On substituting the above into (3.42) we find that

$$\begin{aligned} T(x) &= \sum_{n=-N}^{N}\Big(\frac{1}{q}\sum_{a=1}^{q} e(-an/q)T(a/q)\Big)e(nx) \\ &= \frac{1}{q}\sum_{a=1}^{q} T(a/q)\sum_{n=-N}^{N} e(n(x-a/q)) \qquad (6.2) \\ &= \frac{1}{q}\sum_{a=1}^{q} T(a/q)D_N(x-a/q)\,. \end{aligned}$$

Thus the value of $T(x)$ for general x is determined by the values of T at the special arguments a/q, provided that $q > 2N$. Since $T(x)$ has $2N+1$ coefficients that need to be determined, we need to have $q \ge 2N+1$ in order to determine $T(x)$ uniquely. Thus the condition that $q > 2N$ is necessary, as well as sufficient.

Suppose that the numbers $c(1), c(2), \ldots, c(2N+1)$ are given. Put

$$U(x) = \frac{1}{2N+1}\sum_{a=1}^{2N+1} c(a)D_N(x-a/(2N+1))\,. \qquad (6.3)$$

This is a trigonometric polynomial of degree $\le N$, and $U(b/(2N+1)) = c(b)$. Thus we can construct a trigonometric polynomial $U \in \mathcal{T}_N$ that takes prescribed values at $2N+1$ equally spaced points. Since U has $2N+1$ coefficients, in general we will not be able to force U to take prescribed values at $2N+2$ points.

Suppose that numbers $c_0(1), c_0(2), \ldots, c_0(N)$ are given, and put

$$V_0(x) = \frac{1}{N}\sum_{n=1}^{N} c_0(n)\Delta_N(x - n/N)\,. \tag{6.4}$$

Then $V_0(n/N) = c_0(n)$ and $V_0'(n/N) = 0$ for $n = 1, 2, \ldots, N$. Let numbers $c_1(1), c_1(2), \ldots, c_1(N)$ be given, and put

$$V_1(x) = \frac{1}{2\pi N^2}\sum_{n=1}^{N} c_1(n)\widetilde{D}_N^*(x - n/N) \tag{6.5}$$

where $\widetilde{D}_N^*(x)$ is the modified conjugate Dirichlet kernel,

$$\widetilde{D}_N^*(x) = \sin 2\pi N x + 2\sum_{n=1}^{N-1} \sin 2\pi n x \tag{6.6}$$

$$= 2(\sin \pi N x)^2 \cot \pi x = N\Delta_N(x)\sin 2\pi x, \tag{6.7}$$

as defined in Exercise 3.4.6. Thus $V_1(n/N) = 0$ and $V_1'(n/N) = c_1(n)$ for $n = 1, 2, \ldots, N$. We put $V(x) = V_0(x) + V_1(x)$, and thus obtain a trigonometric polynomial $V \in \mathcal{T}_N$ with prescribed values of $V(n/N)$ and $V_N'(n/N)$. Moreover,

$$\widehat{V}(\pm N) = \frac{\pm 1}{4\pi i N^2}\sum_{n=1}^{N} c_1(n),$$

so $V \in \mathcal{T}_{N-1}$ if $\sum_{n=1}^{N} c_1(n) = 0$. We now use these ideas to construct useful one-sided approximations to the sawtooth function $s(x)$.

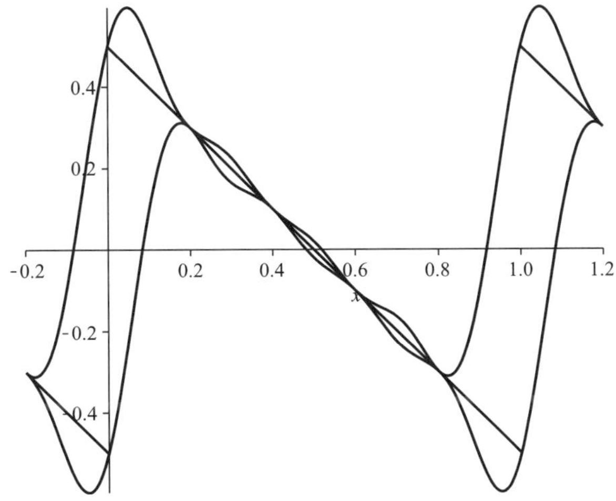

Figure 6.1. Graph of $B_4(x)$, the sawtooth function $s(x)$, and $-B_4(-x)$, for $-0.2 \le x \le 1.2$.

Theorem 6.5. *Let*

$$(6.8)\quad \begin{aligned} B_N(x) = {} & \frac{1}{N+1}\sum_{n=0}^{N}\Big(\frac{1}{2}-\frac{n}{N+1}\Big)\Delta_{N+1}(x-n/(N+1)) \\ & -\frac{1}{2\pi(N+1)}\sin 2\pi(N+1)x+\frac{1}{2\pi}\Delta_{N+1}(x)\sin 2\pi x\,. \end{aligned}$$

Then $B_N \in \mathcal{T}_N$, $B_N(x) \ge s(x)$ for all x, and $\int_0^1 B_N(x)\,dx = 1/(2(N+1))$. Moreover, if $T \in \mathcal{T}_N$ and $T(x) \ge s(x)$ for all x, then $\int_0^1 T(x)\,dx \ge 1/(2(N+1))$, and the equality holds if and only if $T(x) = B_N(x)$.

We call $B_N(x)$ a *Beurling trigonometric polynomial*, which should not be confused with the Bernoulli polynomial $B_k(x)$, as discussed in §9.5. Since $B_N(-x) \ge s(-x) = -s(x)$, it follows that $-B_N(-x) \le s(x)$ for all x. We note that

$$\int_0^1 -B_N(-x)\,dx = \frac{-1}{2(N+1)},$$

so $-B_N(-x)$ minorizes $s(x)$ in the same way that $B_N(x)$ is a majorant.

Proof. The last two terms in (6.8) are trigonometric polynomials of degree $N+1$, but the coefficient of $e(\pm(N+1)x)$ cancels, so the combined contribution of these two terms is a trigonometric polynomial of degree N. Clearly $B_N(n/(N+1)) = 1/2-n/(N+1)$ for $n = 0, 1, \dots, N$. Also, $B_N'(n/(N+1)) = -1$ for $n = 1, 2, \dots, N$. By the Mean Value Theorem of differential calculus applied to the interval $[(n-1)/(N+1), n/(N+1)]$ we see that there is an $x_n \in ((n-1)/(N+1), n/(N+1))$ such that $B_N'(x_n) = -1$, for $n = 1, 2, \dots, N$. Thus we have identified $2N$ roots of the equation $B_N'(x)+1 = 0$, which is the maximum number of roots of a trigonometric polynomial of degree N. If there is an $x' \in ((n-1)/(N+1), n/(N+1))$ such that $B_N(x') = 1/2-x'$, then by two applications of the Mean Value Theorem we would have a root of $B_N'(x)+1 = 0$ in the interval $((n-1)/(N+1), x')$ and also in the interval $(x', n/(N+1))$. This would give at least $2N+1$ roots, which is impossible. Thus we deduce that the only roots of $B_N(x) = 1/2-x$ are at the points $n/(N+1)$ for $n = 0, 1, \dots, N$. To complete the proof that $B_N(x) \ge 1/2-x$ for $0 \le x \le 1$, it suffices to show that

$$(6.9)\qquad B_N''(k/(N+1)) > 0$$

for $k = 1, 2, \dots, N$. Since $\sin \pi(N+1)x \sim (-1)^k\pi(N+1)(x-k/(N+1))$ for x near $k/(N+1)$, it follows that $\big(\sin \pi(N+1)x\big)^2 \sim \pi^2(N+1)(x-k/(N+1))$. Hence

$$\Delta_{N+1}''\Big(\frac{k}{N+1}\Big) = \frac{2\pi^2(N+1)}{\big(\sin \pi\frac{k}{N+1}\big)^2}$$

for $k = 1, 2, \dots, N$. Thus for these k we find that

$$(6.10)\quad B_N''\Big(\frac{k}{N+1}\Big) = \frac{1}{N+1}\sum_{n=0}^{N}\Big(\frac{1}{2}-\frac{n}{N+1}\Big)\Delta_{N+1}''\Big(\frac{k-n}{N+1}\Big)+2\pi\cot\frac{\pi k}{N+1}\,.$$

Since $\Delta_{N+1}(x)$ is a trigonometric polynomial of degree N, it follows that $\Delta''_{N+1}(x)$ is a trigonometric polynomial of degree N, and that $\widehat{\Delta''_{N+1}}(0) = 0$. Thus by Corollary 6.2 it follows that

$$\sum_{n=0}^{N} \Delta''_{N+1}\Big(\frac{k-n}{N+1}\Big) = 0\,.$$

We multiply this by $(1/2 - k/(N+1))/(N+1)$ and subtract from the right hand side of (6.10) to see that

$$B''_N\Big(\frac{k}{N+1}\Big) = \sum_{\substack{n=0\\ n\neq k}}^{N} \frac{k-n}{N+1}\,\frac{2\pi^2}{\sin^2\frac{\pi(k-n)}{N+1}} + 2\pi(N+1)\cot\frac{\pi k}{N+1}\,. \tag{6.11}$$

Suppose first that $2k \le N$. We pair $n = k-j$ with $n = k+j$ for $j = 1, 2, \ldots, k$, and see that the two terms in a pair cancel each other. If $2k = N$, then the sum over n vanishes, and the cosine term is positive. Suppose that $2k < N$. It remains to sum from $n = 2k+1$ to N. By reindexing the sum with $m = n-2k$ we see that

$$\sum_{n=2k+1}^{N} \frac{k-n}{N+1}\,\frac{2\pi^2}{\sin^2\frac{\pi(k-n)}{N+1}} = \sum_{m=1}^{N-2k} \frac{-k-m}{N+1}\,\frac{2\pi^2}{\sin^2\frac{\pi(k+m)}{N+1}}\,.$$

Alternatively, we can sum the same terms in reverse order by setting $m = N+1-n$. Thus the above is

$$= \sum_{m=1}^{N-2k} \frac{k-N-1+m}{N+1}\,\frac{2\pi^2}{\sin^2\frac{k+m}{N+1}}\,.$$

We sum the two right hand sides, and divide by 2 to conclude that

$$B''_N\Big(\frac{k}{N+1}\Big) = 2\pi\cot\frac{\pi k}{N+1} - \pi^2\sum_{m=1}^{N-2k}\frac{1}{\sin^2\frac{\pi(k+m)}{N+1}}$$

when $k < 2N$. Now $\sin^{-2}\pi u$ is convex for $0 < u < 1$, and we know that if f is a convex function then

$$f(x) \le \frac{1}{h}\int_{x-h/2}^{x+h/2} f(u)\,du\,.$$

On taking $x = (k+m)/(N+1)$, $h = 1/(N+1)$, and summing, we deduce that

$$\pi^2\sum_{m=1}^{N-2k}\frac{1}{\sin^2\frac{\pi(k+m)}{N+1}} \le \pi^2(N+1)\int_{(k+1/2)/(N+1)}^{1-(k+1/2)/(N+1)} \sin^{-2}\pi u\,du\,.$$

But $\int \sin^{-2}\pi u\,du = -\frac{1}{\pi}\cot\pi u + C$, and $\cot\pi u$ is an odd function with period 1, so the right hand side above is

$$= 2\pi(N+1)\cot\frac{\pi(k+1/2)}{N+1},$$

and this is

$$< 2\pi(N+1)\cot\frac{\pi k}{N+1}$$

since $\cot \pi x$ is a strictly decreasing function for $0 < x < 1$. Thus we have (6.9) when $2k \le N$.

Now suppose that $N < 2k \le 2N$. We again argue from (6.11). The term for $n = k - j$ is cancelled by the term for $n = k + j$ for $j = 1, 2, \ldots, N - k$. The remaining sum is

$$\sum_{n=0}^{N-2k-1} \frac{k-n}{N+1} \frac{2\pi^2}{\sin^2 \frac{\pi(k-n)}{N+1}}.$$

By reindexing with $m = 2k - N - 1 - n$ we can write the terms in reverse order. Thus the above is

$$= \sum_{m=0}^{N-2k-1} \frac{-k+N+1+m}{N+1} \frac{2\pi^2}{\sin^2 \frac{\pi(k-m)}{N+1}}.$$

We give the two index variables a common name, say n, add the two sums above, and divide by 2 to see that the above sums are

$$= \sum_{n=0}^{N-2k-1} \frac{\pi^2}{\sin^2 \frac{\pi(n-k)}{N+1}}.$$

Now $\sin^2 \pi u$ is a convex function for $0 < u < 1$, and we know that if f is convex, then

$$\frac{1}{h}\int_x^{x+h} f(u)\, du \le \frac{1}{2} f(x) + \frac{1}{2} f(x+h).$$

With $x = (k-n)/(N+1)$ and $h = 1/(N+1)$ we deduce that our sum is

$$> \pi^2(N+1)\int_{-k/(N+1)}^{k/(N+1)-1} \sin^{-2} \pi u\, du = \pi^2 \int_{1-k/(N+1)}^{k/(N+1)} \sin^{-2} \pi u\, du$$
$$= 2\pi(N+1)\cot \frac{\pi k}{N+1}.$$

Thus $B_N''\left(\frac{k}{N+1}\right) > 0$ when $N < 2k \le 2N$. This completes the proof of (6.9) for $k = 1, 2, \ldots, N$.

To complete the proof of the theorem it remains to show that $B_N(x)$ is the unique extremal. Suppose that $T(x)$ is a trigonometric polynomial of degree at most N such that $T(x) \ge s(x)$ for all x. Since $s(0^+) = 1/2$ and T is continuous, it follows that $T(0) \ge 1/2$. By Corollary 6.2 we know that

$$\int_0^1 T(x)\, dx = \widehat{T}(0) = \frac{1}{N+1}\sum_{n=0}^{N} T\left(\frac{n}{N+1}\right)$$
$$\ge \frac{1}{N+1}\left(\frac{1}{2} + \sum_{n=1}^{N} s\left(\frac{n}{N+1}\right)\right) = \frac{1}{2(N+1)}.$$

Moreover, if equality is achieved here, then

$$T(0) = B_N(0) = \frac{1}{2},$$
$$T\left(\frac{n}{N+1}\right) = B_N\left(\frac{n}{N+1}\right) = \frac{1}{2} - \frac{n}{N+1} \quad \text{for } n = 1, 2, \ldots, N,$$

and

$$T'\Big(\frac{n}{N+1}\Big) = B_N'\Big(\frac{n}{N+1}\Big) = -1 \text{ for } n = 1, 2, \ldots, N\,.$$

Put $U(x) = T(x) - B_N(x)$. Then $U(0) = 0$ and $U(x)$ has double zeros at the points $n/(N+1)$, which makes a total of at least $2N+1$ zeros. This implies that $U(x) = 0$ identically, so $T(x) = B_N(x)$. □

Exercises

1. Suppose that $T(x) \in \mathcal{T}_N$. Put $U(x) = T(x + \alpha)$, where α is a real constant. Show that $U(x) \in \mathcal{T}_N$.
2. Suppose that $T_1 \in \mathcal{T}_N$, and that $T_2 \in \mathcal{T}_N$. Show that $c_1T_1(x) + c_2T_2(x) \in \mathcal{T}_N$ for any constants c_1, c_2.
3. Suppose that $T \in \mathcal{T}_M$ and that $U \in \mathcal{T}_N$. Show that $TU \in \mathcal{T}_{M+N}$.
4. Suppose that $T \in \mathcal{T}_N$. Show that $T' \in \mathcal{T}_N$.
5. Suppose that $T(x)$ is given as in (6.1), that q is a positive integer, and that α is a real number. Show that
$$\frac{1}{q}\sum_{a=1}^{q} T(a/q + \alpha) = \sum_{\substack{-N \le n \le N \\ q \mid n}} t_n e(n\alpha)\,.$$
6. Suppose that $T(x)$ is given as in (6.1), that q is a positive integer, k is an integer, and that α is a real number. Show that
$$\frac{1}{q}\sum_{a=1}^{q} e(-ak/q)T(a/q + \alpha) = \sum_{\substack{-N \le n \le N \\ n \equiv k \pmod q}} t_n e(n\alpha)\,.$$
7. Let $x_1, x_2, \ldots, x_{2N+1}$ be distinct points of $\mathbb{T}$. For $1 \le r \le 2N+1$, put
$$T_r(x) = \prod_{\substack{j=1 \\ j \ne r}}^{2N+1} \frac{\sin \pi(x - x_j)}{\sin \pi(x_r - x_j)}\,.$$
   (a) Show that $T_r(x)$ has period 1.
   (b) Show that $T_r(x) \in \mathcal{T}_N$.
   (c) Show that
$$T_r(x_j) = \begin{cases} 1 & (j = r), \\ & (j \ne r). \end{cases}$$
   (d) Let $c(1), c(1), \ldots, c(2N+1)$ be given numbers, and put
$$T(x) = \sum_{r=1}^{2N+1} c(r)T_r(x)\,.$$
   Show that $T \in \mathcal{T}_N$, and that $T(x_r) = c(r)$ for $r = 1, 2, \ldots, 2N+1$.

8. Let $z_1, z_2, \dots, z_N$ be unimodular complex numbers (i.e., complex numbers for which $|z_n| = 1$), and put

$$P(z) = \prod_{n=1}^{N} (z - z_n) = \sum_{n=0}^{N} a_n z^n,$$

$$T(\theta) = P(e(\theta)) = \sum_{n=0}^{N} a_n e(n\theta).$$

(a) Show that $a_N = 1$ and that $|a_0| = 1$.
(b) Show that

$$\frac{1}{N} \sum_{k=1}^{N} T(k/N + \alpha) = a_0 + a_N e(N\alpha)$$

(c) Show that there is an α such that $|a_0 + a_N e(N\alpha)| = 2$.
(d) Show that there exists a k and an α such that $|T(k/N + \alpha)| \ge 2$.
(e) Show that if $z_n = e(n/N)$, then $P(z) = z^N - 1$, and $|T(\theta)| \le 2$ for all θ.
(f) Show that no matter how N points $P_1, P_2, \dots, P_N$ are chosen on a unit circle, there will be a point P on that circle such that the product of the distance from P to P_1, the distance from P to P_2, ..., and the distance from P to P_N is ≥ 2.

9. Suppose that

$$T(x) = \frac{1}{2} a_0 + \sum_{n=1}^{N} a_n \cos 2\pi n x,$$

that $T(x) \ge 0$ for all x, that $T(0) = 1$, and that $a_n = 0$ whenever $3|n$.
(a) Use Corollary 6.2 to show that $a_0 \ge 2/3$.
(b) By considering $\Delta_3(x)/3$, show that a_0 can be as large as $2/3$.

10. Suppose that $P(z)$ is an algebraic polynomial of degree not exceeding N, that $P(0) = 1$, and that $P(1) = 0$. Use Corollary 6.2 to show that

$$\max_{|z| \le 1} |P(z)| \ge 1 + \frac{1}{N+1}.$$

11. Let $B_N(x)$ be defined as in Theorem 6.5.
(a) Show that

$$\sum_{n=0}^{N} B_N'\left(\frac{n}{N+1}\right) = 0.$$

(b) Deduce that $B_N'(0) = N$.
(c) Show that

$$\Delta_{N+1}''(0) = -8\pi^2 \sum_{n=1}^{N} (1 - n/N) n^2.$$

(d) Deduce that $\Delta_{N+1}''(0) = -\frac{2}{3}\pi^2 N(N+1)(N+2)$. (Some of the formulas of (9.28) may be useful here.)

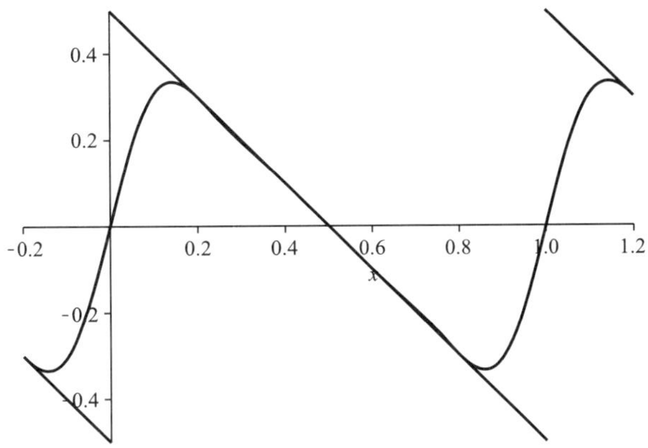

Figure 6.2. Graph of $V_4(x)$ and the sawtooth function $s(x)$, for $-0.2 \le x \le 1.2$.

12. Let

$$\begin{aligned} V_N(x) = {} & \frac{1}{N+1}\sum_{n=1}^{N}\Big(\frac{1}{2} - \frac{n}{N+1}\Big)\Delta_{N+1}(x - n/(N+1)) \\ & - \frac{1}{2\pi(N+1)}\sin 2\pi(N+1)x + \frac{1}{2\pi}\Delta_{N+1}(x)\sin 2\pi x\,. \end{aligned} \tag{6.12}$$

We call this a *Vaaler trigonometric polynomial.*

(a) Show that

$$B_N(x) = \frac{\Delta_{N+1}(x)}{2(N+1)} + V_N(x)\,.$$

(b) In the first term in the definition of $V_N(x)$, write

$$\Delta_{N+1}(x - n/(N+1)) = \sum_{k=-N}^{N}\Big(1 - \frac{|k|}{N+1}\Big)e(kx - kn/(N+1)),$$

invert the order of summation, and use the result of Example 2.4 to show that this first term is

$$\frac{1}{N+1}\sum_{k=1}^{N}\Big(1 - \frac{k}{N+1}\Big)\Big(\cot\frac{\pi k}{N+1}\Big)\sin 2\pi kx\,.$$

(c) Use the closed form formula for the Fejér kernel to show that the combined contribution of the second and third terms in the definition of $V_N(x)$ is

$$\frac{\sin\pi(N+1)x\,\sin\pi Nx}{\pi(N+1)\sin\pi x} = \frac{\widetilde{D}_N(x)}{2\pi(N+1)}$$

where $\widetilde{D}_N(x)$ is the conjugate Dirichlet kernel as in Exercise 3.4.5.

(d) Conclude that

$$V_N(x) = \frac{1}{N+1}\sum_{k=1}^{N} f(k/(N+1))\sin 2\pi kx$$

where $f(u) = (1-u)\cot\pi u + 1/\pi$.

(e) Show that $V_N(n/(N+1)) = s(n/(N+1))$ for $n = 1, 2, \ldots, N$.
(f) Show that $V_N'(n/(N+1)) = -1$ for $n = 1, 2, \ldots, N$. By an argument similar to the proof of Theorem 6.5 it can be shown that $V_N''(n/(N+1)) < 0$ for $2n \le N$, that $V_N''(n/(N+1)) > 0$ for $N < 2n \le 2N$, and hence that $V_N(x) \le s(x)$ for $0 \le x \le 1/2$ and $V_N(x) \ge s(x)$ for $1/2 \le x \le 1$.

6.2. Bernstein's Inequality

We now encounter a wonderful gem, whose proof involves ideas that we have not previously encountered.

Theorem 6.6. (Bernstein) *Suppose that $T(x)$ is a trigonometric polynomial of degree $\le N$. Then*

$$\max_x |T'(x)| \le 2\pi N \max_x |T(x)| .$$

Proof. If we multiply T by a constant c, then both sides of the proposed inequality are multiplied by c. Thus without loss of generality, we may suppose that $\max |T(x)| = 1$. Also, since $T(x+c)$ is also a trigonometric polynomial of the same degree as T, it suffices to prove that

$$|T'(0)| \le 2\pi N$$

provided that $\max_x |T(x)| \le 1$. Our line of attack is to express $T'(0)$ as a linear combination of values of $T(x)$. We observe that equality is attained when $T(x) = \sin 2\pi N x$. With the idea that this is the extremal case, we are challenged to write $T'(0)$ as a linear combination of $T(x)$ at points x where $\sin 2\pi N x$ attains its maximum modulus, i. e., at the points $\frac{2k-1}{4N}$ for $k = 1, 2, \ldots, 2N$. Armed with this insight, we adopt the strategy of expressing $T'(0)$ as a linear combination of T at these points:

$$T'(0) = \sum_{k=1}^{2N} c(k) T\Big(\frac{2k-1}{4N}\Big) . \tag{6.13}$$

Our idea is that this should hold for all T, with the $c(k)$ depending on k and N, but independent of the coefficients of T. Suppose that $T(x)$ is defined as in (6.1). Then

$$T'(x) = \sum_{n=-N}^{N} 2\pi i n t_n e(nx),$$

and hence (6.13) asserts that

$$\sum_{n=-N}^{N} 2\pi i n t_n = \sum_{k=1}^{2N} c(k) \sum_{n=-N}^{N} t_n e\Big(\frac{(2k-1)n}{4N}\Big) .$$

For this to hold for all t_n, it is necessary (and sufficient) for the coefficient of t_n on the left hand side to be the same as that on the right. That is, the coefficients $c(k)$ should be chosen so that

$$2\pi i n = \sum_{k=1}^{2N} c(k) e\Big(\frac{(2k-1)n}{4N}\Big) \tag{6.14}$$

for $-N \le n \le N$. This constitutes $2N+1$ linear equations in the $2N$ variables $c(k)$, which would appear to be a tall order. On closer inspection we find that the case $n = N$ in (6.14) is exactly the same as the case $n = -N$, so we have only $2N$ equations to be solved in $2N$ variables. In the task of solving these equations explicitly, the Discrete Fourier Transform comes to our rescue. The equations (6.14) can be rewritten as

$$\sum_{k=1}^{2N} c(k)e\Big(\frac{kn}{2N}\Big) = 2\pi i n\, e\Big(\frac{n}{4N}\Big) \tag{6.15}$$

for $-N < n \le N$. If we regard $c(k)$ as a sequence with period $2N$, then the above asserts that

$$\widehat{c}(-n) = 2\pi i n\, e\Big(\frac{n}{4N}\Big),$$

which is to say that

$$\widehat{c}(n) = -2\pi i n\, e\Big(\frac{-n}{4N}\Big)$$

for $-N \le n \le N$. By the inversion theorem (Theorem 2.2) it follows that there is a unique set of $c(k)$ for which the above is true, namely

$$c(k) = \frac{-2\pi i}{2N} \sum_{n=-N}^{N} n\, e\Big(\frac{(k-1/2)n}{2N}\Big) + \pi(-1)^k .$$

In the above we have written the sum symmetrically, so that it runs from $-N$ to N. However, the contribution of $n = -N$, namely $\pi(-1)^{k+1}$, is duplicated by that of $n = N$, so to correct for this duplication we have the extra term $\pi(-1)^k$ in the above.

We have established that there is a set of $c(k)$ such that (6.13) holds for all trigonometric polynomials T of degree at most N. By the triangle inequality it follows that

$$|T'(0)| \le \Big(\sum_{k=1}^{2N} |c(k)|\Big) \max_x |T(x)| .$$

To complete the proof, we need to evaluate the above sum. We recall that the Dirichlet kernel is

$$D_N(x) = \sum_{n=-N}^{N} e(nx) = \frac{\sin(2N+1)\pi x}{\sin \pi x} .$$

We differentiate to see that

$$\begin{aligned} D_N'(x) &= \sum_{n=-N}^{N} 2\pi i n e(nx) \\ &= \frac{(2N+1)\pi(\sin \pi x)(\cos(2N+1)\pi x) - \pi(\sin(2N+1)\pi x)(\cos \pi x)}{(\sin \pi x)^2} . \end{aligned}$$

We take $x = (k - 1/2)/(2N) = (2k - 1)/(4N)$. The resulting numerator on the right hand side is

$$(6.16)\qquad \begin{aligned}(2N+1)\pi\sin\Big(\frac{2k-1}{4N}\pi\Big)\cos\Big((2N+1)\frac{2k-1}{4N}\pi\Big)\\ -\pi\sin\Big((2N+1)\frac{2k-1}{4N}\pi\Big)\cos\Big(\frac{2k-1}{4N}\pi\Big).\end{aligned}$$

From the formula (T.14) for $\cos(x+y)$ we find that

$$\cos\Big((2N+1)\frac{2k-1}{4N}\pi\Big) = \cos(k-1/2)\pi\cos\frac{2k-1}{4N}\pi - \sin(k-1/2)\pi\sin\frac{2k-1}{4N}\pi.$$

But $\cos(k-1/2)\pi = 0$ for all k, and $\sin(k-1/2)\pi = (-1)^{k-1}$, so the above simplifies to

$$(-1)^k\sin\frac{2k-1}{4N}\pi.$$

Similarly, by the formula (T.13) for $\sin(x+y)$ we find that

$$\sin\Big((2N+1)\frac{2k-1}{4N}\pi\Big) = (-1)^{k-1}\cos\frac{2k-1}{4N}\pi.$$

Thus our numerator (6.16) is

$$= (-1)^k\pi\Big((2N+1)\sin^2\frac{2k-1}{4N}\pi + \cos^2\frac{2k-1}{4N}\Big) = (-1)^k\pi\Big(1 + 2N\sin^2\frac{2k-1}{4N}\pi\Big)$$

since $\sin^2 + \cos^2 = 1$. Thus we find that

$$\sum_{N=-N}^{N} 2\pi n e\Big(\frac{2k-1}{4N}n\Big) = (-1)^k 2N\pi + \frac{(-1)^k\pi}{\sin^2\frac{2k-1}{4N}\pi}.$$

On multiplying both sides by $-1/(2N)$ we conclude that

$$c(k) = \frac{(-1)^{k-1}\pi}{2N\sin^2\frac{2k-1}{4N}\pi}.$$

Thus

$$\sum_{k=1}^{2N}|c(k)| = \sum_{k=1}^{2N}(-1)^{k-1}c(k).$$

By taking $n = N$ in (6.15) we find that this is $2\pi N$, so the proof is complete. □

A curious by-product of our argument is that

$$\sum_{k=1}^{2N}\frac{1}{\sin^2\frac{\pi(k-1/2)}{2N}} = 4N^2.$$

This is a special case of a more general identity found in Exercise 1.4.4(b).

In view of our remarks in the preceding section to the effect that we need to sample $T(x)$ at least $2N+1$ points in order to determine $T(x)$, it may seem surprising that we are able to express $T'(0)$ using only $2N$ points. The explanation is that t_0 does not contribute to $T'(x)$, so $T'(x)$ has only $2N$ degrees of freedom.

Although the equality in Bernstein's inequality is achieved only when $T(x)$ is of the form $c\sin 2\pi(x+\phi)$, the inequality is useful because the bound provided is usually of the correct order of magnitude, even when T is far from having only one term; see Exercise 1 below and Exercise 6.4.16 for two disparate examples (one with a large peak, and the other with no identifiable peak).

Corollary 6.7. *Let T be a trigonometric polynomial with period 1, and of degree not exceeding N. Then*

$$\max_x |T(x)| \le 2\pi N \int_0^1 |T(x)|\, dx\,.$$

In the case that T is the Fejér kernel, $T(x) = \Delta_{N+1}(x)$, the left hand side above is $N+1$, and the right hand side is $2\pi N$. Thus the bound is sharp apart from the constant.

Proof. Let $M = \max_x |T(x)|$, and let x_0 be a point such that $|T(x_0)| = M$. Clearly

$$T(x) - T(x_0) = \int_{x_0}^{x} T'(u)\, du\,.$$

By Bernstein's inequality, the integrand has absolute value not exceeding $2\pi NM$. Hence

$$|T(x) - T(x_0)| \le 2\pi NM|x - x_0|\,.$$

By the triangle inequality in the form $|a| \ge |b| - |a-b|$ it follows that

$$|T(x)| \ge M(1 - 2\pi N|x - x_0|)\,. \tag{6.17}$$

Let $\delta = 1/(2\pi N)$. Then

$$\int_0^1 |T(x)|\, dx \;\ge\; \int_{x_0-\delta}^{x_0+\delta} |T(x)|\, dx \;\ge\; M\int_{-\delta}^{\delta} 1 - |x|/\delta\, dx \;=\; M\delta,$$

which gives the stated result. □

Exercises

1. (a) Use Bernstein's inequality to show that $|\Delta_N'(x)| \le 2\pi(N-1)N$ for all x.
   (b) Note that $\Delta_N(-1/N) = 0$, and that $\Delta_N(0) = N$. Deduce that there is an $x \in [-1/N, 0]$ such that $\Delta_N'(x) \ge N^2$.
2. For $f \in L^2(\mathbb{T})$ we know by the Cauchy–Schwarz inequality that $\|f\|_1 \le \|f\|_2$. In general, the ratio $\|f\|_2/\|f\|_1$ may be arbitrarily large, but we show now that for trigonometric polynomials this ratio can be bounded in terms of the degree.
   (a) Suppose that $T \in \mathcal{T}_N$. Show that
   $$\|T\|_2 \;\le\; \sqrt{2\pi N}\, \|T\|_1\,.$$
   (b) Give an example of a trigonometric polynomial in $\mathcal{T}_N$ for which $\|T\|_2 > c\sqrt{N}\|T\|_1$ where c is a positive constant.
3. (a) Show that if $T \in \mathcal{T}_N$, then
   $$\max_x |T(x)|^2 \;\le\; (2N+1)\int_0^1 |T(x)|^2\, dx\,.$$
   (Hint: *Do not* use Bernstein's inequality!)

(b) Give an example of a trigonometric polynomial in $\mathcal{T}_N$ for which

$$\max_x |T(x)|^2 = (2N+1)\int_0^1 |T(x)|^2\,dx\,.$$

4. Let $x_1, x_2, \ldots, x_K$ be points of $\mathbb{T}$ such that any interval of $\mathbb{T}$ of length $1/(2\pi N)$ contains at least one of the x_k. Suppose that $T \in \mathcal{T}_N$.
   (a) Show that

$$\max_{1\le k\le K} |T(x_k)| \ge \frac{1}{2}\max_x |T(x)|\,.$$

   (Hint: Use (6.17))
   (b) Show that the above is false if $1/(2\pi N)$ is replaced by $1/N$.
5. Let the numbers $c(k)$ be defined as in the proof of Bernstein's theorem, let $T(x)$ be a trigonometric polynomial as in (6.1), and put $u_k = (2k-1)/(4N)$.
   (a) Show that

$$T'(x) = \sum_{k=1}^{2N} c(k)T(x+u_k)$$

   for all x.
   (b) Show that

$$\int_0^1 |T'(x)|\,dx \le \Big(\sum_{k=1}^{2N} |c(k)|\Big)\int_0^1 |T(x)|\,dx\,.$$

   (c) Deduce that $\|T'\|_1 \le 2\pi N\|T\|_1$.
   (d) Apply Parseval's Identity to T and to T'. Deduce that $\|T'\|_2 \le 2\pi N\|T\|_2$.
   (e) Suppose that $1 < p < \infty$. By an appropriate application of Hölder's inequality, show that

$$|T'(x)|^p \le \Big(\sum_{k=1}^{2N} |c(k)|\Big)^p\Big(\sum_{k=1}^{2N} |c(k)||T(x+u_k)|^p\Big).$$

   (f) Deduce that

$$\int_0^1 |T'(x)|^p\,dx \le \Big(\sum_{k=1}^{2N} |c(k)|\Big)^p\int_0^1 |T(x)|^p\,dx\,.$$

   Conclude that $\|T'\|_p \le 2\pi N\|T\|_p$ for $1 \le p < \infty$.

6.3. Real-valued and nonnegative trigonometric polynomials

Let $T(x)$ be a trigonometric polynomial as in (6.1), and for $z \ne 0$ put

$$Q(z) = \sum_{n=-N}^{N} t_n z^n\,. \tag{6.18}$$

Thus Q is a rational function, and indeed $Q(z)z^N$ is a polynomial. Suppose that $T(x)$ is real for all real x. Then $Q(z)$ is real on the unit circle $|z| = 1$. Write z in polar coordinates, $z = re(\theta)$. Then $1/z = \frac{1}{r}e(-\theta)$, and hence $1/\overline{z} = \frac{1}{r}e(\theta)$. Thus z and $1/\overline{z}$ have the same argument, and $|z||1/\overline{z}| = 1$. We consider $1/\overline{z}$ to be the reflection of z with respect to the unit circle, in the same sense that $\overline{z}$ is the reflection of z with respect to the real axis. In Exercise 2 we found that a polynomial $f(x)$ is real for all real x if and only if all the coefficients of f are real, and in Exercise 3 we discovered that in this case the complex zeros of f come in pairs, which is to say that $f(z) = 0$ if and only if $f(\overline{z}) = 0$. We now establish an analogous relation between $Q(z)$ and $Q(1/\overline{z})$. We note that

$$Q(1/\overline{z}) = \sum_{n=-N}^{N} t_n r^{-n} e(n\theta)\,.$$

Put $k = -n$. Then the above sum is

$$= \sum_{k=-N}^{N} t_{-k} r^{k} e(-k\theta)\,.$$

From Corollary 3.4 we know that $t_{-k} = \overline{t_k}$. Thus the above is $\overline{Q(z)}$. Hence if $Q(z)$ is real for all z on the unit circle, then

$$Q(1/\overline{z}) = \overline{Q(z)} \tag{6.19}$$

for all $z \neq 0$. In particular, $Q(z) = 0$ if and only if $Q(1/\overline{z}) = 0$. Thus the zeros of Q occur in reflected pairs, as do the zeros of a real-valued polynomial $f(x)$. This is useful in proving the following further result.

Theorem 6.8. (Fejér–Riesz) *Let $T(x)$ be a trigonometric polynomial, as in* (6.1). *If $T(x) \geq 0$ for all x, then there exist numbers $\alpha_0, \alpha_1, \ldots, \alpha_N$ such that*

$$T(x) = \left| \sum_{k=0}^{N} \alpha_k e(kx) \right|^2 .$$

In general, the numbers α_k are complex, but in some instances they may all be real. For example, if $\alpha_k = 1/\sqrt{N}$ for all k, then we obtain the Fejér kernel $\Delta_{N+1}(x)$.

Proof. Since $t_{-N} = \overline{t_N}$, if $t_N = 0$, then $t_{-N} = 0$ and the value of N can be reduced. Thus we may assume that t_N and t_{-N} are non-zero. Let $Q(z)$ be defined as in (6.18), and put $P(z) = Q(z)z^N$. Thus $P(z)$ is a polynomial of degree $2N$. The roots of P on the unit circle are of even multiplicity; say there are $2K$ of them, so that $e(\theta_1), \ldots, e(\theta_K)$ each should be repeated two times. Let $L = N - K$. Thus there are L pairs of roots, say $z_1, \ldots, z_L$ with $|z_\ell| < 1$, and their reflections $1/\overline{z_\ell}$. The factorization of $P(z)$ can be written

$$P(z) = c \prod_{k=1}^{K} (z - e(\theta_k))^2 \prod_{\ell=1}^{L} (z - z_\ell)(\overline{z_\ell} z - 1)\,.$$

Put

$$P_1(z) = \sqrt{c}\prod_{k=1}^{K}(z - e(\theta_k))\prod_{\ell=1}^{L}(z - z_\ell),$$

$$P_2(z) = \sqrt{c}\prod_{k=1}^{K}(z - e(\theta_k))\prod_{\ell=1}^{L}(z\overline{z_\ell} - 1).$$

We note that if $|z| = 1$, then

$$|z - z_\ell| = |\overline{z} - \overline{z_\ell}| = |z||\overline{z} - \overline{z_\ell}| = |z\overline{z} - z\overline{z_\ell}| = |1 - z\overline{z_\ell}| = |z\overline{z_\ell} - 1|.$$

Thus if $|z| = 1$, then $|P_1(z)| = |P_2(z)|$, and consequently $T(x) = |P_1(e(x))|^2$. The α_k are determined by expanding:

$$P_1(z) = \sum_{k=0}^{N}\alpha_k z^k.$$

Thus we have the stated result. □

Exercises

1. Let $T(x)$ be a cosine polynomial, say

$$T(x) = a_0 + \sum_{n=1}^{N} a_n \cos 2\pi nx,$$

and suppose throughout this exercise that $T(x) \ge 0$ for all x.
   (a) Note that $a_0 = \int_0^1 T(x)\,dx$, and hence that $a_0 \ge 0$.
   (b) Show that if $1 \le n \le N$, then $\int_0^1 T(x)(1 \pm \cos 2\pi nx)\,dx = a_0 \pm \frac{1}{2}a_n$.
   (c) Deduce that $|a_n| \le 2a_0$ for all n.
   (d) Note that if $T(x) = \Delta_{N+1}(x)$, then $a_0 = 1$ and $a_1 = 2(1 - 1/(N+1))$; thus the constant 2 obtained in (c) is asymptotically best possible.
2. Let $T(x)$ be a nonnegative trigonometric polynomial of degree at most N, and let the α_k be determined as in the Fejér–Riesz theorem (Theorem 6.8).
   (a) Show that $\int_0^1 T(x)\,dx = \sum_{k=0}^{N}|\alpha_k|^2$.
   (b) Show that $T(x) \le \left(\sum_{k=0}^{N}|\alpha_k|\right)^2$ for all x.
   (c) Show that $\max_x T(x) \le (N+1)\int_0^1 T(x)\,dx$.
   (d) Show that equality is achieved in the above inequality when $T(x) = \Delta_{N+1}(x)$.
3. Let

$$C(x) = a_0 + 2\sum_{n=1}^{N-1} a_n \cos 2\pi nx,$$

$$S(x) = \sum_{n=1}^{N} b_n \sin 2\pi nx,$$

and suppose that

$$b_n = \begin{cases} a_{n-1} - a_{n+1} & (n = 1, 2, \dots, N-2), \\ a_{N-2} & (n = N-1), \\ a_{N-1} & (n = N). \end{cases}$$

(a) Show that $S(x) = C(x)\sin 2\pi x$ for all x.
(b) Deduce that $S(x) \ge 0$ for $0 \le x \le 1/2$ if and only if $C(x) \ge 0$ for all x.

6.4. Littlewood polynomials

We call a trigonometric polynomial

$$P(x) = \sum_{n=0}^{N-1} \varepsilon_n e(nx) \tag{6.20}$$

a *Littlewood polynomial* of length N if $\varepsilon = \pm 1$ for all n. For such a polynomial, we note that

$$|P(x)|^2 = \sum_{m=0}^{N-1}\sum_{n=0}^{N-1} \varepsilon_m \varepsilon_n e((m-n)x)\,.$$

We group terms according to the value of $k = m - n$. As arranged below, we have the contribution of $k = 0$, $k > 0$, and $k < 0$, respectively. Thus our expression is

$$\begin{aligned} &= N + \sum_{k=1}^{N-1}\Big(\sum_{n=0}^{N-k-1} \varepsilon_n \varepsilon_{n+k}\Big) e(kx) \\ &\quad + \sum_{k=1}^{N-1}\Big(\sum_{m=0}^{N-k-1} \varepsilon_m \varepsilon_{m+k}\Big) e(-kx)\,. \end{aligned}$$

Here the sum (over n) that forms the coefficient of $e(kx)$ is the same as the sum (over m) defining the coefficient of $e(-k)$. Thus we find that

$$|P(x)|^2 = N + 2\sum_{k=1}^{N-1} a_k \cos 2\pi kx \tag{6.21}$$

where

$$a_k = \sum_{n=0}^{N-k-1} \varepsilon_n \varepsilon_{n+k}\,. \tag{6.22}$$

These a_k are called the *aperiodic autocorrelation coefficients* of the ε_n. Since $\varepsilon_n = \pm 1$ for all n, it follows by Parseval's Identity that

$$\int_0^1 |P(x)|^2\,dx = N\,. \tag{6.23}$$

By applying Parseval's Identity to $|P(x)|^2$ as written in (6.21), we find that

$$\int_0^1 |P(x)|^4\,dx = N^2 + 2\sum_{k=1}^{N-1} a_k^2\,. \tag{6.24}$$

By the Cauchy–Schwarz inequality we know that

$$\Big(\int_0^1 |P(x)|^2\,dx\Big)^2 \le \Big(\int_0^1 1\,dx\Big)\Big(\int_0^1 |P(x)|^4\,dx\Big) = \int_0^1 |P(x)|^4\,dx\,.$$

In view of (6.23), it follows that the integral in (6.24) must be at least as large as N^2. What the a_k provide is a measure of the amount by which the fourth moment (6.24) exceeds this minimum. We are interested in Littlewood polynomials for which this fourth moment is as small as possible, which is to say that the a_k^2 should be as small as possible. We note that a_k is the sum of $N-k$ odd integers. Thus if $N-k$ is odd, then a_k is odd, and hence $a_k^2 \ge 1$. Otherwise, all that is obvious is that $a_k^2 \ge 0$. A *Barker polynomial* is a Littlewood polynomial such that

$$a_k = \begin{cases} \pm 1 & (N-k \text{ odd}),\\ 0 & (N-k \text{ even}).\end{cases} \tag{6.25}$$

If $P(x)$ is a Barker polynomial, then $-P(x)$ is also a Barker polynomial, since it will have the same a_k. When we replace $P(x)$ by $P(x+1/2)$, we are replacing ε_n by $(-1)^n\varepsilon_n$, so a_k becomes $(-1)^k a_k$, and we still have a Barker polynomial. Thus Barker polynomials arise in groups of four: $\pm P(x)$, $\pm P(x+1/2)$, and the pair $(\varepsilon_0, \varepsilon_1)$ will be one of $(1,1), (1,-1), (-1,1), (-1,-1)$, each one once. Thus when we list Barker polynomials, it suffices to list those for which $\varepsilon_0 = \varepsilon_1 = 1$. In Table 6.1 we list the known Barker polynomials of this sort.

Table 6.1. Coefficients of the eight known Barker polynomials with $\varepsilon_0 = \varepsilon_1 = 1$.

N	Signs of the coefficients
2	$+\,+$
3	$+\,+\,-$
4	$+\,+\,+\,-$
4	$+\,+\,-\,+$
5	$+\,+\,+\,-\,+$
7	$+\,+\,+\,-\,-\,+\,-$
11	$+\,+\,+\,-\,-\,-\,+\,-\,-\,+\,-$
13	$+\,+\,+\,+\,+\,-\,-\,+\,+\,-\,+\,-\,+$

Although there seem to be no further Barker polynomials, there are other interesting and useful Littlewood polynomials. Suppose that $P(x)$ and $Q(x)$ are Littlewood polynomials of length N, that $a_1, a_2, \ldots, a_{N-1}$ and $b_1, b_2, \ldots, b_{N-1}$ are the aperiodic autocorrelation coefficients of P and Q, respectively. If $b_k = -a_k$ for all k, then we say that P and Q form a *Golay complementary pair*. For such a pair, we see from (6.21) that

$$|P(x)|^2 + |Q(x)|^2 = 2N \tag{6.26}$$

for all x. Conversely, if the above holds for all x, then by two applications of (6.21) we see that

$$\sum_{k=1}^{N-1} (a_k + b_k)\cos 2\pi k x = 0$$

for all x. This in turn implies (by Theorem 3.15) that $a_k + b_k = 0$ for all k, so P and Q are complementary. The identity (6.26) is quite remarkable, since it means that these quantities compensate for each other, and in particular it means that

$$|P(x)|^2 \le 2N \tag{6.27}$$

for all x, which is only twice the size of the mean square we noted in (6.23). As a first example of a Golay complementary pair, we note that if $P(x) = 1 + e(x)$ and $Q(x) = 1 - e(x)$, with a_1 and b_1 denoting their aperiodic autocorrelations, then $a_1 = 1$ and $b_1 = -1$, so P and Q are complementary. Indeed, if z and w are complex numbers, then $|z+w|^2 + |z-w|^2 = 2|z|^2 + 2|w|^2$ (recall the Parallelogram Law found in Example 1.2), so it is clear that $|1+e(x)|^2 + |1-e(x)|^2 = 4$ for all x. The following lemma is useful in constructing longer Golay pairs.

Lemma 6.9. *Suppose that the Littlewood polynomials $P_1(x)$ and $Q_1(x)$ of length N form a Golay complementary pair. Put $P_2(x) = P_1(x) + e(Nx)Q_1(x)$ and $Q_2(x) = P_1(x) - e(Nx)Q_1(x)$. Then P_2 and Q_2 are Littlewood polynomials of length $2N$, and they form a Golay complementary pair.*

Proof. The first N coefficients of P_2 are those of P_1, and the second set of N coefficients are those of Q_1. Similarly, the first N coefficients of Q_2 are those of P_1 and the final N coefficients are the negatives of the coefficients of Q_1. Also,

$$\begin{aligned} |P_2(x)|^2 + |Q_2(x)|^2 &= |P_1(x) + Q_1(x)|^2 + |P_1(x) - Q_1(x)|^2 \\ &= 2|P_1(x)|^2 + 2|Q_1(x)|^2 = 2N \end{aligned}$$

by (6.26). As we already remarked, this implies that P_2 and Q_2 are complementary. □

By applying this lemma repeatedly we obtain the most famous family of Golay complementary pairs, namely the *Shapiro polynomials*. Let $P_0(x) = Q_0(x) = 1$ for all x. Then P_0 and Q_0 are Littlewood polynomials of length 1, and they form a Golay complementary pair in the sense that $|P_0(x)|^2 + |Q_0(x)|^2 = 2$ for all x. Set

$$P_{k+1}(x) = P_k(x) + e(2^k x)Q_k(x), \tag{6.28}$$

$$Q_{k+1}(x) = P_k(x) - e(2^k x)Q_k(x)\,. \tag{6.29}$$

By using Lemma 6.9 inductively, we see that P_k and Q_k are Golay complementary pairs, so that

$$|P_k(x)|^2 + |Q_k(x)|^2 = 2^{k+1}, \tag{6.30}$$

which implies that $|P_k(x)| \le 2^{(k+1)/2}$ for all x. To summarize: P_k is a Littlewood polynomial of length 2^k such that

$$\|P_k\|_2 = 2^{k/2}, \qquad \|P_k\|_\infty \le 2^{(k+1)/2}\,. \tag{6.31}$$

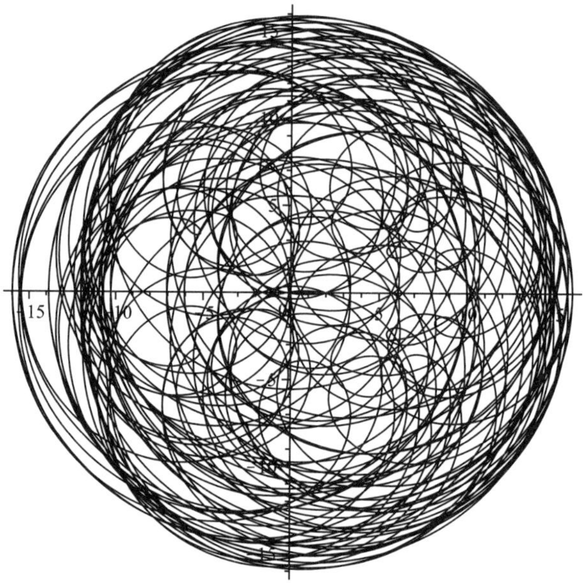

Figure 6.3. Graph of the Shapiro polynomial $P_7(x)$ in the disc $|z| \leq 16$ for $0 \leq x \leq 1$.

Exercises

1. Let $P(x)$ be a Littlewood polynomial as in (6.20). We extend the definition of the ε_n to all integers by asserting that they are periodic with period N. The *symmetric autocorrelation coefficients* of the ε_n are the numbers

$$b_k = \sum_{n=1}^{N} \varepsilon_n \varepsilon_{n+k} \,. \tag{6.32}$$

   (a) Show that the sequence b_k has period N.
   (b) Show that $b_k = a_k + a_{N-k}$ for $1 \leq k \leq N-1$.

2. Let $P(x)$ be a Littlewood polynomial as in (6.20), with aperiodic autocorrelation coefficients as in (6.22). Show that

$$\int_0^1 \left(|P(x)|^2 - N\right)^2 dx = 2 \sum_{k=1}^{N-1} a_k^2 \,.$$

3. Let $P(x)$ be a Barker polynomial of length N.
   (a) Show that

$$\int_0^1 |P(x)|^4 \, dx = \begin{cases} N^2 + N & (N \text{ even}), \\ N^2 + N - 1 & (N \text{ odd}). \end{cases}$$

(b) It has been conjectured that there is a constant $c > 1$ such that if $P(x)$ is a Littlewood polynomial of length N, then

$$\int_0^1 |P(x)|^4 \, dx \geq cN^2 . \tag{6.33}$$

(c) Show that this conjecture implies that there exist only finitely many Barker polynomials.

4. Compute the aperiodic autocorrelation coefficients of the polynomial of length 13 described in Table 6.1, and thus verify that it is indeed a Barker polynomial.

5. Let $P(x)$ be a Littlewood polynomial of length N, and let

$$Q(x) = e((N-1)x)P(-x).$$

(a) Show that $Q(x)$ is a Littlewood polynomial whose coefficients are the same as those of $P(x)$, but in reverse order.
(b) Show that the aperiodic autocorrelation coefficients of Q are the same as those of P. (Hint: Use (6.24), not (6.22).)

6. Suppose that the ε_n are independent random variables with $P(\varepsilon_n = \pm 1/2) = 1/2$ for all n.
(a) Show that $E[a_k^2] = N - k$ for $1 \leq k \leq N-1$.
(b) Deduce that $E[\|P\|_4^4] = 2N^2 - N$.
(c) Note that if the conjecture (6.33) holds, then $c \leq 2$.

7. Suppose that $P_1(x)$ and $Q_1(x)$ is a Golay complementary pair, each of length N. Put $P_2(x) = P_1(2x) + e(x)Q_2(2x)$ and $Q_2(x) = P_1(2x) - e(x)Q_2(2x)$.
(a) Show that P_2 and Q_2 are Littlewood polynomials of length $2N$, and describe how their coefficients are determined in terms of those of P_1 and Q_1.
(b) Show that P_2 and Q_2 are a Golay complementary pair.

8. Put

$$P(x) = 1 + e(x) - e(2x) + e(3x) - e(4x) + e(5x) - e(6x) - e(7x) + e(8x) + e(9x),$$
$$Q(x) = 1 + e(x) - e(2x) + e(3x) + e(4x) + e(5x) + e(6x) + e(7x) - e(8x) - e(9x).$$

Compute the aperiodic correlation coefficients of these polynomials, and thus demonstrate that they form a Golay complementary pair of order 10.

9. Let

$$P(x) = \sum_{n=0}^{N-1} \delta_n e(nx), \qquad Q(x) = \sum_{n=0}^{N-1} \varepsilon_n e(nx)$$

be Littlewood polynomials with autocorrelations $a_1, \ldots, a_{N-1}$ and $b_1, \ldots, b_{N-1}$, respectively. Suppose that P and Q form a Golay pair, so that $a_n + b_n = 0$ for all n. The object of this exercise is to show that if $N > 1$, then N is even. We use only the identities

$$a_1 + b_1 = 0,$$
$$a_{N-1} + b_{N-1} = 0.$$

We note that

$$a_1 = \sum_{n=0}^{N-2} \delta_n \delta_{n+1}, \qquad b_1 = \sum_{n=0}^{N-2} \varepsilon_n \varepsilon_{n+1} .$$

Let s denote the number of $\delta_n\delta_{n+1}$ that are negative, and t the number of $\varepsilon_n\varepsilon_{n+1}$ that are negative.

(a) Show that $a_1 = N - 1 - 2s$ and that $b_1 = N - 1 - 2t$.

(b) Deduce that $N = s + t + 1$.

(c) Show that

$$\prod_{n=0}^{N-2}(\delta_n\delta_{n+1}) = (-1)^s, \qquad \prod_{n=0}^{N-2}(\varepsilon_n\varepsilon_{n+1}) = (-1)^t\,.$$

(d) Show that

$$\prod_{n=0}^{N-2}(\delta_n\delta_{n+1}) = \delta_0\delta_{N-1}, \qquad \prod_{n=0}^{N-2}(\varepsilon_n\varepsilon_{n+1}) = \varepsilon_0\varepsilon_{N-1}\,.$$

(e) Note that $a_{N-1} = \delta_0\delta_{N-1}$, and that $b_{N-1} = \varepsilon_0\varepsilon_{N-1}$.

(f) Deduce that s and t have opposite parity.

(g) Deduce that N is even.

10. Let $P_k(x)$ denote the kth Shapiro polynomial, and let S denote that part of $\mathbb{T}$ on which $|P_k(x)|$ is at least $1/2$ its root mean square size, which is to say that $S = \{x \in \mathbb{T} : |P_k(x)| \ge 2^{k/2-1}\}$. Thus S is the union of finitely many closed intervals. Show that $\operatorname{meas} S \ge 3/7$.

11. Let P_k denote the kth Shapiro polynomial. Show that

$$\frac{1}{\sqrt{2}}\|P_k\|_2 \le \|P_k\|_1 \le \|P_k\|_2\,.$$

12. Let P_k and Q_k denote Shapiro polynomials, and put

$$X_k = \int_0^1 |P_k(x)|^4\,dx, \qquad Y_k = \int_0^1 |P_k(x)Q_k(x)|^2\,dx\,.$$

(a) Write

$$|P_k(x)|^2 = 2^k + 2\sum_{r=1}^{2^k-1} a_r \cos 2\pi r x$$

as in (6.21). Explain why

$$|Q_k(x)|^2 = 2^k - 2\sum_{r=1}^{2^k-1} a_r \cos 2\pi r x$$

for all x.

(b) Deduce that

$$X_k = 2^{2k} + 2\sum_{r=1}^{2^k-1} a_r^2 = \int_0^1 |Q_k(x)|^4\,dx\,.$$

(c) Explain why

$$\int_0^1 |P_{k-1}(x) + e(2^{k-1}x)Q_{k-1}(x)|^4 |P_{k-1}(x) - e(2^{k-1}x)Q_{k-1}(x)|^4\,dx = 2X_k\,.$$

(d) Show that if z and w are complex numbers, then

$$|z+w|^4 + |z-w|^4 = 2|z|^4 + 2|w|^4 + 8|zw|^2 + 4\operatorname{Re} z^2\overline{w}^2\,.$$

(e) When $P_{k-1}(x)^2$ is expanded so that it is a linear combination of exponentials $e(rx)$, the frequencies r lie in an interval $I_k = [\alpha_k, \beta_k]$. Determine the values of α_k and β_k.

(f) When $\big(e(-2^{k-1}x)Q_{k-1}(-x)\big)^2$ is expanded so that it is a linear combination of exponentials $e(rx)$, the frequencies r lie in an interval $J_k = [\gamma_k, \delta_k]$. Determine γ_k and δ_k.

(g) Note that $\beta_k + \delta_k = -2$. Deduce that

$$\int_0^1 P_{k-1}(x)^2\big(e(-2^{k-1}x)Q_{k-1}(x)\big)^2\,dx = 0\,.$$

(h) Deduce that $2X_k = 4X_{k-1} + 8Y_{k-1}$.

(i) Explain why $\big(|P_k(x)|^2 + |Q_k(x)|^2\big)^2 = 2^{2k+2}$ for all x. Integrate this, and deduce that $2X_k + 2Y_k = 2^{2k+2}$.

(j) Deduce that $X_k + 2X_{k-1} = 2^{2k+1}$.

(k) Put $x_k = X_k - \frac{4}{3}2^{2k}$. Show that $x_k + 2x_{k-1} = 0$.

(l) Note that $X_0 = 1$; deduce that $x_0 = -1/3$.

(m) Show that $x_k = (-1)^{k+1}2^k/3$.

(n) Conclude that

$$\int_0^1 |P_k(x)|^4\,dx = \frac{4}{3}2^{2k} - (-1)^k\frac{2^k}{3}\,.$$

Thus

$$\int_0^1 |P_k(x)|^4\,dx \sim \frac{4}{3}\Big(\int_0^1 |P_k(x)|^2\,dx\Big)^2\,.$$

(o) Note that if the conjecture (6.33) holds, then $c \le 4/3$.

(p) Show that

$$\int_0^1 |P_k(x)Q_k(x)|^2\,dx = \frac{2}{3}2^{2k} + (-1)^k\frac{2^k}{3}\,.$$

(q) The left hand side above is an integer, because (by Parseval) it is the sum of the squares of the coefficients of $P_k(x)Q_k(x)$, which are integers. Give a direct proof that the right hand side above is an integer. That is, show that 3 divides $2\cdot 4^k + (-2)^k$.

13. In (3.15) we noted that $\max_n |\widehat{f}(n)| \le \|f\|_1$ for all $f \in L^1(\mathbb{T})$. While this holds with equality if $f(x) = e(nx)$ for some n, in general the ratio between the two sides of the inequality may be arbitrarily large. We now show that for trigonometric polynomials this ratio is bounded in terms of the degree.

(a) Suppose that $T \in \mathcal{T}_N$. Show that

$$\max_n |\widehat{T}(n)| \;\ge\; \frac{1}{\sqrt{2N+1}}\int_0^1 |T(x)|\,dx.$$

(b) Suppose that N is given, and choose k so that $2^{k-1} \le N < 2^k$. Put

$$T(x) = \sum_{n=0}^{N}(1 - n/N)\widehat{P}_k(n)e(nx)\,. \tag{6.34}$$

Explain why $\|T\|_\infty \le \|P_k\|_\infty \le 2^{(k+1)/2} \le 2\sqrt{N}$.

(c) Show that

$$\|T\|_2^2 = \sum_{n=0}^{N}(1 - n/N)^2 > N/3\,.$$

(Hint: Use an appropriate formula on page 225.)

(d) Show that $\|T\|_2^2 \le \|T\|_1 \|T\|_\infty$.

(e) Conclude that

$$\max_n |\widehat{T}(n)| \le \frac{6}{\sqrt{N}} \int_0^1 |T(x)|\,dx\,.$$

(Thus the inequality in part (a) is sharp, apart from the constant.)

14. Let $T \in \mathcal{T}_N$.

(a) Show that

$$\max_x |T(x)| \le \sum_{n=-N}^{N} |\widehat{T}(n)|$$

for all x. (This is an analogue of (3.15), but with the roles reversed.)

(b) Show that

$$\max_x |T(x)| \ge \frac{1}{\sqrt{2N+1}} \sum_{n=-N}^{N} |\widehat{T}(n)|$$

(c) Let $T(x)$ be defined as in (6.34). Show that

$$\sum_{n=-N}^{N} |\widehat{T}(n)| > N/2\,.$$

(d) Conclude that for this T,

$$\max_x |T(x)| \le \frac{4}{\sqrt{N}} \sum_{n=-N}^{N} |\widehat{T}(n)|\,.$$

(Thus the inequality in part (b) is sharp, apart from the constant.)

15. Put

$$f(x) = \sum_{k=1}^{\infty} \frac{e(2^k x) P_k(x)}{2^{k/2} k^2}\,.$$

(a) Let

$$f_K(x) = \sum_{k=1}^{K} \frac{e(2^k x) P_k(x)}{2^{k/2} k^2}\,.$$

Explain why f_K is continuous.

(b) Show that $f(x) = f_K(x) + O(1/K)$.

(c) Deduce that $f(x)$ is continuous.

(d) Show that $|\widehat{f}(n) - \widehat{f_K}(n)| = O(1/K)$.

(e) Note that if $2^k \le n < 2^{k+1}$, then

$$\widehat{f}(n) = \frac{\pm 1}{2^{k/2} k^2}\,.$$

(f) Deduce that

$$\sum_{2^k \le n < 2^{k+1}} |\widehat{f}(n)|^{2-\varepsilon} = \frac{2^{\varepsilon k/2}}{k^{4-2\varepsilon}}.$$

(g) Conclude that

$$\sum_{n=1}^{\infty} |\widehat{f}(n)|^{2-\varepsilon} = \infty.$$

(By Bessel's inequality we know that $\|f\|_2 < \infty$ implies that $\|\widehat{f}\|_2 < \infty$. However, by this example, we see that even the much stronger hypothesis that $f \in C(\mathbb{T})$ does not imply that $\|\widehat{f}\|_p < \infty$ if $p < 2$. Thus in a sense Parseval's Identity cannot be improved.)

16. Let $T(x) = e(x)P_k(x)$.
   (a) Use Bernstein's inequality to show that $\max_x |T'(x)| \le (2\pi\sqrt{2})2^{3k/2}$.
   (b) Show that

$$\int_0^1 |T'(x)|^2 \, dx > (2\pi)^2 2^{3k}/3.$$

   (c) Deduce that $\max_x |T'(x)| \ge (2\pi/\sqrt{3})2^{3k/2}$, which means that the bound provided by Bernstein's inequality is within a factor $\sqrt{6}$ of the truth in this case.

17. Suppose that a nonnegative integer n is written in binary,

$$n = \sum_{r=0}^{R} b_r 2^r$$

where each b_r is 0 or 1, and set

$$c(n) = \sum_{r=0}^{R-1} b_r b_{r+1}.$$

Thus $c(n)$ is the number of pairs of consecutive 1's in the binary expansion of n.
   (a) Show that $c(2n) = c(n)$.
   (b) Show that if n is even, then $c(2n+1) = c(n)$.
   (c) Show that if n is odd, then $c(2n+1) = c(n) + 1$.
   (d) Show that if $0 \le n < 2^k$, then $c(2^{k+1} + n) = c(n)$.
   (e) Show that if $2^k \le n < 2^{k+1}$, then $c(2^{k+1} + n) = c(n) + 1$.
   (f) For $n \ge 0$ put

$$a(n) = (-1)^{c(n)}. \tag{6.35}$$

   Show that $a(0) = 0$, $a(2n) = a(n)$, and that $a(2n+1) = (-1)^n a(n)$.

18. For nonnegative integers k put

$$\begin{aligned} R_k(x) &= \sum_{0 \le n < 2^k} a(n)e(nx), \\ S_k(x) &= \sum_{0 \le n < 2^k} a(2^k + n)e(nx). \end{aligned} \tag{6.36}$$

   (a) Show that $R_k(x) + e(2^k x)S_k(x) = R_{k+1}(x)$.

(b) Write

$$S_{k+1}(x) = \sum_{0 \le n < 2^k} a(2^{k+1}+n)e(nx) + \sum_{2^k \le n < 2^{k+1}} a(2^{k+1}+n)e(nx)\,.$$

Use properties of $a(n)$ to deduce that the above is

$$= R_k(x) - e(2^k x)S_k(x)\,.$$

(c) Conclude that since R_k and S_k satisfy the same initial conditions and same recurrences as P_k and Q_k, they must be the same: $R_k(x) = P_k(x)$ and $S_k(x) = Q_k(x)$. In particular,

$$P_k(x) = \sum_{0 \le n < 2^k} a(n)e(nx),$$

and we call the $a(n)$ the *Shapiro coefficients.*

19. (a) By considering separately the contributions of even and odd n in the above sum, and using properties of the $a(n)$, show that

$$P_k(x) = P_{k-1}(2x) + e(x)P_{k-1}(2x+1/2)\,.$$

(b) By replacing x by $x+1/2$ in the above, deduce that

$$P_k(x+1/2) = P_{k-1}(2x) - e(x)P_{k-1}(2x+1/2)\,.$$

(c) By inducting on k, deduce that $P_k(x)$ and $P_k(x+1/2)$ form a Golay complementary pair:

$$|P_k(x)|^2 + |P_k(x+1/2)|^2 = 2^{k+1}$$

for all x.

20. (a) Suppose that $P_{2k}(0) = P_{2k}(1/2) = 2^k$. Show that $P_{2k+1}(0) = 2^{k+1}$ and $P_{2k+1}(1/2) = 0$.
(b) Show that if $P_{2k+1}(0) = 2^{k+1}$ and $P_{2k+1}(1/2) = 0$, then $P_{2k+2}(0) = P_{2k+2}(1/2) = 2^{k+1}$.
(c) Note that the first identity in (a) holds when $k = 0$. Deduce by induction that

$$P_{2k}(0) = P_{2k}(1/2) = 2^k,$$
$$P_{2k+1}(0) = 2^{k+1},$$
$$P_{2k+1}(1/2) = 0$$

for all $k \ge 0$.

21. (a) Show that $P_{2k}(1/4) = i2^k$.
(b) Show that $P_{2k+1}(1/4) = (1+i)2^k$.

22. For $|z| < 1$, put

$$f(z) = \sum_{n=0}^{\infty} a(n)z^n.$$

(a) Show that $f(z) = f(z^2) + zf(-z^2)$.
(b) Show that $f(-z) = f(z^2) - zf(-z^2)$.

23. Put

$$P(N;x) = \sum_{n=0}^{N-1} a(n)e(nx)$$

where the $a(n)$ are the Shapiro coefficients. The object of this exercise is to show that

$$|P(N;x)| \le (2+\sqrt{2})\sqrt{N} \tag{6.37}$$

for all x. To this end we prove a little more: A trigonometric polynomial formed by summing the first N terms of either $P_k(x)$ or $Q_k(x)$ has the property that its absolute value is uniformly $\le (2+\sqrt{2})\sqrt{N}$. Call this Assertion (k). We argue by induction on k.

(a) Verify that the assertion is true for $k = 0$. This is the basis of the induction.
(b) From now on, assume that the assertion has been proved for k.
(c) Suppose we have a sum of the first N terms of $P_k(x)$ or of $Q_k(x)$. Note that if $N \le 2^k$ then the sum is a partial sum of $P_k(x)$, and so the bound has already been established.
(d) Assume that $2^k < N \le 2^{k+1}$ and that we are summing the first N terms of $P_{k+1}(x)$. Thus our sum consists of $P_k(x)$ plus $e(2^k x)$ times the first $N - 2^k$ terms of $Q_k(x)$. Use the inductive hypothesis and the triangle inequality to show that this sum of N terms has absolute value

$$\begin{aligned} &\le 2^{(k+1)/2} + (2+\sqrt{2})(N-2^k)^{1/2} \\ &= \left(\sqrt{2}(2^k/N)^{1/2} + (2+\sqrt{2})(1-2^k/N)^{1/2}\right)\sqrt{N}\,. \end{aligned} \tag{6.38}$$

(e) Show that $1/2 \le 2^k/N \le 1$.
(f) Let $f(x) = \sqrt{2x} + C\sqrt{1-x}$ where C is a positive constant. Show that if $C \ge \sqrt{2}$, then f is decreasing for $1/2 \le x \le 1$. Note that $f(1/2) = 1 + C/\sqrt{2}$.
(g) Take $C = 2+\sqrt{2}$ to see that the coefficient of $\sqrt{N}$ in (6.38) is $\le 2+\sqrt{2}$.
(h) Show that steps (d) through (g) work equally well if we are summing the first N terms of $Q_{k+1}(x)$. Deduce that the induction is complete, so that the assertion has been proved for all k.

6.5. Quantitative approximation of continuous functions

From Theorem 4.1 we know that if $f \in C(\mathbb{T})$, then $|f(x) - \sigma_N(x)| < \varepsilon$ uniformly in x, if $N > N_0(\varepsilon)$. Our object now is to describe how quickly ε tends to 0, as a function of N. This of course depends on how smooth f is. For $f \in C(\mathbb{T})$, let

$$\omega(\delta) = \max_{\substack{x,y \\ |x-y|\le\delta}} |f(x) - f(y)|\,.$$

This quantity is known as the *modulus of continuity* of f. Since f is uniformly continuous, it follows that $\omega(\delta) \to 0$ as $\delta \to 0$. The rate at which it tends to 0 is a quantitative measure of how continuous f is. If $0 < \delta_1 \le \delta_2$, then the pairs

x, y for which $|x - y| \le \delta_1$ is a subset of the pairs x, y for which $|x - y| \le \delta_2$, so $\omega(\delta_1) \le \omega(\delta_2)$. That is, ω is an increasing function:

$$\omega(\delta) \nearrow . \tag{6.39}$$

We continue by establishing a further property concerning the modulus of continuity.

Lemma 6.10. *Suppose that $f \in C(\mathbb{T})$, and let $\omega(\delta)$ denote its modulus of continuity. Then $\omega(n\delta) \le n\omega(\delta)$.*

Proof. Suppose that $x \le y \le x + n\delta$. For $k = 0, 1, \ldots, n$ put $x_k = x + k(y - x)/n$. Thus $x_0 = x$, and $x_n = y$. Clearly

$$f(y) - f(x) = \sum_{k=1}^{n} f(x_k) - f(x_{k-1}),$$

so by the triangle inequality,

$$|f(x) - f(y)| \le \sum_{k=1}^{n} |f(x_k) - f(x_{k-1})| .$$

But $x_k - x_{k-1} = (y - x)/n \le \delta$, so $|f(x_k) - f(x_{k-1})| \le \omega(\delta)$ by (6.39). Thus the above sum is $\le n\omega(\delta)$, and hence $\omega(n\delta) \le n\omega(\delta)$. □

We recall that $\sigma_N(x) = (f * \Delta_N)(x)$. While the Fejér kernel is great for many things, in the present situation we need a summability kernel that decays even more rapidly than the Fejér kernel.

Lemma 6.11. *Let $c_N = 3/(2N^3 + N)$, and put*

$$J_N(x) = c_N \Big(\frac{\sin \pi N x}{\sin \pi x}\Big)^4 .$$

Then $J_N(x)$ is a summability kernel.

The kernel $J_N(x)$ is known as the *Jackson kernel.* Since $J_N(x) = c_N N^2 \Delta_N(x)^2$, we know that J_N is a trigonometric polynomial of degree $2N - 2$.

Proof. Of the necessary properties (4.76), (4.77), (4.78), all are evident except for the first one, so our task is to show that $\int_0^1 J_N(x)\,dx = 1$. Clearly

$$\int_0^1 J_N(x)\,dx = c_N N^2 \int_0^1 \Delta_N(x)^2\,dx .$$

By Parseval's Identity (which is trivial for trigonometric polynomials), this is

$$= c_N N^2 \sum_{|n|<N} \Big(1 - \frac{|n|}{N}\Big)^2 = c_N N^2 \Big(1 + 2 \sum_{n=1}^{N-1} (1 - n/N)^2\Big)$$

$$= c_N \Big(N^2 + 2 \sum_{n=1}^{N-1} (N - n)^2\Big) .$$

Put $m = N - n$. Then the above is

$$= c_N\Big(N^2 + 2\sum_{m=1}^{N-1} m^2\Big)\,. \tag{6.40}$$

Now by induction it is easy to show that

$$\sum_{m=1}^{N-1} m^2 = \frac{1}{3}N^3 - \frac{1}{2}N^2 + \frac{1}{6}N\,. \tag{6.41}$$

This is also a special case of Corollary 9.10, which we shall encounter later. (See especially (9.27).) On inserting this, we conclude that the expression (6.40) is

$$= c_N\Big(\frac{2}{3}N^3 + \frac{1}{3}N\Big) = 1\,.$$

□

Theorem 6.12. (Jackson) *Suppose that $f \in C(\mathbb{T})$ and that $\omega(\delta)$ is the modulus of continuity of f. Then there is a trigonometric polynomial $T \in \mathcal{T}_N$ such that*

$$|f(x) - T(x)| \le 3\omega(1/N)$$

for all x.

Proof. Let M be the largest integer such that $2M - 2 \le N$. That is,

$$2M - 2 \le N < 2M\,. \tag{6.42}$$

We take $T(x) = (f * J_M)(x)$. Thus

$$T(x) = \sum_{m=-2M+2}^{2M-2} \widehat{J}_M(m)\widehat{f}(m)e(mx)$$

is a trigonometric polynomial of degree at most N. Also,

$$f(x) - T(x) = \int_{-1/2}^{1/2} (f(x) - f(x-u))J_M(u)\,du$$

and $|f(x) - f(x-u)| \le \omega(|u|)$, so by the triangle inequality

$$|f(x) - T(x)| \le 2\int_0^{1/2} \omega(u)J_M(u)\,du \le 2\sum_{n=1}^{(N+1)/2} \int_{(n-1)/N}^{n/N} \omega(u)J_M(u)\,du\,. \tag{6.43}$$

(If N is odd, then the integral in the last term should run not from $1/2 - 1/(2N)$ to $1/2 + 1/(2N)$ but rather from $1/2 - 1/(2N)$ to $1/2$.) For $n = 1$ we observe that $\omega(u) \le \omega(1/N)$ and that $\int_0^{1/N} J_M(u)\,du \le \int_0^{1/2} J_M(u)\,du = 1/2$. For $n > 1$ we see by (6.39) and Lemma 6.10 that $\omega(u) \le n\omega(1/N)$. Since $\sin \pi u \ge 2u$ for $0 \le u \le 1/2$, it follows that

$$\begin{aligned}\int_{(n-1)/N}^{n/N} \omega(u)J_M(u)\,du &\le \frac{1}{16}n\omega(1/N)c_M \int_{(n-1)/N}^{n/N} u^{-4}\,du \\ &= \frac{1}{48}\omega(1/N)c_M N^3\Big(\frac{n}{(n-1)^3} - \frac{1}{n^2}\Big)\,.\end{aligned}$$

Since $M \ge N/2$, it follows that $2M^3 + M \ge N^3/4$, so $c_M N^3 \le 12$. On inserting these estimates in (21) we deduce that

$$|f(x) - T(x)| \le \omega(1/N)\Big(1 + \frac{1}{2}\sum_{n=2}^{\infty}\Big(\frac{n}{(n-1)^3} - \frac{1}{n^2}\Big)\Big).$$

By writing $n = 1 + (n-1)$ in the numerator we see that the sum over n is

$$= \sum_{n=2}^{\infty} \frac{1}{(n-1)^3} + \sum_{n=2}^{\infty}\Big(\frac{1}{(n-1)^2} - \frac{1}{n^2}\Big).$$

Here the second sum is telescoping and has the value 1. By the integral test the first sum is

$$\le 1 + \int_1^\infty u^{-3}\, du = 3/2,$$

so

$$|f(x) - T(x)| \le \frac{9}{4}\omega(1/N),$$

and the proof is complete. □

Let $f(x) = \|x\|$. Thus f has period 1 and $f(x) = |x|$ for $-1/2 \le x \le 1/2$. Thus $\omega(\delta) = \delta$ for $0 \le \delta \le 1/2$, and so by Theorem 6.12 there is a trigonometric polynomial $T \in \mathcal{T}_N$ such that $\big|\|x\| - T(x)\big| \le 3/N$ for all x. We now show that this is the best one can do, apart possibly from improving the constant.

Theorem 6.13. *Let $f(x) = \|x\|$. There is an absolute constant $c > 0$ such that if $T \in \mathcal{T}_N$, then*

$$\max_x |f(x) - T(x)| \ge \frac{c}{N}.$$

Proof. Let $m = \max_x |f(x) - T(x)|$. Then

$$\int_0^1 |f(x) - T(x)|^2 \Delta_N(x)\, dx \le m^2 \int_0^1 \Delta_N(x)\, dx = m^2. \tag{6.44}$$

We recall that

$$\Delta_N(x) = \frac{1}{N}\Big|\sum_{n=1}^{N} e(nx)\Big|^2.$$

Let

$$g(x) = (f(x) - T(x))\sum_{n=1}^{N} e(nx).$$

Then the left hand side of (6.44) is

$$= \frac{1}{N}\int_0^1 |g(x)|^2\, dx = \frac{1}{N}\sum_{k=-\infty}^{\infty} |\widehat{g}(k)|^2 \tag{6.45}$$

by Parseval's Identity. But

$$\widehat{g}(k) = \sum_{n=1}^{N}\int_0^1 (f(x) - T(x))e(nx)e(-kx)\, dx = \sum_{n=1}^{N}(\widehat{f}(k-n) - \widehat{T}(k-n)).$$

In Example 3.5 we found that $\widehat{f}(0) = 1/4$, that $\widehat{f}(k) = 0$ for even non-zero k, and that $\widehat{f}(k) = -1/(\pi k)^2$ for k odd. If $2N < k \le 3N$, then $k - n > N$ for all $n \in [1, N]$, and hence $\widehat{T}(k-n) = 0$ for all n in the above sum. On the other hand, there are $\sim N/2$ values of n for which $k-n$ is odd, and for these $\widehat{f}(k-n) \asymp -1/N^2$. Thus $|\widehat{g}(k)| \geq c/N$ when $2N < k \leq 3N$, so the right hand side of (6.45) is $\geq c^2/N^2$. On combining this with (6.44) we obtain the stated result. □

Exercises

1. Let $P(M)$ be the proposition that the identity (6.41) holds for the integer M.
   (a) Verify that $P(1)$ is true.
   (b) Verify that $\frac{1}{3}M^3 + \frac{1}{2}M^2 + \frac{1}{6}M = \frac{1}{3}(M+1)^3 - \frac{1}{2}(M+1)^2 + \frac{1}{6}(M+1)$.
   (c) Deduce that $P(M+1)$ is true if $P(M)$ is true.
   (d) By the principle of mathematical deduction, deduce that $P(M)$ is true for all positive integers M.
2. Let $J_N(x)$ be defined as in Lemma 6.11.
   (a) Explain why
   $$\widehat{J_N}(k) = c_N \sum_{\substack{-N\leq m\leq N \\ -N\leq m-k\leq N}} (N-|m|)(N-|m-k|)\,.$$
   (b) Show that if $N \leq k \leq 2N$, then
   $$\widehat{J_N}(k) = c_N \sum_{m=-N+k}^{N} (N-m)(N+m-k)\,.$$
   (c) Deduce that if $N \leq k \leq 2N$, then
   $$\widehat{J_N}(k) = c_N(2N-k+1)(2N-k)(2N-k-1)/6\,.$$
   (d) Show that if $0 \leq k \leq N$, then
   $$\widehat{J_N}(k) = c_N \sum_{m=-N+k}^{0} (N+m)(N+m-k) + c_N \sum_{m=1}^{k} (N-m)(N+m-k)$$
   $$+ c_N \sum_{m=k+1}^{N} (N-m)(N-m+k)\,.$$
   (e) Deduce that if $0 \leq k \leq N$, then
   $$\widehat{J_N}(k) = c_N\Big(\frac{1}{2}k^3 - Nk^2 + \frac{2}{3}N^3 - \frac{1}{2}k + \frac{1}{3}N\Big)\,.$$
   (f) Deduce that $\widehat{J_N}(0) = 1$.
   (g) The value of $\widehat{J_N}(N)$ can be deduced from (c) and also from (e). Do the two formulæ agree?
   (h) Explain why $\widehat{J_N}(-k) = \widehat{J_N}(k)$ for all k.
3. Let K_N be a summability kernel, and let C be a constant such that
   $$\int_0^1 |K_N(x)|\,dx \leq C$$

for all N. Suppose that $f \in C(\mathbb{T})$ with modulus of continuity $\omega(\delta)$. Show that $f * K_N$ has modulus of continuity not exceeding $C\omega(\delta)$. (Note that if K_N is non-negative, then we may take $C = 1$.)

4. Suppose that $f = \|x\|$.
   (a) Show that $\sigma_N(x) - f(x)$ is an even function.
   (b) Show that $\sigma_N(x) - f(x) = -(f(1/2 - x) - \sigma_N(1/2 - x))$.
   (c) Show that $\sigma_N(-1/4) = f(-1/4) = 1/4$.
   (d) By Exercise 3, or otherwise, show that $|\sigma_N'(x)| \le 1$ for all x.
   (e) Show that $\sigma_N(x) - f(x)$ is an increasing function on the interval $[-1/4, 0]$.
   (f) Conclude that

$$\max_x |f(x) - \sigma_N(x)| = \sigma_N(0) = 2\int_0^{1/2} x\Delta_N(x)\,dx\,.$$

   (With a little more work it can be shown that the above is $= c(\log N)/N + O(1/N)$ with $c = 1/\pi^2$.)

5. (a) The kernel $J_N(x)$ shares certain properties with $\Delta_N(x)$ that allows the arguments in Exercise 4 to be repeated with $\sigma_N(x)$ replaced by $(f * J_N)(x)$. What properties are those?
   (b) Conclude that if $f(x) = \|x\|$, then

$$\max_x |f(x) - (f * J_N)(x)| = (f * J_N)(0) = 2\int_0^{1/2} xJ_N(x)\,dx\,.$$

   (With a little work it can be shown that this is $O(1/N)$.)

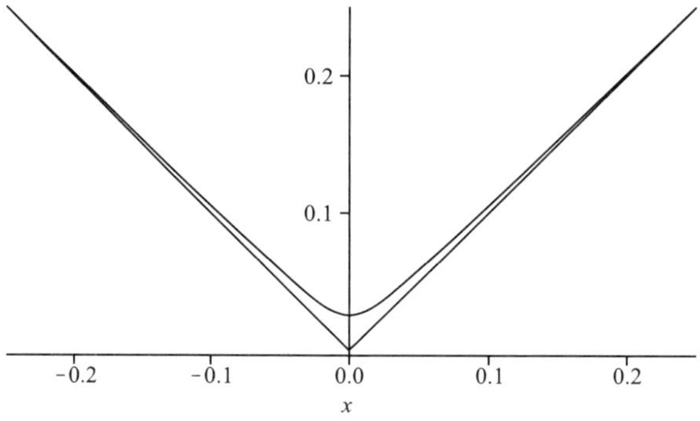

Figure 6.4. Graph of $f(x)$ vs. $(f * \Delta_{21})(x)$ for $-1/4 \le x \le 1/4$.

Note that $\deg \Delta_{11} = \deg J_{11} = 20$, so that $f * \Delta_{21}$ and $f * J_{11}$ have the same degree.

6. Let b be the unique real number, $1/6 < b < 1/4$, such that

$$b + b\sqrt{1 - \frac{1}{(2\pi b)^2}} + \frac{1}{2\pi}\arcsin\frac{1}{2\pi b} = 1/2\,.$$

Thus $b = 0.2196418$. Let $T_1(x) = 1/4 - b\cos 2\pi x$, and take $x_0 = \frac{1}{2\pi}\arcsin\frac{1}{2\pi b}$. Let f have period 1 and let $f(x) = T_1(x) - |x|$ for $-1/2 \le x \le 1/2$. Put $M = f(0) = 1/4 - b = 0.030358$.

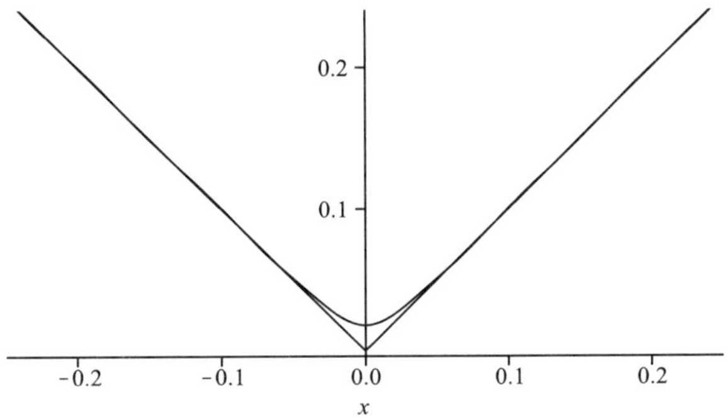

Figure 6.5. Graph of $f(x)$ vs. $(f * J_{11})(x)$ for $-1/4 \le x \le 1/4$.

(a) Show that $f(-1/2) = -M$, that $f(-1/2 + x_0) = M$, that $f(-x_0) = -M$, that $f(0) = M$, that $f(x_0) = -M$, that $f(1/2 - x_0) = M$, and that $f(1/2) = -M$.
(b) Let $T(x)$ be a trigonometric polynomial of degree at most 2, and suppose that $\big|T(x) - |x|\big| \le M$ for $-1/2 \le x \le 1/2$. Put $T_2(x) = T(x) - T_1(x)$. Show that $T_2(-1/2) \le 0$, that $T_2(-1/2 + x_0) \ge 0$, that $T_2(-x_0) \le 0$, that $T_2(0) \ge 0$, that $T_2(x_0) \le 0$, that $T_2(1/2 - x_0) \ge 0$, and that $T_2(1/2) \le 0$.
(c) Show that $T_2(x) = 0$ has at least one solution in each of the six intervals $(-1/2, -1/2 + x_0]$, $-1/2 + x_0, -x_0]$, $(-x_0, 0]$, $(0, x_0]$, $(x_0, 1/2 - x_0]$, and $(1/2 - x_0, 1/2]$.
(d) Deduce that $T_2(x)$ is identically 0.
(e) Conclude that

$$\min_{T \in \mathcal{T}_2} \max_{-1/2 \le x \le 1/2} \big|T(x) - |x|\big| = M = 0.030358,$$

and that the only T that achieves this minimum is T_1.

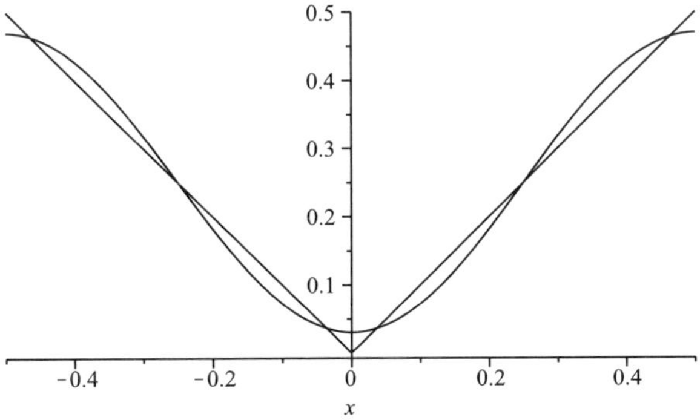

Figure 6.6. Graph of $f(x)$ vs. $T_1(x)$ for $-1/4 \le x \le 1/4$.

Notes

6.1. The identity (6.2) goes back to Cauchy (1841), who derived it by Lagrange interpolation. It is the analogue for $L^1(\mathbb{T})$ of the cardinal series for $L^1(\mathbb{R})$; see (10.35). For more on one-sided trigonometric approximation, see Graham and Vaaler (1981).

6.2. Bernstein's inequality originates in Bernstein (1912). The generalization in Exercise 5 (f) is due to Zygmund (1932). R. P. Boas (1969) gave a very simple and elementary proof of Bernstein's theorem totally different from ours, but it lacks the identity (6.13), which we find to be quite useful. For further reading in this area, see the excellent expository article of Schaeffer (1941).

6.3. The Fejér–Riesz theorem was conjectured by Fejér (1916), and immediately proved by F. Riesz (1916). It is very useful in the study of extremal problems for nonnegative trigonometric polynomials, but it is perhaps the only topic we cover that does not extend gracefully to higher dimensions.

6.4. Hardy and Littlewood (1952) mentioned such trigonometric polynomials as

$$\sum_{n=1}^{N} e\Big(\frac{n^2}{N}\Big)e(nx) \quad \text{and} \quad \sum_{n=1}^{N} e(n\log n)e(nx) \tag{6.46}$$

that are uniformly $O(\sqrt{N})$, but it came as quite a shock in the world of harmonic analysis (as done by pure mathematicians) when H. S. Shapiro (1951) produced, in his master's thesis at MIT, an example of trigonometric polynomials that achieve this with coefficients ± 1. We now realize that Golay (1949) had achieved this earlier, although these were independent discoveries. The pure and applied literatures have since gradually become aware of each other. What we call a Littlewood polynomial is sometimes called a PM polynomial in the applied literature. While it is still not known that there are no further Barker polynomials, Borwein and Mossinghoff (2009) showed that if there is a further Barker code, then its length is $> 10^{31}$. Turyn (1974) showed that if one has a pair of complementary codes, one of length m and the other of length n, then one can construct a complementary pair of length mn. Borwein and Ferguson (2004) have determined all Golay pairs of length ≤ 100.

Shapiro never published his work on the Shapiro polynomials. Later, Rudin (1959) needed them, and included the construction of the Shapiro polynomials in his paper. As a result, the Shapiro polynomials have often been called the Rudin–Shapiro polynomials, although Rudin's role was that of publicizer, not inventor.

It is still not known whether there exist absolute constants c, C with $0 < c < 1 < C < \infty$ with the property that there exist Littlewood polynomials of arbitrarily great length N such that $c\sqrt{N} \le |P(x)| \le C\sqrt{N}$ for all x. However, Kahane (1980) showed that one can achieve $(1-\varepsilon)\sqrt{N} \le |P(x)| \le (1+\varepsilon)\sqrt{N}$ uniformly in x for $N > N_0(\varepsilon)$ if one allows complex unimodular coefficients. Such polynomials are said to be *ultra-flat*.

6.5. Jackson's inequality (our Theorem 6.12) is just one of several theorems of this sort proved by Jackson. Bernstein established converses: If a function can be very well approximated by a trigonometric sum, then it (and its derivatives to an appropriate order) are smooth. For an account of this area, see Achieser (1956).

Chapter 7

Absolutely Convergent Fourier Series

7.1. Convergence

We let $A(\mathbb{T})$ denote the set of those continuous functions with period 1 whose Fourier Series is absolutely convergent. As with other spaces of functions that we have studied, for $f \in A(\mathbb{T})$ we have a norm, namely

$$\|f\|_A = \sum_{n=-\infty}^{\infty} |\widehat{f}(n)| \,. \tag{7.1}$$

In Corollary 4.5 we showed that if $f \in A(\mathbb{T})$, then

$$f(x) = \sum_{n=-\infty}^{\infty} \widehat{f}(n)e(nx)$$

for all x. By the triangle inequality it follows that

$$\|f\|_\infty = \max_x |f(x)| \le \sum_{n=-\infty}^{\infty} |\widehat{f}(n)| = \|f\|_A \,. \tag{7.2}$$

When the Fourier coefficients of a function are known explicitly, it is usually a simple matter to determine whether the sum of the coefficients is absolutely convergent. However, we can sometimes show that $f \in A(\mathbb{T})$ without determining its coefficients, just by knowing that f is sufficiently smooth. We give two useful examples of this.

Theorem 7.1. *Suppose that $f \in C(\mathbb{T})$, that f is piecewise differentiable, and that $f' \in L^2(\mathbb{T})$. Then $f \in A(\mathbb{T})$.*

Proof. By Corollary 3.8 we know that $\widehat{f'}(n) = 2\pi in\widehat{f}(n)$, so by Parseval's Identity (Theorem 5.7) it follows that

$$\int_0^1 |f'(x)|^2\,dx = \sum_{n=-\infty}^{\infty} |2\pi n\widehat{f}(n)|^2\,.$$

These quantities are finite, since $f' \in L^2(\mathbb{T})$. By Cauchy's inequality,

$$\sum_{n\neq 0} |\widehat{f}(n)| \le \Big(\sum_{n\neq 0} \frac{1}{n^2}\Big)^{1/2}\Big(\sum_{n\neq 0} |n\widehat{f}(n)|^2\Big)^{1/2}.$$

Since the right hand side above is finite, it follows that $f \in A(\mathbb{T})$. □

For a function f with period 1, we say that f is *Lipschitz*, and write $f \in \mathrm{Lip}(\mathbb{T})$, if there is a constant C such that $|f(x) - f(y)| \le C|x - y|$ for all x and y. For example, if f is differentiable and $|f'(x)| \le C$ for all x, then

$$|f(x) - f(y)| = \Big|\int_y^x f'(u)\,du\Big| \le \int_y^x |f'(u)|\,du \le C|x - y|,$$

so such a function is Lipschitz. But a function can be Lipschitz without being everywhere differentiable; an example of this is found in the function $\|x\|$ considered in Theorem 6.13. More generally, if $0 < \alpha \le 1$, then we say that $f \in \mathrm{Lip}_\alpha(\mathbb{T})$ if f has period 1 and there is a constant C such that $|f(x) - f(y)| \le C|x - y|^\alpha$ for all x and y. Here x and y are considered modulo 1, so the distance from x to y is never more than 1/2. If $0 < \alpha < \beta \le 1$, then $|x - y|^\beta \le |x - y|^\alpha$, so $\mathrm{Lip}_\beta(\mathbb{T}) \subseteq \mathrm{Lip}_\alpha(\mathbb{T})$. Thus the classes Lip_α become progressively larger as α becomes smaller, since progressively more jagged functions are allowed. Note that all these classes are subclasses of the set of continuous functions.

In terms of the modulus of continuity introduced in §6.5, to say that f is Lipschitz is to say that there is a constant C such that $\omega(\delta) \le C\delta$. That is, $\omega(\delta) = O(\delta)$. Similarly, to say that $f \in \mathrm{Lip}_\alpha(\mathbb{T})$ is to say that $\omega(\delta) \le C\delta^\alpha$, i.e. $\omega(\delta) = O(\delta^\alpha)$.

Theorem 7.2. (Bernstein) *If* $1/2 < \alpha \le 1$ *and* $f \in \mathrm{Lip}_\alpha(\mathbb{T})$, *then* $f \in A(\mathbb{T})$.

Proof. For a given δ, let $g(x) = f(x + \delta) - f(x)$. Then

$$\widehat{g}(n) = \int_0^1 (f(x + \delta) - f(x))e(-nx)\,dx = (e(n\delta) - 1)\widehat{f}(n)\,.$$

By Parseval's Identity it follows that

$$\sum_{n=-\infty}^{\infty} |(e(n\delta) - 1)\widehat{f}(n)|^2 = \int_0^1 |f(x + \delta) - f(x)|^2\,dx\,.$$

Since $f \in \mathrm{Lip}_\alpha(\mathbb{T})$, there is a constant C such that $|f(x + \delta) - f(x)| \le C\delta^\alpha$. Thus the right hand side above is $\le C^2\delta^{2\alpha}$. On the other hand,

$$|e(x) - 1| = |e(x/2)(e(x/2) - e(-x/2))| = |e(x/2) - e(-x/2)| = |2\sin\pi x|\,.$$

Hence with $x = n\delta$ we find that

$$\sum_{n=-\infty}^{\infty} (2\sin\pi n\delta)^2|\widehat{f}(n)|^2 \le C^2\delta^{2\alpha}\,.$$

Take $\delta = 2^{-r-1}$. If $2^{r-1} < n \le 2^r$, then $\sin \pi/4 \le \sin \pi n\delta \le \sin \pi/2$, which is to say that $1/2 \le (\sin \pi n \delta)^2 \le 1$. Thus

$$\sum_{2^{r-1}<|n|\le 2^r} |\widehat{f}(n)|^2 = O\big(2^{-2r\alpha}\big).$$

By Cauchy's inequality,

$$\begin{aligned}\sum_{2^{r-1}<|n|\le 2^r} |\widehat{f}(n)| &\le \Big(\sum_{2^{r-1}<|n|\le 2^r} 1\Big)^{1/2}\Big(\sum_{2^{r-1}<|n|\le r} |\widehat{f}(n)|^2\Big)^{1/2}\\ &= O\big(2^{r/2}\cdot 2^{-r\alpha}\big) = O\big(2^{(1/2-\alpha)r}\big).\end{aligned}$$

Here our bound is a geometric progression indexed by r, and it tends to 0 since $\alpha > 1/2$. Thus the sum over r of the right hand side is convergent, and hence

$$\sum_{n=-\infty}^{\infty} |\widehat{f}(n)| < \infty,$$

as desired. □

When α is given, $1/2 < \alpha \le 1$, and we have a value of C for which $|f(x+\delta) - f(x)| \le C\delta^\alpha$. Then by the above method we can derive an explicit bound for

$$\sum_{n\ne 0} |\widehat{f}(n)|$$

in terms of C and α. However, there is no way to bound $|\widehat{f}(0)|$, since a change in $\widehat{f}(0)$ changes $f(x+\delta)$ by the same amount as it changes $f(x)$, and so the value of $f(x+\delta) - f(x)$ is independent of $\widehat{f}(0)$.

Example 7.1. The *Cantor function* $c(x)$ maps the interval $[0, 1]$ to itself, as depicted in Figure 7.1. It can be defined by an iterative procedure in which for an interval on which it has not yet been defined, it is set to be constant in the middle one-third of that interval. The constant value in question is the average of the values at the two end points of the interval that is being dissected. We start with $c(0) = 0$ and $c(1) = 1$. Then $c(x) = 1/2$ for $1/2 \le x \le 2/3$. Next we set $c(x) = 1/4$ for $1/9 \le x \le 2/9$ and $c(x) = 3/4$ for $7/9 \le x \le 8/9$. At this point there are four intervals on which $c(x)$ has not yet been defined. These intervals are of length $1/9$, and on the middle one-third of these intervals we take $c(x)$ to be $1/8, 3/8, 5/8$, and $7/8$, respectively. Alternatively, we can define the Cantor function as follows. Set $c(1) = 1$, and for $0 \le x < 1$ write x in base 3:

$$x = \sum_{k=1}^{\infty} \frac{d_k}{3^k}$$

where each d_k is $0, 1$, or 2. Let $K = K(x)$ be the least integer k such that $d_k = 1$, if there is such a k. Put $K = \infty$ otherwise. Thus $d_k = 0$ or 2 for all $k < K$. Set

$$c(x) = \frac{1}{2^K} + \sum_{k=1}^{K-1} \frac{d_k/2}{2^k}. \tag{7.3}$$

From either of our two descriptions of the Cantor function it is clear that if a is an integer, then $c(a/3^k)$ is a number of the form $b/2^k$ where b is an integer depending on a. Also, $c((a+1)/3^k) - c(a/3^k)$ is either 0 or $1/2^k$ when a is an integer. Thus $c(x)$ is not only continuous, but also Lip_α for suitable α.

Lemma 7.3. *Let the Cantor function $c(x)$ be described as above, and let $\alpha = \log 2/\log 3$. Then $c \in \mathrm{Lip}_\alpha[0,1]$.*

Proof. Suppose that $0 \le x_0 < x_1 \le 1$, and choose k so that $3^{-k-1} < x_1 - x_0 \le 3^{-k}$. Of the intervals $[a/3^k, (a+1)/3^k]$, either x_0 and x_1 lie in the same interval, or they lie in adjacent intervals. Thus

$$|c(x_1) - c(x_0)| \le 2/2^k = 4/2^{k-1} = 4\big(3^{-k+1}\big)^\alpha \le 4|x_1 - x_0|^\alpha . \tag{7.4}$$

□

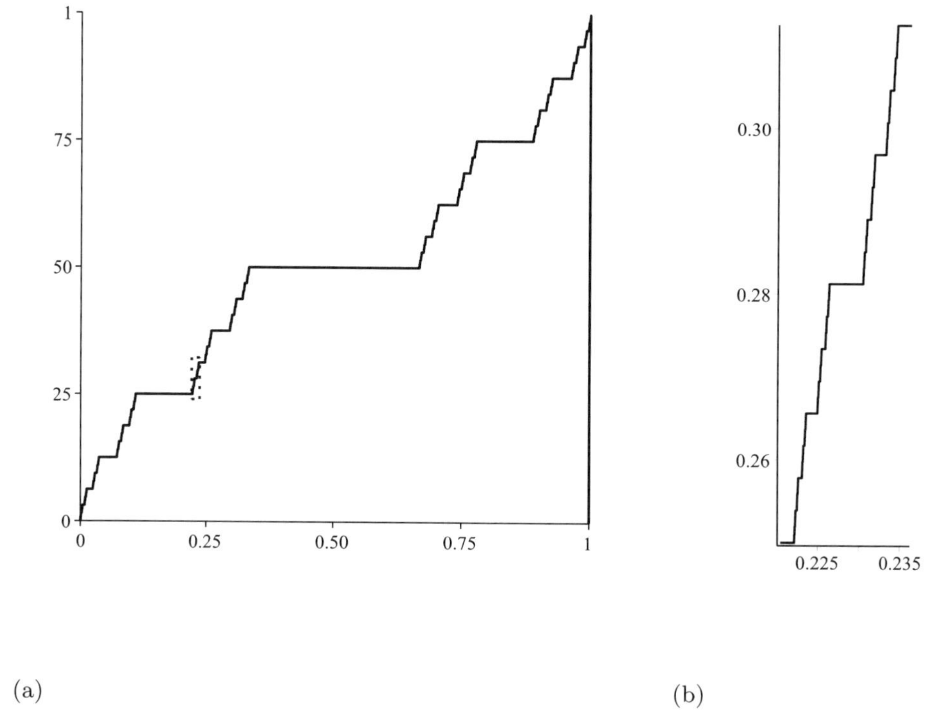

Figure 7.1. (a) The Cantor function; (b) Detail for $2/9 - \delta \le x \le 19/81 + \delta$.

We note that $c'(x) = 0$ on an interval of length 1/3, also on two intervals of length 1/9, on 4 intervals of length 1/27, and so on. The sum of the lengths of these intervals is

$$\sum_{k=1}^{\infty} \frac{2^{k-1}}{3^k} = \frac{1}{3} \cdot \frac{1}{1 - 2/3} = 1 .$$

Thus $c'(x) = 0$ for almost all x, but $c(x)$ is not constant. In Lebesgue measure theory, such a function is said to be *singular*.

If we were to extend the definition of $c(x)$, making it periodic with period 1, we would have a jump discontinuity at 0, since $c(0) = 0$ and $c(1) = 1$. Instead, we let $C(x)$ have period 1 and we put

$$C(x) = c(x) - x \tag{7.5}$$

for $0 \le x \le 1$. Clearly $C \in \mathrm{Lip}_\alpha(\mathbb{T})$ where $\alpha = \frac{\log 2}{\log 3} = 0.63093 > 1/2$, so by Bernstein's theorem (Theorem 7.2) it follows that $C \in A(\mathbb{T})$.

Theorem 7.4. *Let $C(x)$ be defined as above. Then $\widehat{C}(0) = 0$, and*

$$\widehat{C}(n) = \frac{(-1)^n}{2\pi i n} \prod_{k=1}^{\infty} \cos \frac{2\pi n}{3^k} \tag{7.6}$$

for $n \ne 0$.

Since $C \in A(\mathbb{T})$, we can infer that the product above is usually small. Here we see the value of Bernstein's theorem, since a direct proof that $C \in A(\mathbb{T})$ using the above formula might be difficult to construct.

Proof. Since $c(x) + c(1-x) = 1$ for all x, it is clear that $\widehat{c}(0) = 1/2$. The average of $-x$ is $-1/2$, so $\widehat{C}(0) = 0$. To treat nonzero n we set

$$h(u) = e(-u) + 2\pi i u \int_0^1 c(x)e(-ux)\,dx$$

where u is a real number. We first show that

$$h(u) = e(-u/3)h(u/3)\cos\frac{2\pi u}{3}\,. \tag{7.7}$$

Since $c(2/3 + x) = 1/2 + c(x)$ for $0 \le x \le 1/3$, we see that

$$\int_0^1 c(x)e(-ux)\,dx = \int_0^{1/3} c(x)e(-ux)\,dx + \frac{1}{2}\int_{1/3}^{2/3} e(-ux)\,dx$$
$$+ \int_{2/3}^1 c(x)e(-ux)\,dx = (1 + e(-2u/3))\int_0^{1/3} c(x)\,dx + \frac{1}{2}\int_{1/3}^1 e(-ux)\,dx\,.$$

Now $c(x) = c(3x)/2$ for $0 \le x \le 1/3$, so the above is

$$= e(-u/3)cos\frac{2\pi u}{3}\int_0^{1/3} c(3x)e(-ux)\,dx + \frac{e(-u) - e(-u/3)}{-4\pi i u}$$
$$= \frac{1}{3}e(-u/3)cos\frac{2\pi u}{3}\int_0^1 c(x)e(-ux/3)\,dx + \frac{e(-u) - e(-u/3)}{-4\pi i u}.$$

Thus

$$h(u) = \frac{2\pi i u}{3}e(-u/3)\cos\frac{2\pi u}{3}\int_0^1 c(x)e(-ux/3)\,dx + e(-2u/3)\cos\frac{2\pi u}{3}$$
$$= e(-u/3)\cos\frac{2\pi u}{3}\Big(e(-u/3) + \frac{2\pi i u}{3}\int_0^1 c(x)e(-ux/3)\,dx\Big),$$

which is the desired identity (7.7). We note that $h(u) \to 1$ as $u \to 0$. Also, $\sum_{k=1}^{\infty} 1/3^k = 1/2$, so when we use (7.7) repeatedly we find that

$$h(u) = e(-u/2) \prod_{k=1}^{\infty} \cos \frac{2\pi u}{3^k} .$$

On taking $u = n \neq 0$, we deduce that

$$\widehat{c}(n) = \frac{-1}{2\pi i n} + \frac{(-1)^n}{2\pi i n} \prod_{k=1}^{\infty} \cos \frac{2\pi n}{3^k} .$$

Let $s(x) = 1/2 - x$ be the sawtooth function. Then $C(x) = c(x) + s(x) - 1/2$. We know that $\widehat{s}(n) = 1/(2\pi i n)$ for $n \neq 0$. Thus we obtain (7.6), and the proof is complete. □

Exercises

1. Suppose that $f \in A(\mathbb{T})$ and that $g \in L^1(\mathbb{T})$.
   (a) By quoting Theorems 3.12 and 3.13, or otherwise, show that $f * g \in A(\mathbb{T})$ and that $\|f * g\|_A \le \|f\|_A \|g\|_1$.
   (b) Deduce that

$$(f * g)(x) = \sum_{n=-\infty}^{\infty} \widehat{f}(n)\widehat{g}(n)e(nx)$$

   for all x.

2. Suppose that $f \in C(\mathbb{T})$, and write the Fourier Series of f in the old-fashioned form

$$\frac{1}{2}a_0 + \sum_{n=1}^{\infty} (a_n \cos 2\pi n x + b_n \sin 2\pi n x) .$$

   (a) Show that if $n \ge 0$, then $|a_n| \le |\widehat{f}(n)| + |\widehat{f}(-n)|$.
   (b) Show that if $n > 0$, then $|b_n| \le |\widehat{f}(n)| + |\widehat{f}(-n)|$.
   (c) Show that if

$$\sum_{n=-\infty}^{\infty} |\widehat{f}(n)| < \infty, \tag{7.8}$$

   then

$$\sum_{n=0}^{\infty} |a_n| < \infty \quad \text{and} \quad \sum_{n=1}^{\infty} |b_n| < \infty . \tag{7.9}$$

   (d) Show that if $n \neq 0$, then $|\widehat{f}(\pm n)| \le (|a_n| + |b_n|)/2$.
   (e) Show that (7.9) implies (7.8).

3. Suppose that $f \in A(\mathbb{T})$ and that $\sum_{n=-\infty}^{\infty} |n\widehat{f}(n)| < \infty$. Show that f has a continuous first derivative f', that $f' \in A(\mathbb{T})$, and that $\widehat{f'}(n) = 2\pi i n \widehat{f}(n)$.

In the next exercise we construct an example of a function $f \in \mathrm{Lip}_{1/2}(\mathbb{T})$ whose Fourier Series is not absolutely convergent. Thus Bernstein's Theorem (Theorem 7.2) would become false if we allowed $\alpha = 1/2$.

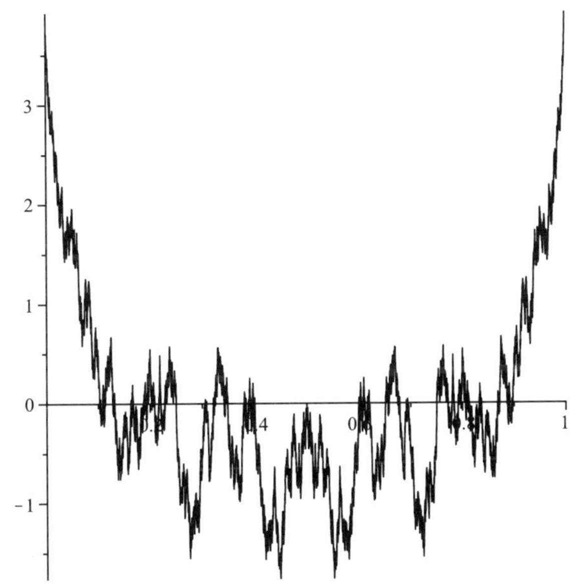

Figure 7.2. Graph of $\mathrm{Re}\, f(x)$ as defined in Exercise 4.

4. Let $P_k(x)$ denote the kth Shapiro polynomial, as defined in §6.4, put

$$T_k(x) = \frac{e(2^k x)}{2^k} P_k(x),$$

and set

$$f(x) = \sum_{k=0}^{\infty} T_k(x)\,.$$

A graph of the real part of this function is found in Figure 7.2.

(a) T_k is a trigonometric polynomial whose coefficients are either $\pm 1/2^k$ or 0. For what n is the coefficient of $e(nx)$ non-zero?

(b) Show that $\|T_k\|_A = 1$ for all k.

(c) Explain why $\|f\|_A = \sum_{k=0}^{\infty} \|T_k\|_A$.

(d) Deduce that $\|f\|_A = \infty$.

(e) Let K be an integer to be chosen later. Explain why

$$|f(x+\delta) - f(x)| \le \sum_{k=1}^{K} |T_k(x+\delta) - T_k(x)| \;+\; \sum_{k>K} \left(|T_k(x+\delta)| + |T_k(x)|\right).$$

Let S_1 and S_2 denote these two sums.

(f) Use results from two different sections of Chapter 6 to show that $T_k'(x) = O(2^{k/2})$.

(g) Explain why

$$T_k(x+\delta) - T_k(x) = \int_x^{x+\delta} T_k'(u)\,du\,.$$

(h) Show that $T_k(x+\delta) - T_k(x) = O\big(2^{k/2}\delta\big)$.
(i) Deduce that $S_1 = O(2^{K/2}\delta)$.
(j) Explain why $T_k(x) = O\big(2^{-k/2}\big)$ for all k and x.
(k) Show that $S_2 = O\big(2^{-K/2}\big)$.
(l) Make a suitable choice of K so that the estimates in (h) and (k) imply that

$$|f(x+\delta) - f(x)| = O\big(\delta^{1/2}\big),$$

which is to say that $f \in \mathrm{Lip}_{1/2}(\mathbb{T})$.
(m) Show that $f(0) = 4$.
(n) Show that $f(1/2) = 0$.
(o) Show that $\widehat{f}(n) = O(1/|n|)$ for all $n \ne 0$.
(p) Deduce that $s_N(x) \to f(x)$ as $N \to \infty$, for all x.

5. In Theorems 7.1 and 7.2 we saw that if f is sufficiently smooth in a suitable sense, then $f \in A(\mathbb{T})$. We now show that there is no converse to this: For $f \in A(\mathbb{T})$, its modulus of continuity $\omega(\delta)$ may tend to 0 arbitrarily slowly.
(a) By recalling (3.20), or otherwise, show that $|\widehat{f}(n)| \le \omega(1/n)$.
(b) By quoting Example 4.2, show that $\omega(1/n)$ can tend to 0 arbitrarily slowly.

6. Let $c(x)$ denote the Cantor function, as defined in (7.3).
(a) Show that $c(1/4) = 1/3$.
(b) Show that $c(1/5) = 1/4$.
(c) Show that $c(1/13) = 1/7$.

7. Use (7.3) to establish the following identities:
(a) $c(x) + c(1-x) = 1$ for $0 \le x \le 1$;
(b) $c(2/3 + x) = 1/2 + c(x)$ for $0 \le x \le 1/3$;
(c) $c(x) = c(3x)/2$ for $0 \le x \le 1/3$.

8. Let $C(x)$ be defined as in Theorem 7.4. Show that C is an odd function.

9. The object of this exercise is to show that the arc length of the curve $(x, c(x))$, $0 \le x \le 1$ is 2. We employ the usual definition (0.9) of arc length.
(a) Apply the triangle inequality (the triangle inequality for triangles!) to the right triangle with vertices $(0,0)$, $(x,0)$, (x,y) to show that $\sqrt{x^2+y^2} \le |x| + |y|$.
(b) Let $0 = x_0 < x_1 < \cdots < x_n = 1$ be a partition of the interval $[0,1]$. Show that

$$\sum_{j=1}^{n} \sqrt{(x_j - x_{j-1})^2 + (c(x_j) - c(x_{j-1}))^2} \le x_n - x_0 + c(x_n) - c(x_0) = 2\,.$$

(c) Show that $\sqrt{x^2+y^2} \ge \max(|x|, |y|)$.

(d) Consider the partition $x_j = j/3^k$ of the interval $[0,1]$ where $j = 0, 1, \ldots, 3^k$. Explain why the arc length of the curve $(x, c(x))$ is at least

$$\sum_{j=1}^{3^k} \sqrt{\frac{1}{3^k} + \left(c(j/3^k) - c((j-1)/3^k)\right)^2}\,.$$

(e) Note that $c(j/3^k) = c((j-1)/3^k)$ if $3^{k-1} < j \le 2 \cdot 3^{k-1}$, a total of 3^{k-1} values of j. Similarly count the j for which $1/9 < j/3^k \le 2/9$ and the j for which $7/9 < j/3^k \le 8/9$. By continuing in this way, show that $c(j/3^k) = c((j-1)/3^k)$ for exactly $3^k - 2^k$ values of j.

(f) Explain why $c(j/3^k) = c((j-1)/3^k) + 1/2^k$ for exactly 2^k values of j.

(g) Deduce that

$$\sum_{j=1}^{3^k} \sqrt{\frac{1}{3^{2k}} + \left(c(j/3^k) - c((j-1)/3^k)\right)^2} \ge \frac{3^k - 2^k}{3^k} + \frac{2^k}{2^k} = 2 - (2/3)^k\,.$$

(h) Conclude that the arc length of the curve $(x, c(x))$ is 2.

7.2. Wiener's theorem

Suppose that $f \in A(\mathbb{T})$. If there is an x for which $f(x) = 0$, say $f(x_0) = 0$, then $|1/f(x)|$ will be unbounded for x near x_0, and hence $1/f \notin A(\mathbb{T})$. Thus in order for $1/f$ to be in $A(\mathbb{T})$ it is necessary that the equation $f(x) = 0$ have no solution. Our object in this section is to show that this necessary condition is also sufficient: If $f \in A(\mathbb{T})$ and $f(x) = 0$ has no solution, then $1/f \in A(\mathbb{T})$. We begin by making some simple observations.

Lemma 7.5. *Suppose that $f \in A(\mathbb{T})$ and that $g \in A(\mathbb{T})$. Then $f + g \in A(\mathbb{T})$, and $\|f+g\|_A \le \|f\|_A + \|g\|_A$.*

Proof. Clearly $f + g$ is continuous, has period 1, and $\widehat{f+g}(n) = \widehat{f}(n) + \widehat{g}(n)$, so by the triangle inequality

$$\|f+g\|_A = \sum_n |\widehat{f}(n) + \widehat{g}(n)| \le \sum_n (|\widehat{f}(n)| + |\widehat{g}(n)|) = \|f\|_A + \|g\|_A\,.$$

□

Lemma 7.6. *Suppose that $f \in A(\mathbb{T})$ and that $g \in A(\mathbb{T})$. Then $fg \in A(\mathbb{T})$,*

$$\widehat{fg}(k) = \sum_{m=-\infty}^{\infty} \widehat{f}(m)\widehat{g}(k-m),$$

and $\|fg\|_A \le \|f\|_A \|g\|_A$.

Proof. Clearly fg is continuous and has period 1. We observe that

$$\begin{aligned} f(x)g(x) &= \Big(\sum_m \widehat{f}(m)e(mx)\Big)\Big(\sum_n \widehat{g}(n)e(nx)\Big) = \sum_k \Big(\sum_m \widehat{f}(m)\widehat{g}(k-m)\Big)e(kx) \\ &= \sum_k c_k e(kx), \end{aligned}$$

say. Here the rearrangement of the series is justified by their absolute convergence. We also see that

$$\sum_k |c_k| \le \sum_k \sum_m |\widehat{f}(m)|\,|\widehat{g}(k-m)| = \sum_m |\widehat{f}(m)| \sum_k |\widehat{g}(k-m)| = \|f\|_A \|g\|_A < \infty,$$

so by Theorem 4.4 it follows that $\widehat{fg}(k) = c_k$. We could also derive this conclusion from Theorem 5.9, but for absolutely convergent series this is not needed. The left hand side above is $\|fg\|_A$, so we have the desired inequality, and the proof is complete. □

Next we establish a quantitative version of Theorem 7.1.

Lemma 7.7. *Suppose that* $f \in L^1(\mathbb{T})$, *and that* $f'(x)$ *is a bounded function. Then*

$$\max_x |f(x)| \le \|f\|_A \le \max_x |f(x)| + \max_x |f'(x)|\,.$$

Proof. We prove the second inequality first. We note that

$$|\widehat{f}(0)| \;=\; \left|\int_0^1 f(x)\,dx\right| \;\le\; \int_0^1 |f(x)|\,dx \;\le\; \max_x |f(x)|\,.$$

By Cauchy's inequality,

$$\sum_{n\ne 0} |\widehat{f}(n)| \le \Big(\sum_{n\ne 0} \frac{1}{n^2}\Big)^{1/2} \Big(\sum_{n\ne 0} |n\widehat{f}(n)|^2\Big)^{1/2}.$$

By Example 3.9 we know that the first sum on the right hand side is $\pi^2/3$. Thus the right hand side is

$$\begin{aligned}\Big(\frac{1}{12}\sum_n |2\pi n\widehat{f}(n)|^2\Big)^{1/2} &= \Big(\frac{1}{12}\sum_n |\widehat{f'}(n)|^2\Big)^{1/2} \\ &= \Big(\frac{1}{12}\int_0^1 |f'(x)|^2\,dx\Big)^{1/2} \le \max_x |f'(x)|\end{aligned}$$

by Theorem 3.8 and Parseval's Identity. This gives the second inequality, which establishes in particular that $f \in A(\mathbb{T})$. Hence (7.2) applies, so we have the first inequality, and the proof is complete. □

We are now ready to prove

Theorem 7.8. (Wiener) *If* $f \in A(\mathbb{T})$ *and the equation* $f(x) = 0$ *has no solution, then* $1/f \in A(\mathbb{T})$.

While this is a very pleasing result, it should not come as a total shock. We know that the rate of decay of Fourier coefficients $\widehat{f}(n)$ is dependent on the extent to which f is smooth. If f is differentiable, then the smoothness of f is expressed by the size of $f'(x)$. Suppose that f is never 0, and let $m = \min_x |f(x)|$, $M = \max_x |f(x)|$. Since $(1/f(x))' = -f'(x)/f(x)^2$ it follows that

$$\frac{|f'(x)|}{M^2} \le \Big|\Big(\frac{1}{f(x)}\Big)'\Big| \le \frac{|f'(x)|}{m^2}$$

for all x. Thus $1/f$ is about as smooth as f.

Proof. Let $m = \min_x |f(x)|$. Since $f(x) = 0$ has no solution, it follows that $m > 0$. Put $g(x) = f(x)/m$. We prove the theorem for g. Once we know that $1/g \in A(\mathbb{T})$, it will follow that $1/f \in A(\mathbb{T})$, since $1/f = 1/(mg)$, and so $\|1/f\|_A = \|1/g\|_A/m$.

Let

$$s_N(x) = \sum_{n=-N}^{N} \widehat{g}(n)e(nx)$$

with N chosen so large that

$$\sum_{|n|>N} |\widehat{g}(n)| \le 1/3\,. \tag{7.10}$$

Then $|g(x) - s_N(x)| \le 1/3$ for all x. But $|g(x)| \ge 1$ for all x, so by the triangle inequality $|s_N(x)| \ge 2/3$ for all x. Consider

$$\sum_{n=1}^{\infty} \frac{(s_N(x) - g(x))^{n-1}}{s_N(x)^n} = \frac{1}{s_N(x)} \sum_{n=1}^{\infty} \Big(\frac{s_N(x) - g(x)}{s_N(x)}\Big)^{n-1}.$$

Since $|(s_N(x) - g(x))/s_N(x)| \le 1/2$, this is a convergent geometric series, and the above is

$$= \frac{1}{s_N(x)} \frac{1}{1 - \frac{s_N(x)-g(x)}{s_N(x)}} = \frac{1}{s_N(x) - (s_N(x) - g(x))} = \frac{1}{g(x)}\,.$$

By Lemmas 7.5 and 7.6 it follows that

$$\Big\|\frac{1}{g}\Big\|_A \le \sum_{n=1}^{\infty} \Big\|\frac{(s_N(x) - g(x))^{n-1}}{s_N(x)^n}\Big\| \le \sum_{n=1}^{\infty} \|s_N(x) - g(x)\|_A^{n-1} \Big\|\frac{1}{s_N(x)^n}\Big\|_A\,. \tag{7.11}$$

From (7.10) we know that $\|s_N(x) - g(x)\|_A \le 1/3$. From Lemma 7.7 we see that

$$\Big\|\frac{1}{s_N(x)^n}\Big\|_A \le \max_x \Big|\frac{1}{s_N(x)^n}\Big| + \max_x \Big|\Big(\frac{1}{s_N(x)^n}\Big)'\Big|\,.$$

Since $|s_N(x)| \ge 2/3$ for all x, the first term on the right hand side above is $\le (3/2)^n$. Now

$$\Big(\frac{1}{s_N(x)^n}\Big)' = -n\frac{s_N'(x)}{s_N(x)^{n+1}}\,.$$

Let $C = \max_x |s_N'(x)|$. Then the above has absolute value $\le Cn(3/2)^{n+1}$. Thus the series in (7.11) converges by comparison with the series

$$\sum_{n=1}^{\infty} \frac{3/2 + 9Cn/4}{2^n}\,.$$

This completes the proof. □

Exercises

1. Show that if $f \in A(\mathbb{T})$, then $e^f \in A(\mathbb{T})$, and that

$$\|e^f\|_A \le \exp(\|f\|_A)\,.$$

2. Suppose that $f \in A(\mathbb{T})$. Show that $\|\cos f\|_A \le \cosh \|f\|_A$.

Notes

7.1. Cantor (1884) was the first to define the Cantor function $c(x)$, which is sometimes referred to as the devil's staircase. That the arc length of the Cantor function is 2 was proved by Scheeffer (1884); see also Darst (1972). The formula in Theorem 7.4 is a surprise as we approached it, but in the context of probability theory it is quite natural. For $k = 1, 2, \ldots$ let X_k be independent identically distributed random variables with the distribution $P(X_k = 0) = P(X_k = 2) = 1/2$. The Fourier Transform (in probability theory it is called "characteristic function") of this is

$$\int_{-\infty}^{\infty} e(-tx)\,d\mu(x) = \frac{1}{2} + \frac{1}{2}e(-2t) = e(-t)\cos 2\pi t\,.$$

We put

$$X = \sum_{k=1}^{\infty} \frac{X_k}{3^k}\,.$$

The cumulative distribution function of X is 0 for $-\infty < x \le 0$, $c(x)$ for $0 \le x \le 1$, and 1 for $1 \le x < \infty$. The distribution of the sum of two independent random variables is the convolution of the two distributions, and the Fourier transform of that convolution is the product of the Fourier transforms of the distributions. Thus if μ_X is the measure for which

$$c(x) = \int_0^x 1\,d\mu_X,$$

then

$$\widehat{\mu_X}(t) = \int_0^1 e(-tx)\,d\mu_X(x) = e(-t/2)\prod_{k=1}^{\infty} \cos\frac{2\pi t}{3^k}\,.$$

The measure μ_X has support $[0, 1]$. If we were to use this to define a measure with period 1, the resulting measure would have Fourier coefficients

$$e(-n/2)\prod_{k=1}^{\infty} \cos\frac{2\pi n}{3^k},$$

and then the Fourier coefficients of the periodization of $c(x)$ would be obtained by integration, which explains the $2\pi i n$ in the denominator in (7.6).

7.2. The original proof of Wiener (1932, pp. 10–14) was somewhat complicated. Zygmund (1968, pp. 245–246) simplified the proof, and Newman (1972) simplified it further. Lévy (1934) generalized Wiener's theorem by showing that if $f \in A(\mathbb{T})$ and F is analytic at every value taken by f, then $F(f) \in A(\mathbb{T})$.

Chapter 8

Convergence of Fourier Series

8.1. Conditions ensuring convergence

We begin by recalling from §4.1 that for a function $f \in L^1(\mathbb{T})$ with Fourier coefficients $\widehat{f}(n)$, the symmetric partial sums of its Fourier Series are

$$s_N(x) = \sum_{n=-N}^{N} \widehat{f}(n)e(nx).$$

If we need to discuss partial sums of several functions, say f and g, then to distinguish among them we would write $s_N(f;x)$ and $s_N(g;x)$.

The Dirichlet kernel is

$$D_N(x) = \sum_{n=-N}^{N} e(nx) = \begin{cases} \dfrac{\sin(2N+1)\pi x}{\sin \pi x} & (x \notin \mathbb{Z}), \\ 2N+1 & (x \in \mathbb{Z}). \end{cases} \tag{8.1}$$

Thus

$$s_N(x) = (f * D_N)(x) = \int_0^1 f(u)D_N(x-u)\,du = \int_0^1 f(x-u)D_N(u)\,du\,.$$

Since $\sin \pi x \ge 2x$ for $0 \le x \le 1/2$, it follows from (8.1) that

$$|D_N(x)| \le \min\left(2N+1, \frac{1}{2\|x\|}\right). \tag{8.2}$$

Riemann's *Principle of Localization* asserts that whether the sequence $s_N(x)$ converges to $f(x)$ (or to any other number) depends entirely on the behavior of f in a neighborhood $(x-\delta, x+\delta)$ of x. To understand why this is so, we establish

Lemma 8.1. *If $f \in L^1(\mathbb{T})$ and $0 < a < b < 1$, then*

$$\lim_{N\to\infty} \int_a^b f(u)D_N(u)\,du = 0\,.$$

Proof. Our idea is to apply the Riemann–Lebesgue Lemma (Theorem 3.6) to the function $f_1(u)$, which is defined to be $f(u)/\sin \pi u$ on the interval $[a, b]$, and 0 otherwise. However, this leads us to consider $\int_0^1 f_1(u)e(\pm(N + 1/2)u)\,du$, so the Riemann–Lebesgue Lemma does not quite apply, since we established it only for integer frequencies, and $N + 1/2$ is not an integer. To avoid this problem, we take a clue from the modified Dirichlet kernel $D_N^*(x)$ defined in (3.60), and write $D_N(u) = \cos 2\pi Nu + \sin 2\pi Nu \cot \pi u$. Accordingly, we put

$$f_1(u) = \begin{cases} f(u) & (a \le u \le b), \\ 0 & (\text{otherwise}), \end{cases} \qquad f_2(u) = \begin{cases} f(u)\cot \pi u & (a \le u \le b), \\ 0 & (\text{otherwise}). \end{cases}$$

We note that $f_2(u) \in L^1(\mathbb{T})$, since $\cot \pi u$ is bounded in the interval $[a, b]$. Thus

$$\int_a^b f(u)D_N(u)\,du = \int_0^1 f_1(u)\cos 2\pi Nu\,du + \int_0^1 f_2(u)\sin 2\pi Nu\,du,$$

and now the desired result follows upon four applications of the Riemann–Lebesgue Lemma. □

Write

(8.3)
$$s_N(x) = \int_0^1 f(x-u)D_N(u)\,du = \int_{-\delta}^{\delta} f(x-u)D_N(u)\,du + \int_{\delta}^{1-\delta} f(x-u)D_N(u)\,du\,.$$

By our lemma, the second integral tends to 0 as N tends to infinity. Thus the limiting behavior of $s_N(x)$ depends entirely on the first integral, which in turn depends on f only in the interval $[x - \delta, x + \delta]$. We use this to establish a helpful sufficient condition for convergence.

Theorem 8.2. (Dini's Test) *Suppose that $f \in L^1(\mathbb{T})$. If c is a number such that*

$$\int_{-1/2}^{1/2} \left| \frac{f(x-u) - c}{u} \right| du < \infty, \tag{8.4}$$

then $s_N(x)$ tends to c as N tends to infinity.

Proof. From (8.3) we see that

$$s_N(x) = c + \int_{-\delta}^{\delta} (f(x-u) - c)D_N(u)\,du + \int_{\delta}^{1-\delta} (f(x-u) - c)D_N(u)\,du$$

since $\int_0^1 D_N(u)\,du = 1$. By (8.2) and our hypothesis that the integral (8.4) converges, it follows that the first integral above has modulus $< \varepsilon$ if δ is small enough. With δ fixed in this way, the second integral tends to 0 as $N \to \infty$, by Lemma 8.1. Hence the result. □

Dini's Test succeeds for most continuous functions that we encounter.

Corollary 8.3. *If $0 < \alpha \le 1$ and $f \in \mathrm{Lip}_\alpha(\mathbb{T})$, then $s_N(x)$ tends to $f(x)$ as $N \to \infty$, for all x.*

Proof. By hypothesis, there is a C such that $|f(x-u) - f(x)| \le C|u|^\alpha$. Thus it suffices to recall that the integral $\int_0^1 1/u^p\,du$ converges when $p < 1$. □

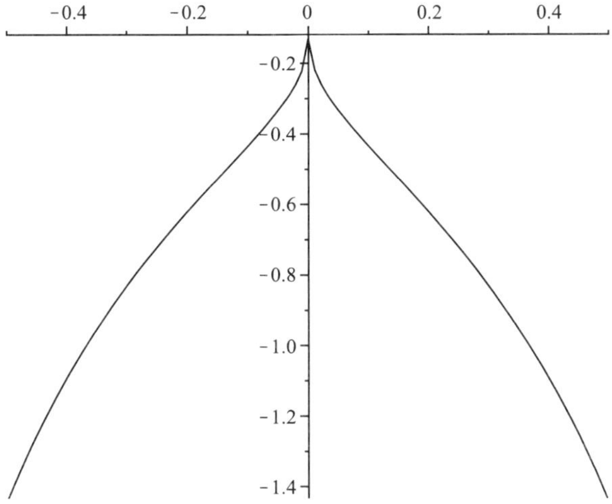

Figure 8.1. Graph of $f(x)$ defined in Example 8.1.

Example 8.1. Let f have period 1, and put

$$f(x) = \begin{cases} \dfrac{1}{\log|x|} & (0 < |x| \le 1/2), \\ 0 & (x = 0). \end{cases}$$

Then f is continuous, but $\int_0^{1/2} 1/(x \log x)\, dx$ diverges, so in this case Dini's Test fails when $x = 0$.

In §8.2 we shall develop tools that enable us to show that indeed $s_N(0)$ converges to 0 in the above Example. However, in §8.3 we shall see that a Fourier Series does not necessarily converge at a point of continuity.

Exercises

1. Suppose that $f \in L^1(\mathbb{T})$.
   (a) Explain why

$$s_N(x) = \frac{L^- + L^+}{2} + \int_0^{1/2} (f(x-u) - L^-) D_N(u)\, du + \int_{-1/2}^{0} (f(x-u) - L^+) D_N(u)\, du\,.$$

   (b) Show that if

$$\int_0^{1/2} \left| \frac{f(x-u) - L^-}{u} \right| du < \infty \quad \text{and} \quad \int_{-1/2}^{0} \left| \frac{f(x-u) - L^+}{u} \right| du < \infty,$$

   then $s_N(x)$ tends to $(L^- + L^+)/2$ as $N \to \infty$.
   (c) Suppose that there is a $\delta > 0$ and an $\alpha > 0$ and a constant C such that $|f(u) - L^-| \le C(x-u)^\alpha$ for $x - \delta \le u < x$ and $|f(u) - L^+| \le C(u-x)^\alpha$ for $x < u \le x + \delta$. Show that $s_N(x)$ converges to $(L^- + L^+)/2$.

8.2. Functions of bounded variation

Let f have period 1. In order to measure how much $f(x)$ varies as x runs through a period, we first choose points at which f is to be sampled:

$$x_0 \le x_1 \le \ldots \le x_{J-1} \le x_J = x_0 + 1. \tag{8.5}$$

We call this a *partition* of $\mathbb{T}$. For brevity we put $\boldsymbol{\pi} = \{x_j\}$. This is the same sort of partition that we form when defining Riemann sums for a Riemann integral. With such a partition in place, we put

$$S(\boldsymbol{\pi}) = \sum_{j=1}^{J} |f(x_j) - f(x_{j-1})| \,. \tag{8.6}$$

Then the *variation* of f, written $\operatorname{Var}(f)$ or $\|f\|_{BV}$, is the least upper bound (i.e. supremum) of the sums $S(\boldsymbol{\pi})$ taken over all partitions $\boldsymbol{\pi}$ of $\mathbb{T}$:

$$\operatorname{Var}(f) = \|f\|_{\mathrm{BV}} = \sup_{\boldsymbol{\pi}} S(\boldsymbol{\pi}) \,. \tag{8.7}$$

If $\operatorname{Var}(f) < \infty$, then we say that f is a function of *Bounded Variation*, and we let $BV(\mathbb{T})$ denote the set of all such functions. If f is a bounded piecewise monotonic function with finitely many jump discontinuities, then $\operatorname{Var}(f)$ is the sum of the amounts of the increases, plus the sum of the amounts of the decreases, plus the sum of the heights of the jumps. If f has a continuous first derivative, then

$$\operatorname{Var}(f) = \int_0^1 |f'(x)| \, dx \,. \tag{8.8}$$

While we have formulated this for the circle group $\mathbb{T}$, this has little to do with periodicity, and could equally well be formulated for functions defined on an interval $[a, b]$ or on the real line $\mathbb{R}$.

First we establish that these are rather tame functions.

Theorem 8.4. *Suppose that $f \in BV(\mathbb{T})$. Then for any $x \in \mathbb{T}$, both one-sided limits $f(x^-)$ and $f(x^+)$ exist and are finite.*

In other words, while a BV function may have jump discontinuities or removable discontinuities, it may not have an essential discontinuity.

Proof. Let $\{x_j\}$ be a strictly increasing sequence in $[x - 1/2, x)$ tending to x. Then

$$\sum_{j=1}^{\infty} |f(x_j) - f(x_{j-1})| < \infty \tag{8.9}$$

since every partial sum is bounded above by $\operatorname{Var}(f)$. Suppose that $J < K$. Then

$$\big|f(x_K) - f(x_J)\big| = \Big| \sum_{n=J+1}^{K} \big(f(x_j) - f(x_{j-1})\big) \Big| \le \sum_{n=J+1}^{K} \big|f(x_j) - f(x_{j-1})\big| \,.$$

This last quantity is $< \varepsilon$ if J is large, since it is a piece of the tail of the convergent sum (8.9). Thus $\{f(x_j)\}$ is a Cauchy sequence, so $\lim_{j\to\infty} f(x_j)$ exists. Since this is the case for all such sequences $\{x_j\}$, it follows that $f(x^-)$ exists (and is finite). The argument for the limit from above is similar, so the proof is complete. □

Theorem 8.5. *If f is of Bounded Variation on $\mathbb{T}$, then f is Riemann-integrable.*

Proof. We follow the notation established in §0.2. Let $f \in BV(\mathbb{T})$ be given, and suppose we have a partition $\boldsymbol{\pi}$ with $\text{mesh}(\boldsymbol{\pi}) = \delta$, and interspersing points $\boldsymbol{\xi}$ and $\boldsymbol{\xi}'$. Then

$$S(\boldsymbol{\pi}, \boldsymbol{\xi}) - S(\boldsymbol{\pi}, \boldsymbol{\xi}') = \sum_{j=1}^{J} \big(f(\xi_j) - f(\xi_j')\big)(x_j - x_{j-1}) .$$

By the triangle inequality,

$$\begin{aligned} |S(\boldsymbol{\pi}, \boldsymbol{\xi}) - S(\boldsymbol{\pi}, \boldsymbol{\xi}')| \ &\le \ \sum_{j=1}^{J} \big|f(\xi_j) - f(\xi_j')\big|(x_j - x_{j-1}) \\ &\le \ \delta \sum_{j=1}^{J} \big|f(\xi_j) - f(\xi_j')\big| \ \le \ \delta \text{Var}(f) . \end{aligned}$$

This suffices. □

For purposes of Fourier Series, functions of Bounded Variation are amenable because of the following important property.

Theorem 8.6. *If $f \in BV(\mathbb{T})$, then*

$$|\widehat{f}(n)| \le \frac{\text{Var}(f)}{2\pi|n|}$$

for all $n \neq 0$.

If we were to make the stronger hypothesis that f has a continuous first derivative, then to achieve the above we would merely integrate by parts:

$$\widehat{f}(n) = \int_0^1 f(x)e(-nx)\,dx = \frac{1}{2\pi i n}\int_0^1 f'(x)e(-nx)\,dx . \tag{8.10}$$

Then by the triangle inequality,

$$|\widehat{f}(n)| \le \frac{1}{2\pi|n|}\int_0^1 |f'(x)|\,dx = \frac{\text{Var}(f)}{2\pi|n|}, \tag{8.11}$$

in view of (8.8). In the proof that follows, we mimic this, using Riemann sums.

Proof. By Theorem 8.5 we know that $f(x)e(-nx)$ is Riemann-integrable. Let

$$S = \sum_{j=1}^{J} f(\xi_j)e(-n\xi_j)(x_j - x_{j-1}) . \tag{8.12}$$

This will be close to $\widehat{f}(n)$ if $\text{mesh}(\boldsymbol{\pi})$ is small. We note that

$$\begin{aligned} e(-nx_j) - e(-nx_{j-1}) &= -2\pi i n \int_{x_{j-1}}^{x_j} e(-nu)\,du \\ &= -2\pi i n e(-n\xi_j)(x_j - x_{j-1}) - 2\pi i n \int_{x_{j-1}}^{x_j} \big(e(-nu) - e(-n\xi_j)\big)\,du . \end{aligned} \tag{8.13}$$

The derivative of $e(-nu)$ has modulus $2\pi|n|$, so

$$|e(-nu) - e(-n\xi_j)| \le 2\pi|n||u - \xi_j| \le 2\pi|n|(x_j - x_{j-1}).$$

Thus the second term in (8.13) is $O\big(n^2(x_j - x_{j-1})^2\big)$. That is,

$$e(-nx_j) - e(-nx_{j-1}) = -2\pi i n e(-n\xi_j)(x_j - x_{j-1}) + O\big(n^2(x_j - x_{j-1})^2\big).$$

On dividing both sides of this by $-2\pi i n$, and rearranging, we conclude that

$$e(-n\xi_j)(x_j - x_{j-1}) = \frac{e(-nx_{j-1}) - e(-nx_j)}{2\pi i n} + O\big(|n|(x_j - x_{j-1})^2\big).$$

On inserting this in (8.12), we deduce that

$$(8.14)\quad S = \frac{1}{2\pi i n}\sum_{j=1}^{J} f(\xi_j)\big(e(-nx_{j-1}) - e(-nx_j)\big) + O\Big(|n|\sum_{j=1}^{J}|f(\xi_j)|(x_j - x_{j-1})^2\Big).$$

Let $\delta = \operatorname{mesh}(\boldsymbol{\pi})$. Thus the error term above is

$$\le C|n|\delta\sum_{j=1}^{J}|f(\xi_j)|(x_j - x_{j-1}).$$

Here the sum is a Riemann sum for $\int_0^1 |f(x)|\,dx$, which is finite, and so the sum will be close to this integral when δ is small. Thus this error term tends to 0 as $\delta \to 0$. The sum in the main term in (8.14) is

$$\sum_{j=1}^{J} f(\xi_j)e(-nx_{j-1}) - \sum_{j=1}^{J} f(\xi_j)e(-nx_j).$$

In the first sum we reindex, setting $k = j - 1$. Thus the above is

$$= \sum_{k=0}^{J-1} f(\xi_{k+1})e(-nx_k) - \sum_{j=1}^{J} f(\xi_j)e(-nx_j).$$

For j and k running from 1 to $J - 1$ we combine terms into one sum; the term for $k = 0$ in the first sum and the term for $j = J$ in the second must be accounted for separately. Thus the above is

$$= \sum_{j=1}^{J-1}\big(f(\xi_{j+1}) - f(\xi_j)\big)e(-nx_j) \quad + f(\xi_1)e(-x_0) - f(\xi_J)e(-nx_J).$$

We recall that $x_0 = x_J$, and put $\xi_{J+1} = \xi_1 + 1$. Thus the above is

$$= \sum_{j=1}^{J}\big(f(\xi_{j+1}) - f(\xi_j)\big)e(-nx_j).$$

The manipulations just completed amount to summation by parts, which is the discrete analogue of integration by parts as in (8.10). Thus far we have shown that

$$\widehat{f}(n) = \frac{1}{2\pi i n}\sum_{j=1}^{J}\big(f(\xi_{j+1}) - f(\xi_j)\big)e(-nx_j) \quad + o(1)$$

as $\delta \to 0$. To complete our argument we use the triangle inequality in the same way it was used in (8.11). The sum above has absolute value not exceeding

$$\sum_{j=1}^{J} |f(\xi_{j+1}) - f(\xi_j)| \le \operatorname{Var}(f)$$

since $\xi_1 \le \xi_2 \le \cdots \le \xi_M \le \xi_{M+1} = \xi_1 + 1$ is partition of $\mathbb{T}$. Thus we have the stated inequality on letting $\delta \to 0$. □

Corollary 8.7. (Jordan) *If $f \in BV(\mathbb{T})$, then*

$$\lim_{N\to\infty} s_N(x) = \frac{f(x^-) + f(x^+)}{2} \tag{8.15}$$

for all x.

Proof. By Fejér's theorem (Theorem 4.14) we know that

$$\lim_{N\to\infty} \sigma_N(x) = \frac{f(x^-) + f(x^+)}{2} .$$

The bound of Theorem 8.6 supplies the Tauberian hypothesis (4.58) of Hardy's Theorem (Theorem 4.32), so the result follows. □

For $f \in BV(\mathbb{T})$, the convergence of $s_N(x)$ to $f(x)$ may not be uniform, and indeed cannot be uniform in the neighborhood of a jump discontinuity. However, it is *boundedly convergent*, as we see from the following.

Corollary 8.8. *If $f \in BV(\mathbb{T})$, then*

$$|s_N(x)| \le \sup_x |f(x)| + \frac{1}{\pi}\operatorname{Var}(f) .$$

Proof. We observe that

$$s_N(x) = \sigma_N(x) + \frac{1}{N} \sum_{n=-N}^{N} n\widehat{f}(n)e(nx) . \tag{8.16}$$

As we already observed in (4.10),

$$\begin{aligned} |\sigma_N(x)| &= \Big| \int_0^1 f(x-u)\Delta_N(u)\,du \Big| \le \int_0^1 |f(x-u)|\Delta_N(u)\,du \\ &\le \big(\sup_x f(x)\big) \int_0^1 \Delta_N(u)\,du = \sup_x |f(x)|. \end{aligned}$$

For $n \ne 0$ in the second term in (8.16) we apply Theorem 8.6, and we find by the triangle inequality that this term has modulus no more than $\frac{1}{\pi}\operatorname{Var}(f)$. Thus the proof is complete. □

By the Lebesgue Dominated Convergence Theorem, the bounded convergence of the $s_N(x)$ to $f(x)$ has the following consequence.

Corollary 8.9. *If $f \in BV(\mathbb{T})$, then*

$$\lim_{N\to\infty} \int_0^1 |s_N(x) - f(x)|\,dx = 0 .$$

We have encountered a number of BV functions (square wave, sawtooth, etc.) with jump discontinuities, and in every case some of the Fourier coefficients were as large as $1/n$. We now show that this is universally the case.

Theorem 8.10. *Suppose that $f \in L^1(\mathbb{T})$, that*

$$\lim_{x\to x_0^-} f(x) = L^-, \qquad \lim_{x\to x_0^+} f(x) = L^+,$$

and that $L^- \neq L^+$. If $0 < c < |L^+ - L^-|$, then there exist infinitely many n for which

$$|\widehat{f}(n)| > \frac{c}{2\pi|n|}\,.$$

In Example 3.1 concerning the sawtooth function we have $x_0 = 0$, $L^- = -1/2$, $L^+ = 1/2$, and $\widehat{s}(n) = 1/(2\pi i n)$ for $n \neq 0$, so for this function $\widehat{s}(n) = |L^+ - L^-|/2\pi|n|$ for all non-zero n.

Proof. We establish the contrapositive, which asserts that if $|\widehat{f}(n)| \le c/|2\pi n|$ for all sufficiently large n, then $c \ge |L^+ - L^-|$. For $0 < \delta < 1/2$ put $g_\delta(x) = 1/\delta$ if $\|x\| \le \delta/2$, and $g_\delta(x) = 0$ otherwise. Then

$$\widehat{g_\delta}(n) = \begin{cases} 1 & (n = 0), \\ \dfrac{\sin \pi n\delta}{\pi n\delta} & (n \neq 0). \end{cases} \tag{8.17}$$

Put $h_\delta = f * g_\delta$. Then h_δ is continuous by Theorem 3.13. By Theorem 3.12 we know that $\widehat{h_\delta}(n) = \widehat{f}(n)\widehat{g_\delta}(n)$. Suppose that $|\widehat{f}(n)| \le c/|2\pi n|$ for all n with $|n|$ sufficiently large. Then $\sum_n |\widehat{h_\delta}(n)| < \infty$, so $h_\delta \in A(\mathbb{T})$. Thus by Corollary 4.5

$$h_\delta(x) = 1 + \sum_{n\neq 0} \widehat{f}(n)\frac{\sin \pi n\delta}{\pi n\delta} e(nx)$$

for all x. We take $x = x_0 + \delta/2$ and $x = x_0 - \delta/2$, and difference, to see that

$$\begin{aligned} h_\delta(x_0 + \delta/2) - h_\delta(x_0 - \delta/2) &= \sum_{n\neq 0} \widehat{f}(n)\frac{\sin \pi n\delta}{\pi n\delta} e(nx_0)(e(n\delta/2) - e(-n\delta/2)) \\ &= 2i\sum_{n\neq 0} \widehat{f}(n)\frac{(\sin \pi n\delta)^2}{\pi n\delta} e(nx_0)\,. \end{aligned} \tag{8.18}$$

The quantity $h_\delta(x_0 - \delta/2)$ is the average of the values of f over the interval $(x_0 - \delta, x_0)$. Since these values are all near L^-, it follows (as in Theorem 4.28) that $h_\delta(x_0 - \delta/2) \to L^-$ as $\delta \to 0^+$. Similarly, $h_\delta(x_0 + \delta/2) \to L^+$ as $\delta \to 0^+$. Thus the left hand side above tends to $L^+ - L^-$ as δ tends to 0. Suppose that N is so large that $|\widehat{f}(n)| \le c/|2\pi n|$ for all $n > N$. Then the expression (8.18) has absolute value not exceeding

$$2\sum_{0<|n|\le N} |\widehat{f}(n)|\Big|\frac{\sin \pi n\delta}{\pi n\delta}\Big||\sin \pi n\delta| + c\delta \sum_{|n|>N} \Big(\frac{\sin \pi n\delta}{\pi n\delta}\Big)^2 = T_1 + T_2,$$

say.

We bound T_1 first. For each n, $|\widehat{f}(n)| \le \|f\|_1$. The fraction inside the large absolute value signs is $\widehat{g_\delta}(n)$, and $|\widehat{g_\delta}(n)| \le \|g_\delta\|_1 = 1$. Finally, $|\sin \pi n\delta| \le \pi|n|\delta$. On combining these estimates, we deduce that

$$T_1 \le 4\pi\delta\|f\|_1 \sum_{n=1}^{N} n = 2\pi\delta\|f\|_1 N(N+1)\,.$$

To bound T_2, we apply Parseval's Identity to g_δ to see that

$$1 + \sum_{n\ne 0} \Big(\frac{\sin \pi n\delta}{\pi n\delta}\Big)^2 = \int_0^1 |g_\delta(x)|^2\,dx = \frac{1}{\delta}\,.$$

Thus

$$\sum_{|n|>N} \Big(\frac{\sin \pi n\delta}{\pi n\delta}\Big)^2 < \frac{1}{\delta},$$

and hence

$$T_2 = c\delta \sum_{|n|>N} \Big(\frac{\sin \pi n\delta}{\pi n\delta}\Big)^2 \le c\,.$$

On combining our estimates, we find that

$$|h_\delta(x_0+\delta/2) - h_\delta(x_0-\delta/2)| \le c + O(\delta)\,.$$

As already noted, the left hand side here tends to $|L^+ - L^-|$ as $\delta \to 0$. Thus we conclude that $c \ge |L^+ - L^-|$, and the proof is complete. □

For a continuous function F of the form

$$F(x) = \int_0^x f(u)\,du - \widehat{f}(0)x,$$

as in Theorem 3.17 where $f \in L^1(\mathbb{T})$, we have $\widehat{f}(n) = o(1)$, so $\widehat{F}(n) = o(1/|n|)$. However, in §7.1 we encountered a continuous function $C(x)$ whose Fourier coefficients we computed in Theorem 7.4. From that calculation we see that $\widehat{C}(3n) = \widehat{C}(n)/3$. Thus $\widehat{C}(3^k) = \widehat{C}(1)/3^k$. Since $\widehat{C}(1) = -0.059116i \ne 0$, it follows that $\widehat{C}(n) = \Omega(1/|n|)$, which is to say that $|\widehat{C}(n)| \ge c/|n|$ for infinitely many n. Of course there is a distinction, since $\widehat{C}(n) = o(1/|n|)$ for most n, while $\widehat{f}(n) \ge c/|n|$ for a positive proportion of n if $f \in \mathrm{BV}(\mathbb{T})$ and f has a jump discontinuity.

Exercises

1. Suppose that $f \in BV(\mathbb{T})$. Show that $\mathrm{Var}(cf) = |c|\mathrm{Var}(f)$.
2. Suppose that $f \in BV(\mathbb{T})$ and $g \in BV(\mathbb{T})$. Show that $\mathrm{Var}(f+g) \le \mathrm{Var}(f) + \mathrm{Var}(g)$.
3. Let $C(x)$ be defined as in Theorem 7.4.
   (a) Show that $\mathrm{Var}(C) \le 2$.
   (b) Show that $\mathrm{Var}(C) = 2$.
4. Suppose that $f \in BV(\mathbb{T})$. Show that $\mathrm{Var}(\mathrm{Re}\, f) \le \mathrm{Var}(f)$ and that $\mathrm{Var}(\mathrm{Im}\, f) \le \mathrm{Var}(f)$.

5. Suppose that $u(x)$ and $v(x)$ are real-valued functions, and that $w(x) = u(x) + iv(x)$. Show that $\max(\mathrm{Var}(u), \mathrm{Var}(v)) \le \mathrm{Var}(w) \le \mathrm{Var}(u) + \mathrm{Var}(v)$.

6. Suppose that $f \in BV(\mathbb{T})$, that $g \in BV(\mathbb{T})$, that $|f(x)| \le M$ for all x, and that $|g(x)| \le N$ for all x. Show that $\mathrm{Var}(fg) \le M\mathrm{Var}(g) + N\mathrm{Var}(f)$.

7. Prove that if f has period 1 and has a continuous first derivative, then (8.8) holds.

8. Let $f(x)$ be the function defined in Example 8.1.
   (a) Show that $\mathrm{Var}(f) = 2/\log 2$.
   (b) Deduce that $s_N(x)$ converges to $f(x)$ for all x, and in particular when $x = 0$.

9. Let f have period 1 and for $0 \le x \le 1$ put $f(x) = x \sin \pi/x$.
   (a) Show that $f \in C(\mathbb{T})$.
   (b) Show that $\mathrm{Var}(f) = \infty$.

10. Suppose that $f \in L^1(\mathbb{T})$, that $x \in \mathbb{T}$ is given, and that $\delta > 0$. Show that if f is of Bounded Variation in the interval $[x - \delta, x + \delta]$, then (8.15) holds.

11. Recall that in §3.4 step functions with jumps of height h_k at points x_k were discussed, and that the Fourier Series of such functions was determined in (3.55).
   (a) Show that for any integer M and any positive integer N,

$$\sum_{n=M-N}^{M+N} \Big|\sum_{k=1}^{K} h_k e(-nx_k)\Big|^2 = \sum_{j=1}^{K} h_j e(-Mx_j) \sum_{k=1}^{K} \overline{h_k} e(Mx_k) D_N(x_j - x_k)\,. \tag{8.19}$$

   (b) Explain why the contribution of those terms in the above for which $j = k$ is

$$(2N+1)\sum_{k=1}^{K} |h_k|^2\,.$$

   (c) Let R denote the sum of all those terms on the right hand side of (8.19) for which $j \ne k$. Show that

$$|R| \le \sum_{1\le j<k\le K} \frac{|h_j h_k|}{\|x_j - x_k\|}\,.$$

   (d) Conclude that for any $\varepsilon > 0$ there is an N_0 such that if $N > N_0$ then for every M there is an n in the interval $M - N \le n \le M + N$ such that

$$\Big|\sum_{k=1}^{K} h_k e(-nx_k)\Big| \ge (1-\varepsilon)\Big(\sum_{k=1}^{K} |h_k|^2\Big)^{1/2}\,.$$

12. Let $f \in L^1(\mathbb{T})$ with $\widehat{f}(0) = 0$, and put $F(x) = \int_0^x f(u)\,du$. Thus $F \in C(\mathbb{T})$. Let $x_0 < x_1 < \cdots < x_J$ be a partitioning of $\mathbb{T}$.
   (a) Explain why

$$\sum_{j=1}^{J} |F(x_j) - F(x_{j-1})| = \sum_{j=1}^{J} \Big|\int_{x_{j-1}}^{x_j} f(u)\,du\Big|\,.$$

(b) Deduce that F has bounded variation and that $\|F\|_{BV} \le \|f\|_1$.

8.3. Examples of divergence

We show below that the Fourier Series of a continuous function can diverge at a point. We contend that this can happen because $\int_0^1 |D_N(x)|\,dx$ is not bounded. In preparation for our discussion, we examine this integral in greater detail.

Theorem 8.11. *For $N \ge 1$,*

$$\int_0^1 |D_N(x)|\,dx = \frac{4}{\pi^2}\log N + O(1)\,.$$

The numbers $L_N = \int_0^1 |D_N(x)|\,dx$ are known as the *Lebesgue constants.*

Proof. We write

$$\begin{aligned}\int_0^1 |D_N(x)|\,dx &= 2\int_0^{1/2} |D_N(x)|\,dx \\ &= 2\int_0^{1/(2N+1)} |D_N(x)|\,dx + 2\int_{1/(2N+1)}^{1/2} |D_N(x)|\,dx = T_1 + T_2,\end{aligned}$$

say. Now $|D_N(x)| \le 2N+1$ for all x, so

$$0 \le T_1 \le 2\int_0^{1/(2N+1)} 2N+1\,dx = 2\,.$$

We know that $\int_0^1 |\sin \pi u|\,du = 2/\pi$. Thus we write

$$I(x) = \int_0^x |\sin \pi u|\,du = \frac{2}{\pi}x + R(x)\,.$$

Here $R(x) \in C(\mathbb{T})$ and $R(0) = R(1/2) = 0$. As $I'((2N+1)x) = |\sin(2N+1)\pi x|$, we see that

$$T_2 = 2\int_{1/(2N+1)}^{1/2} \frac{I'((2N+1)x)}{\sin \pi x}\,dx\,.$$

But $I'(x) = 2/\pi + R'(x)$, so this is

$$= 2\int_{1/(2N+1)}^{1/2} \frac{R'((2N+1)x)}{\sin \pi x}\,dx + \frac{4}{\pi}\int_{1/(2N+1)}^{1/2} \frac{1}{\sin \pi x}\,dx = T_{21} + T_{22},$$

say. In T_{21} we integrate by parts to see that

$$T_{21} = 2\left[\frac{R((2N+1)x)}{(2N+1)\sin \pi x}\right|_{1/(1N+1)}^{1/2} + \frac{2\pi}{2N+1}\int_{1/(2N+1)}^{1/2} \frac{R((2N+1)x)\cos \pi x}{(\sin \pi x)^2}\,dx\,.$$

Here the first term vanishes because $R((2N+1)x) = 0$ at both endpoints. The integral is

$$O\Big(\int_{1/(2N+1)}^{1/2} \frac{\cos \pi x}{(\sin \pi x)^2}\,dx\Big) = O\Big(\frac{1}{\sin \pi/(2N+1)}\Big) = O(N),$$

so $T_{21} = O(1)$. As for T_{22}, we note that

$$\frac{1}{\sin \pi x} - \frac{1}{\pi x} = \frac{\pi x - \sin \pi x}{\pi x \sin \pi x}.$$

Here the numerator is $O(x^3)$, so the above is $O(x)$. Hence

$$\begin{aligned} T_{22} &= \frac{4}{\pi^2}\int_{1/(2N+1)}^{1/2} \frac{1}{x}\,dx + \frac{4}{\pi}\int_{1/(2N+1)}^{1/2} \frac{\pi x - \sin \pi x}{\pi x \sin \pi x}\,dx \\ &= \frac{4}{\pi^2}\log\frac{2N+1}{2} + O(1) = \log N + O(1). \end{aligned}$$

The stated result now follows on combining our estimates. □

We have seen that if $f \in C(\mathbb{T})$, then the Cesàro partial sums $\sigma_N(x)$ of the Fourier Series of f converge to $f(x)$ uniformly in x. We now show that this is not true for the unweighted symmetric partial sums $s_N(x)$.

Theorem 8.12. *There is a function $f \in C(\mathbb{T})$ such that $s_N(0)$ does not tend to a limit.*

Proof. We first construct continuous functions f_N such that $\int_0^1 D_N(x) f_N(x)\,dx$ is large. We note that if

$$g_N(x) = \operatorname{sgn} D_N(x) = \begin{cases} 1 & (D_N(x) > 0), \\ 0 & (D_N(x) = 0), \\ -1 & (D_N(x) < 0), \end{cases}$$

then $\int_0^1 D_N(x) g_N(x)\,dx = \int_0^1 |D_N(x)|\,dx = L_N$ is large. However, this $g_N(x)$ has jump discontinuities, so we take a continuous $f_N(x)$ that mimics g_N: Let f_N be the function with period 1 such that $f_N(x) = \sin(2N+1)\pi x$ for $0 \le x \le 1$. (Note that $\sin(2N+1)\pi x$ has period 2; so we are not saying that $f_N(x) = \sin(2N+1)\pi x$ for all x.) Then

$$\begin{aligned} \widehat{f_N}(n) &= \int_0^1 \sin(2N+1)\pi x)e(-nx)\,dx = \frac{1}{2i}\int_0^1 e^{i\pi(2N+1-2n)x} - e^{-i\pi(2N+1+2n)x}\,dx \\ &= \frac{1}{2i}\Big[\frac{e^{i\pi(2N+1-2n)x}}{i\pi(2N+1-2n)}\Big|_0^1 - \frac{1}{2i}\Big[\frac{e^{-i\pi(2N+1+2n)x}}{-i\pi(2N+1+2n)}\Big|_0^1 \\ &= \frac{1}{\pi}\Big(\frac{1}{2N+1-2n} + \frac{1}{2N+1+2n}\Big) = \frac{2}{\pi}\,\frac{2N+1}{(2N+1)^2-(2n)^2} \end{aligned}$$

Hence

$$\begin{aligned} \int_0^1 D_N(x) f_N(x)\,dx &= \sum_{n=-N}^{N} \widehat{f}(n) = \frac{1}{\pi}\sum_{n=-N}^{N}\Big(\frac{1}{2N+1-2n} + \frac{1}{2N+1+2n}\Big) \\ &= \frac{2}{\pi}\Big(1 + \frac{1}{3} + \frac{1}{5} + \cdots + \frac{1}{4N+1}\Big). \end{aligned}$$

Now $\int_1^{2N} 1/(2x+1)\,dx = \frac{1}{2}\big[\log(2x+1)\big|_0^1 = \frac{1}{2}\log N + O(1)$, so by the integral test we conclude that

$$\int_0^1 D_N(x) f_N(x)\,dx = \frac{1}{\pi}\log N + O(1). \tag{8.20}$$

We now use the f_N to construct our example: We take

$$f(x) = \sum_{k=1}^{\infty} \frac{1}{k^2} f_{N_k}(x)$$

where the N_k tend to infinity sufficiently rapidly. Since the f_{N_k} are continuous and the series is uniformly convergent, it follows that f is continuous. As

$$\int_0^1 D_{N_k}(x) f_{N_k}(x)\, dx \asymp \log N_k,$$

and we want this to be large enough to overwhelm the coefficient $1/k^2$, we take $N_k = 3^{k^3}$. With this choice of the N_k, we have

$$s_{N_k}(0) = \int_0^1 D_{N_k}(x) f(x)\, dx = \sum_{j=1}^{\infty} \frac{1}{j^2} \int_0^1 D_{N_k}(x) f_{N_j}(x)\, dx\,. \tag{8.21}$$

The term $j = k$ gives a big contribution, of the order of magnitude k. We need to show that the contributions of the terms $j \ne k$ are small enough so that they do not interfere with this. To this end we show that

$$\int_0^1 D_N(x) f_M(x)\, dx = \begin{cases} O(M/N) & (M \le N/2), \\ O(N/M) & (M \ge 2N). \end{cases} \tag{8.22}$$

To prove this, we observe that

$$\int_0^1 D_N(x) f_M(x)\, dx = \frac{1}{\pi} \sum_{n=-N}^{N} \Big(\frac{1}{2M+1-2n} + \frac{1}{2M+1+2n}\Big) = \frac{2}{\pi} \sum_{n=M-N}^{M+N} \frac{1}{2n+1}$$

for any positive integers M and N. If $M \ge 2N$, then we have $2N+1$ summands, each of which is $O(1/M)$, so the expression is $O(N/M)$. If $N \ge 2M$, then the summand for $n = j$ is the negative of the summand when $n = -j-1$. Thus the terms for $M - N \le n \le M - M - 1$ cancel, and all that is left is

$$\frac{2}{\pi} \sum_{n=N-M}^{N+M} \frac{1}{2n+1}\,.$$

Here we have $2M+1$ terms, each one $O(N)$, so the expression is $O(M/N)$. This completes the proof of (8.22). With this in hand, we observe that

$$\sum_{j=1}^{k-1} \frac{1}{j^2} \int_0^1 D_{N_k}(x) f_{N_j}(x)\, dx = O\Big(\sum_{j=1}^{k-1} \frac{N_j}{j^2 N_k}\Big) = O\big(3^{-k^2}\big)$$

since $N_j/N_k \le N_{k-1}/N_k = O\big(3^{-k^2}\big)$ for $j < k$. Similarly,

$$\sum_{j=k+1}^{\infty} \frac{1}{j^2} \int_0^1 D_{N_k}(x) f_{N_j}(x)\, dx = O\Big(\sum_{j=k+1}^{\infty} \frac{N_k}{j^2 N_j}\Big) = O\big(3^{-k^2}\big)$$

since $N_k/N_j \le (N_k/N_{k+1}) = O\big(3^{-k^2}\big)$ for $j > k$. Thus $s_{N_k}(0) \asymp k$, as desired.

To complete the proof we show that it is not the case that $s_N(0) \to \infty$. To see why this is so, recall that

$$\sigma_N(x) = \frac{1}{N} \sum_{n=0}^{N-1} s_n(x) .$$

Thus if it were the case that $s_N(0) \to \infty$, it would follow that $\sigma_N(0) \to \infty$. However, f is continuous, and $f(0) = 0$, so $\sigma_N(0) \to 0$ as $N \to \infty$. □

For $f \in L^1(\mathbb{T})$ we know that $\int_0^1 |\sigma_N(x) - f(x)|\, dx \to 0$ as $N \to \infty$. We now show that this is not necessarily the case if $\sigma_N(x)$ is replaced by $s_N(x)$. Our approach is based on Theorem 4.27, where we showed that if $a_0, a_1, a_2, \ldots$ are real numbers such that $a_n \to 0$ as $n \to \infty$ and $a_{n-1} - 2a_n + a_{n+1} \ge 0$ for all $n \ge 1$, then there is a function $f \in L^1(\mathbb{T})$ that is continuous for $0 < x < 1$, nonnegative for $0 < x < 1$, with $s_N(x) \to f(x)$ for $0 < x < 1$, and whose Fourier Series is

$$\frac{1}{2}a_0 + \sum_{n=1}^{\infty} a_n \cos 2\pi n x .$$

Theorem 8.13. *Let the numbers a_n and the function $f(x)$ be as described above. Then*

$$\int_0^1 |s_N(x) - f(x)|\, dx = \frac{1}{2}|a_N| L_N + o(1)$$

as $N \to \infty$.

Thus for functions f of this type, $\|s_N - f\|_1 \to 0$ if and only if $a_n = o(1/\log n)$, and this need not be the case. Indeed, in (4.53) we considered the possibility that $a_n = 1/\log(n+2)$.

Proof. By Theorem 4.27 (a) and (c) we see that

$$\begin{aligned} s_N(x) - f(x) &= \frac{1}{2}a_N D_N(x) + \frac{1}{2}(a_{N-1} - a_N) N \Delta_N(x) \\ &\quad - \frac{1}{2}\sum_{n=N}^{\infty} (a_{n-1} - 2a_n + a_{n+1}) n \Delta_n(x) \\ &= \frac{1}{2}a_N D_N(x) + E_N(x), \end{aligned}$$

say. Since $L_N = \|D_N\|_1$, it follows by the triangle inequality that

$$\frac{1}{2}|a_N| L_N - \|E_N\|_1 \le \|s_N - f\|_1 \le \frac{1}{2}|a_N| L_N + \|E_N\|_1 .$$

By further use of the triangle inequality we see that

$$\|E_N\|_1 \le \frac{1}{2}|a_{N-1} - a_N| N + \frac{1}{2}\sum_{n=N}^{\infty} (a_{n-1} - 2a_n + a_{n+1}) n$$

since $\|\Delta_n\|_1 = 1$ for all n. By Lemma 4.26 (a) and (c) it follows that the above is

$$= (a_{N-1} - a_N) N + \frac{1}{2}a_N .$$

By Lemma 4.26 (b), this tends to 0 as $N \to \infty$. □

Notes

8.1. Dini's test originates in Dini (1880). For the more elaborate Dini–Lipschitz and Lebesgue tests see for example Hardy and Rogosinski (2013, pp. 41–46) or Zygmund (1968, Vol. 1, pp. 62–66).

8.2. In order to keep our exposition as elementary as possible, we have not introduced the concept of Riemann–Stieltjes integration. Consequently the proof of Theorem 8.6 is rather tortuous. More advanced readers may prefer to contemplate the basic properties of the Riemann–Stieltjes integral, as it not only makes the proof of Theorem 8.6 simple, it also has many other useful applications. The Riemann–Stieltjes integral $\int_a^b f(x)\,dg(x)$ is said to exist and have the value I if the Riemann–Stieltjes sums

$$\sum_{j=1}^{J} f(\xi_j)\big(g(x_j) - g(x_{j-1})\big) \tag{8.23}$$

all tend to I as the mesh of the partition tends to 0. Here the x_j and ξ_j are as in (0.7). It is easy to show that $\int_a^b f(x)\,dg(x)$ exists if f is continuous and g is of bounded variation. In the negative direction, this Riemann–Stieltjes integral does not exist if f and g have a common discontinuity in (a, b). By writing $\sum f(\xi_j)g(x_j)$ and $\sum f(\xi_j)g(x_{j-1})$ separately, reindexing one of the sums, and recombining, we obtain a Riemann–Stieltjes sum for $\int_a^b g(x)\,df(x)$ with the roles of the x_j and ξ_j reversed. Thus if $\int_a^b f(x)\,dg(x)$ exists, then $\int_a^b g(x)\,df(x)$ exists, and

$$\int_a^b f(x)\,dg(x) = f(b)g(b) - f(a)g(a) - \int_a^b g(x)\,df(x)\,. \tag{8.24}$$

These manipulations of the Riemann–Stieltjes sums amount to summation by parts, so summation by parts yields integration by parts for the Riemann–Stieltjes integral. If g' is continuous on $[a, b]$, then

$$\operatorname{Var}_{[a,b]} g = \int_a^b |g'(x)|\,dx\,. \tag{8.25}$$

If in addition f is Riemann-integrable, then

$$\int_a^b f(x)\,dg(x) = \int_a^b f(x)g'(x)\,dx\,. \tag{8.26}$$

Here the integral on the right is a Riemann integral. Finally, suppose that g has bounded variation, and put $g^*(x) = \operatorname{Var}_{[a,x]} g$. Since the triangle inequality applies to the respective Riemann–Stieltjes sums, it follows that

$$\Big|\int_a^b f(x)\,dg(x)\Big| \le \int_a^b |f(x)|\,dg^*(x) \tag{8.27}$$

if these integrals exist. For the details of what we have sketched here, see Montgomery and Vaughan (2007, pp. 486–490).

Suppose that $f \in BV(\mathbb{T})$ and that $n \neq 0$. Then

$$\widehat{f}(n) = \int_0^1 f(x)e(-nx)\,dx = \frac{-1}{2\pi i}\int_0^1 f(x)de(-nx) = \frac{1}{2\pi i n}\int_0^1 e(-nx)\,df(x)$$

by (8.26) and (8.24), and then Theorem 8.6 follows from (8.27).

Suppose that $f \in BV(\mathbb{T})$. Clearly

$$s_N(x) = \int_0^1 f(x-u)D_N(u)\,du = \int_{0+}^{1^-} f(x-u)D_N(u)\,du\,.$$

Let $E_N(x)$ be defined as in (3.52). Thus $E'(x) = -D_N(x)$. By (8.26) and (8.24) the above is

$$= -\int_{0+}^{1^-} f(x-u)\,dE_N(u) = \frac{1}{2}f(x^+) + \frac{1}{2}f(x^-) + \int_{0+}^{1^-} E_N(u)\,df(x-u)\,.$$

For any $\varepsilon > 0$ there is a $\delta > 0$ such that

$$\mathrm{Var}_{[x-\delta,x)} < \varepsilon, \quad \text{and} \quad \mathrm{Var}_{(x,x+\delta]} < \varepsilon\,.$$

Thus from the estimates of Theorem 3.16 we obtain a simple and direct proof of Jordan's theorem (our Corollary 8.7).

8.3. That a continuous function with period 1 could have a Fourier Series that diverges at a point was first proved by du Bois-Reymond (1873). Further to the negative results of Theorems 8.12 and 8.13, Kolmogorov (1926)constructed a function $f \in L^1(\mathbb{T})$ whose Fourier Series diverges everywhere. All such negative results are specific to L^1. If $f \in L^p(\mathbb{T})$ with $1 < p < \infty$, then

$$\lim_{N\to\infty} \int_0^1 |f(x) - s_N(x)|^p\,dx = 0\,.$$

Without equations we would say that s_N converges to f in the L^p norm. Also, if $1 < p < \infty$, then $s_N(x) \to f(x)$ as $N \to \infty$ for almost all x. This last result, known as the Carleson–Hunt theorem, is one of the most spectacular achievements of 20th century harmonic analysis. Most of the great harmonic analysts of the time thought that this would be false, so it came as a surprise when Carleson (1966) established the case $p = 2$. Prior to that, this result was not even known for continuous functions. Hunt (1968) generalized Carleson's result to all p, $1 < p < \infty$.

Chapter 9

Applications of Fourier Series

9.1. The heat equation

Let u be a function of the real variables x and y. Then the *laplacian* of u, denoted Δu, is

$$\Delta\, u(x,y) = \frac{\partial^2 u}{\partial x^2} + \frac{\partial^2 u}{\partial y^2} = u_{xx} + u_{yy}\,.$$

A function for which $\Delta\, u(x,y) = 0$ throughout a region is said to be *harmonic.* Suppose that $\mathcal{R}$ is a region in the plane with boundary C. The *Dirichlet problem* is to find a function that is harmonic in the interior of $\mathcal{R}$ and has prescribed values on C. For reasons of physics that we do not explore, if $u(x,y,t)$ denotes the temperature of a planar object at the position (x,y) at time t, then

$$\Delta\, u = c\frac{\partial u}{\partial t}\,.$$

Here it is assumed that the planar object is homogeneous, so it conducts heat at a rate that is independent of position. This is the *time-dependent heat equation.* If the temperature on C is independent of time, then as $t \to \infty$ the temperature inside $\mathcal{R}$ will tend to limiting values $u(x,y)$ such that

$$\Delta u = 0\,. \tag{9.1}$$

This is the *steady-state heat equation.* Following Fourier, we consider the case in which $\mathcal{R}$ is a disc of radius 1 centered at the origin. We switch to polar coordinates, so that $u = u(r,\theta)$. The values of u on the unit circle are prescribed, which means that we are given a function $f(\theta)$ with period 2π, and our task is to find a function $u(r,\theta)$ such that

$$\begin{aligned} u(1,\theta) &= f(\theta) \qquad (0 \le \theta \le 2\pi),\\ \Delta\, u &= 0 \qquad (|r| < 1)\,. \end{aligned}$$

By the chain rule, the laplacian in polar coordinates is

$$\Delta u = \frac{\partial^2 u}{\partial r^2} + \frac{1}{r}\frac{\partial u}{\partial r} + \frac{1}{r^2}\frac{\partial^2 u}{\partial \theta^2}.$$

We set this to 0, multiply by r^2, and rearrange to find that

$$r^2\frac{\partial^2 u}{\partial r^2} + r\frac{\partial u}{\partial r} = -\frac{\partial^2 u}{\partial \theta^2}.$$

In search of solutions of this equation, we adopt a strategy known as *separation of variables*. That is, we seek solutions of the form $u(r,\theta) = g(r)h(\theta)$. Such a solution is very special, but by taking linear combinations of such solutions we can obtain more general solutions—hopefully all solutions. For u of this kind, our equation reads

$$\big(r^2 g''(r) + rg'(r)\big)h(\theta) = -g(r)h''(\theta).$$

On dividing by $g(r)h(\theta)$ we deduce that

$$\frac{r^2 g''(r) + rg'(r)}{g(r)} = -\frac{h''(\theta)}{h(\theta)}.$$

If we hold θ fixed and allow r to vary, we find that the left hand side is a constant function of r. Similarly, if we hold r fixed and allow θ to vary, then the right hand side is a constant function of θ. Let λ denote the common value of the two sides. Then

$$\begin{aligned} r^2 g''(r) + rg'(r) - \lambda g(r) &= 0, \\ h''(\theta) + \lambda h(\theta) &= 0. \end{aligned} \tag{9.2}$$

The second of these equations has constant coefficients. We apply the standard theory for dealing with such equations, as discussed in §F.2. The associated polynomial is $z^2 + \lambda$, whose roots are $\pm\sqrt{-\lambda}$. Thus h is of the form

$$h(\theta) = ae^{\sqrt{-\lambda}\theta} + be^{-\sqrt{-\lambda}\theta}.$$

But h must have period 2π, so λ must be of the form $\lambda = n^2$, and thus

$$h(\theta) = ae^{in\theta} + be^{-in\theta}.$$

We observe that for $n > 0$ the equation

$$r^2 g''(r) + rg'(r) - n^2 g(r) = 0 \tag{9.3}$$

has the solutions $g(r) = r^n$, $g(r) = r^{-n}$. However, u must be continuous at the origin, and a term of the form $r^{-n}e^{\pm in\theta}$ would create a discontinuity as $r \to 0^+$, so we discard the second solution, and keep only the first. For $n = 0$ the equation (36) has the solutions 1 and $\log r$. We reject $\log r$ for the same reason that we rejected r^{-n}. The equation (9.1) is linear in u, in the sense that if u_1 and u_2 are two solutions, then so also is $\alpha u_1 + \beta u_2$ for any choice of α and β. This is called *superposition* of solutions. Thus we have solutions of the form

$$u(r,\theta) = \frac{1}{2}a_0 + \sum_{n=1}^{\infty} r^n(a_n \cos n\theta + b_n \sin n\theta).$$

The coefficients a_n and b_n need to be chosen so that

$$\frac{1}{2}a_0 + \sum_{n=1}^{\infty}(a_n \cos n\theta + b_n \sin n\theta) = f(\theta).$$

Fourier was of the opinion that this could be done for an 'arbitrary' function f. Of course whether this is so or not depends on exactly how "arbitrary" the function f is allowed to be. By Abel's theorem (Theorem 4.30) we see that

$$\lim_{r \to 1^-} u(r, \theta) = f(\theta)$$

if the Fourier series of f converges at θ. Also, we recall the Poisson kernel $P_r(\theta)$ defined in (4.79), and note that u is the convolution of f with P_r (with suitable adjustments for the period 2π instead of 1).

Exercises

In general we expect that a homogeneous linear differential equation of degree 2 will have 2 linearly independent solutions. In the following two exercises we show that this is the case for the differential equation (9.3).

1. Suppose that $n > 0$, that $g(r)$ satisfies (36), and define $G(r)$ so that $g(r) = r^n G(r)$.
   (a) Show that $(2n+1)G'(r) + rG''(r) = 0$.
   (b) Write this as $2nG'(r) + (rG'(r))' = 0$. Deduce that $2nG(r) + rG'(r) = c$ for some constant c.
   (c) Deduce that G is of the form $G(r) = ar^{-2n} + b$.
2. When $n = 0$ in (9.3), the differential equation implies that $rg''(r) = -g'(r)$. Let $g_1(r) = g(r)$.
   (a) Show that $g_1(r)$ is of the form $g_1 = K/r$.
   (b) Show that g is of the form $K \log r + L$.
3. Confirm that
$$\frac{\partial^2 u}{\partial x^2} + \frac{\partial^2 u}{\partial y^2} = \frac{\partial^2 u}{\partial r^2} + \frac{1}{r}\frac{\partial u}{\partial r} + \frac{1}{r^2}\frac{\partial^2 u}{\partial \theta^2} .$$
4. Show that
$$\left|\frac{\partial u}{\partial x}\right|^2 + \left|\frac{\partial u}{\partial y}\right|^2 = \left|\frac{\partial u}{\partial r}\right|^2 + \frac{1}{r^2}\left|\frac{\partial u}{\partial \theta}\right|^2 .$$

9.2. The wave equation

The laplacian of a function u of n real variables $x_1, \ldots, x_n$ is

$$\Delta u = \frac{\partial^2 u}{\partial x_1^2} + \frac{\partial^2 u}{\partial x_2^2} + \cdots + \frac{\partial^2 u}{\partial x_n^2} = u_{x_1 x_1} + u_{x_2 x_2} + \cdots + u_{x_n x_n} .$$

The *wave equation* in n dimensions asserts that

$$c^2 \Delta u = \frac{\partial^2 u}{\partial t^2} .$$

This is used as a model for

(1) The vibration of a stretched string ($n = 1$).

(2) The vibration of a column of air (as in an organ pipe, $n = 1$).

(3) A stretched membrane (as in a drumhead, $n = 2$).
(4) A wave in an incompresible fluid (such as water, $n = 2$).
(5) Sound waves ($n = 3$).
(6) Electromagnetic waves ($n = 3$).

All of these situations are amenable to Fourier analysis, but for now we consider a plucked string of length ℓ, with fixed endpoints. Thus

$$u_{tt} = c^2 u_{xx} \tag{9.4}$$

with

$$u(0, t) = u(\ell, t) = 0 \tag{9.5}$$

for all t. As in the previous section we try the method of separation of variables. That is, we search for special solutions of the form $u(x, t) = X(x)T(t)$. This gives

$$XT'' = c^2 X'' T$$

with $X(0) = X(\ell) = 0$. Thus

$$\frac{X''(x)}{X(x)} = \frac{1}{c^2} \frac{T''(t)}{T(t)} .$$

If we allow x to vary with t fixed, we find that the left hand side of the equation is a constant function of x. Similarly, if we allow t to vary with x fixed, we find that the right hand side of the equation is a constant function of t. Let the common value of these two constants be denoted $-\lambda^2$. Then

$$\begin{aligned} X''(x) &= -\lambda^2 X(x), \\ T''(t) &= -(\lambda c)^2 T(t) . \end{aligned}$$

These are second order homogeneous linear differential equations with constant coefficients. From the theory of such equations (as outlined in §.F.2) we find that

$$\begin{aligned} X(x) &= a_1 \cos \lambda x + b_1 \sin \lambda x \\ T(t) &= a_2 \cos c\lambda t + b_2 \sin c\lambda t . \end{aligned}$$

From the condition $X(0) = 0$ we deduce that $a_1 = 0$. If $b_1 = 0$, then $u = 0$ identically. Otherwise, the condition that $X(\ell) = 0$ implies that $\lambda = \pi n/\ell$ for some integer n. On combining such solutions we obtain solutions $u(x, t)$ of the form

$$u(x, t) = \sum_{n=1}^{\infty} \Big(\sin \frac{\pi n x}{\ell} \Big) \Big(a_n \cos \frac{\pi n c t}{\ell} + b_n \sin \frac{\pi n c t}{\ell} \Big) . \tag{9.6}$$

Let $U_0(x)$ and $U_1(x)$ be two given functions defined on $[0, \ell]$ with $U_0(0) = U_0(\ell) = 0$ and $U_1(0) = U_1(\ell) = 0$. Suppose that we are told that $U_0(x)$ describes the initial position of the string, and that $U_1(x)$ is the initial velocity of the string. So we want solutions of the wave equation (9.4) that satisfy not only the constraint (9.5), but also the initial conditions

$$u(x, 0) = U_0(x), \tag{9.7}$$

$$u_t(x, 0) = U_1(x) \tag{9.8}$$

for $0 \le x \le \ell$. From (9.6) we see that

$$u(x,0) = \sum_{n=1}^{\infty} a_n \sin \frac{\pi n x}{\ell},$$

so in order to satisfy (9.7) it suffices to take the a_n to be the Fourier coefficients of U_0. On differentiating (9.6) with respect to t we find that

$$u_t(x,0) = \frac{\partial u(x,t)}{\partial t}\bigg|_{t=0} = \frac{\pi c}{\ell} \sum_{n=1}^{\infty} n b_n \sin \frac{\pi n x}{\ell}.$$

The second initial condition (9.8) will be satisfied if we choose the b_n appropriately in terms Thus we have constructed a solution of the wave equation (9.4) that satisfies the constraint (9.5) and the initial conditions (9.7) and (9.8). This solution is unique, but proving the uniqueness is an issue in the theory of partial differential equations (PDE).

9.3. Continuous, nowhere differentiable functions

We recall our notation: $\|x\| = \min_{n \in \mathbb{Z}} |x - n|$ is the distance from x to the nearest integer. This function has period 1, and for $-1/2 \le x \le 1/2$ is $|x|$.

Theorem 9.1. *Let*

$$f(x) = \sum_{k=1}^{\infty} \frac{\|10^k x\|}{10^k}.$$

Then f is continuous but nowhere differentiable.

Proof. That f is continuous is clear, since it is a uniformly convergent sum of continuous functions (see §0.2). To see that f is nowhere differentiable, suppose that x is given. We consider difference quotients $(f(x + \delta_r) - f(x))/\delta_r$ where $\delta_r = \pm 1/10^r$. The choice of the sign here depends on the decimal expansion of x. Suppose that $x = 0.d_1 d_2 \ldots$ in base 10. The choice of the sign of δ_r depends on d_r as follows: Consider the contribution of the term

d_r	0	1	2	3	4	5	6	7	8	9
sign	$+$	$\pm$	$\pm$	$\pm$	$-$	$+$	$\pm$	$\pm$	$\pm$	$-$

$$\frac{\|10^k(x + \delta_r)\| - \|10^k x\|}{10^k \delta_r}.$$

If $k \ge r$, then $10^k(x + \delta_r)$ differs from $10^k x$ by an integer. The function $\|x\|$ has period 1, so the numerator is 0 in this case. Now suppose that $k < r$. Our choice of the sign of δ_r was made to ensure that when u moves from x to $x + \delta_r$, the function $\|10^k u\|$ does not pass any of its corners. That is, $\|10^k u\|$ is linear, with slope $\pm 10^k$. Hence the numerator displayed above is $\pm 10^k \delta_r$, and so the entire expression is ± 1. Our difference quotient is therefore a sum of $r - 1$ ± 1's. This is alternately odd and even, so it cannot be constant. Hence the limit does not exist. □

One could complain that in the above construction, the individual summand is not everywhere differentiable. We now give a second construction, in which the summands are smooth.

Theorem 9.2. *Let*

$$f(x) = \sum_{k=1}^{\infty} \frac{\cos 2^k \pi x}{c^k}$$

where $1 < c < 2$. *Then* f *has period* 1, *is continuous, but is nowhere differentiable.*

Proof. That f is continuous follows because the sum is uniformly convergent and each summand is continuous (see §0.2). Consider

(9.9)

$$I(\alpha) = \int_0^1 f(\alpha - x)\Delta_N(x)e(Hx)\,dx = \sum_{n=-N+1}^{N-1} (1 - |n|/N)\widehat{f}(n+H)e((n+H)\alpha)$$

where N and H will be chosen later with $H > N$. We will argue by contradiction, but rather than assume that f is differentiable at α we will assume only that it is Lipschitz near α in the sense that there is a $\delta > 0$ and a constant C such that $|f(\alpha - x) - f(\alpha)| \le C|x|$ when $|x| \le \delta$. We note that

$$\int_0^1 \Delta_N(x)e(Hx)\,dx = 0$$

since $H > N$. We muliply both sides of this by $f(\alpha)$, and subtract the result from $I(\alpha)$ to see that

$$I(\alpha) = \int_0^1 (f(\alpha - x) - f(\alpha))\Delta_N(x)e(Hx)\,dx\,.$$

Write this integral as $\int_{-\delta}^{\delta} + \int_{\delta}^{1-\delta} = I_1 + I_2$, say. We recall that

$$\Delta_N(x) \le \begin{cases} N & (|x| \le 1/N), \\ \dfrac{1}{4Nx^2} & (1/N \le |x| \le 1/2)\,. \end{cases}$$

Hence

$$|I_1| \le 2C\int_0^{1/N} Nx\,dx + \frac{C}{N}\int_{1/N}^{\delta} \frac{1}{x}\,dx \le \frac{C\log 4N}{N}\,.$$

The function f is bounded, say $|f(\alpha)| \le M$ for all α. Hence $|f(\alpha - x) - f(\alpha)| \le 2M$ for all x, and so

$$|I_2| \le \frac{M}{N}\int_{\delta}^{1/2} \frac{1}{x^2}\,dx \le \frac{M}{N\delta}\,.$$

Take $H = 2^k$ for some k. Then the sum on the right hand side of (9.9) includes the term $\widehat{f}(2^k)e(2^k\alpha)$, which has absolute value $\frac{1}{2}c^{-k}$. If we take $N = 2^{k-1}$, then this is the only non-zero term in the sum. On combining our estimates, we find that we have shown that $c^{-k} \le Ak/2^k$. Here A is a constant whose value may depend on C, δ, and M, but is independent of k. We thus have a contradiction, since $c < 2$, and hence c^{-k} tends to 0 more slowly than $k/2^k$. □

9.4. Inequalities

We consider several useful and interesting (but unconnected) inequalities that are easily proved using properties of Fourier Series (mainly Parseval's Identity). For the more basic inequalities relating to various kinds of mean values, see Appendix I. Properties of norms of polynomials are discussed in §9.8.

Theorem 9.3. (Hilbert's Inequality) *Let* $a_1, a_2, \ldots, a_N$ *and* $b_1, b_2, \ldots, b_N$ *be arbitrary real or complex numbers. Then*

$$\Bigg|\sum_{\substack{1\le m\le N\\ 1\le n\le N\\ m\ne n}} \frac{a_m b_n}{m-n}\Bigg| \le \pi\Big(\sum_{m=1}^N |a_m|^2\Big)^{1/2}\Big(\sum_{n=1}^N |b_n|^2\Big)^{1/2}.$$

Proof. Put

$$A(x) = \sum_{m=1}^N a_m e(mx), \qquad B(x) = \sum_{n=1}^N b_n e(nx).$$

If $f \in L^1(\mathbb{T})$, then

$$\begin{aligned}\int_0^1 f(x)A(-x)B(x)\,dx &= \sum_{m=1}^N a_m \sum_{n=1}^N b_n \int_0^1 f(x)e((n-m)x)\,dx \\ &= \sum_{m=1}^N\sum_{n=1}^N a_m b_n \widehat{f}(m-n).\end{aligned}$$

In Example 3.1 we saw that if $s(x) = 1/2 - \{x\}$ is the sawtooth function, then $\widehat{s}(k) = 1/(2\pi i k)$ for all $k \ne 0$. Hence if $f(x) = 2\pi i s(x)$, then $\widehat{f}(k) = 1/k$ for all $k \ne 0$, and $\widehat{f}(0) = 0$. Thus

$$\sum_{\substack{1\le m\le N\\ 1\le n\le N\\ m\ne n}} \frac{a_m b_n}{m-n} = \int_0^1 f(x)A(-x)B(x)\,dx.$$

Now $|f(x)| \le \pi$ for all x, so

$$\Big|\int_0^1 f(x)A(-x)B(x)\,dx\Big| \le \pi\int_0^1 |A(-x)B(x)|\,dx.$$

By the Cauchy–Schwartz inequality we know that

$$\int_0^1 |A(-x)B(x)|\,dx \le \Big(\int_0^1 |A(-x)|^2\,dx\Big)^{1/2}\Big(\int_0^1 |B(x)|^2\,dx\Big)^{1/2},$$

which by Parseval's Identity is

$$= \Big(\sum_{m=1}^M |a_m|^2\Big)^{1/2}\Big(\sum_{n=1}^N |b_n|^2\Big)^{1/2}.$$

Thus we have the stated result. □

From the above argument we see further that equality is attained only in the trivial case when one of $A(x)$, $B(x)$ is identically 0. On the other hand, the constant π is optimal, since we could have $A(-x) = \overline{B(x)}$ and choose $B(x)$ so that almost all of its mean square mass is in the interval $(0, \delta)$ where $f(x)$ is approximately $i\pi$.

Our method generalizes to give bounds for sums of the form

$$\sum_{\substack{1\le m\le N\\ 1\le n\le N}} a_m b_n c(m-n),$$

provided that the numbers $c(k)$ are the Fourier coefficients of a bounded function.

Theorem 9.4. (Wirtinger's Inequality) *Suppose that $f(x)$ has a continuous derivative on the interval $[0,1]$, and that $\int_0^1 f(x)\,dx = 0$. Then*

$$\int_0^1 |f'(x)|^2\,dx \ge 4\pi^2 \int_0^1 |f(x)|^2\,dx\,.$$

Proof. We may suppose that f has period 1. Thus $\widehat{f}(0) = 0$, so by Parseval's Identity,

$$\int_0^1 |f(x)|^2\,dx = \sum_{\substack{-\infty<n<\infty\\ n\ne 0}} |\widehat{f}(n)|^2,$$

$$\int_0^1 |f'(x)|^2\,dx = \sum_{n=-\infty}^{\infty} |2\pi i n \widehat{f}(n)|^2\,.$$

Thus

$$\int_0^1 |f'(x)|^2\,dx - 4\pi^2\int_0^1 |f(x)|^2\,dx = 4\pi^2 \sum_{\substack{-\infty<n<\infty\\ |n|\ge 2}} (n^2-1)|\widehat{f}(n)|^2 \ge 0\,.$$

□

From the above argument we see, moreover, that equality holds precisely when f is of the form $f(x) = ae(-x) + be(x)$.

Finally, we consider the classical result that a simple closed curve of given length encloses the largest area when it is a circle.

Theorem 9.5. (The Isoperimetric Inequality) *Let $\mathcal{C}$ be a simple closed curve in the plane of arc length L that encloses an area A. Then*

$$A \le \frac{L^2}{4\pi} \tag{9.10}$$

with equality if and only if $\mathcal{C}$ is a circle of radius $L/(2\pi)$.

Proof. Let the point $(x(t), y(t))$ move around the curve in the positive sense (i.e., counter-clockwise), at a constant speed L, so that the curve is traversed as t runs from 0 to 1. Here the velocity vector is $(x'(t), y'(t))$, so our speed is $L = \sqrt{x'(t)^2 + y'(t)^2}$ for all t. Hence

$$L^2 = x'(t)^2 + y'(t)^2. \tag{9.11}$$

On integrating this, we see in particular that

$$L^2 = \int_0^1 x'(t)^2 +' (t)^2 \, dt \,. \tag{9.12}$$

For the moment we assume that

$$\int_0^1 y(t)\, dt = 0. \tag{9.13}$$

By taking $P(x, y) = y$, $Q(x, y) = 0$ in Green's Theorem (see Theorem 0.20), we find that

$$A = -\int_{\mathcal{C}} y\, dx = -\int_0^1 y(t)x'(t)\, dt \,.$$

On multiplying this by 4π and subtracting the result from (9.12) we find that

$$L^2 - 4\pi A = \int_0^1 (x'(t) + 2\pi y(t))^2 \, dt + \int_0^1 y'(t)^2 \, dt - 4\pi^2 \int_0^1 y(t)^2 \, dt \,.$$

Here the first integral contributes a nonnegative amount because the integrand is the square of a real number. The combined contributions of the second and third integral is nonnegative, by Wirtinger's Inequality. Hence $4\pi A \le L^2$, as claimed.

In order to achieve equality, we must have $y(t) = ae(-x) + be(x)$ for some a and b. Since y is real-valued, we must have $a = \overline{b}$, and so $y(t) = R\sin(2\pi t + \phi)$ for some $R > 0$ and real ϕ. In order for the first integral above to vanish, we must have $x'(t) = -2\pi R \sin(2\pi t + \phi)$. Thus $x(t) = R\cos(2\pi t + \phi) + c$, and $y'(t) = 2\pi R\cos(2\pi t + \phi)$. On returning to (9.11), we deduce that $L^2 = 4\pi^2 R^2$, so $R = L/(2\pi)$ and our curve is the circle $(c + R\cos(2\pi t + \phi), R\sin(2\pi t + \phi))$. In case $\mu = \int_0^1 y(t)\, dt \ne 0$, we apply the above argument to $y(t) - \mu$, and note that

$$A = -\int_0^1 y(t)x'(t)\, dt = -\int_0^1 (y(t) - \mu)x'(t)\, dt - \mu \int_0^1 x'(t)\, dt \,.$$

This last integral is $x(1) - x(0) = 0$ because the curve is closed. That is, translation by μ in the y-coordinate does not change the area. □

Exercises

1. Let $A(x)$, $B(x)$, and $f(x)$ be as in the proof of Hilbert's Inequality (Theorem 9.3).
   (a) Show that
   $$\sum_{\substack{1\le m\le N \\ 1\le n\le N}} a_m b_m \widehat{f}(m+n) = \int_0^1 f(x)A(-x)B(-x)\, dx \,.$$
   (b) Deduce that
   $$\Bigg| \sum_{\substack{1\le m\le N \\ 1\le n\le N}} \frac{a_m b_n}{m+n} \Bigg| \le \pi \Big(\sum_{m=1}^{M} |a_m|^2 \Big)^{1/2} \Big(\sum_{n=1}^{N} |b_n|^2 \Big)^{1/2} .$$

2. Suppose that f has period 1 and a continuous second derivative on the interval $[0,1]$. Show that
$$\int_0^1 |f''(x)|^2 \geq 4\pi^2 \int_0^1 |f'(x)|^2\,dx\,.$$
3. Suppose that f has continuous first derivative on the interval $[0,c]$ where $c>0$. For $0 \leq x \leq 1$, put $g(x) = f(cx)$.
   (a) Explain why $g'(x) = cf'(cx)$ for $0 \leq x \leq 1$.
   (b) Show that
$$\int_0^1 g(x)\,dx = \frac{1}{c}\int_0^c f(y)\,dy,$$
$$\int_0^1 |g(x)|^2\,dx = \frac{1}{c}\int_0^c |f(y)|^2\,dy,$$
$$\int_0^1 |g'(x)|^2\,dx = c\int_0^c |f'(y)|^2\,dy\,.$$
   (c) Show that if $\mu = \int_0^c f(y)\,dy$, then
$$\int_0^c |f'(y)|^2 \geq \frac{4\pi^2}{c^2}\int_0^c |f(y)-\mu|^2\,dy\,.$$
4. Suppose that f has continuous first derivative on $[0,1]$. Show that
$$\int_0^1 |f(x)|^2\,dx \leq \left|\int_0^1 f(x)\,dx\right|^2 + \frac{1}{4\pi^2}\int_0^1 |f'(x)|^2\,dx\,.$$
5. Suppose that as $(x(t),y(t))$ is the parameterization of a simple closed curve of length L that encloses an area A, and that $x'(t)^2 + y'(t)^2 = L^2$ for all t.
   (a) Explain why
$$A = \int_{\mathcal{C}} x\,dy = \int_0^1 x(t)y'(t)\,dt\,.$$
   (b) Show that
$$L^2 - 4\pi A = \int_0^1 \big(y'(t) - 2\pi x(t)\big)^2\,dt + \Big(\int_0^1 x'(t)^2\,dt - 4\pi^2\int_0^1 x(t)^2\,dt\Big)\,.$$

9.5. Bernoulli polynomials

The *Bernoulli polynomials* are generated recursively by the following three properties:

(9.14) $$B_0(x) = 1,$$

(9.15) $$B_k'(x) = kB_{k-1}(x) \qquad (k \geq 1),$$

(9.16) $$\int_0^1 B_k(x)\,dx = 0 \qquad (k \geq 1)\,.$$

For example, from (9.14) and (9.15) we find that $B_1'(x) = 1$, so $B_1(x) = x + c$ for some constant c. Since $\int_0^1 x + c\,dx = 1/2 + c$, we deduce from (9.16) that $c = -1/2$.

That is, $B_1(x) = x - 1/2$. From (9.15) we have $B_2'(x) = 2B_1(x) = 2x - 1$, so $B_2(x) = x^2 - x + c$ for some c. Since $\int_0^1 x^2 - x + c\,dx = 1/3 - 1/2 + c$, we deduce that $c = 1/6$, so $B_2(x) = x^2 - x + 1/6$. The kth *Bernoulli number* is simply the constant term of the kth Bernoulli polynomial:

$$B_k = B_k(0)\,. \tag{9.17}$$

Thus $B_0 = 1$, $B_1 = -1/2$, and $B_2 = 1/6$.

Theorem 9.6. *For each nonnegative integer k,*

$$B_k(x) = \sum_{j=0}^{k} \binom{k}{j} B_{k-j} x^j\,.$$

Proof. We proceed by induction. When $k = 0$, the above asserts that $B_0(x) = B_0 = 1$, which is (9.14). Suppose the identity holds for $k - 1$. Then by (9.15) we have

$$B_k'(x) = kB_{k-1}(x) = k\sum_{j=0}^{k-1} \binom{k-1}{j} B_{k-1-j} x^j\,.$$

We integrate both sides to see that

$$B_k(x) = k\sum_{j=0}^{k-1} \binom{k-1}{j} B_{k-1-j}\frac{x^{j+1}}{j+1} + C$$

for some constant C. Since $B_k(0) = C$, by (9.17) we see that $C = B_k$. Put $i = j+1$. Then the above is

$$\sum_{i=1}^{k} \frac{k}{i}\binom{k-1}{i-1} B_{k-i} x^i + B_k\,.$$

For nonnegative integers m and r, the binomial coefficient $\binom{m}{r}$ is

$$\binom{m}{r} = \begin{cases} \dfrac{m!}{r!(m-r)!} & (0 \le r \le m), \\ 0 & (r > m)\,. \end{cases}$$

Hence

$$\frac{k}{i}\binom{k-1}{i-1} = \frac{k(k-1)!}{i(i-1)!(k-i)!} = \frac{k!}{i!(k-i)!} = \binom{k}{i}.$$

Thus we have the desired identity for k, and the proof is complete. □

Suppose that $k \ge 2$. By (9.15) and (9.16) we see that

$$0 = \int_0^1 kB_{k-1}(x)\,dx = \int_0^1 B_k'(x)\,dx = B_k(1) - B_k(0).$$

That is,

$$B_k(1) = B_k(0) = B_k \qquad (k \ge 2). \tag{9.18}$$

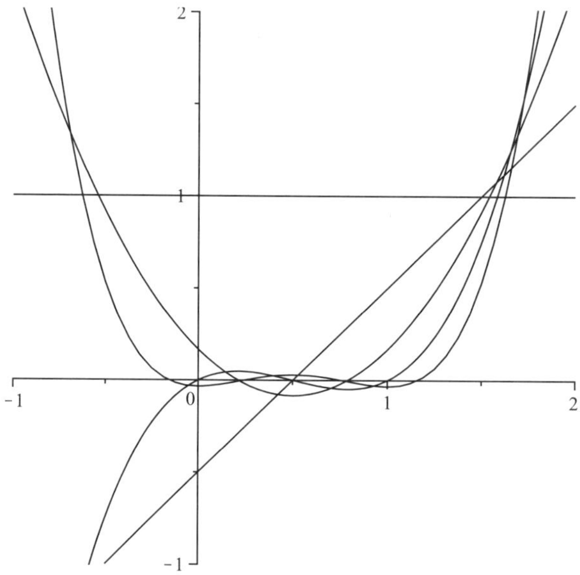

Figure 9.1. The Bernoulli polynomials $B_k(x)$ for $k = 0, 1, 2, 3, 4$ and $-1 \le x \le 2$.

The restriction to $k \ge 2$ is essential here, since $B_1(x) = x - 1/2$, and hence $B_1(1) = 1/2 \ne B_1(0) = -1/2$. On setting $x = 1$ in Theorem 9.6, and using (9.18), we discover that

$$B_k = \sum_{j=0}^{k} \binom{k}{j} B_{k-j}$$

for $k \ge 2$. When $j = 0$, the summand is B_k, so on subtracting this from both sides we find that

$$\sum_{j=1}^{k} \binom{k}{j} B_{k-j} = 0 \qquad (k \ge 2). \tag{9.19}$$

This serves to define B_{k-1} as a linear combination of B_{k-2}, B_{k-3}, $\ldots$, B_0. Since the coefficients in the linear combination are all rational, it follows that B_k is a rational number for all k.

For positive integers k let $P_k(x)$ have period 1, $P_k(x) = B_k(x)$ for $0 < x < 1$, and $P_k(0) = \big(B_k(0^+)+B_k(1^-)\big)/2$. Thus $P_1(x) = -s(x)$ where $s(x)$ is the sawtooth function as defined in Example 3.1. For $k \ge 2$, the function $P_k(x)$ is continuous, in view of (9.18). The periodic restriction $P_k(x)$ of $B_k(x)$ is interesting because of its striking Fourier Series.

Theorem 9.7. *Let k be a positive integer, and let P_k be defined as above. Then*

$$P_k(x) = -k! \sum_{n \ne 0} \frac{e(nx)}{(2\pi i n)^k} \tag{9.20}$$

for all x.

Proof. We argue by induction. In Example 3.1 we found that the sawtooth function $s(x) = 1/2 - x$ $(0 < x < 1)$ has $\widehat{s}(m) = 1/(2\pi i m)$. That the resulting Fourier Series converges to $s(x)$ for $0 < x < 1$ follows from Corollary 8.7. Since $B_1(x) = x - 1/2 = -s(x)$, we obtain (9.20) when $k = 1$. Now suppose that $k > 1$ and that (9.20) holds for $k - 1$. By (9.16) we know that $\widehat{P}_k(0) = 0$. Suppose that $n \neq 0$. By Corollary 3.8 we know that $\widehat{P'_k}(n) = 2\pi i n \widehat{P}_k(n)$ for all n. However, by (9.15) we have $P'_k(x) = kP_{k-1}(x)$. Since $\widehat{P}_{k-1}(n) = -(k-1)!/(2\pi i n)^{k-1}$ by (9.20) (for $k-1$), it follows that $\widehat{P}_k(n) = -k!/(2\pi i n)^k$. Hence the right hand side of (9.20) is the Fourier Series of $P_k(x)$. The function P_k is continuous for $k \geq 2$, and its Fourier Series is absolutely convergent, and so it converges to P_k, in view of Corollary 4.5. □

By taking $x = 0$ in (9.20), we find that if $k \geq 2$, then

$$B_k = \frac{-k!}{(2\pi i)^k} \sum_{n=1}^{\infty} \Big(\frac{1}{n^k} + \frac{1}{(-n)^k}\Big).$$

If k is odd, then the summands are all 0, and hence

(9.21) $$B_k = 0 \qquad (k \text{ odd}, k \geq 3).$$

On the other hand, when k is even we discover that

(9.22) $$\sum_{n=1}^{\infty} \frac{1}{n^{2k}} = \frac{(-1)^{k-1} 2^{2k-1} \pi^{2k} B_{2k}}{(2k)!}.$$

Since the sum on the left is positive, it follows that $\operatorname{sgn} B_{2k} = (-1)^{k-1}$. Also, when k is large, the above sum is almost exactly 1, and hence

(9.23) $$(-1)^{k-1} B_{2k} \sim \frac{(2k)!}{2^{2k-1}\pi^{2k}} \text{ as } k \to \infty.$$

We have noted that $B_1 = -1/2$, and that $B_k = 0$ if k is odd, $k > 1$. The first few nonzero Bernoulli numbers are given in Table 9.1.

Corollary 9.8. *IF k is a nonnegative integer, then $B_k(x) = (-1)^k B_k(1-x)$ identically.*

Proof. Suppose that $0 < x < 1$. Then also $0 < 1 - x < 1$. Thus (9.20) holds for x and for $1 - x$. But when x is replaced by $1 - x$, the terms corresponding to $\pm n$ are exchanged. Since $(-n)^k = (-1)^k n^k$, the effect is to multiply the right hand side by $(-1)^k$. Thus $B_k(x) = (-1)^k B_k(1-x)$ when $0 < x < 1$. That is,

$$B_k(x) - (-1)^k B_k(1-x) = 0$$

for $0 < x < 1$. However, the left hand side here is a polynomial in x, and a polynomial can have no more roots than its degree unless it is identically 0. Hence the polynomial above is identically 0, and we have the result. □

Theorem 9.9. *Let k be a nonnegative integer. Then*

(9.24) $$\frac{B_{k+1}(x+1) - B_{k+1}(x)}{k+1} = x^k$$

identically.

Table 9.1. Bernoulli numbers

k	B_k	
0	$1/1 =$	$1.00000\,00000$
1	$-1/2 =$	$-0.50000\,00000$
2	$1/6 =$	$0.16666\,66667$
4	$-1/30 =$	$-0.03333\,33333$
6	$1/42 =$	$0.02380\,95238$
8	$-1/30 =$	$-0.03333\,33333$
10	$5/66 =$	$0.07575\,75758$
12	$-691/2730 =$	$-0.25311\,35531$
14	$7/6 =$	$1.16666\,66667$
16	$-3617/510 =$	$-7.09215\,68627$
18	$43867/798 =$	$54.97117\,79449$
20	$-174611/330 =$	$-529.12424\,24242$
22	$854513/138 =$	$6192.12318\,84058$
24	$-236364091/2730 =$	$-86580.25311\,35531$

Proof. We induct on k. When $k = 0$, we find that $B_1(x+1) - B_1(x) = (x+1) - 1/2 - (x - 1/2) = 1 = x^0$, so the assertion holds when $k = 0$. Now suppose that $k > 0$ and that the identity holds for $k - 1$. That is,

$$\frac{B_k(x+1) - B_k(x)}{k} = x^{k-1}\,. \tag{9.25}$$

By (9.15) we see that the derivative of the left hand side of (9.24) is $B_k(x+1) - B_k(x)$. The derivative of the right hand side of (9.24) is kx^{k-1}. These two quantities are equal, in view of (9.25). Since the two sides of (9.24) have the same derivative, it follows that they differ at most by a constant. However, when we take $x = 0$, the right hand side is 0, and the left hand side is $B_{k+1}(1) - B_{k+1}(0) = 0$ because of (9.18). Thus the constant in question is 0, and so we have (9.24). □

Corollary 9.10. *Let k be a positive integer. Then*

$$\sum_{m=1}^{n-1} m^k = \frac{B_{k+1}(n) - B_{k+1}}{k+1}\,. \tag{9.26}$$

Proof. In (9.24), take $x = 0, 1, 2, \ldots, n-1$, and sum. The left hand side telescopes, and we obtain the stated result. □

We note the initial instances of (9.26):

(9.27)
$$\begin{aligned}
\sum_{m=1}^{n-1} m &= \frac{1}{2}n^2 - \frac{1}{2}n, \\
\sum_{m=1}^{n-1} m^2 &= \frac{1}{3}n^3 - \frac{1}{2}n^2 + \frac{1}{6}n, \\
\sum_{m=1}^{n-1} m^3 &= \frac{1}{4}n^4 - \frac{1}{2}n^3 + \frac{1}{4}n^2, \\
\sum_{m=1}^{n-1} m^4 &= \frac{1}{5}n^5 - \frac{1}{2}n^4 + \frac{1}{3}n^3 - \frac{1}{30}n, \\
\sum_{m=1}^{n-1} m^5 &= \frac{1}{6}n^6 - \frac{1}{2}n^5 + \frac{5}{12}n^4 - \frac{1}{12}n^2, \\
\sum_{m=1}^{n-1} m^6 &= \frac{1}{7}n^7 - \frac{1}{2}n^6 + \frac{1}{2}n^5 - \frac{1}{6}n^3 + \frac{1}{42}n,
\end{aligned}$$

It might seem more natural to sum from 1 to n, but the alteration required is predictable, in view of Theorem 9.9:

(9.28)
$$\begin{aligned}
\sum_{m=1}^{n} m &= \frac{1}{2}n^2 + \frac{1}{2}n, \\
\sum_{m=1}^{n} m^2 &= \frac{1}{3}n^3 + \frac{1}{2}n^2 + \frac{1}{6}n, \\
\sum_{m=1}^{n} m^3 &= \frac{1}{4}n^4 + \frac{1}{2}n^3 + \frac{1}{4}n^2, \\
\sum_{m=1}^{n} m^4 &= \frac{1}{5}n^5 + \frac{1}{2}n^4 + \frac{1}{3}n^3 - \frac{1}{30}n, \\
\sum_{m=1}^{n} m^5 &= \frac{1}{6}n^6 + \frac{1}{2}n^5 + \frac{5}{12}n^4 - \frac{1}{12}n, \\
\sum_{m=1}^{n} m^6 &= \frac{1}{7}n^7 + \frac{1}{2}n^6 + \frac{1}{2}n^5 - \frac{1}{6}n^3 + \frac{1}{42}n.
\end{aligned}$$

Suppose that $f(x)$ has continuous first derivative on $[0, 1]$. Then by integrating by parts we find that

$$\begin{aligned}
\int_0^1 f(x)\,dx &= \Big[(x-1/2)f(x)\Big|_0^1 - \int_0^1 (x-1/2)f'(x)\,dx \\
&= \frac{1}{2}f(0) + \frac{1}{2}f(1) - \int_0^1 (x-1/2)f'(x)\,dx\,.
\end{aligned}$$

For the interval $[n, n+1]$ we have similarly

$$\int_n^{n+1} f(x)\,dx = \frac{1}{2}f(n) + \frac{1}{2}f(n+1) - \int_n^{n+1} (x-n-1/2)f'(x)\,dx\,.$$

Here $x - n - 1/2 = B_1(\{x\})$. On summing this, we find that

$$\sum_{n=M+1}^{N} f(n) = \int_M^N f(x)\,dx - \frac{1}{2}f(M) + \frac{1}{2}f(N) + \int_M^N B_1(\{x\})f'(x)\,dx\,.$$

If f has a continuous second derivative, then we may integrate by parts again:

$$\int_M^N B_1(\{x\})f'(x)\,dx = \Big[\frac{1}{2}B_2(\{x\})f'(x)\Big|_M^N - \frac{1}{2}\int_M^N B_2(\{x\})f''(x)\,dx\,.$$

If f has continuous derivatives through the order K, then after K integrations by parts we find that

$$\begin{aligned}\sum_{n=M+1}^{N} f(n) = \int_M^N f(x)\,dx + \sum_{k=1}^{K}\frac{(-1)^k B_k}{k!}\big(f^{(k-1)}(N) - f^{(k-1)}(M)\big) \\ + \frac{(-1)^{K-1}}{K!}\int_M^N B_K(\{x\})f^{(K)}(x)\,dx\,.\end{aligned} \tag{9.29}$$

This is the *Euler–Maclaurin summation formula.* Here we are assuming that M and N are integers with $M < N$.

Example 9.1. Take $f(x) = 1/x$, $M = 1$, and $K = 1$ in the above. Then we find that

$$\sum_{n=2}^{N}\frac{1}{n} = \log N + \frac{1}{2N} - \frac{1}{2} - \int_1^N \frac{B_1(\{x\})}{x^2}\,dx\,.$$

Put

$$\gamma = \frac{1}{2} - \int_1^\infty \frac{B_1(\{x\})}{x^2}\,dx = 0.5772156649\ldots\,.$$

This is known as *Euler's constant.* We see that

$$\sum_{n=1}^{N}\frac{1}{n} = \log N + \gamma + R_N$$

where

$$R_N = \frac{1}{2N} + \int_N^\infty \frac{B_1(\{x\})}{x^2}\,dx\,.$$

Since $-1/2 \le B_1(\{x\}) \le 1/2$ for all x, we have

$$\frac{-1}{2N} = \frac{-1}{2}\int_N^\infty \frac{1}{x^2}\,dx \le \int_N^\infty \frac{B_1(\{x\})}{x^2}\,dx \le \frac{1}{2}\int_N^\infty \frac{1}{x^2}\,dx = \frac{1}{2N}\,.$$

Thus $0 \le R_N \le 1/N$, which is to say that

$$\log N + \gamma \le \sum_{n=1}^{N}\frac{1}{n} \le \log N + \gamma + \frac{1}{N}\,.$$

If more precision were desired, we could integrate by parts to find that

$$R_N = \frac{1}{2N} + O\Big(\frac{1}{N^2}\Big)\,.$$

If still more precision were desired, then we could integrate by parts more times, which would amount to applying the Euler–Maclaurin summation formula with a larger value of K.

Exercises

1. Show that $B_k(1/2) = 0$ if k is odd.
2. Show that
$$\sum_{n=0}^{\infty} \frac{(-1)^n}{(2n+1)^3} = \frac{\pi^3}{32}\,.$$
3. Show that $(-1)^k B_k(-x) = B_k(x) + kx^{k-1}$ for all $k \geq 1$.
4. Let k be a nonnegative integer. Show that for all x and y,
$$B_k(x+y) = \sum_{j=0}^{k} \binom{k}{j} B_{k-j}(x) y^j\,.$$
5. (a) Use (9.23) to show that the power series
$$\sum_{k=0}^{\infty} \frac{B_k}{k!} z^k$$
converges for $|z| < 2\pi$.
   (b) Write
$$(e^z - 1)\sum_{k=0}^{\infty} \frac{B_k}{k!} z^k = \Big(\sum_{\ell=1}^{\infty} \frac{1}{\ell!} z^\ell\Big)\Big(\sum_{k=0}^{\infty} \frac{B_k}{k!} z^k\Big) = \sum_{m=1}^{\infty} \frac{c_m}{m!} z^m\,.$$
Show that
$$c_m = \sum_{j=1}^{m} \binom{m}{j} B_{m-j}.$$
   (c) Deduce that $c_1 = 1$ and that $c_m = 0$ for all $m > 1$.
   (d) Conclude that
$$\frac{z}{e^z - 1} = \sum_{k=0}^{\infty} \frac{B_k}{k!} z^k$$
for $|z| < 2\pi$.
6. Let j and k be positive integers. Use Theorem 9.7 to show that
$$\int_0^1 B_j(x) B_k(x)\, dx = (-1)^{k-1} \frac{j!k!}{(j+k)!} B_{j+k}\,.$$
7. Let k be a nonnegative integer and q a positive integer.
   (a) Use Theorem 9.7 to show that
$$P_k(qx) = q^{k-1} \sum_{a=0}^{q-1} P_k(x + a/q)$$
for all x.

(b) Deduce that

$$B_k(qx) = q^{k-1}\sum_{a=0}^{q-1} B_k(x + a/q)$$

for $0 < x < 1/q$.

(c) Deduce that the above holds for all x.

8. Show that $B_k(1/2) = -(1 - 2^{1-k})B_k$ for all $k \ge 1$.

9. Let m and n denote positive integers. Show that

$$\int_0^1 P_1(mx)P_1(nx)\,dx = \frac{(m,n)^2}{12mn}$$

where (m, n) denotes the greatest common divisor of m and n.

10. (a) Use the Euler–Maclaurin summation formula to show that

$$\log(N!) = N\log N - N + \frac{1}{2}\log N + c + \frac{1}{12N} - \frac{1}{2}\int_N^\infty \frac{P_2(x)}{x^2}\,dx$$

where

$$c = \frac{11}{12} + \frac{1}{2}\int_1^\infty \frac{P_2(x)}{x^2}\,dx\,.$$

(b) Show that $-1/12 \le P_2(x) \le 1/6$ for all x.

(c) Deduce that

$$\frac{-1}{12N} \le \int_N^\infty \frac{P_2(x)}{x^2}\,dx \le \frac{1}{6N}\,.$$

(d) Deduce that

$$N\log N - N + \frac{1}{2}\log N + c \le \log(N!) \le N\log N - N + \frac{1}{2}\log N + c + \frac{1}{8N}\,.$$

(e) Put $C = e^c$. Show that

$$C\sqrt{N}\Big(\frac{N}{e}\Big)^N \le N! \le C\sqrt{N}\Big(\frac{N}{e}\Big)^N \exp(1/(8N))\,.$$

(f) Show that $1 \le \exp(1/(8N)) \le 1 + 1/N$ for all positive integers N.

(g) Conclude that

$$C\sqrt{N}\Big(\frac{N}{e}\Big)^N \le N! \le C\sqrt{N}\Big(\frac{N}{e}\Big)^N (1 + 1/N)\,. \tag{9.30}$$

Stirling's formula asserts that

$$N! \sim \sqrt{2\pi N}\Big(\frac{N}{e}\Big)^N \quad \text{as } N \to \infty\,. \tag{9.31}$$

In (9.30) we have a quantitative form of this, apart from the need to show that $C = \sqrt{2\pi}$. This issue is settled in the next exercise.

11. For nonnegative integers n, put $I_n = \int_0^1 (\sin \pi x)^n\,dx$.

(a) Show that $I_0 = 1$ and that $I_1 = 2/\pi$.

(b) Show that

$$I_{n+2} = I_n - \int_0^1 (\sin \pi x)^n (\cos \pi x)^2\,dx\,.$$

(c) Note that if $f(x) = (\sin \pi x)^{n+1}$, then $f'(x) = \pi(n+1)(\sin \pi x)^n \cos \pi x$. Use integration by parts to show that

$$\int_0^1 (\sin \pi x)^n (\cos \pi x)^2 \, dx = \frac{1}{n+1} \int_0^1 f(x) \sin \pi x \, dx = I_{n+2}/(n+1) .$$

(d) Deduce that

$$I_{n+2} = \frac{n+1}{n+2} I_n .$$

(e) Show that $I_0 \ge I_1 \ge I_2 \ge \ldots$.

(f) Show that $I_{n+2}/I_n \to 1$ as $n \to \infty$.

(g) Deduce that $I_{n+1}/I_n \to 1$ as $n \to \infty$. Show that

$$\frac{I_{2n}}{I_{2n+1}} = \Big(1 - \frac{1}{4n^2}\Big) \frac{I_{2n-2}}{I_{2n-1}} .$$

(h) Use induction to show that

$$\frac{I_{2n}}{I_{2n+1}} = \frac{I_0}{I_1} \prod_{k=1}^{n} \Big(1 - \frac{1}{4k^2}\Big) .$$

(i) Deduce that

$$\prod_{k=1}^{\infty} \Big(1 - \frac{1}{4k^2}\Big) = \frac{2}{\pi} .$$

This is *Wallis's formula*; it dates from 1656.

(j) Show that

$$\prod_{k=1}^{n} \Big(1 - \frac{1}{4k^2}\Big) = (2n+1) \frac{(2n)!^2}{n!^4 2^{4n}} .$$

(k) Use Stirling's formula in the form of 10 (g) to show that the above tends to $4/C^2$ as $n \to \infty$. Deduce that $C = \sqrt{2\pi}$.

12. Use appropriate software to complete the following tasks.
(a) For $n = 10, 20, \ldots, 100$, compute the decimal value of $n!/\big(\sqrt{2\pi n}(n/e)^n\big)$.
(b) For $n = 10, 20, \ldots, 100$, compute the decimal value of $n! - \sqrt{2\pi n}(n/e)^n$.
(c) Let c_n be determined by the identity $n! = \sqrt{2\pi n}(n/e)^n\big(1 + c_n/n\big)$. For $n = 10, 20, \ldots,$ 100, compute the decimal value of c_n. Compare these values with the decimal expansion of $1/12$.

9.6. Uniform distribution

We say that a sequence $u_1, u_2, \ldots$ of real numbers is *uniformly distributed modulo* 1 if

$$\lim_{N \to \infty} \frac{1}{N} \sum_{\substack{1 \le n \le N \\ 0 \le \{u_n\} \le b}} 1 = b \tag{9.32}$$

for all real numbers $b \in [0,1]$. Suppose that f has period 1. It is reasonable to ask whether

$$\lim_{N\to\infty} \frac{1}{N}\sum_{n=1}^{N} f(u_n) = \int_0^1 f(x)\,dx\,. \tag{9.33}$$

If f has period 1 and

$$f(x) = \begin{cases} 1 & (0 \le x \le b), \\ 0 & (b < x < 1), \end{cases} \tag{9.34}$$

then (9.33) is exactly the same as (9.32). Thus to say that a sequence is uniformly distributed modulo 1 is to say that (9.33) holds for functions of the type (9.34). Suppose that $f_1, f_2, \ldots, f_K$ are functions with period 1, and that (9.33) holds for each f_k. Then (9.33) holds also for any linear combination $f(x) = a_1 f_1(x) + \cdots + a_K f_K(x)$ of the f_k. When we take a linear combination of functions of the type (9.34) we obtain a step function, and any step function can be expressed as a linear combination of functions of the type (9.34). Hence a sequence is uniformly distributed modulo 1 if and only if (9.33) holds for all step functions with period 1.

Let $\mathcal{F}$ denote a family of functions in $L^1(\mathbb{T})$ for which (9.33) holds, and suppose that g is a further function with period 1. If for every $\varepsilon > 0$ there are functions f_+ and f_- in $\mathcal{F}$ with the two properties

$$f_-(x) \le g(x) \le f_+(x) \text{ for all } x, \tag{9.35a}$$

$$\int_0^1 f_+(x)\,dx \le \int_0^1 g(x)\,dx + \varepsilon, \qquad \int_0^1 f_-(x)\,dx \ge \int_0^1 g(x)\,dx - \varepsilon, \tag{9.35b}$$

then (9.33) holds also for g. To see why this is so, note that from 9.35a it follows that

$$\frac{1}{N}\sum_{n=1}^{N} f_-(u_n) \le \frac{1}{N}\sum_{n=1}^{N} g(u_n) \le \frac{1}{N}\sum_{n=1}^{N} f_+(u_n)\,. \tag{9.36}$$

From (9.35b) we also see that

$$\int_0^1 f_+(x)\,dx - \varepsilon \le \int_0^1 g(x)\,dx \le \int_0^1 f_-(x)\,dx + \varepsilon\,.$$

We multiply all members by -1, which has the effect of reversing the inequalities. Thus we find that

$$-\int_0^1 f_-(x)\,dx - \varepsilon \le -\int_0^1 g(x)\,dx \le -\int_0^1 f_+(x)\,dx + \varepsilon\,.$$

We add these inequalities to the inequalities (9.36) to find that

$$\begin{aligned} \frac{1}{N}\sum_{n=1}^{N} f_-(u_n) - \int_0^1 f_-(x)\,dx - \varepsilon &\le \frac{1}{N}\sum_{n=1}^{N} g(u_n) - \int_0^1 g(x)\,dx \\ &\le \frac{1}{N}\sum_{n=1}^{N} f_+(u_n) - \int_0^1 f_+(x)\,dx + \varepsilon\,. \end{aligned}$$

Since (9.33) holds for f_-, the left hand member above tends to $-\varepsilon$ as $N \to \infty$. Hence it is $> -2\varepsilon$ for all sufficiently large N. Since (9.33) also holds for f_+, the quantity on the lower right above is tending to ε as $N \to \infty$. Hence it is $< 2\varepsilon$ for all sufficiently large N. Thus the middle member above lies in the interval $[-2\varepsilon, 2\varepsilon]$ for all sufficiently large N. Since this is true no matter how small ε is, it follows that this middle member tends to 0 as $N \to \infty$. That is, (9.33) holds for g.

If g is a real-valued Riemann-integrable function with period 1, then for any $\varepsilon > 0$ there exist step functions f_- and f_+ such that 9.35a and 9.35b hold. Indeed, this is essentially what it means for a function to be Riemann-integrable. We have seen that if the sequence $\{u_n\}$ is uniformly distributed modulo 1, then (9.33) holds for all step functions. We now see that (9.33) therefore holds also for any real-valued Riemann-integrable function with period 1. If f is complex valued, then we may write $f(x) = f_r(x) + if_i(x)$, and f is Riemann-integrable if f_r and f_i are. Thus (9.33) holds also for complex-valued Riemann-integrable functions with period 1. A continuous function is Riemann-integrable, and the functions $e(nx)$ are continuous and have period 1, so if $\{u_n\}$ is uniformly distributed modulo 1, then (9.33) holds for $f(x) = e(kx)$. That is,

$$\lim_{N\to\infty} \frac{1}{N} \sum_{n=1}^{N} e(ku_n) = \begin{cases} 1 & (k = 0), \\ 0 & (k \neq 0). \end{cases} \tag{9.37}$$

We continue our discussion by proving that the converse of the above is also true. That is, if (9.37) holds for all integers k, then the sequence $\{u_n\}$ is uniformly distributed modulo 1. We have observed that if (9.33) holds for several functions, then it holds for any linear combination of them. Thus if (9.33) holds for the functions $e(kx)$, then (9.33) holds for any trigonometric polynomial. Let f be as in (9.34). We want to show that (9.33) holds for f. To this end, we show that there exist trigonometric polynomials f_- and f_+ such that 9.35a and 9.35b hold. We do this in two steps. First we approximate to f above and below by continuous functions, and then we approximate these continuous functions uniformly by trigonometric polynomials. Let g_- and g_+ have period 1, with

$$g_-(x) = \begin{cases} -\varepsilon/2 + 2x/\varepsilon & (0 \le x \le \varepsilon/2), \\ 1 - \varepsilon/2 & (\varepsilon/2 \le x \le b - \varepsilon/2), \\ -\varepsilon/2 - 2(x-b)/\varepsilon & (b - \varepsilon/2 \le x \le b), \\ -\varepsilon/2 & (b \le x \le 1). \end{cases}$$

$$g_+(x) = \begin{cases} 1 + \varepsilon/2 & (0 \le x \le b), \\ 1 + \varepsilon/2 - 2(x-b)/\varepsilon & (b \le x \le b + \varepsilon/2), \\ \varepsilon/2 & (b + \varepsilon/2 \le x \le 1 - \varepsilon/2), \\ 1 + \varepsilon/2 + 2(x-1)/\varepsilon & (1 - \varepsilon/2 \le x \le 1). \end{cases}$$

These are continuous piece-wise linear functions, and it is easy to verify that

$$g_-(x) \le f(x) - \varepsilon/2, \qquad g_+(x) \ge f(x) + \varepsilon/2 \tag{9.38}$$

for all x, as depicted in Figure 9.2. Also, it is easy to show that

$$\int_0^1 g_-(x)\,dx = b - \varepsilon/2, \qquad \int_0^1 g_+(x)\,dx = b + \varepsilon/2\,. \tag{9.39}$$

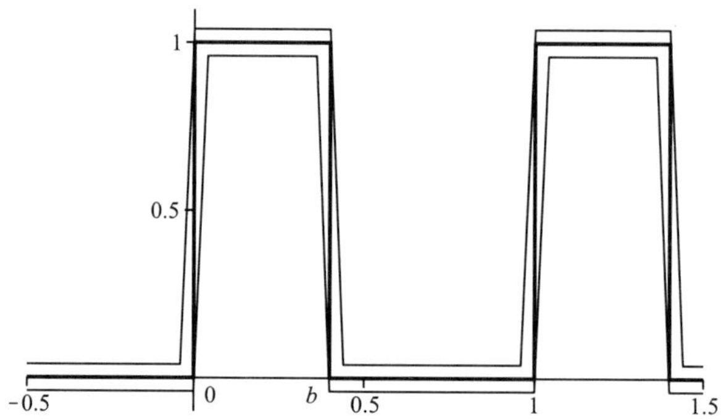

Figure 9.2. Approximations $g_{\pm}(x)$ to $f(x)$.

Since $g_{\pm}(x)$ are continuous functions, by Theorem 4.1 there exist trigonometric polynomials $f_{\pm}(x)$ such that $|g_{\pm}(x) - f_{\pm}(x)| \le \varepsilon/2$ for all x. From (9.38) it follows that $f_(x) \le g(x) \le f_+(x)$ for all x. That is, we have (9.35a). Since

$$\left| \int_0^1 h_{\pm}(x)\,dx - \int_0^1 f_{\pm}(x)\,dx \right| < \varepsilon/2,$$

from (9.39) it follows that

$$\int_0^1 f_-(x)\,dx \ge b - \varepsilon, \qquad \int_0^1 f_+(x)\,dx \le b + \varepsilon\,.$$

That is, we have (9.35b). Since (9.33) holds for f_- and f_+, and (9.35a) and (9.35b) hold for arbitrarily small ε, it follows that (9.33) holds for f. That is, the sequence $\{u_n\}$ is uniformly distributed modulo 1. Thus we have proved

Theorem 9.11. (Weyl's Criterion) *Let $\{u_n\}$ be an infinite sequence of real numbers. Then the following are equivalent:*

(a) *The sequence $\{u_n\}$ is uniformly distributed modulo* 1;

(b) $\lim\limits_{N\to\infty} \frac{1}{N} \sum\limits_{n=1}^{N} f(u_n) = \int_0^1 f(x)\,dx$ *for all Riemann-integrable functions f with period* 1;

(c) $\lim\limits_{N\to\infty} \frac{1}{N} \sum\limits_{n=1}^{N} e(ku_n) = 0$ *for all non-zero integers k.*

By the above, we have reduced the problem of establishing uniform distribution to that of estimating an exponential sum. However, estimating exponential sums is not always an easy matter. We begin with a simple case.

Example 9.2. If α is an irrational number, then the sequence $u_n = n\alpha$ is uniformly distributed modulo 1. To see why this is so, note by (4.9) that

$$\left| \sum_{n=1}^{N} e(kn\alpha) \right| \le \frac{1}{|\sin \pi k\alpha|} \le \frac{1}{2\|k\alpha\|}\,.$$

Since α is irrational, the number $k\alpha$ is not an integer, and hence the denominator above is not 0. Since this is a bounded function of N, part (c) of Weyl's Criterion holds, and so the sequence is uniformly distributed.

In order to estimate more general exponential sums, we next prove a useful inequality.

Lemma 9.12. (van der Corput's Inequality) *Let $z_1, z_2, \ldots, z_N$ be arbitrary complex numbers. Then for any integer H with $1 \le H \le N$ we have*

$$H^2\left|\sum_{n=1}^{N} z_n\right|^2 \le H(N+H-1)\sum_{n=1}^{N}|z_n|^2 + 2(N+H-1)\sum_{h=1}^{H-1}(H-h)\left|\sum_{n=1}^{N-h} z_{n+h}\overline{z_n}\right|.$$

Proof. It suffices to show that

$$H^2\left|\sum_{n=1}^{N} z_n\right|^2 \le H(N+H-1)\sum_{n=1}^{N}|z_n|^2 + 2(N+H-1)\operatorname{Re}\sum_{h=1}^{H-1}(H-h)\sum_{n=1}^{N-h} z_{n+h}\overline{z_n}, \tag{9.40}$$

since then the stated result follows by the triangle inequality and the inequality $|\operatorname{Re} z| \le |z|$.

In order to ease the task of computing the ranges of summation, we suppose that $z_n = 0$ when $n < 1$ or $n > N$. Our argument follows our derivation of the closed form formula for the Fejér kernel $\Delta_N(x)$ (recall (4.8)). We write the sum $\sum_{n=1}^{N} z_n$ with H different shifts of the indices:

$$H\sum_{n=1}^{N} z_n = \sum_{0\le r<H}\ \sum_{0<n<N+H} z_{n-r}\,.$$

Next we invert the order of summation to see that the above is

$$= \sum_{0<n<N+H}\ \sum_{0\le r<H} z_{n-r}\,.$$

By Cauchy's inequality it follows that

$$H^2\left|\sum_{n=1}^{N} z_n\right|^2 \le (N+H-1)\sum_{0<n<N+H}\left|\sum_{0\le r<H} z_{n-r}\right|^2.$$

On multiplying out the square on the right, and inverting the order of summation we see that this is

$$= (N+H-1)\sum_{\substack{0\le r<H\\ 0\le s<H}}\ \sum_{n} z_{n-r}\overline{z_{n-s}}.$$

The inner sum depends only on $r-s$, and a given value h of $r-s$ occurs for $H-|h|$ different pairs r, s. Thus the above is the right hand side of (9.40), so the desired result now follows. □

Theorem 9.13. (van der Corput) *Let $\{u_n\}$ be a sequence of real numbers with the property that, for each positive integer h, the sequence $\{u_{n+h}-u_n\}$ is uniformly distributed. Then the sequence $\{u_n\}$ is uniformly distributed.*

From Example 9.2 we see that the converse of the above is false.

Proof. By Weyl's criterion (Theorem 9.11) we know that it suffices to show that

$$\lim_{N\to\infty} \frac{1}{N}\sum_{n=1}^{N} e(ku_n) = 0$$

for any positive integer k. Taking $z_n = e(ku_n)$ in van der Corput's Inequality (Lemma) 9.12), we see that

$$H^2\Big|\sum_{n=1}^{N} e(ku_n)\Big|^2 \le H(N+H-1)N+2(N+H-1)\sum_{h=1}^{H-1}(H-h)\Big|\sum_{n=1}^{N-h} e(k(u_{n+h}-u_n))\Big|.$$

We divide both sides by H^2N^2 to see that

$$\Big|\frac{1}{N}\sum_{n=1}^{N} e(ku_n)\Big|^2 \le \frac{N+H-1}{NH} + 2\frac{N+H-1}{NH}\sum_{h=1}^{H-1}\Big|\frac{1}{N}\sum_{n=1}^{N-h} e(k(u_{n+h}-u_n))\Big|.$$

Since the sequence $\{u_{n+h} - u_n\}$ is uniformly distributed, we find by a second application of Weyl's criterion that the innermost sum on the right is small compared with N. Thus the right hand side above is $< \frac{C}{N} + \varepsilon$ for all large N. Since H may be taken arbitrarily large, it follows that the sum on the left hand side is small compared with N. Thus the proof is complete. □

Corollary 9.14. (Weyl) *Let $P(x) = \sum_{j=0}^{k} c_j x^j$ be a polynomial with real coefficients, $k > 0$, and suppose that c_k is irrational. Then the sequence $\{P(n)\}$ is uniformly distributed modulo 1.*

Proof. We induct on the degree. The basis of our induction is established in Example 9.2. Let $P_h(x) = P(x+h) - P(x)$. Then P_h has degree $k-1$ and leading coefficient hkc_k, which is irrational. Hence by the inductive hypothesis, the sequence $\{P_h(n)\}$ is uniformly distributed, for every positive integer h. Thus $\{P(n)\}$ is uniformly distributed. □

Some uniformly distributed sequences are more uniformly distributed than others, in the sense that the limiting uniformity is more quickly approached. In Figure 9.3(a) we have 1000 points $(n/N, \{\sqrt{2}n\})$ that are nearly as well-distributed as a sequence can be. However, in Figure 9.3(b), we have 1000 points $(n/N, \{\sqrt{2}n^2\})$ that seem to be much more irregularly distributed, although by Corollary 9.14 we know that in the limit such points tend toward uniform distribution.

If we choose an a, $0 < a < 1$, and consider that part of Figure 9.3(a) with the x-coordinate between 0 and a, then $n \le aN$. As is visually suggestive, the number of these n for which $\{\sqrt{2}n\} \le b$ is very close to the expected number, abN, as depicted in Figure 9.4(a). However, it appears in Figure 9.4(b) that if we were to count points in a rotated square, then we might encounter a larger deviation from the expected number of points. Similarly, in Figure 9.3(b) it seems that we might have a relatively large error if we were to count points under certain circular arcs.

Thus far we have discussed Weyl's criterion as a qualitative principle, but we are in a position to construct quantitative bounds. For a given infinite sequence u_n, let $Z(N; a, b)$ denote the number of n, $1 \le n \le N$, for which $u_n \in [a, b]$. Here we are assuming that $a \le b \le a+1$. If the sequence is tending toward uniform

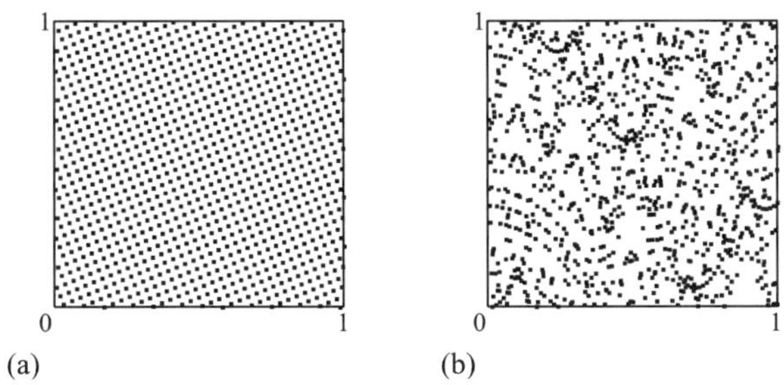

Figure 9.3. (a) Points $(n/N, \{\sqrt{2}n\})$; (b) points $(n/N, \{\sqrt{2}n^2\})$, both with $N = 1000$.

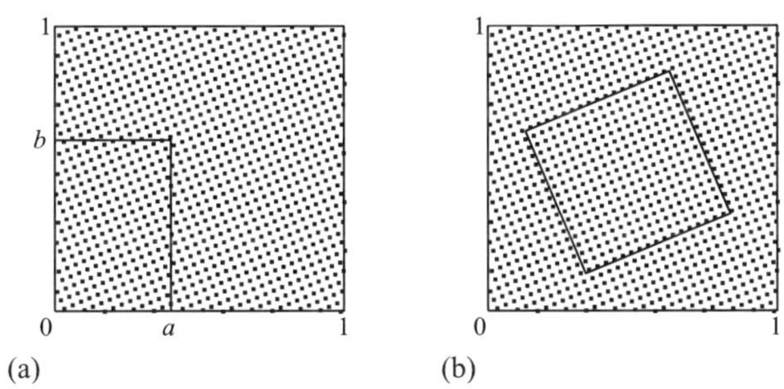

Figure 9.4. (a) Points in a rectangle. (b) Points in a rotated square.

distribution, then $Z(N; a, b)$ will be near $(b-a)N$ when N is large. The *discrepancy* of the sequence is

$$D(N) = \sum_{\substack{a,b \\ a \le b \le a+1}} |Z(N; a, b) - (b-a)N| \,. \tag{9.41}$$

By a simple compactness argument it can be shown that the sequence u_n is uniformly distributed (mod 1) if and only if $D(N) = o(N)$ as $N \to \infty$. When we have quantitative upper bounds for the exponential sums that arise in Weyl's criterion, we can derive a quantitative bound for the discrepancy. Let

$$f(x) = b - a + s(x-a) + s(b-x)$$

where $s(x)$ is the sawtooth function. We note that

$$f(x) = \begin{cases} 1 & a < x < b (\text{mod } 1), \\ 1/2 & x = a \text{ or } x = b, \\ 0 & \text{otherwise.} \end{cases}$$

Thus $f(x)$ is almost the characteristic function of the interval $[a, b]$. Let $B_K(x)$ be the Beurling trigonometric polynomial defined in (6.8), and put

$$S_K^+(x) = b - a + B_K(x-a) + B_K(b-x), \tag{9.42}$$

$$S_K^-(x) = b - a - B_K(a-x) - B_K(x-b)\,. \tag{9.43}$$

These are the *Selberg trigonometric polynomials.* From Theorem 6.5 we know that $S_K^-(x) \le f(x) \le S_K^+(x)$ for all x. Since the functions $S_K^\pm(x)$ are continuous, it follows that $S_K^-(x) \le \chi_{[a,b]} \le S_K^+(x)$ for all x. Also from Theorem 6.5 we know that

$$\widehat{S_K^\pm}(0) = \int_0^1 S_K^\pm(x)\,dx = b - a \pm \frac{1}{K+1}\,. \tag{9.44}$$

These inequalities allow us to bound the discrepancy in terms of exponential sums.

Theorem 9.15. *Let $\{u_n\}_{n=1}^N$ be a sequence of points in $\mathbb{T}$, and let $\mathcal{I} = [a,b]$ be an interval of $\mathbb{T}$ with $a \le b \le a+1$. Then*

(9.45)

$$|Z(N;a,b) - (b-a)N| \le \frac{N}{K+1} + 2\sum_{k=1}^K \left(\frac{1}{K+1} + \min\left(b-a, \frac{1}{\pi k}\right)\right)\left|\sum_{n=1}^N e(ku_n)\right|$$

for positive integers N and K.

Proof. We see that

$$\begin{aligned} Z(N;a,b) &= \sum_{n=1}^N \chi_{\mathcal{I}}(u_n) \le \sum_{n=1}^N S_K^+(u_n) \\ &= \sum_{n=1}^N \sum_{k=-K}^K \widehat{S_K^+}(k) e(ku_n) = \sum_{k=-K}^K \widehat{S_K^+}(k) \sum_{n=1}^N e(ku_n)\,. \end{aligned}$$

Hence by (9.44),

$$Z(N;a,b) - (b-a)N \le \frac{N}{K+1} + \sum_{\substack{k=-K \\ k\ne 0}}^K \left|\widehat{S_K^+}(k)\right| \left|\sum_{n=1}^N e(ku_n)\right|.$$

We note that

$$|\widehat{\chi_{\mathcal{I}}}(k)| = \left|\frac{\sin \pi k(b-a)}{\pi k}\right|$$

for $k \ne 0$. Since $|\sin x| \le \min(|x|, 1)$, it follows that the above is

$$\le \min\left(b-a, \frac{1}{\pi k}\right).$$

By (3.15) applied to the function $B_K^+(x) - \chi_{\mathcal{I}}(x)$ we deduce that

$$\left|\widehat{B_K^+}(k) - \widehat{\chi_{\mathcal{I}}}(k)\right| \le \int_0^1 B_K^+(x) - \chi_{\mathcal{I}}(x)\,dx = \frac{1}{K+1}\,.$$

Thus

$$\left|\widehat{B_K^+}(k)\right| \le \frac{1}{K+1} + \min\left(b-a, \frac{1}{\pi k}\right),$$

so

$$Z(N;a,b) - (b-a)N \le \frac{N}{K+1} + 2\sum_{k=1}^K \left(\frac{1}{K+1} + \min\left(b-a, \frac{1}{\pi k}\right)\right)\left|\sum_{n=1}^N e(ku_n)\right|.$$

The lower bound is proved similarly, using $S_K^-(x)$, so the proof is complete. □

Since

$$\frac{1}{K+1}+\frac{1}{\pi k}\le\frac{3}{2k},$$

the following corollary is immediate.

Corollary 9.16. (The Erdős–Turán Inequality) *For any sequence* $\{u_n\}_{n=1}^N$ *and any positive integer* K,

$$D(N)\le\frac{N}{K+1}+3\sum_{k=1}^K\Big|\sum_{n=1}^N e(ku_n)\Big|.$$

A sequence that is tending toward uniform distribution will not have long gaps, but if our object is to bound gaps rather than the discrepancy, then the theorem above gives a stronger estimate than the corollary.

Corollary 9.17. *Suppose that* K *is a positive integer such that*

$$\sum_{k=1}^K\Big|\sum_{n=1}^N e(ku_n)\Big|<\frac{N}{10}. \tag{9.46}$$

Then every interval $\mathcal{I}=[a,b]$ *of* $\mathbb{T}$ *of length* $b-a>4/(K+1)$ *contains at least* $(b-a)N/2$ *of the points* u_n, $1\le n\le N$.

Proof. It suffices to prove that $D(N;a,b)\le(b-a)N/2$. From Theorem 9.15 we see that

$$D(N;a,b)\le\frac{N}{K+1}+2\Big(\frac{1}{K+1}+b-a\Big)\sum_{k=1}^K\Big|\sum_{n=1}^N e(ku_n)\Big|,$$

which by hypothesis is

$$\le\Big(\frac{1}{K+1}+\frac{1}{5}\Big(\frac{1}{K+1}+b-a\Big)\Big)N.$$

Since

$$\frac{1}{K+1}\le\frac{b-a}{4},$$

this gives the desired bound. □

We now return to the subject of Example 9.2, where α is irrational and $u_n=n\alpha$. Suppose that instead of considering the first N members of this sequence we consider N consecutive members of this sequence, starting at some arbitrary initial point, say $n=M+1,M+2,\dots,M+N$. The exponential sum that arises in Weyl's criterion is now

$$\sum_{n=M+1}^{M+N}e(ku_n)=\sum_{n=M+1}^{M+N}e(kn\alpha)=e(kM\alpha)\sum_{n=1}^N e(kn\alpha),$$

and hence

$$\Big|\sum_{n=M+1}^{M+N}e(ku_n)\Big|\le\frac{1}{2\|k\alpha\|}.$$

Note that this upper bound is independent of M. Thus the bounds provided by Corollaries 9.16 and 9.17 depend on α, K, and N, but are independent of M. In particular, if α is a fixed irrational number, and $\mathcal{I}$ is a fixed interval of positive length in $\mathbb{T}$, and if $0 < n_1 < n_2 < \cdots$ are the integers for which $n_j\alpha \in \mathcal{I}$ (mod 1), then the gaps $n_{j+1} - n_j$ between these n_j are bounded in length.

Exercises

1. Suppose that $\{u_n\}$ is uniformly distributed modulo 1, and that c is a real number. Put $v_n = u_n + c$. Show that $\{v_n\}$ is uniformly distributed modulo 1.
2. Let $f(x)$ be a given real-valued function with period 1. Suppose that for every $\varepsilon > 0$ there exist trigonometric polynomials $T_\pm(x)$ such that $T_-(x) \le f(x) \le T_+(x)$ for all x, and $\int_0^1 T_+(x) - T_-(x)\,dx < \varepsilon$. Show that f is properly Riemann-integrable on $[0, 1]$.
3. (a) Use the Euler–Maclaurin summation formula with $K = 1$ to show that if k is a positive integer, then
$$\sum_{n=1}^{N} e\big(k\sqrt{n}\big) = \int_0^N e\big(kx^{1/2}\big)\,dx + O\big(kN^{1/2}\big)\,.$$
   (b) Make the change of variable $y = x^{1/2}$, and then integrate by parts, to show that
$$\int_0^N e\big(kx^{1/2}\big)\,dx = \frac{e(k\sqrt{N})\sqrt{N}}{\pi i k} + \frac{e(k\sqrt{N}) - 1}{4\pi^2 k^2}\,.$$
   (c) Deduce that the sequence $\{\sqrt{n}\}$ is uniformly distributed modulo 1.
4. (a) Use the Euler–Maclaurin summation formula (9.29) with $K = 1$ to show that if k is a positive integer, then
$$\sum_{n=1}^{N} e(k\log n) = \int_0^N e(k\log x)\,dx + O(k\log 2N)\,.$$
   (b) Show that
$$\int_0^N e(k\log x)\,dx = \frac{Ne(k\log N)}{1 + 2\pi i k}\,.$$
   (c) Deduce that the sequence $\{\log n\}$ is not uniformly distributed modulo 1.
5. (a) Show that
$$\limsup_{N\to\infty} \frac{1}{N} \sum_{\substack{1\le n\le N\\ \{\log n\}\in[0,1/2]}} 1 = \frac{e - e^{1/2}}{e - 1}\,.$$
   (b) Show that
$$\liminf_{N\to\infty} \frac{1}{N} \sum_{\substack{1\le n\le N\\ \{\log n\}\in[0,1/2]}} 1 = \frac{e^{1/2} - 1}{e - 1}\,.$$
   (c) Deduce that the sequence $\{\log n\}$ is not uniformly distributed modulo 1.

9.7. Positive definite kernels

We say that a function $K \in C(\mathbb{T})$ is *positive definite* if K is real-valued, even, and if

$$\int_0^1 \int_0^1 f(x)K(x-y)\overline{f(y)}\,dy\,dx \ge 0 \tag{9.47}$$

for all $f \in L^1(\mathbb{T})$. Note that the complex conjugate of the above is equal to itself, so this expression is a real number. If we take $f(x) = e(-nx)$, then

$$\begin{aligned}\int_0^1 \int_0^1 f(x)K(x-y)\overline{f(y)}\,dy\,dx &= \int_0^1 \int_0^1 K(x-y)e(-n(x-y))\,dx\,dy \\ &= \int_0^1 \widehat{K}(n)\,dy = \widehat{K}(n)\,.\end{aligned}$$

Thus for K to be positive definite, it is necessary that $\widehat{K}(n) \ge 0$ for all n. By Theorem 4.1 we know that

$$\lim_{N\to\infty} \sum_{n=-N}^{N} \Big(1 - \frac{|n|}{N}\Big)\widehat{K}(n) = K(0)\,.$$

Since $\widehat{K}(n) \ge 0$ for all n, the sum on the left is $\ge (1-\varepsilon)\sum_{n=-M}^{M} \widehat{K}(n)$ if $N > M/\varepsilon$. Since this is uniformly bounded above by $K(0)$, it follows that $K \in A(\mathbb{T})$.

Suppose conversely that K is even, that $K \in A(\mathbb{T})$, and that $\widehat{K}(n) \ge 0$ for all n. Then for $f \in L^1(\mathbb{T})$,

$$\begin{aligned}\int_0^1 \int_0^1 f(x)K(x-y)\overline{f(y)}\,dy\,dx &= \sum_{n=-\infty}^{\infty} \widehat{K}(n) \int_0^1 \int_0^1 f(x)e(n(x-y))\overline{f(y)}\,dy\,dx \\ &= \sum_{n=-\infty}^{\infty} \widehat{K}(n)\widehat{f}(-n)\overline{\widehat{f}(-n)} \\ &= \sum_{n=-\infty}^{\infty} \widehat{K}(n)|\widehat{f}(-n)|^2 \ge 0,\end{aligned}$$

so K is positive definite. Thus we have proved

Theorem 9.18. *Suppose that K has period 1, is continuous, even, and real-valued. Then K is positive definite if and only if $\widehat{K}(n) \ge 0$ for all n, and in that case $\sum_{n=-\infty}^{\infty} \widehat{K}(n) < \infty$.*

Corollary 9.19. *Suppose that K is positive definite. For any positive integer N and any real number θ,*

$$\sum_{n=1}^{N} \Big(1 - \frac{n}{N+1}\Big)K(n\theta) \ge \frac{N+1}{2}\widehat{K}(0) - \frac{1}{2}K(0)\,. \tag{9.48}$$

Proof. The left hand side is

$$\begin{aligned}\widehat{K}(0)\sum_{n=1}^{N}\Big(1-\frac{n}{N+1}\Big)&+\sum_{k=1}^{\infty}2\widehat{K}(k)\sum_{n=1}^{N}\Big(1-\frac{n}{N+1}\Big)\cos 2\pi nk\theta\\ &=\frac{N}{2}\widehat{K}(0)+\sum_{k=1}^{\infty}\widehat{K}(k)(\Delta_{N+1}(k\theta)-1)\\ &\ge\frac{N}{2}\widehat{K}(0)-\sum_{k=1}^{\infty}\widehat{K}(k)=\frac{N}{2}\widehat{K}(0)-\frac{1}{2}(K(0)-\widehat{K}(0))\end{aligned}$$

since $\widehat{K}(k)\ge 0$ and $\Delta_{N+1}(x)\ge 0$. This gives the stated result. □

Corollary 9.20. (Dirichlet) *For any positive integer N and any real number θ, there is an integer n, $1\le n\le N$, such that $\|n\theta\|\le 1/(N+1)$.*

This is best possible, since equality is achieved when $\theta=1/(N+1)$.

Proof. For $0<\delta\le 1/2$ put $K_\delta(x)=\max(0,1-\|x\|/\delta)$. Then

$$\widehat{K}_\delta(k)=\begin{cases}\dfrac{1}{\delta}\Big(\dfrac{\sin\pi k\delta}{\pi k}\Big)^2 & (k\ne 0),\\ \delta & (k=0).\end{cases}$$

Thus K_δ is positive definite, $K(0)=1$ and $\widehat{K}(0)=\delta$, so the lower bound provided by Corollary 9.19 is $\frac{1}{2}((N+1)\delta-)$. Thus if $\delta>1/(K+1)$, then the sum in (9.48) is positive, so there must be at least one n, $1\le n\le N$, such that $K_\delta(n\theta)>0$. For such an n we have $\|n\theta\|<\delta$. This suffices, since δ can be arbitrarily close to $1/(N+1)$. □

Corollary 9.21. *If θ is an irrational real number, then there exist infinitely many rational numbers a/q with $(a,q)=1$, such that*

$$\Big|\theta-\frac{a}{q}\Big|<\frac{1}{q^2}\,.$$

Proof. Suppose that R such solutions a_r/q_r have been found (where possibly $R=0$). Each of the differences $|q_r\theta-a_r|$ is non-zero, so if N is large enough, then $\|q_r\theta\|>1/(N+1)$ for all r. Suppose that $\|n\theta\|\le 1/(N+1)$, say $|n\theta-a|\le 1/(N+1)$. Then $|\theta-a/n|\le 1/(n(N+1)<1/n^2$. If a and n have a common factor, then this can be divided out, to give a new solution a_r/q_r. □

Exercises

1. Let θ and N be given, as in Corollary 9.20. Arrange the $N+1$ fractional parts $\{0\theta\},\{1\theta\},\ldots,\{N\theta\}$ in increasing order, and let $\delta_0,\delta_1,\ldots,\delta_N$ be the distance from one to the next, so that δ_N is the distance from the largest fractional part to 1.
   (a) Explain why $\sum_{n=0}^{N}\delta_n=1$.
   (b) Explain why there is an n, $0\le n\le N$, such that $\delta_n\le 1/(N+1)$.

(c) Deduce that there exist integers m and n, $0 \le m < n \le N$, such that $\|(n-m)\theta\| \le 1/(N+1)$.
(d) Observe that this gives a second proof of Corollary 9.20.

2. A sequence $\{a_n\}_{n=-\infty}^{\infty}$ is said to be *positive definite* if $a_{-n} = \overline{a_n}$ for all n, and if
$$\sum_{m=1}^{N}\sum_{n=1}^{N} z_m a_{m-n}\overline{z_n} \ge 0$$
for any complex numbers $z_1, z_2, \ldots, z_N$. Show that if $g \in L^1(\mathbb{T})$ and $g(x) \ge 0$ for all x, then $\{\widehat{g}(n)\}$ is a positive definite sequence.

9.8. Norms of polynomials

For a polynomial $P(z) = a_N z^N + a_{N-1}z^{N-1} + \cdots + a_1 z + a_0$ we let $\|P\|_1$ denote the little ℓ_1 norm of its coefficients:

$$\|P\|_1 = |a_0| + |a_1| + \cdots + |a_{N-1}| + |a_N|\,. \tag{9.49}$$

For the norm of the product of two polynomials we have a simple upper bound.

Theorem 9.22. *If P and Q are polynomials with real or complex coefficients, then $\|PQ\|_1 \le \|P\|_1\|Q\|_1$.*

Proof. Suppose that
$$\begin{aligned} P(x) &= a_M z^M + a_{M-1}z^{M-1} + \cdots + a_0,\\ Q(x) &= b_N z^N + b_{N-1}z^{N-1} + \cdots + b_0. \end{aligned}$$

Then
$$P(z)Q(z) = c_{M+N}z^{M+N} + c_{M+N-1}z^{M+N-1} + \cdots + c_0$$

where
$$c_r = \sum_{\substack{0\le m\le M\\ r-N\le m\le r}} a_m b_{r-m}$$

is the *Cauchy convolution* of the sequences a_m and b_n. Hence
$$\begin{aligned} \|PQ\|_1 = \sum_{r=0}^{M+N}|c_r| &= \sum_{r=0}^{M+N}\Big|\sum_{\substack{0\le m\le M\\ r-N\le m\le r}} a_m b_{r-m}\Big| \le \sum_{r=0}^{M+N}\sum_{\substack{0\le m\le M\\ r-N\le m\le r}} |a_m b_{r-m}|\\ &= \Big(\sum_{m=0}^{M}|a_m|\Big)\Big(\sum_{n=0}^{N}|b_n|\Big) = \|P\|_1\|Q\|_1\,. \end{aligned}$$

□

Our next objective is to establish an inequality in the reverse direction: $\|PQ\|_1 \ge c_{M,N}\|P\|_1\|Q\|_1$. In this connection it is useful to note that of the various norms that we can define for polynomials, there is one that is exactly multiplicative, namely the *Mahler measure*, which is defined to be

$$M(P) = \exp\left(\int_0^1 \log|P(e(\theta))|\,d\theta\right). \tag{9.50}$$

We note that by the identities $\log ab = \log a + \log b$, $e^{a+b} = e^a e^b$ it is evident that

$$M(PQ) = M(P)M(Q). \tag{9.51}$$

The quantity $M(P)$ is the geometric mean of $|P(z)|$ for $|z| = 1$, which means that it is a kind of average. Indeed, it is obvious that if $\mu = \max_{|z|\le 1}|P(z)|$, then

$$M(P) \le \exp\left(\int_0^1 \log\mu\,d\theta\right) = \exp(\log\mu) = \mu \le \|P\|_1. \tag{9.52}$$

In order to derive a corresponding lower bound for the Mahler measure, we use Fourier Series.

Lemma 9.23. *If $0 \le r < 1$, then*

$$\log|1 - re(\theta)| = -\sum_{n=1}^{\infty} \frac{r^n}{n}\cos 2\pi n\theta \tag{9.53}$$

for all θ. Also,

$$\log|2\sin\pi\theta| = -\sum_{n=1}^{\infty} \frac{1}{n}\cos 2\pi n\theta \tag{9.54}$$

for $0 < \theta < 1$. In both cases, the series on the right is the Fourier Series of the function on the left.

Proof. Let r be fixed, with $0 \le r < 1$. From the formula for the sum of a geometric series we know that

$$\sum_{n=0}^{\infty} r^n e(n\theta) = \frac{1}{1 - re(\theta)} = \frac{1 - re(-\theta)}{(1 - re(\theta))(1 - re(-\theta))}.$$

On taking imaginary parts we find that

$$\sum_{n=1}^{\infty} r^n \sin 2\pi n\theta = \frac{r\sin 2\pi\theta}{1 - 2r\cos 2\pi\theta + r^2}.$$

Since this series is absolutely convergent, and since the function on the right is a continuous function of θ, it follows by Theorem 4.4 that the series on the left is the Fourier Series of the function on the right. Write this identity as $f_r(\theta) = g_r(\theta)$. We put

$$F_r(\theta) = \sum_{n=1}^{\infty} \frac{-r^n}{n}\cos 2\pi n\theta, \qquad G_r(\theta) = \frac{1}{2}\log(1 - 2r\cos 2\pi\theta + r^2).$$

Then

$$\frac{\partial}{\partial\theta}F_r(\theta) = 2\pi\sum_{n=1}^{\infty} r^n \sin 2\pi n\theta = 2\pi f_r(\theta), \quad \frac{\partial}{\partial\theta}G_r(\theta) = \frac{2\pi r\sin 2\pi\theta}{1 - 2r\cos 2\pi\theta + r^2} = 2\pi g_r(\theta).$$

Since F_r and G_r have the same derivative, it follows that there is a constant c (possibly depending on r) such that $F_r(\theta) = G_r(\theta) + c$ for all θ. However,

$$F_r(0) = -\sum_{n=1}^{\infty} \frac{r^n}{n} = \log(1-r) = G_r(0),$$

so in fact $F_r(\theta) = G_r(\theta)$ for all θ, and the series F_r is the Fourier Series of the function $G_r(\theta)$.

To see that the series $F_1(\theta)$ is the Fourier Series of $G_1(\theta)$, we note that $|\widehat{G}_r(n) - \widehat{G}_1(n)| \le \|G_r - G_1\|_1$ by (3.15). Thus it suffices to show that this latter quantity tends to 0 as $r \to 1^-$. To this end we show that

$$G_r(\theta) - G_1(\theta) = \begin{cases} O\Big(\log \dfrac{1}{\theta}\Big) & (|\theta| \le 1-r), \\ O\Big(\dfrac{1-r}{|\theta|}\Big) & (1-r \le |\theta| \le 1/2). \end{cases} \tag{9.55}$$

For the first of these estimates we note that

$$G_r(\theta) = O\Big(\log \frac{1}{1-r}\Big) = O\Big(\log \frac{1}{|\theta|}\Big), \qquad G_1(\theta) = O\Big(\log \frac{1}{|\theta|}\Big).$$

We note that $1 - re(\theta) = 1 - e(\theta) + (1-r)e(\theta)$. By two applications of the triangle inequality it follows that

$$|1-e(\theta)| - (1-r) \le |1 - re(\theta)| \le |1-e(\theta)| + (1-r).$$

On dividing through by $|1-e(\theta)|$, we deduce that

$$1 - \frac{1-r}{|1-e(\theta)|} \le \frac{|1-re(\theta)|}{|1-e(\theta)|} \le 1 + \frac{1-r}{|1-e(\theta)|}.$$

Now $|1-e(\theta)| = 2|\sin \pi\theta| \ge 4|\theta|$ and $\log(1+u) = O(|u|)$ for $-1/2 \le u \le 1/2$, so by taking logarithms of the three quantities above we obtain the second estimate in (9.55). By integrating this bound we find that

$$\|G_r - G_1\|_1 = O\Big((1-r)\log \frac{1}{1-r}\Big)$$

since $\int \log u \, du = u \log u - u + C$. Thus $\|G_r - G_1\|_1 \to 0$ as $r \to 1^-$. The function $G_1(\theta)$ is continuous for $0 < \theta < 1$, so the series in (9.54) converges to $G_1(\theta)$, by Corollary 4.33. □

Lemma 9.24. *For $r > 0$,*

$$\int_0^1 \log |e(\theta) - r| \, d\theta = \max(0, \log r).$$

Proof. Suppose that $0 \le r < 1$. Let $F_r(\theta) = \log |1 - re(\theta)|$. We note that $|e(\theta) - r| = |1 - re(\theta)|$ for all θ. Thus the integral in question is $\widehat{F}_r(0)$. Since the series (9.53) is the Fourier Series of F_r, and since there is no constant term in that series, we know that $\widehat{F}_r(0) = 0$. If $r = 1$, then we argue in the same way, but we use (9.54). Finally, if $r > 1$, we note that $\log |e(\theta) - r| = \log r + \log |e(\theta) - 1/r|$ and the integral of this last term is 0 since $1/r < 1$. □

Theorem 9.25. (Jensen's Formula) *Suppose that the polynomial* $P(z) = a_N z^N + \cdots + a_0$ *has the factored form* $P(z) = a_N(z - z_1)\cdots(z - z_N)$. *Then*

$$M(P) = \exp\left(\int_0^1 \log|P(e(\theta))|\,d\theta\right) = |a_N|\prod_{n=1}^{N}\max(1, |z_n|)\,.$$

Proof. Clearly

$$\log|P(e(\theta))| = \log|a_N| + \sum_{n=1}^{N}\log|e(\theta) - z_n|.$$

Write $z_n = r_n e(\phi_n)$ with $r_n \ge 0$. Then

$$\int_0^1 \log|e(\theta) - z_n|\,d\theta = \int_0^1 \log|e(\theta - \phi_n) - r_n|\,d\theta = \int_{-\phi_n}^{1-\phi_n} \log|e(\theta) - r_n|\,d\theta$$
$$= \int_0^1 \log|e(\theta) - r_n|\,d\theta\,.$$

That is, this integral may depend on r_n, but it is independent of ϕ_n. By applying Lemma 9.24 N times, we find that

$$\int_0^1 \log|P(e(\theta))|\,d\theta = \log|a_N| + \sum_{n=1}^{N}\max(0, \log r_n)\,.$$

The stated result now follows by exponentiating. □

Corollary 9.26. (Mahler) *Let* P *be a polynomial of degree* N. *Then*

$$M(P) \ge \|P\|_1/2^N\,.$$

Proof. For any complex number z it is clear that

$$\frac{1 + |z|}{2} \le \max(1, |z|)\,.$$

Hence by Theorem 9.25,

$$M(P) \ge \frac{|a_N|}{2^N}\prod_{n=1}^{N}(1 + |z_n|)\,.$$

Now $\|aP\|_1 = |a|\|P\|_1$, and $\|z - z_n\|_1 = 1 + |z_n|$, so by N applications of Theorem 9.22 we find that

$$\|P\|_1 = |a_N|\|(z - z1)\cdots(z - z_N)\|_1 \le |a_N|\prod_{n=1}^{N}(1 + |z_n|)\,.$$

Thus we have the result. □

Corollary 9.27. *Let* P *be a polynomial of degree* M, *and* Q *a polynomial of degree* N. *Then*

$$2^{-M-N}\|P\|_1\|Q\|_1 \le \|PQ\|_1 \le \|P\|_1\|Q\|_1\,.$$

Proof. The second inequality is in Theorem 9.22. For the first inequality we note by (9.52) and (9.51) that

$$\|PQ\|_1 \geq M(PQ) = M(P)M(Q) \geq \frac{\|P\|_1}{2^M} \cdot \frac{\|Q\|_1}{2^N} .$$

□

For a polynomial $P(z) = a_N z^N + \cdots + a_0$ we also have other norms, such as

$$\|P\|_2 = \Big(\sum_{n=0}^{N} |a_n|^2\Big)^{1/2}, \qquad H(P) = \|P\|_\infty = \max_{0 \leq n \leq N} |a_n|,$$

and we call $H(P)$ the *height* of P. All our norms are comparable to within constants depending on N. In particular, it is obvious that $H(P) \leq \|P\|_1 \leq (N+1)H(P)$.

By the Fundamental Theorem of Algebra we know that a polynomial can be factored into linear factors. Thus over $\mathbb{C}$ the only irreducible polynomials are linear. However, if we are considering only polynomials with integer coefficients, then some (such as $z^2 - 5z + 6$) are reducible, while others (such as $z^2 + 1$) are irreducible. We now show that over the integers, almost all polynomials are irreducible.

Theorem 9.28. *Let N be fixed, $N \geq 2$. For given d, the number of pairs Q, R of polynomials with integer coefficients and $\deg Q = d$, $\deg R = N - d$ such that $H(QR) \leq H$ is $O(H^{N+1-d})$ if $0 < d < N/2$, and is $O(H^{N/2+1} \log H)$ if $d = N/2$.*

In the above, the implicit constant may depend on N, but this dependence is not displayed since we said that N is fixed.

Let $H \geq 2$ be an integer. The number of polynomials P of degree at most N and height not exceeding H is exactly $(2H+1)^{N+1}$. If we want to count only those polynomials of degree exactly N with positive leading coefficient, and height not exceeding H, then we have $H(2H+1)^N$ polynomials, which is also $> H^{N+1}$. From the theorem we see that the number of reducible polynomials of height $\leq H$ is $O(H^N)$ if $N < 2$, and is $O(H \log H)$ if $N = 2$. Moreover, of the $\asymp H^N$ reducible polynomials of degree $N > 2$, most of them factor into a linear polynomial and a polynomial of degree $N - 1$.

Proof. Suppose that Q has degree d and that $2^{k-1} < H(Q) \leq 2^k$. Since $H(QR) \asymp H(Q)H(R)$, in order that $H(QR) \leq H$, we must have $H(R) \leq cH/2^k$. The number of possible such Q is $O\big(2^{(d+1)k}\big)$, and the number of possible such R is $O\big((cH/2^k)^{N+1-d}\big)$. Hence the number of such products is $O\big(H^{N+1-d} 2^{(2d-N)k}\big)$. We sum this estimate over k, but only for k such that $2^k \leq cH$, since past that point there is no possible R. If $2d < N$, then our estimates decrease geometrically with k. If $2d = N$, then we have $O(\log H)$ terms, each of size $O\big(H^{N/2+1}\big)$. □

Exercises

1. (a) Show that
$$\int_0^1 \log(2\sin \pi x)\, dx = 0\,.$$
(b) Show that
$$\int_0^{1/2} x\cot \pi x\, dx = \frac{\log 2}{2\pi}\,.$$

2. Show that
$$\int_0^1 (\log 2\sin \pi x)^2\, dx = \frac{\pi^2}{12}\,.$$

3. The object of this exercise is to give an alternative derivation of (9.54) from (9.53).
(a) Use Lemma 4.20 to show that the series $F_1(\theta)$ converges for $0 < \theta < 1$.
(b) Use Abel's Theorem (Theorem 4.30) to show that if $0 < \theta < 1$, then $\lim_{r\to 1^-} F_r(\theta) = F_1(\theta)$.
(c) Note that if $0 < \theta < 1$, then
$$\lim_{r\to 1^-} G_r(\theta) = G_1(\theta) = \frac{1}{2}\log(2 - 2\cos 2\pi\theta) = \frac{1}{2}\log(2\sin \pi\theta)^2 = \log|2\sin \pi\theta|\,.$$
(d) To show that $F_1(\theta)$ is the Fourier Series of $G_1(\theta)$, apply Theorem 4.27 to $3/4 - G_1(\theta)$.

4. Let $P(z) = (z+1)^N$.
(a) Show that $M(P) = 1$.
(b) Show that $\|P\|_1 = 2^N$.
(c) Deduce that Mahler's inequality in Corollary 9.26 is best possible.
(d) Put $Q(z) = (z-1)^N$. Show that $\|Q\|_1 = 2^N$.
(e) Show that $\|PQ\|_1 = 2^N = \|P\|_1\|Q\|_1/2^N$.

Notes

9.1. On December 21, 1807, Joseph Fourier presented his 234-page paper *On the propagation of heat in solid bodies* to the French Academy of Sciences. Fourier had determined the partial differential equation that governs the propagation of heat, and had solved the equation in several circumstances, including the conducting disc. He could treat the disc when the temperature on the boundary is sinusoidal, and by superimposing solutions he could treat linear combinations of sines and cosines. He made the audacious claim that any periodic function could be expressed by such a series. A committee of four (Lacroix, Lagrange, Laplace, and Monge) was created to evaluate his manuscript. Their skepticism of his claim was so great that they rejected the paper. Fourier continued to work on heat, and in doing so discovered the greenhouse effect through a combination of mathematical reasoning and physical experiments. Eventually Fourier (1822) was able to publish his findings.

We have written Δu for the laplacian. While this is a common notation, some authors write $\nabla^2 u$ for the laplacian. As always, one must be alert to the conventions in place when reading the work of an unfamiliar author.

9.2. The wave motion of a vibrating string was for the most part derived by Euler and Lagrange.

9.5. The Bernoulli numbers and polynomials were invented by Jakob Bernoulli (1655–1705). Jakob Bernoulli, in the course of studying a question concerning compound interest, also discovered the number e, although the symbol e for this number was introduced later by Euler.

9.6. Weyl's criterion originates in Weyl (1914), (1916). Its generalization to the weak convergence of measures is an important tool in probability.

9.7. At a somewhat more advanced level, a sequence $\{a_n\}_{n=-\infty}^{\infty}$ is said to be *positive definite* if for any finite collection of complex numbers $z_1, z_2, \ldots, z_N$ we have

$$\sum_{m=1}^{N}\sum_{n=1}^{N} a_{m-n} z_m \overline{z_n} \ge 0\,.$$

Herglotz's theorem asserts that $\{a_n\}$ is positive definite if and only if there is a positive measure μ on $\mathbb{T}$ such that

$$a_n = \int_0^1 e(-nx)\,d\mu(x)$$

for all n. Here the integral on the right is considered to be the Fourier coefficient of μ, denoted $\widehat{\mu}(n)$. See Katznelson (2004, pp. 40–43). The analogue of this for Fourier transforms on the real line is known as Bochner's Theorem. See Goldberg (1970, pp. 59–65).

Chapter 10

The Fourier Transform

10.1. Definition and basic properties

Let $L^1(\mathbb{R})$ denote the set of real or complex-valued functions $f(x)$ defined for $x \in \mathbb{R}$ for which

$$\int_{-\infty}^{\infty} |f(x)|\, dx < \infty.$$

This integral is called the L^1 norm of f; in symbols we write $\|f\|_{L^1(\mathbb{R})}$, or sometimes more briefly just $\|f\|_1$. For $f \in L^1(\mathbb{R})$, we define its *Fourier Transform* to be

$$\widehat{f}(t) = \int_{-\infty}^{\infty} f(x)e(-tx)\, dx\,. \tag{10.1}$$

Here t is an arbitrary real number. Our hope is to establish a Fourier inversion (or expansion) so that we can write

$$f(t) = \int_{-\infty}^{\infty} \widehat{f}(t)e(xt)\, dt\,.$$

As we develop the theory, we find much that is analogous to what we have seen of Fourier Series, but there are some surprises, not all of them welcome.

By applying the triangle inequality in the definition (10.1) of the Fourier Transform it is immediate that

$$|\widehat{f}(t)| \le \int_{-\infty}^{\infty} |f(x)|\, dx \tag{10.2}$$

for all t. This is analogous to a similar inequality, (3.15), for Fourier coefficients.

We note a few manipulative properties of the Fourier Transform, which are analogous to corresponding properties of the Discrete Fourier Transform and of Fourier coefficients. Suppose that $f \in L^1(\mathbb{R})$ and that $g \in L^1(\mathbb{R})$. Set $h(x) = af(x) + bg(x)$ for all x. The Fourier Transform is linear in the sense that $\widehat{h}(t) = a\widehat{f}(t) + b\widehat{g}(t)$ for all t. We note also how the Fourier Transform is altered by translation and rescaling.

Theorem 10.1. *Suppose that $f \in L^1(\mathbb{R})$.*

(a) *If a is a real number, and $g(x) = f(x+a)$ for all x, then $g \in L^1(\mathbb{R})$ and $\widehat{g}(t) = e(at)\widehat{f}(t)$ for all t;*

(b) *If b is a real number and $h(x) = f(x)e(bx)$ for all x, then $h \in L^1(\mathbb{R})$ and $\widehat{h}(t) = \widehat{f}(t-b)$ for all t;*

(c) *If c is a non-zero real number, and $j(x) = f(cx)$ for all x, then $j \in L^1(\mathbb{R})$ and $\widehat{j}(t) = \widehat{f}(t/c)/|c|$ for all t;*

(d) *If $\ell(x) = \overline{f(x)}$, then $\ell \in L^1(\mathbb{R})$ and $\widehat{\ell}(t) = \overline{\widehat{f}(-t)}$.*

Proof. (a) By the change of variable $y = x + a$ we see that

$$\int_{-\infty}^{\infty} g(x)e(-tx)\,dx = \int_{-\infty}^{\infty} f(x+a)e(-tx)\,dx = e(at)\int_{-\infty}^{\infty} f(x+a)e(-t(x+a))\,dx$$
$$= e(at)\int_{-\infty}^{\infty} f(y)e(-ty)\,dy = e(at)\widehat{f}(t)\,.$$

(b) Clearly

$$\int_{-\infty}^{\infty} h(x)e(-tx)\,dx = \int_{-\infty}^{\infty} f(x)e(-(t-b)x)\,dx = \widehat{f}(t-b)\,.$$

(c) Some care is required here, because c might be negative.

$$\widehat{j}(t) = \int_{-\infty}^{\infty} j(x)e(-tx)\,dx = \int_{-\infty}^{\infty} f(cx)e(-tx)\,dx\,.$$

By the change of variable $y = cx$, this is

$$= \frac{1}{c}\int_{-c\infty}^{c\infty} f(y)e(-(t/c)y)\,dy\,.$$

If $c > 0$, then this is $\widehat{f}(t/c)/c$. However, if $c < 0$, then a minus sign is introduced when we reverse the endpoints, so the above is

$$= \frac{-1}{c}\int_{-\infty}^{\infty} f(y)e(-(t/c)y)\,dy = \frac{-1}{c}\widehat{f}(t/c)\,.$$

(d) We know that

$$\widehat{f}(-t) = \int_{-\infty}^{\infty} f(x)e(tx)\,dx\,.$$

The complex conjugate of an integral is the integral of the complex conjugate, and the complex conjugate of a product is the product of the complex conjugates, so

$$\overline{\widehat{f}(-t)} = \int_{-\infty}^{\infty} \ell(x)e(-tx)\,dx = \widehat{\ell}(t)\,.$$

□

Corollary 10.2. *Suppose that $f \in L^1(\mathbb{R})$. If f is even, then $\widehat{f}$ is even. If f is odd, then $\widehat{f}$ is odd.*

Corollary 10.3. *Suppose that $f \in L^1(\mathbb{R})$. If f is real-valued, then $\widehat{f}(-t) = \overline{\widehat{f}(t)}$ for all t.*

Next we consider the smoothness of the Fourier Transform. This has no analogue for Fourier coefficients, since $\widehat{f}(n)$ is defined only when n is an integer.

Theorem 10.4. *Suppose that $f \in L^1(\mathbb{R})$. Then $\widehat{f}(t)$ is uniformly continuous.*

Proof. Let $\varepsilon > 0$ be given. Then there is an X such that

$$\int_{-\infty}^{-X} |f(x)|\, dx < \varepsilon, \quad \text{and} \quad \int_X^{\infty} |f(x)|\, dx < \varepsilon\,.$$

We observe that

$$\widehat{f}(t_1) - \widehat{f}(t_2) = \int_{-\infty}^{\infty} f(x)(e(-t_1x) - e(-t_2x))\, dx\,.$$

The difference of complex exponentials has absolute value at most 2, so

$$|\widehat{f}(t_1) - \widehat{f}(t_2)| \le 4\varepsilon + \int_{-X}^{X} |f(x)||(e(-t_1x) - e(-t_2x))|\, dx\,.$$

We write

$$\begin{aligned} e(-t_1x) - e(-t_2x) &= e(-(t_1+t_2)x/2)(e((t_2-t_1)x/2) - e(-(t_2-t_1)x/2)) \\ &= 2ie(-(t_1+t_2)x/2)\sin \pi(t_2-t_1)x\,. \end{aligned}$$

Hence

$$|e(-t_1x) - e(-t_2x)| \le 2|\sin \pi(t_2-t_1)x| \le 2\pi|t_2-t_1||x| \le 2\pi|t_2-t_1|X$$

for $-X \le x \le X$. Thus

$$\int_{-X}^{X} |f(x)||(e(-t_1x)-e(-t_2x))|\, dx \le 2\pi|t_2-t_1|X \int_{-X}^{X} |f(x)|\, dx \le 2\pi|t_2-t_1|X\|f\|_1.$$

Consequently, there is a $\delta > 0$ so small that if $|t_2 - t_1| < \delta$, then the above is $< \varepsilon$. Since $|\widehat{f}(t_2) - \widehat{f}(t_1)| < 5\varepsilon$ whenever $|t_2 - t_1| < \delta$, it follows that $\widehat{f}$ is uniformly continuous. □

For Fourier coefficients we also had the Riemann–Lebesgue Lemma, which also has an analogue in our new setting. We first extend Lemma 3.5 to the real line.

Lemma 10.5. *If $f \in L^1(\mathbb{R})$, then*

$$\lim_{\delta \to 0} \int_{-\infty}^{\infty} |f(x+\delta) - f(x)|\, dx = 0\,. \tag{10.3}$$

Proof. Let $\varepsilon > 0$ be given, and choose X so large that

$$\int_{|x|>X-1} |f(x)|\, dx < \varepsilon\,.$$

Then for $\delta \le 1$ it follows by the triangle inequality that

$$\int_{-\infty}^{\infty} |f(x+\delta) - f(x)|\, dx \le \int_{-X}^{X} |f(x+\delta) - f(x)|\, dx + 2\varepsilon\,.$$

By principles of Lebesgue measure theory (see Theorem 0.22) we know that continuous functions are dense in the L^1 norm on compact intervals. Thus there is a continuous function g such that

$$\int_{-X-1}^{X+1} |f(x) - g(x)|\, dx < \varepsilon\,.$$

By the triangle inequality it follows that

$$\int_{-X}^{X} |f(x+\delta) - f(x)|\, dx \le \int_{-X}^{X} |g(x+\delta) - g(x)|\, dx + 2\varepsilon\,.$$

Since g is continuous on $[-X-1, X+1]$ it follows that g is uniformly continuous on this interval, so the integrand in this last integral tends to 0 uniformly in x. Hence the integral tends to 0 as $\delta \to 0$, and the proof is complete. □

Theorem 10.6. (Riemann–Lebesgue Lemma) *Suppose that $f \in L^1(\mathbb{R})$. Then $\widehat{f}(t)$ tends to 0 as t tends to infinity, and also as t tends to minus infinity.*

Proof. By definition,

$$\widehat{f}(t) = \int_{-\infty}^{\infty} f(x)e(-tx)\, dx\,. \tag{10.4}$$

By the change of variable $x = y + 1/(2t)$, we see that the above is

$$= \int_{-\infty}^{\infty} f(y + 1/(2t))e(-t(y + 1/(2t)))\, dy\,.$$

We write x in place of y. Thus

$$\widehat{f}(t) = -\int_{-\infty}^{\infty} f(x + 1/(2t))e(-tx)\, dx\,.$$

(10.5)

We add this to (10.4) to see that

$$2\widehat{f}(t) = \int_{-\infty}^{\infty} (f(x + 1/(2t)) - f(x))e(-tx)\, dx\,.$$

Hence by the triangle inequality,

$$|\widehat{f}(t)| \le \frac{1}{2}\int_{-\infty}^{\infty} |f(x + 1/(2t)) - f(x)|\, dx\,. \tag{10.6}$$

The desired result now follows from Lemma 10.5. □

Example 10.1. Suppose that $c > 0$, and put $f(x) = e^{-c|x|}$. Then

$$\begin{aligned}\widehat{f}(t) &= \int_{-\infty}^{0} e^{cx}e(-tx)\, dx + \int_{0}^{\infty} e^{-cx}e(-tx)\, dx\\ &= \left[\frac{e^{(-2\pi it+c)x}}{-2\pi it + c}\right]_{-\infty}^{0} + \left[\frac{e^{(-2\pi it-c)x}}{-2\pi it - c}\right]_{0}^{\infty}\\ &= \frac{1}{c - 2\pi it} + \frac{1}{c + 2\pi it} = \frac{2c}{c^2 + 4\pi^2t^2}\,.\end{aligned}$$

The behavior here is typical in that if f is even and convex upwards on $(0, \infty)$, then $\widehat{f}(t) > 0$ for all t (recall Theorem 4.27).

Example 10.2. Suppose that $f(x) = 1$ for $-1/2 \le x \le 1/2$, and $f(x) = 0$ otherwise. Then

$$\widehat{f}(t) = \int_{-1/2}^{1/2} e(-tx)\, dx = \left[\frac{e(-tx)}{-2\pi i t}\right|_{-1/2}^{1/2} = \frac{e(-t/2) - e(t/2)}{-2\pi i t} = \frac{\sin \pi t}{\pi t}\,.$$

Lemma 10.7. (Sobolev) *Suppose that f has a continuous first derivative on the interval $[0,1]$. Then*

$$|f(x)| \le \int_0^1 |f(u)|\, du + \int_0^1 |f'(u)|\, du$$

for all $x \in [0,1]$.

Proof. By integrating by parts we find that

$$\int_0^x uf'(u)\, du = \Big[uf(u)\Big|_0^x - \int_0^x f(u)\, du = xf(x) - \int_0^x f(u)\, du,$$

and that

$$\int_x^1 (u-1)f'(u)\, du = \Big[(u-1)f(u)\Big|_x^1 - \int_x^1 f(u)\, du = (1-x)f(x) - \int_x^1 f(u)\, du\,.$$

We add these two identities, and rearrange to find that

$$f(x) = \int_0^1 f(u)\, du + \int_0^x uf'(u)\, du + \int_x^1 (u-1)f'(u)\, du\,.$$

Hence by the triangle inequality,

$$\begin{aligned} |f(x)| &\le \int_0^1 |f(u)|\, du + \int_0^x u|f'(u)|\, du + \int_x^1 (1-u)|f'(u)|\, du \\ &\le \int_0^1 |f(u)|\, du + \int_0^1 |f'(u)|\, du\,. \end{aligned}$$

□

Theorem 10.8. *If $f \in L^1(\mathbb{R})$, $f'(x)$ exists and is continuous, and $f' \in L^1(\mathbb{R})$, then $\widehat{f'}(t) = 2\pi i t \widehat{f}(t)$.*

By applying the inequality (10.2) to f', we obtain the quantitative estimate

$$|\widehat{f}(t)| \le \frac{\|f'\|_1}{2\pi|t|}, \tag{10.7}$$

and from Theorem 10.6 the stronger (but qualitative) estimate

$$\widehat{f}(t) = o(1/|t|) \qquad \text{as } |t| \to \infty\,. \tag{10.8}$$

Proof. By integration by parts we find that

$$\begin{aligned}\int_{-X}^{X} f'(x)e(-tx)\,dx &= \Big[f(x)e(-tx)\Big|_{-X}^{X} - \int_{-X}^{X} f(x)(-2\pi it)e(-tx)\,dx \\ &= f(X)e(-tX) - f(-X)e(tX) + 2\pi it\int_{-X}^{X} f(x)e(-tx)\,dx\,. \qquad (10.9)\end{aligned}$$

By Sobolev's inequality (Lemma 10.7) we find that

$$|f(X)| \le \int_{X-1/2}^{X+1/2} |f(u)|\,du + \int_{X-1/2}^{X+1/2} |f'(u)|\,du\,.$$

Thus $f(X) \to 0$ as $X \to \infty$. Similarly, $f(-X) \to 0$ as $X \to \infty$. We obtain the stated result by letting X tend to infinity in (10.9). □

For functions f and g defined on the real line, we define their *convolution* to be

$$(f * g)(x) = \int_{-\infty}^{\infty} f(x-u)g(u)\,du\,. \qquad (10.10)$$

It may happen that $\int_{-\infty}^{\infty} |f(x-u)g(u)|\,du = \infty$, in which case the convolution does not exist. However, it will exist if $f \in L^1(\mathbb{R})$ and g is bounded. If both f and g are in $L^1(\mathbb{R})$, then the convolution will exist for almost all x, as we now show.

Theorem 10.9. *If $f \in L^1(\mathbb{R})$ and $g \in L^1(\mathbb{R})$, then $f * g \in L^1(\mathbb{R})$, and $\widehat{f * g}(t) = \widehat{f}(t)\widehat{g}(t)$.*

Proof. By the triangle inequality,

$$\begin{aligned}\int_{-\infty}^{\infty} |(f*g)(x)|\,dx &= \int_{-\infty}^{\infty}\left|\int_{-\infty}^{\infty} f(x-u)g(u)\,du\right|dx \\ &\le \int_{-\infty}^{\infty}\int_{-\infty}^{\infty} |f(x-u)g(u)|\,du\,dx\,.\end{aligned}$$

We exchange the order of integration, and see that the above is

$$= \int_{-\infty}^{\infty} |g(u)|\Big(\int_{-\infty}^{\infty} |f(x-u)|\,dx\Big)\,du\,.$$

Here the inner integral is $\|f\|_1$, which is independent of u, so we conclude that

$$\|f*g\|_1 \le \|f\|_1\|g\|_1\,.$$

As for the Fourier transform, we find that

$$\begin{aligned}\widehat{f*g}(t) &= \int_{-\infty}^{\infty}\int_{-\infty}^{\infty} f(x-u)g(u)\,du\,e(-tx)\,dx \\ &= \int_{-\infty}^{\infty} g(u)e(-tu)\Big(\int_{-\infty}^{\infty} f(x-u)e(-t(x-u))\,dx\Big)\,du\,.\end{aligned}$$

Here the exchange of the order of integration is justified because the double integral is absolutely convergent. The inner integral above is $\widehat{f}(t)$, which is independent of u, and the remaining integral with respect to u is $\widehat{g}(t)$. □

Example 10.3. Suppose that $g(x) = 1 - |x|$ for $|x| \le 1$, and that $g(x) = 0$ otherwise. Then $g(x) = (f * f)(x)$ where f is the function of Example 10.2. Hence

$$\widehat{g}(t) = \widehat{f}(t)^2 = \Big(\frac{\sin \pi t}{\pi t}\Big)^2 .$$

We can confirm this by direct calculation:

$$\begin{aligned}\widehat{g}(t) &= \int_{-1}^{1} (1 - |x|)e(-tx)\, dx = 2\int_0^1 (1-x)\cos 2\pi tx\, dx \\ &= 2\Big[(1-x)\frac{\sin 2\pi tx}{2\pi t}\Big|_0^1 + \int_0^1 \frac{\sin 2\pi tx}{\pi t}\, dx \\ &= \Big[\frac{-\cos 2\pi tx}{2(\pi t)^2}\Big|_0^1 = \frac{1 - \cos 2\pi t}{2(\pi t)^2} = \Big(\frac{\sin \pi t}{\pi t}\Big)^2 .\end{aligned}$$

For $T > 0$ put $g_T(x) = Tg(Tx)$. Then g_T has a peak of width $1/T$, height T, and $\int_{-\infty}^{\infty} g_T(x)\, dx = 1$ for all T. Thus the family of functions g_T can be used as a kernel as $T \to \infty$. This applies not just to this g, but to any $K \in L^1(\mathbb{R})$ with $\int_{-\infty}^{\infty} K(x)\, dx = 1$, since it suffices to take $K_T(x) = TK(Tx)$. From Theorem 10.1(c) we see that $\widehat{K_T}(t) = \widehat{K}(t/T)$. Hence kernels are more easily constructed on the real line than when working modulo 1.

10.2. The inversion formula

Up to this point, our development of the Fourier Transform has followed our discussion of Fourier Series quite closely. In the same way that we wish to use the Fourier Series of a function as a formula for that function, for $f \in L^1(\mathbb{R})$ we would like to write f in terms of its Fourier Transform. Indeed, we might hope that

$$f(x) = \int_{-\infty}^{\infty} \widehat{f}(t)e(tx)\, dt .$$

Of course we anticipate that we may encounter the same sort of convergence difficulties that we had with Fourier Series. Depending on the hypotheses imposed, one might have to settle for a formula such as

$$f(x) = \lim_{T\to\infty} \int_{-T}^{T} \widehat{f}(t)e(tx)\, dt$$

or

$$f(x) = \lim_{T\to\infty} \int_{-T}^{T} (1 - |t|/T)\widehat{f}(t)e(tx)\, dt .$$

In Chapter 4, our ability to approximate to a general $f \in L^1(\mathbb{T})$ by a trigonometric polynomial was of immense value. For the Fourier Transform, the analogue of a trigonometric polynomial is a function of the sort

$$B(x) = \int_{-T}^{T} b(t)e(tx)\, dt . \tag{10.11}$$

We begin by establishing the natural analogue of Lemma 3.14.

Lemma 10.10. *Suppose that $B(x)$ is given as in (10.11), and also that $\int_{-T}^{T} |b(t)|\, dt < \infty$. If $f \in L^1(\mathbb{R})$, then*

$$(f * B)(x) = \int_{-T}^{T} b(t)\widehat{f}(t)e(tx)\, dt\,.$$

Proof. By the triangle inequality

$$|B(x)| \le \int_{-T}^{T} |b(t)|\, dt < \infty\,.$$

Thus B is a bounded function, and so the convolution $(f * B)(x)$ exists for all x, and indeed $|(f * B)(x)| \le \|f\|_1 \|b\|_1$ for all x. Clearly

$$\begin{aligned}(f * B)(x) &= \int_{-\infty}^{\infty} f(x-u)B(u)\, du = \int_{-\infty}^{\infty} f(x-u)\int_{-T}^{T} b(t)e(tu)\, dt\, du \\ &= \int_{-T}^{T} b(t)\Big(\int_{-\infty}^{\infty} f(x-u)e(tu)\, du\Big)\, dt\,.\end{aligned}$$

Here the exchange of the integrals is justified by their absolute convergence. On making the change of variable $y = x - u$ in the inner integral, we see that it is

$$= \int_{-\infty}^{\infty} f(y)e(t(x-y))\, dy = \widehat{f}(t)e(tx)\,.$$

Thus we have the stated result. □

As we continue to emulate our treatment of Fourier Series, we recall our derivation (3.43) of the Fourier coefficients of a trigonometric polynomial. Unfortunately the analogue of this for a function $B(x)$ as given in (10.11) is in general false. Indeed, in Example 10.2 we saw that $B(x)$ is not necessarily a member of $L^1(\mathbb{R})$, even when $\int_{-T}^{T} |b(t)|\, dt < \infty$. (When reading Example 2 in the present context, one needs to interchange the variables x and t.) The relation (3.43) is so obvious that we may use it without being very much aware of it. For example, it contributed to our proof of (4.7), which states that $\int_0^1 \Delta_N(x)\, dx = 1$. We now establish the natural analogue of this, but of course by a rather different argument.

Lemma 10.11. *We have*

$$\int_{-\infty}^{\infty} \Big(\frac{\sin \pi x}{\pi x}\Big)^2\, dx = 1\,.$$

Proof. From (4.7) and (4.8) we know that

$$\int_{-1/2}^{1/2} \frac{1}{N}\Big(\frac{\sin \pi N x}{\sin \pi x}\Big)^2\, dx = 1\,.$$

Write this integral as

$$\int_{-1/2}^{1/2} \frac{1}{N}\Big(\frac{\sin \pi N x}{\pi x}\Big)^2\, dx + \int_{-1/2}^{1/2} \frac{(\sin \pi N x)^2}{N}\Big(\frac{1}{(\sin \pi x)^2} - \frac{1}{(\pi x)^2}\Big)\, dx = I_1 + I_2,$$

say. By rescaling the variable of integration in I_1, we see immediately that

$$I_1 = \int_{-N/2}^{N/2} \Big(\frac{\sin \pi x}{\pi x}\Big)^2\, dx\,.$$

In I_2, the quantity in large parentheses is

$$\frac{(\pi x)^2 - (\sin \pi x)^2}{(\pi x \sin \pi x)^2} = \frac{(\pi x - \sin \pi x)(\pi x + \sin \pi x)}{(\pi x \sin \pi x)^2}\,.$$

Here the denominator has a zero of order 4 at $x = 0$. Since $\sin \pi x = \pi x - (\pi x)^3/6 + \cdots$, we see that $\sin \pi x - \pi x$ has a zero of order 3 at 0, and that $\sin \pi x + \pi x$ has a zero of order 1 at 0. Thus the numerator also has a zero at 0 of multiplicity 4, and hence this expression is a bounded function for $-1/2 \le x \le 1/2$. Consequently, $|I_2| \le C/N$ for some constant C. As N tends to infinity, I_1 tends to the desired integral, I_2 tends to 0, but $I_1 + I_2 = 1$ for all N. Hence the integral in question has value 1. □

For real numbers $T > 0$ let

$$\Delta_T(x) = \begin{cases} \dfrac{1}{T}\Big(\dfrac{\sin \pi T x}{\pi x}\Big)^2 & (x \ne 0), \\ T & (x = 0)\,. \end{cases} \tag{10.12}$$

This is a slight abuse of notation, since Δ_N denotes the Fejér kernel. However, the above is analogous to the Fejér kernel, and we want to emphasize the similarities. From Lemma 10.11 we find that

$$\int_{-\infty}^{\infty} \Delta_T(x)\, dx = 1\,. \tag{10.13}$$

Let g be the function in Example 10.3. By Theorem 10.1 (c) with $c = 1/T$, we find that

$$\int_{-T}^{T} (1 - |t|/T) e(tx)\, dt = \Delta_T(x)\,. \tag{10.14}$$

By Lemma 10.10 it follows that

$$(f * \Delta_T)(x) = \int_{-T}^{T} (1 - |t|/T) \widehat{f}(t) e(tx)\, dt\,. \tag{10.15}$$

Moreover, from (10.13) we see that

$$(f * \Delta_T)(x) - f(x) = \int_{-\infty}^{\infty} (f(x-u) - f(x)) \Delta_T(u)\, du\,. \tag{10.16}$$

How we use these relations depends on our goal. We begin by establishing an analogue of Fejér's theorem (Theorem 4.14).

Theorem 10.12. *Suppose that $f \in L^1(\mathbb{R})$. If f is continuous at x, then*

$$\lim_{T\to\infty} \int_{-T}^{T} (1 - |t|/T) \widehat{f}(t) e(tx)\, dt = f(x)\,.$$

From the proof below it will be clear also that if f is uniformly continuous throughout the real line, then the above limit is attained uniformly.

Proof. Suppose that $\varepsilon > 0$ has been given, and that $|f(x-u) - f(x)| < \varepsilon$ when $|u - x| < \delta$. We write the integral in (10.16) as $I_1 + I_2 + I_3$ where I_1 is the integral over $-\infty < u \le -\delta$, I_2 is the integral over $-\delta \le u \le \delta$, and I_3 is the integral over $\delta \le u < \infty$. From the definition (10.12), we see that $\Delta_T(u) \le 1/(T\pi^2\delta^2)$ when $u \le -\delta$. Thus

$$|I_1| \le \frac{1}{T\pi^2\delta^2}\int_{-\infty}^{-\delta} |f(x-u)|\,du + |f(x)|\int_{-\infty}^{-\delta} \Delta_T(u)\,du\,.$$

Here the first integral is $\le \|f\|_1$, which by hypothesis is finite, and the second integral is $\le 1/(T\pi^2\delta)$, by (10.12). Hence $|I_1| \le C/T$ where C is a constant that may depend on f, x, ε, and δ, but is independent of T. The integral I_3 is bounded similarly. As for I_2, we see that

$$|I_2| < \varepsilon\int_{-\delta}^{\delta} \Delta_T(u)\,du \le \varepsilon\int_{-\infty}^{\infty} \Delta_T(u)\,du = \varepsilon$$

by (10.12) and (10.13). Since $|(f * \Delta_T)(x) - f(x)| < 2\varepsilon$ for all sufficiently large T, we have the stated result, in view of (10.15). □

We continue with analogues of Theorem 4.28, Corollary 4.29, Theorem 4.32, and Corollary 4.33.

Theorem 10.13. *Suppose that f is Riemann–integrable in intervals of the form $[-T, T]$, and put*

$$I(T) = \int_{-T}^{T} f(t)\,dt, \qquad J(T) = \int_{-T}^{T} (1 - |t|/T)f(t)\,dt\,.$$

If $\lim_{T\to\infty} I(T) = a$, then $\lim_{T\to\infty} J(T) = a$.

Proof. We observe that

$$\frac{1}{T}\int_0^T I(u)\,du = \frac{1}{T}\int_0^T\int_{-u}^{u} f(t)\,dt\,du = \frac{1}{T}\int_{-T}^{T} f(t)\Big(\int_{|t|}^{T} 1\,du\Big)\,dt = J(T)\,.$$

Given $\varepsilon > 0$, choose U so that $|I(u) - a| < \varepsilon$ for all $u \ge U$. Write

$$J(T) - a = \frac{1}{X}\int_0^T I(u) - a\,du = \frac{1}{T}\int_0^U I(u) - a\,du + \frac{1}{T}\int_U^T I(u) - a\,du = E_1 + E_2,$$

say. Clearly $E_1 \to 0$ as $T \to \infty$. Also,

$$|E_2| \le \frac{1}{T}\int_U^T |I(u) - a|\,du < \frac{1}{T}\int_U^T \varepsilon\,du = \frac{T-U}{T}\varepsilon < \varepsilon\,.$$

Thus $|J(T) - a| < 2\varepsilon$ when T is sufficiently large, so we have the result. □

Corollary 10.14. *Suppose that $f \in L^1(\mathbb{R})$. If f is continuous at x, and if*

$$\lim_{T\to\infty}\int_{-T}^{T} \widehat{f}(t)e(xt)\,dt$$

exists, then this limit is $f(x)$.

Proof. Let a denote the limit, and take this integral to be $I(T)$ in Theorem 10.13. By Theorem 10.12 we find that $J(T) \to f(x)$ as $T \to \infty$. Thus $a = f(x)$ by Theorem 10.13. □

Theorem 10.15. *Suppose that f is Riemann–integrable in intervals of the form $[-T,T]$, and put*

$$I(T)=\int_{-T}^{T} f(t)\,dt,\qquad J(T)=\int_{-T}^{T}(1-|t|/T)f(t)\,dt\,.$$

If $\lim_{T\to\infty} J(T)=a$, and if $f(t)=O(1/|t|)$ for $|t|\ge 1$, then $\lim_{T\to\infty} I(T)=a$.

Proof. Since

$$\max(0,T+H-|t|)-\max(0,T-|t|)=\begin{cases}0 & (t>T+H),\\ T+H-t & (T\le t\le T+H),\\ H & (-T\le t\le T),\\ T+H+t & (-T-H\le t\le -T),\\ 0 & (t<-T),\end{cases}$$

it follows that

$$\begin{aligned}&(T+H)(J(T+H)-a)-T(J(T)-a)\\ &\quad = H(I(T)-a)+\int_T^{T+H}(T+H-t)f(t)\,dt+\int_{-T-H}^{-T}(T+H+t)f(t)\,dt\,.\end{aligned}$$

We divide through by H, rearrange, and apply the triangle inequality to deduce that

$$\begin{aligned}|I(T)-a|&\le \frac{T+H}{H}|J(T+H)-a|+\frac{T}{H}|J(T)-a|\\ &\quad+\int_T^{T+H}\frac{T+H-t}{H}|f(t)|\,dt+\int_{-T-H}^{-T}\frac{T+H+t}{H}|f(t)|\,dt\\ &= E_1+E_2+E_3+E_4,\end{aligned}$$

say. Suppose that $|J(T)-a|<\varepsilon$ for $T\ge T_0(\varepsilon)$, and that $T>T_0$. Take $H=\sqrt{\varepsilon}T$. Thus $E_1\le 2\sqrt{\varepsilon}$ and $E_2\le\sqrt{\varepsilon}$. As for E_3, we note that $0\le (T+H-t)/H\le 1$ and that $|f(t)|=O(1/T)$, so $E_3=O(\sqrt{\varepsilon})$. Similarly, $E_4=O(\sqrt{\varepsilon})$. Since ε can be taken arbitrarily small, it follows that $I(T)\to a$. □

Corollary 10.16. *If $f\in L^1(\mathbb{R})$, if $\widehat{f}(t)=O(1/|t|)$ uniformly for $|t|\ge 1$, and if f is continuous at x, then*

$$f(x)=\lim_{T\to\infty}\int_{-T}^{T}\widehat{f}(t)e(xt)\,dt\,.$$

Proof. This is immediate on combining Theorem 10.15 with Theorem 10.12. □

In §7.1 we showed that if $f\in A(\mathbb{T})$, then the Fourier Series of f converges to $f(x)$ for all x. We now establish an analogue of this for the Fourier Transform.

Lemma 10.17. *Suppose that f is Riemann–integrable in intervals of the form $[-T,T]$, and put*

$$I(T)=\int_{-T}^{T} f(t)\,dt,\qquad J(T)=\int_{-T}^{T}(1-|t|/T)f(t)\,dt\,.$$

If $f \in L^1(\mathbb{R})$, then $\lim_{T\to\infty} I(T)$ and $\lim_{T\to\infty} J(T)$ both exist, and are equal.

Proof. That $I(T)$ tends to a limit is clear, because an integral will converge if it is absolutely convergent. This is a continuous analogue of the familiar fact that $\sum_1^\infty a_n$ converges if $\sum_1^\infty |a_n| < \infty$. Let a denote the limit of $I(T)$. Then $J(T)$ tends to a by Theorem 10.13. □

Theorem 10.18. *If $f \in L^1(\mathbb{R})$, if $f(x)$ is everywhere continuous, and if $\widehat{f} \in L^1(\mathbb{R})$, then*

$$\int_{-\infty}^{\infty} \widehat{f}(t)e(tx)\,dt = f(x) \tag{10.17}$$

for all x.

Proof. This is immediate on combining Lemma 10.17 with Theorem 10.12. □

We are now in a position to recover an analogue of (3.43), which gives the Fourier coefficients of a trigonometric polynomial.

Corollary 10.19. *Let $B(x)$ be given as in (10.11). If $B \in L^1(\mathbb{R})$ and $b(t)$ is continuous for $-T \le t \le T$ with $b(\pm T) = 0$, then*

$$\widehat{B}(t) = \begin{cases} b(t) & (|t| \le T), \\ 0 & (|t| > T). \end{cases}$$

Proof. Suppose, as we may, that $b(t) = 0$ for $|t| > T$. The integral (10.11) can thus be considered to run over the entire real line, and so we see that $B(x) = \widehat{b}(-x)$. By Theorem 10.18 applied to $b(t)$ we see that

$$b(t) = \int_{-\infty}^{\infty} \widehat{b}(x)e(xt)\,dx = \int_{-\infty}^{\infty} B(-x)e(xt)\,dx = \int_{-\infty}^{\infty} B(x)e(-xt)\,dx\,.$$

That is, $b(t) = \widehat{B}(t)$. □

Corollary 10.20. *We have*

$$\int_{-\infty}^{\infty} \Big(\frac{\sin \pi t}{\pi t}\Big)^2 e(xt)\,dt = \begin{cases} 1 - |x| & (|x| \le 1), \\ 0 & (|x| > 1). \end{cases} \tag{10.18}$$

Lemma 10.11—whose proof involved some work—is the special case $x = 0$ of this.

Proof. In Example 10.3 we found that

$$\int_{-1}^{1} (1 - |x|)e(-tx)\,dx = \Big(\frac{\sin \pi t}{\pi t}\Big)^2$$

for all t. (When $t = 0$, the right hand side above is interpreted to be 1, by continuity.) The desired result now follows from Corollary 10.19. □

Suppose that $f \in L^1(\mathbb{R})$. We recognize that the integral in (10.15) is the continuous analogue of the Cesàro partial sum $\sigma_N(x)$ of a Fourier series. With a slight abuse of notation, we now set

$$\sigma_T(x) = (f * \Delta_T)(x) = \int_{-T}^{T} (1 - |t|/T)\widehat{f}(t)e(tx)\, dt\,. \tag{10.19}$$

Theorem 10.21. *If $f \in L^1(\mathbb{R})$, then*

$$\lim_{T\to\infty} \int_{-\infty}^{\infty} |f(x) - \sigma_T(x)|\, dx = 0\,.$$

Proof. By (10.16) we see that

$$\int_{-\infty}^{\infty} |f(x) - \sigma_T(x)|\, dx = \int_{-\infty}^{\infty} \left| \int_{-\infty}^{\infty} (f(x) - f(x-u))\Delta_T(u)\, du \right| dx\,.$$

By the triangle inequality this is

$$\leq \int_{-\infty}^{\infty} \int_{-\infty}^{\infty} |(f(x) - f(x-u))|\Delta_T(u)\, du \Big| \, dx\,.$$

We invert the order of integration to see that the above is

$$\int_{-\infty}^{\infty} \Delta_T(u) \Big(\int_{-\infty}^{\infty} |f(x) - f(x-u)|\, dx \Big)\, du\,. \tag{10.20}$$

Let $I(u)$ denote the integral inside the large parentheses. In the first place,

$$I(u) \leq \int_{-\infty}^{\infty} |f(x)| + |f(x-u)|\, dx = 2\|f\|_1$$

for all u. Secondly, from Lemma 10.5 we know that $\lim_{u\to 0} I(u) = 0$. Suppose that $\varepsilon > 0$ is given. Choose δ so that $I(u) < \varepsilon$ when $|u| < \delta$. Write the integral with respect to u in (10.20) as

$$\int_{-\infty}^{-\delta} + \int_{-\delta}^{\delta} + \int_{\delta}^{\infty} = I_1 + I_2 + I_3\,.$$

Then

$$I_1 \leq 2\|f\|_1 \int_{-\infty}^{-\delta} \Delta_T(u)\, du \leq \frac{2\|f\|_1}{T\pi^2\delta}$$

by (10.12), and I_3 is estimated similarly. On the other hand,

$$I_2 \leq \varepsilon \int_{-\delta}^{\delta} \Delta_T(u)\, du < \varepsilon \int_{-\infty}^{\infty} \Delta_T(u)\, du = \varepsilon$$

by (10.13). Since the expression (10.20) is $< 3\varepsilon$ for all sufficiently large T, it follows that the limit is 0. □

Corollary 10.22. *If $f \in L^1(\mathbb{T})$ and $\widehat{f}(t) = 0$ for all t, then*

$$\int_{-\infty}^{\infty} |f(x)|\, dx = 0,$$

which is to say that $f(x) = 0$ for almost all x.

Proof. Since $\widehat{f}(t) = 0$ for all t, it follows from the definition (10.19) that $\sigma_T(x) = 0$ for all x. Thus Theorem 10.21 asserts that

$$\lim_{T\to\infty} \int_{-\infty}^{\infty} |f(x)|\, dx = 0\,.$$

Here the integral is independent of T, so the only way this can hold is for the integral to be 0. □

Corollary 10.23. *Suppose that $f \in L^1(\mathbb{R})$ and that $g \in L^1(\mathbb{R})$. If $\widehat{f}(t) = \widehat{g}(t)$ for all t, then*

$$\int_{-\infty}^{\infty} |f(x) - g(x)|\, dx = 0,$$

which is to say that $f(x) = g(x)$ for almost all x.

Proof. Apply the corollary above to the function $f(x) - g(x)$. □

Exercises

1. Suppose that f is continuous on the real line, and that $\lim_{x\to\pm\infty} f(x) = 0$. Show that f is uniformly continuous on the real line.
2. (a) Give an example of a function $f \in L^1(\mathbb{R})$ such that f is everywhere continuous, but not uniformly continuous.
   (b) Show that for the function constructed in (a), $\widehat{f} \notin L^1(\mathbb{R})$.
3. Suppose that $f \in L^1(\mathbb{R})$ and that $g \in L^1(\mathbb{R})$. Show that

$$\int_{-\infty}^{\infty} f(x)\widehat{g}(x)\, dx = \int_{-\infty}^{\infty} \widehat{f}(t) g(t)\, dt\,.$$

4. Let

$$f(x) = \begin{cases} (\cos \pi x/2)^2 & (|x| \le 1), \\ 0 & (|x| > 1). \end{cases}$$

   (a) Show that

$$\widehat{f}(t) = \frac{\sin 2\pi t}{2\pi t(1 - 4t^2)}\,.$$

   (b) The above needs to be interpreted by continuity as t approaches 0 or $\pm 1/2$. What are the values of $\widehat{f}$ at these three points?
   (c) Show that

$$\int_{-\infty}^{\infty} \widehat{f}(t) e(xt)\, dt = \begin{cases} (\cos \pi x/2)^2 & (|x| \le 1), \\ 0 & (|x| > 1). \end{cases}$$

   (d) With the aid of appropriate software, graph these functions.
5. Suppose that

$$f(x) = \begin{cases} 1 - |x| + \dfrac{\sin 2\pi |x|}{2\pi} & (|x| \le 1), \\ 0 & (\text{otherwise}). \end{cases}$$

(a) Show that

$$\widehat{f}(t) = \frac{(\sin \pi t)^2}{\pi^2 t^2 (1 - t^2)}$$

(b) Show that

$$\int_{-\infty}^{\infty} \frac{(\sin \pi t)^2}{\pi^2 t^2 (1-t^2)} e(tx)\, dt = \begin{cases} 1 - |x| + \dfrac{\sin 2\pi |x|}{2\pi} & (|x| \le 1), \\ 0 & \text{(otherwise)}. \end{cases}$$

(c) With the aid of appropriate software, graph these functions.

6. Suppose that $f \in L^1(\mathbb{R})$ and that $g \in L^1(\mathbb{R})$. Show that

$$\int_{-\infty}^{\infty} f(x)\widehat{g}(x)\, dx = \int_{-\infty}^{\infty} \widehat{f}(t) g(t)\, dt\,.$$

7. Let

$$f(x) = \max(0, 1 - |x|) + \Big(\frac{\sin \pi x}{\pi x}\Big)^2.$$

Show that $\widehat{f}(t) = f(t)$ for all t.

8. Suppose that $f \in L^1(\mathbb{R})$ and that f is everywhere continuous. Put $g(t) = \widehat{f}(t)$. Show that if $\widehat{g} \in L^1(\mathbb{R})$, then $\widehat{g}(x) = f(-x)$.

10.3. Fourier transforms in mean square

For $p \ge 1$, we let $L^p(\mathbb{R})$ denote the space of those functions f for which

$$\int_{-\infty} |f(x)|^p\, dx < \infty\,.$$

For such functions we have a norm,

$$\|f\|_p = \|f\|_{L^p(\mathbb{R})} = \Big(\int_{-\infty}^{\infty} |f(x)|^p\, dx\Big)^{1/p}. \tag{10.21}$$

We consider now the important case $p = 2$. The first thing that we need to understand is that $L^2(\mathbb{R})$ is not a subset of $L^1(\mathbb{R})$. Indeed, there is no inclusion in either direction. This is made clear by considering the following examples: If

$$f(x) = \begin{cases} x^{-p} & (x \ge 1), \\ 0 & \text{(otherwise)}, \end{cases}$$

then $f \in L^1(\mathbb{R})$ for $p > 1$, and not otherwise, and $f \in L^2(\mathbb{R})$ for $p > 1/2$, but not otherwise. Hence if $p = 3/4$, then $f \in L^2(\mathbb{R})$, but $f \notin L^1(\mathbb{R})$. On the other hand, if

$$f(x) = \begin{cases} x^{-p} & (0 < x \le 1), \\ 0 & \text{(otherwise)}, \end{cases}$$

then $f \in L^1(\mathbb{R})$ for $p < 1$, but not for $p \ge 1$, and $f \in L^2(\mathbb{R})$ for $p < 1/2$ but not for $p \ge 1/2$. Hence if $p = 3/4$, then $f \in L^1(\mathbb{R})$, but $f \notin L^2(\mathbb{R})$. We need f to be in $L^1(\mathbb{R})$ in order to define $\widehat{f}(t)$, and we need f to be in $L^2(\mathbb{R})$ if we are to have any hope of establishing an analogue of the Parseval Identity. Thus much of our discussion will involve functions that are in both $L^1(\mathbb{R})$ and $f \in L^2(\mathbb{R})$.

As was the case with $L^2(\mathbb{T})$, the space $L^2(\mathbb{R})$ is an inner product space. For $f, g \in L^2(\mathbb{R})$ we put

(10.22)
$$\langle f, g\rangle = \int_{-\infty}^{\infty} f(x)\overline{g(x)}\,dx\,.$$

Thus we have the Cauchy–Schwarz inequality

(10.23)
$$|\langle f, g\rangle| \le \|f\|_2\|g\|_2,$$

and a triangle inequality

(10.24)
$$\|f+g\|_2 \le \|f\|_2 + \|g\|_2\,.$$

Lemma 10.24. *If $f \in L^2(\mathbb{R})$, then*

(10.25)
$$\lim_{\delta\to 0}\int_{-\infty}^{\infty} |f(x+\delta) - f(x)|^2\,dx = 0\,.$$

This is analogous to Lemma 5.4, and the proof follows the same lines.

Proof. Let $\varepsilon > 0$ be given, and choose X so large that

$$\Big(\int_{|x|>X-1} |f(x)|^2\,dx\Big)^{1/2} < \varepsilon\,.$$

Then for $\delta \le 1$ it follows by the triangle inequality that

$$\Big(\int_{-\infty}^{\infty} |f(x+\delta) - f(x)|^2\,dx\Big)^{1/2} \le \Big(\int_{-X}^{X} |f(x+\delta) - f(x)|^2\,dx\Big)^{1/2} + 2\varepsilon\,.$$

By principles of Lebesgue measure theory (see Theorem 0.22) we know that continuous functions are dense in the L^2 norm on compact intervals. Thus there is a continuous function g such that

$$\Big(\int_{-X-1}^{X+1} |f(x) - g(x)|^2\,dx\Big)^{1/2} < \varepsilon\,.$$

By the triangle inequality it follows that

$$\Big(\int_{-X}^{X} |f(x+\delta) - f(x)|^2\,dx\Big)^{1/2} \le \Big(\int_{-X}^{X} |g(x+\delta) - g(x)|\,dx\Big)^{1/2} + 2\varepsilon\,.$$

Since g is continuous on $[-X-1, X+1]$ it follows that g is uniformly continuous on this interval, so the integrand in this last integral tends to 0 uniformly in x. Hence the integral tends to 0 as $\delta \to 0$, and the proof is complete. □

Lemma 10.25. *If $f \in L^2(\mathbb{R})$ and $g \in L^2(\mathbb{R})$, then $(f*g)(x)$ is uniformly continuous.*

Proof. Let $h(x) = (f*g)(x)$. Then

$$h(x+\delta) - h(x) = \int_{-\infty}^{\infty} (h(x+\delta-u) - h(x-u))g(u)\,du\,.$$

Hence by the Cauchy–Schwarz inequality,

$$|h(x+\delta) - h(x)|^2 \le \int_{-\infty}^{\infty} |f(x+\delta-u) - h(x-u)|^2\,du \int_{-\infty}^{\infty} |g(u)|^2\,du\,.$$

Here the second integral is a fixed (finite) number, independent of x and δ. The first integral on the right is

$$= \int_{-\infty}^{\infty} |f(v+\delta) - f(v)|^2 \, dv,$$

which tends to 0 as δ tends to 0, by Lemma 10.24. Hence we have the result. □

In $L^2(\mathbb{R})$, the analogue of Parseval's Identity is known as *Plancherel's Identity.*

Theorem 10.26. (Plancherel) *If* $f \in L^1(\mathbb{R}) \cap L^2(\mathbb{R})$, *then*

$$\int_{-\infty}^{\infty} |f(x)|^2 \, dx = \int_{-\infty}^{\infty} |\widehat{f}(t)|^2 \, dt \, .$$

Proof. Let $g(x) = \overline{f(-x)}$, and put $h(x) = (f * g)(x)$. By Theorem 10.9 we know that $h \in L^1(\mathbb{R})$, and by Lemma 10.25 we know that h is uniformly continuous. Hence by Theorem 10.12 it follows that

$$h(x) = \lim_{T \to \infty} \int_{-T}^{T} (1 - |t|/T) \widehat{h}(t) e(tx) \, dt$$

for all x. In particular, on taking $x = 0$ we see that

$$h(0) = \lim_{T \to \infty} \int_{-T}^{T} (1 - |t|/T) \widehat{h}(t) \, dt \, .$$

From Theorem 10.9 we know that $\widehat{h}(t) = \widehat{f}(t)\widehat{g}(t)$. From the definition of g we find that

$$\widehat{g}(t) = \int_{-\infty}^{\infty} g(x) e(-xt) \, dx = \int_{-\infty}^{\infty} \overline{f(-x)} e(-xt) \, dx = \int_{-\infty}^{\infty} \overline{f(x)} e(xt) \, dx$$

$$= \overline{\int_{-\infty}^{\infty} f(x) e(-xt) \, dx} = \overline{\widehat{f}(t)} \, .$$

Since

$$h(0) = (f * g)(0) = \int_{-\infty}^{\infty} f(u) g(-u) \, du = \int_{-\infty}^{\infty} |f(u)|^2 \, du,$$

we deduce that

$$\int_{-\infty}^{\infty} |f(x)|^2 \, dx = \lim_{T \to \infty} \int_{-T}^{T} (1 - |t|/T) |\widehat{f}(t)|^2 \, dt \, .$$

Thus for any ε, $0 < \varepsilon < 1$, we have

$$(1-\varepsilon) \int_{-\varepsilon T}^{\varepsilon T} |\widehat{f}(t)|^2 \, dt \le \int_{-T}^{T} (1 - |t|/T) |\widehat{f}(t)|^2 \, dt \le \int_{-\infty}^{\infty} |\widehat{f}(t)|^2 \, dt \, .$$

On letting T tend to infinity we deduce that

$$(1-\varepsilon) \int_{-\infty}^{\infty} |\widehat{f}(t)|^2 \, dt \le \int_{-\infty}^{\infty} |f(x)|^2 \, dx \le \int_{-\infty}^{\infty} |\widehat{f}(t)|^2 \, dt \, .$$

Since ε is arbitrarily small, this gives the result. □

Just as the Parseval Identity has an asymmetric form, the Plancherel Identity has an asymmetric form:

Theorem 10.27. (The Extended Plancherel Identity) *Suppose that* $f \in L^1(\mathbb{R}) \cap L^2(\mathbb{R})$ *and that* $g \in L^1(\mathbb{R}) \cap L^2(\mathbb{R})$. *Then*

$$\int_{-\infty}^{\infty} f(x)\overline{g(x)}\,dx = \int_{-\infty}^{\infty} \widehat{f}(t)\overline{\widehat{g}(t)}\,dt\,.$$

Proof. We observe that

$$f(x)\overline{g(x)} = \frac{1}{4}\big(|f(x)+g(x)|^2 + i|f(x)+ig(x)|^2 - |f(x)-g(x)|^2 - i|f(x)-ig(x)|^2\big)\,.$$

We integrate with respect to x, and apply Theorem 10.26 four times. By the identity

$$\frac{1}{4}\big(|\widehat{f}(t)+\widehat{g}(t)|^2 + i|\widehat{f}(t)+i\widehat{g}(t)|^2 - |\widehat{f}(t)-\widehat{g}(t)|^2 - i|\widehat{f}(t)-i\widehat{g}(t)|^2\big) = \widehat{f}(t)\overline{\widehat{g}(t)}$$

we obtain the stated result. □

Example 10.4. Suppose one has a telescope mounted in a satellite in space. If the telescope is aimed at a distant star, then the signal is very weak. If the telescope is aimed at the sun, then the signal is very strong. The weak signal would be lost in noise, and the strong signal is too strong to cope with. Consequently, the satellite is designed not to transmit the actual signal $f(x)$, but rather $F(f(x))$ where F is an increasing function that emphasizes small values (those near 0) and attenuates large values. For example, one might take $F(y) = y^{1/9}$. With this choice of F, if x is a point at which $f(x) = 10^{-9}$, then $F(f(x)) = 0.1$, while if x is a point at which $f(x) = 10^9$, then $F(f(x)) = 10$. The disparity between 10^{-9} and 10^9 is thus reduced to values in $[0.1, 10.0]$. By the extended form of Plancherel's Identity, we recognize that

$$\int_{-\infty}^{\infty} \big(f_1(x) - f_2(x)\big)\big(F(f_1(x)) - F(f_2(x))\big)\,dx$$
$$= \int_{-\infty}^{\infty} \big(\widehat{f_1}(t) - \widehat{f_2}(t)\big)\big(\overline{\widehat{F(f_1)}(t)} - \overline{\widehat{F(f_2)}(t)}\big)\,dt\,.$$

If $\widehat{f_1}(t) = \widehat{f_2}(t) = 0$ for $|t| \ge T$, and if $\widehat{F(f_1)}(t) = \widehat{F(f_2)}(t)$ for $-T \le t \le T$, then the right hand side above is 0 for all t. However, since F is increasing, the sign of $f_1(x) - f_2(x)$ is the same as that of $F(f_1(x)) - F(f_2(x))$. Thus the integrand on the left hand side above is everywhere nonnegative. For this integral to be 0, the integrand must be 0 almost everywhere. But in practice f_1 and f_2 are continuous, and F also, so $f_1(x) = f_2(x)$ for all x. Thus there is only one function $f(x)$ that can produce the received $\widehat{F(f)}(t)$.

To prepare for the discussion below, we establish a variant of the Sobolev inequality (Lemma 10.7).

Lemma 10.28. (Sobolev) *If* $g(x)$ *has a continuous first derivative on the interval* $[a, b]$, *then*

$$|g(x)|^2 \le \frac{1}{b-a}\int_a^b |g(u)|^2\,du + 2\Big(\int_a^b |g(u)|^2\,du\Big)^{1/2}\Big(\int_a^b |g'(u)|^2\,du\Big)^{1/2}$$

for all $x \in [a, b]$.

Proof. We apply Lemma 10.7 with $f(u) = g(a+(b-a)u)^2$. Thus,

$$|g(x)|^2 \le \int_0^1 |g(a+(b-a)u)|^2\,du + 2(b-a)\int_0^1 |g(a+(b-a)u)g'(a+(b-a)u)|\,du\,.$$

In these integrals we make the change of variable $v = a + (b-a)u$ to see that the above is

$$= \frac{1}{b-a}\int_a^b |g(v)|\,dv + 2\int_a^b |g(v)g'(v)|\,dv\,.$$

We obtain the stated result by applying the Cauchy–Schwarz inequality to the second integral above. □

The function $\max(0, 1-|t|/T)$ has a peak of width T. In (10.14) we see that its Fourier transform is $\Delta_T(x)$, which has a peak of width $1/T$. Here T may be large or small, but the product of the widths of the peaks is ≥ 1. This is true not just in this case, but more generally. In physics, this is known as the *Heisenberg uncertainty principle*. In the following we have a quantitative form of this.

Theorem 10.29. (The Heisenberg Uncertainty Principle) *If* $f \in L^1(\mathbb{R})$, $xf(x) \in L^2(\mathbb{R})$, *and* $f' \in L^1(\mathbb{R}) \cap L^2(\mathbb{R})$, *then*

$$\Big(\int_{-\infty}^{\infty} |xf(x)|^2\,dx\Big)\Big(\int_{-\infty}^{\infty} |t\widehat{f}(t)|^2\,dt\Big) \ge \frac{1}{16\pi^2}\Big(\int_{-\infty}^{\infty} |f(x)|^2\,dx\Big)^2\,.$$

The point is that if f has most of its L^2 mass concentrated in the the interval $[-A, A]$, then the first integral on the left is comparable to $A^2\int|f|^2$. If $\widehat{f}$ has most of its L^2 mass concentrated in the interval $[-B, B]$, then the second integral on the left is comparable to $B^2 \int |\widehat{f}|^2 = B^2 \int |f|^2$. Thus from the above we learn that AB is bounded below by a positive absolute constant.

Proof. If any of these integrals vanish, then $f(x) = 0$ for almost all x, and hence all these integrals vanish. Thus we may assume that all these integrals are positive. If either of the integrals on the left hand side is infinite, then there is nothing to prove. Hence we may assume that they are finite.

By integrating by parts we see that

$$\int_{-X}^{X} |f(x)|^2\,dx = X|f(X)|^2 + X|f(-X)|^2 - \int_{-X}^{X} xf'(x)\overline{f(x)} + xf(x)\overline{f'(x)}\,dx\,.$$

By the Cauchy–Schwarz inequality we see that

$$\int_{-X}^{X} \left|xf'(x)\overline{f(x)}\right|dx \le \Big(\int_{-\infty}^{\infty} |xf(x)|^2\,dx\Big)^{1/2}\Big(\int_{-\infty}^{\infty} |f'(x)|^2\,dx\Big)^{1/2}.$$

By Theorem 10.8 we know that $\widehat{f'}(t) = 2\pi it\widehat{f}(t)$. By Plancherel's Identity it follows that

$$\int_{-\infty}^{\infty} |f'(x)|^2\,dx = \int_{-\infty}^{\infty} |2\pi it\widehat{f}(t)|^2\,dt\,. \tag{10.26}$$

Thus

$$\Big|\int_{-X}^{X} xf'(x)\overline{f(x)}+xf(x)\overline{f'(x)}\,dx\Big| \le 4\pi\Big(\int_{-\infty}^{\infty} |xf(x)|^2\,dx\Big)^{1/2}\Big(\int_{-\infty}^{\infty} |t\widehat{f}(t)|^2\,dt\Big)^{1/2}.$$

By Lemma 10.28 applied to the interval $[X, 2X]$, we see that

$$\begin{aligned}|f(x)|^2 &\le \frac{1}{X}\int_X^{2X} |f(u)|^2\,du + 2\Big(\int_X^{2X} |f(u)|^2\,du\Big)^{1/2}\Big(\int_X^{2X} |f'(u)|^2\,du\Big)^{1/2}\\ &\le \frac{1}{X^3}\int_X^{2X} |uf(u)|^2\,du + \frac{2}{X}\Big(\int_X^{2X} |uf(u)|^2\,du\Big)^{1/2}\Big(\int_X^{2X} |f'(u)|^2\,du\Big)^{1/2}.\end{aligned}$$

By (10.26) we know that $\int_{-\infty}^{\infty} |f'(u)|^2\,du < \infty$. Hence the integrals above are tails of convergent integrals, and thus tend to 0 as $X \to \infty$. Consequently $X|f(X)|^2 \to 0$ as $X \to \infty$. Similarly, $X|f(-X)|^2 \to 0$, so we obtain the stated result by letting X tend to infinity. □

We know by Theorem 10.9 that if $f \in L^1(\mathbb{R})$ and $g \in L^1(\mathbb{R})$, then $f * g \in L^1(\mathbb{R})$, and that $\widehat{f*g}(t) = \widehat{f}(t)\widehat{g}(t)$ for all t. We conclude this section with some remarks concerning convolutions of functions in $L^2(\mathbb{R})$.

Theorem 10.30. *If $f \in L^2(\mathbb{R})$ and $g \in L^2(\mathbb{R})$, then $(f * g)(x)$ exists, and*

$$|(f * g)(x)| \le \|f\|_2\|g\|_2$$

for all x.

Proof. By the Cauchy–Schwarz inequality,

$$\Big|\int_{-\infty}^{\infty} f(u)g(x-u)\,du\Big| \le \Big(\int_{-\infty}^{\infty} |f(u)|^2\,du\Big)^{1/2}\Big(\int_{-\infty}^{\infty} |g(x-u)|^2\,du\Big)^{1/2} = \|f\|_2\|g\|_2.$$

□

Example 10.5. Let

$$f(x) = \begin{cases} 1 & (|x| \le 1), \\ |x|^{-2/3} & (|x| > 1). \end{cases}$$

Thus $f \in L^2(\mathbb{R})$ but $f \notin L^1(\mathbb{R})$. We determine the rate of decay of $f * f$. For x large and positive, we break the integral

$$\int_{-\infty}^{\infty} f(u) f(x-u)\, du$$

that defines $f * f$ into four ranges: $u \le -x/2$, $-x/2 \le u \le x/2$, $x/2 \le u \le 3x/2$, and $u > 3x/2$. In the first range, $f(u) \asymp |u|^{-2/3}$ and $f(x-u) \asymp |u|^{-2/3}$, so the integrand is comparable to $|u|^{-4/3}$, and hence

$$\int_{-\infty}^{-x/2} f(u) f(x-u)\, du \asymp x^{-1/3}\,. \tag{10.27}$$

In the second range we have $f(x-u) \asymp x^{-2/3}$, while $\int_{-x/2}^{x/2} f(u)\, du \asymp x^{1/3}$, so

$$\int_{-x/2}^{x/2} f(u) f(x-u)\, du \asymp x^{-1/3}\,. \tag{10.28}$$

The third range is the same as the one just treated, but with the roles of $f(u)$ and of $f(x-u)$ reversed. Similarly, the fourth range is the same as the first. Thus on combining (10.27) and (10.28) we conclude that

$$(f * f)(x) \asymp x^{-1/3}$$

for $x \ge 1$. The integral of this over $1 \le x < \infty$ diverges, and the same is true of the square. Thus $f * f \notin L^1(\mathbb{R})$, and $f * f \notin L^2(\mathbb{R})$.

Theorem 10.31. *If $f \in L^1(\mathbb{R})$ and $g \in L^2(\mathbb{R})$, then $f * g$ exists, and $f * g \in L^2(\mathbb{R})$.*

Proof. By the triangle inequality,

$$\int_{-\infty}^{\infty} \Big| \int_{-\infty}^{\infty} f(x-u) g(u)\, du \Big|^2 dx \le \int_{-\infty}^{\infty} \Big(\int_{-\infty}^{\infty} |f(x-u) g(u)|\, du \Big)^2 dx\,.$$

Let $F(u) = |f(u)|$ and $G(u) = |g(u)|$. Thus the right hand side above is

$$= \int_{-\infty}^{\infty} \Big(\int_{-\infty}^{\infty} F(x-u) G(u)\, du \Big)^2 dx$$

$$= \int_{-\infty}^{\infty} \Big(\int_{-\infty}^{\infty} F(x-u) G(u)\, du \Big) \Big(\int_{-\infty}^{\infty} F(x-v) G(v)\, dv \Big)\, dx\,.$$

We now take the integrals with respect to u and v outside, and the integral with respect to x inside. This interchange is justified since the integrands are nonnegative. Thus the above is

$$= \int_{-\infty}^{\infty} \int_{-\infty}^{\infty} G(u) G(v) \Big(\int_{-\infty}^{\infty} F(x-u) F(x-v)\, dx \Big)\, du\, dv\,.$$

By the arithmetic–geometric mean inequality we know that $G(u)G(v) \le (G(u)^2 + G(v)^2)/2$. Thus the above is

$$\le \int_{-\infty}^{\infty} \int_{-\infty}^{\infty} \Big(\frac{1}{2} G(u)^2 + \frac{1}{2} G(v)^2 \Big) \Big(\int_{-\infty}^{\infty} F(x-u) F(x-v)\, dx \Big)\, du\, dv\,.$$

The contribution of $G(u)^2/2$ is

$$\frac{1}{2}\int_{-\infty}^{\infty} G(u)^2\Big(\int_{-\infty}^{\infty}\int_{-\infty}^{\infty} F(x-u)F(x-v)\,dx\,dv\Big)\,du\,. \tag{10.29}$$

The two inner integrals are

$$\int_{-\infty}^{\infty}\int_{-\infty}^{\infty} F(x-u)F(x-v)\,dx\,dv = \int_{-\infty}^{\infty} F(x-u)\int_{-\infty}^{\infty} F(x-v)\,dv\,dx\,.$$

On the right hand side, the integral with respect to v is equal to $\int_{-\infty}^{\infty} F(v)\,dv = \|f\|_1$, which is independent of x. On factoring this out, we see that the remaining integral of $F(x-u)$ is also $\|f\|_1$, which is independent of u. Thus the expression (10.29) is

$$= \frac{1}{2}\|f\|_1^2\int_{-\infty}^{\infty} G(u)^2\,du = \frac{1}{2}\|f\|_1\|g\|_2^2.$$

The contribution of $G(v)^2/2$ is the same, with the roles of u and v reversed, so altogether we have shown not only that $\|f*g\|_2 < \infty$, but also that

$$\|f*g\|_2 \le \|f\|_1\|g\|_2\,. \tag{10.30}$$

□

Theorem 10.32. *If f and g are both in $L^1(\mathbb{R})\cap L^2(\mathbb{R})$, then $f*g$ is also in $L^1(\mathbb{R})\cap L^2(\mathbb{R})$, and*

$$\int_{-\infty}^{\infty} |(f*g)(x)|^2\,dx = \int_{-\infty}^{\infty} |\widehat{f}(t)|^2|\widehat{g}(t)|^2\,dt\,.$$

Proof. From the information that f and g are in $L^1(\mathbb{R})$ we know that $f*g$ is also in $L^1(\mathbb{R})$, and also that $\widehat{f*g}(t) = \widehat{f}(t)\widehat{g}(t)$. From the information that $f\in L^1(\mathbb{R})$ and $g\in L^2(\mathbb{R})$ it follows by Theorem 10.31 that $f*g\in L^2(\mathbb{R})$. Since $f*g\in L^1(\mathbb{R})\cap L^2(\mathbb{R})$, the last identity follows by applying Plancherel's Identity to $f*g$. □

10.4. The Poisson summation formula

Roughly, the proposition is that

$$\sum_{n=-\infty}^{\infty} f(n) = \sum_{k=-\infty}^{\infty} \widehat{f}(k)\,. \tag{10.31}$$

Of course one immediately wonders why, and under what hypotheses.

Suppose that $f\in L^1(\mathbb{R})$, and put

$$F(\alpha) = \sum_{n=-\infty}^{\infty} f(n+\alpha)\,. \tag{10.32}$$

It is possible that this sum does not converge for all α, but

$$\int_0^1 \sum_{n=-\infty}^{\infty} |f(n+\alpha)|\,d\alpha = \sum_{n=-\infty}^{\infty}\int_n^{n+1} |f(\alpha)|\,d\alpha = \int_{-\infty}^{\infty} |f(\alpha)|\,d\alpha < \infty\,.$$

Thus we see that the sum in (10.32) is absolutely convergent for almost all α. Moreover, $F(\alpha)$ has period 1, and by the triangle inequality,

$$\int_0^1 |F(\alpha)|\,d\alpha \le \int_0^1 \sum_{n=-\infty}^{\infty} |f(n+\alpha)|\,d\alpha \tag{10.33}$$

$$= \sum_{n=-\infty}^{\infty} \int_n^{n+1} |f(\alpha)|\,d\alpha = \int_{-\infty}^{\infty} |f(\alpha)|\,d\alpha < \infty\,. \tag{10.34}$$

That is, $\|F\|_{L^1(\mathbb{T})} \le \|f\|_{L^1(\mathbb{R})}$. Since $F \in L^1(\mathbb{T})$, our first interest is to determine its Fourier coefficients. We find that

$$\begin{aligned}\widehat{F}(k) &= \int_0^1 F(\alpha)e(-k\alpha)\,d\alpha = \int_0^1 \Big(\sum_{n=-\infty}^{\infty} f(n+\alpha)\Big)e(-ka)\,d\alpha \\ &= \sum_{n=-\infty}^{\infty} \int_0^1 f(n+\alpha)e(-k\alpha)\,d\alpha = \sum_{n=-\infty}^{\infty} \int_n^{n+1} f(\alpha)e(-k\alpha)\,d\alpha \\ &= \int_{-\infty}^{\infty} f(\alpha)e(-k\alpha)\,d\alpha = \widehat{f}(k)\,.\end{aligned}$$

Here the exchange of summation and integration is justified by the absolute convergence established in (10.33). Hence the Fourier Series of $F(\alpha)$ is

$$\sum_{k=-\infty}^{\infty} \widehat{f}(k)e(k\alpha)\,.$$

On taking $\alpha = 0$ we expect that the above should be $F(0) = \sum_n f(n)$, which is to say that under reasonable hypotheses we should expect that (10.31) is true. Of course, we are painfully aware that the Fourier Series of a function need not converge, and even when it does converge might not converge to the anticipated value. We consider several combinations of hypotheses that ensure that the formula holds.

Theorem 10.33. *If $f \in L^1(\mathbb{R})$, f is continuous, and $\sum_{n=-\infty}^{\infty} f(n+\alpha)$ is uniformly convergent for $0 \le \alpha \le 1$, then*

$$\sum_{n=-\infty}^{\infty} f(n+\alpha) = \lim_{K\to\infty} \sum_{k=-K}^{K} (1-|k|/K)\widehat{f}(k)e(k\alpha)$$

for all α.

Proof. A uniformly convergent sum of continuous functions is continuous, so $F(\alpha)$ is continuous. It suffices to apply Theorem 4.1 to F. □

Theorem 10.34. *Suppose that $f \in L^1(\mathbb{R})$, let $F(\alpha)$ be defined as in (10.32), and suppose that*

$$\sum_{k=-\infty}^{\infty} |\widehat{f}(k)| < \infty\,.$$

Then

$$F(\alpha) = \sum_{k=-\infty}^{\infty} \widehat{f}(k)e(k\alpha)$$

for any α at which F is continuous.

Proof. We know that the series on the right hand side above is the Fourier Series of F. The result thus follows from Corollary 4.12. □

The following formulation, with slightly stronger hypotheses, often suffices.

Corollary 10.35. *Suppose that $f \in L^1(\mathbb{R})$, that f is continuous, that*

$$|f(x)| \le \frac{C}{1+|x|^\lambda} \quad \text{and} \quad |\widehat{f}(t)| \le \frac{C}{1+|t|^\lambda}$$

for all real x and all real t where $\lambda > 1$, then

$$\sum_{n=-\infty}^{\infty} f(n+\alpha) = \sum_{k=-\infty}^{\infty} \widehat{f}(k)e(k\alpha)$$

for all α.

Proof. The bound on f ensures that the sum defining $F(\alpha)$ in (10.32) is locally uniformly convergent; since f is continuous it follows that F is continuous. The bound on $\widehat{f}(t)$ implies that $\sum_k |\widehat{f}(k)| < \infty$, so the result follows from Theorem 10.34. □

Finally, we note that if f is of bounded variation on the real line, then the function F defined in (10.32) is of bounded variation on $\mathbb{T}$. This allows us to quote Jordan's theorem (Corollary 8.7).

Theorem 10.36. *Suppose that $f \in L^1(\mathbb{R})$, that f is of bounded variation on $\mathbb{R}$, and that $F(\alpha)$ is defined as in (10.32). Then*

$$\frac{F(\alpha^-)+F(\alpha^+)}{2} = \lim_{N\to\infty} \sum_{k=-N}^{N} \widehat{f}(k)e(k\alpha)$$

for all α.

Theorem 10.37. *If $f \in L^1(\mathbb{R})$, if there is an $A > 0$ such that $f(x) = 0$ whenever $|x| > A$, and if there is a $B > 0$ such that $\widehat{f}(t) = 0$ whenever $|t| > B$, then $f(x) = 0$ for almost all x, and $\widehat{f}(t) = 0$ for all t.*

The *support* of a function $f(x)$, written $\operatorname{supp} f$, is the set of those x in the domain of f for which $f(x) \ne 0$. Thus the above could be more succinctly phrased by saying that if $f \in L^1(\mathbb{R})$ and $\widehat{f}$ both have compact support, then $f = 0$ a. e.

Proof. Put $g(x) = f(4Ax)$. Thus $g(x) = 0$ when $|x| > 1/4$, and

$$\widehat{g}(t) = \frac{\widehat{f}\left(\frac{t}{4A}\right)}{4A}$$

by Theorem 10.1 (c). Hence $\widehat{g}(t) = 0$ when $|t| > 4AB$. Put

$$G(\alpha) = \sum_{n=-\infty}^{\infty} g(n+\alpha).$$

Then $G \in L^1(\mathbb{T})$ and its Fourier Series is

$$\sum_{k=-\infty}^{\infty} \widehat{g}(k)e(k\alpha).$$

Now $\widehat{g}(k) = 0$ when $|k|$ is sufficiently large, so this Fourier Series is actually a trigonometric polynomial. Hence it is absolutely convergent, so by Theorem 10.34 we have

$$G(\alpha) = \sum_{k=-\infty}^{\infty} \widehat{g}(k)e(k\alpha)$$

whenever G is continuous at α. But $G(\alpha) = 0$ for $1/4 < \alpha < 3/4$, since every term in the sum defining G vanishes there. Thus

$$\sum_{k=-\infty}^{\infty} \widehat{g}(k)e(k\alpha) = 0$$

for $1/4 < \alpha < 3/4$. But by Theorem 3.15 we know that a trigonometric polynomial has infinitely many zeros only when all its coefficients are 0. That is, $\widehat{g}(k) = 0$ for all k. Since G is a function in $L^1(\mathbb{T})$ all of whose Fourier coefficients vanish, it follows by Corollary 4.9 that $g(x) = 0$ for almost all x. Hence $f(x) = 0$ for almost all x, so $\|f\|_1 = 0$, and thus $\widehat{f}(t) = 0$ for all t by (10.2). □

Example 10.6. Suppose that $\delta > 0$. By applying the Poisson summation formula to $f(\delta x)$, we see that

$$\sum_{n=-\infty}^{\infty} f(n\delta) = \frac{1}{\delta} \sum_{k=-\infty}^{\infty} \widehat{f}(k/\delta)\,.$$

Suppose that $\widehat{f}(t) = 0$ when $|t| \ge T$, and that $\delta \le 1/T$. Then $\widehat{f}(k/\delta) = 0$ for all k except possibly $k = 0$. Thus in this situation,

$$\int_{-\infty}^{\infty} f(x)\,dx = \delta \sum_{n=-\infty}^{\infty} f(n\delta)\,.$$

By applying the Poisson summation formula to $f(\delta x)e(-\delta\alpha x)$ we find that

$$\delta \sum_{n=-\infty}^{\infty} f(n\delta)e(-n\delta\alpha) = \sum_{k=-\infty}^{\infty} \widehat{f}(k/\delta + \alpha)\,.$$

Suppose that $\widehat{f}(t) = 0$ when $|t| \ge T$, and that $\delta \le 1/(2T)$. If $|\alpha| \le T$, then $\widehat{f}(k/\delta + \alpha) = 0$ except possibly when $k = 0$. Thus

$$\widehat{f}(t) = \delta \sum_{n=-\infty}^{\infty} f(n\delta)e(-n\delta t)$$

for $|t| \le T$. Suppose that f is continuous. Then by the Fourier inversion formula,

$$\begin{aligned} f(x) &= \int_{-T}^{T} \widehat{f}(t)e(tx)\,dt = \int_{-T}^{T} \delta \sum_{n=-\infty}^{\infty} f(n\delta)e((x-n\delta)t)\,dt \\ &= \delta \sum_{n=-\infty}^{\infty} f(n\delta) \int_{-T}^{T} e((x-n\delta)t)\,dt\,. \end{aligned}$$

This last integral is

$$\left[\frac{e((x-n\delta)t)}{2\pi i(x-n\delta)}\right]_{-T}^{T} = \frac{\sin 2\pi(x-n\delta)T}{\pi(x-n\delta)}\,.$$

Thus we conclude that

$$f(x) = \delta \sum_{n=-\infty}^{\infty} f(n\delta)\frac{\sin 2\pi(x - n\delta)T}{\pi(x - n\delta)}\,. \tag{10.35}$$

In summary, we have found that if $\widehat{f}(t) = 0$ when $|t| \ge T$, and if $\delta \le 1/(2T)$, then we can reconstruct $f(x)$ for all x from the numbers $f(n\delta)$. This interpolation formula, which is important in signal processing, is known as the *cardinal series.* The number $1/(2T)$ is the *Nyquist sampling rate.*

Example 10.7. Let

$$f(x) = \Big(\frac{\sin \pi x}{\pi x}\Big)^2.$$

By Corollary 10.20 we know that $\widehat{f}(t) = \max(0, 1 - |t|)$. We apply the Poisson summation formula in the form

$$\sum_{n=-\infty}^{\infty} f(n + \alpha) = \sum_{k=-\infty}^{\infty} \widehat{f}(k)e(k\alpha)\,.$$

Since $\widehat{f}(k) = 0$ for all k except $k = 0$, and since $\widehat{f}(0) = 1$, we deduce that

$$\sum_{n=-\infty}^{\infty} \Big(\frac{\sin \pi(n + \alpha)}{\pi(n + \alpha)}\Big)^2 = 1\,.$$

Since $(\sin \pi(n + \alpha))^2 = (\sin \pi\alpha)^2$ for all n, we conclude that

$$\Big(\frac{\pi}{\sin \pi\alpha}\Big)^2 = \sum_{n=-\infty}^{\infty} \frac{1}{(n + \alpha)^2} \tag{10.36}$$

when α is not an integer.

Example 10.8. Let f be defined as in the preceding example, and put

$$f_T(x) = Tf(Tx) = T\Big(\frac{\sin \pi Tx}{\pi x}\Big)^2.$$

Suppose that T is a positive integer, say $T = N$. Since $(\sin \pi N(n + \alpha))^2 = (\sin \pi N\alpha)^2$, it follows that

$$\sum_{n=-\infty}^{\infty} f_N(n + \alpha) = N\Big(\frac{\sin \pi\alpha}{\pi}\Big)^2 \sum_{n=-\infty}^{\infty} \frac{1}{(n + \alpha)^2},$$

which by (10.36) is

$$= N\Big(\frac{\sin \pi N\alpha}{\sin \pi\alpha}\Big)^2.$$

On the other hand, by Theorem 10.1 (c) we know that $\widehat{f_T}(t) = \widehat{f}(t/T) = \max(0, 1-|t|/T)$, so

$$\sum_{k=-\infty}^{\infty} \widehat{f_N}(k)e(k\alpha) = \sum_{k=-N}^{N} \Big(1 - \frac{|k|}{N}\Big)e(k\alpha) = \Delta_N(\alpha),$$

the Fejér kernel. Thus by applying the Poisson summation formula to the Fourier Transform pair of Example 10.3 and Corollary 10.20 we recover the closed form formula for the Fejér kernel.

Example 10.9. Suppose that $f(x) = e^{-\pi x^2}$. We show that

$$\widehat{f}(t) = e^{-\pi t^2}. \tag{10.37}$$

To this end we first note that

$$\widehat{f}(t) = \int_{-\infty}^{\infty} e^{-\pi x^2} e(-xt)\, dx = \int_{-\infty}^{\infty} e^{-\pi x^2 - 2\pi i t x}\, dx\,.$$

The exponent here is a quadratic polynomial in x. We complete the square, which is to say that we observe that $-\pi x^2 - 2\pi i t x = -\pi(x+it)^2 - \pi t^2$. Thus

$$\widehat{f}(t) = e^{-\pi t^2} \int_{-\infty}^{\infty} e^{-\pi(x+it)^2}\, dx = e^{-\pi t^2} I(t),$$

say. To complete our derivation of (10.37), it remains to show that $I(t) = 1$ for all t. We first show that it is constant. By Leibniz's rule (see Theorem 0.17) we see that

$$I'(t) = \int_{-\infty}^{\infty} \frac{\partial}{\partial t} e^{-\pi(x+it)^2}\, dx\,.$$

As

$$\frac{\partial}{\partial t} e^{-\pi(x+it)^2} = -2\pi i(x+it)e^{-\pi(x+it)^2}, \tag{10.38}$$

we find that

$$I'(t) = -2\pi i \int_{-\infty}^{\infty} (x+it)e^{-\pi(x+it)^2}\, dx\,. \tag{10.39}$$

We note that

$$\frac{\partial}{\partial x} e^{-\pi(x+it)^2} = -2\pi(x+it)e^{-\pi(x+it)^2}\,. \tag{10.40}$$

That is, for any given t, $-e^{-\pi(x+it)^2}/(2\pi)$ is an antiderivative of the integrand in (10.39). Hence

$$I'(t) = i\Big[e^{-\pi(x+it)^2}\Big|_{x=-\infty}^{\infty} = 0\,.$$

Thus $I(t)$ is a constant, so we have shown that $\widehat{f}(t) = ce^{-\pi t^2}$ where $c = I(0) = \int_{-\infty}^{\infty} e^{-\pi x^2} dx$. We give two proofs that $c = 1$. In the first place, by the Fourier inversion theorem (e.g., Corollary 10.14), we know that

$$e^{-\pi x^2} = c \int_{-\infty}^{\infty} e^{-\pi t^2} e(tx)\, dt$$

for all x. When we take $x = 0$, the left hand side is 1, and the right hand side is $c\int_{-\infty}^{\infty} e^{-\pi t^2}\, dt = c^2$. Since $c^2 = 1$, it follows that $c = \pm 1$. In the definition of $I(0)$, the integrand is positive, so it is clear that $c > 0$. Hence $c = 1$. Secondly,

$$\begin{aligned} c^2 &= \Big(\int_{-\infty}^{\infty} e^{-\pi x^2}\, dx\Big)^2 = \Big(\int_{-\infty}^{\infty} e^{-\pi x^2}\, dx\Big)\Big(\int_{-\infty}^{\infty} e^{-\pi y^2}\, dy\Big) \\ &= \int_{-\infty}^{\infty}\int_{-\infty}^{\infty} e^{-\pi(x^2+y^2)}\, dx\, dy\,. \end{aligned}$$

We switch to polar coordinates, and recall that $dx\,dy = r\,dr\,d\theta$, to see that the above is

$$\int_0^{2\pi}\int_0^{\infty} e^{-\pi r^2} r\,dr\,d\theta = 2\pi \int_0^{\infty} e^{-\pi r^2} r\,dr = \Big[- e^{-\pi r^2} \Big|_0^{\infty} = 1\,.$$

Thus $c^2 = 1$, and so $c = 1$ as before. This completes the proof of (10.37).

In the computations (10.38) and (10.40), we have tacitly assumed that the chain rule still holds for complex-valued functions. This could be left as a detail to be dealt with when one studies functions of a complex variable, but we can give a rigorous proof of these identities based on what we know (and without using the chain rule). In Chapter 1 we showed that the identity $e^{z+w} = e^z e^w$ continues to hold when z and w are complex. Moreover, we also showed that $(e^{ax})' = ae^{ax}$ holds even when a is complex. We observe that

$$e^{-\pi(x+it)^2} = e^{-\pi x^2} e^{-2\pi i t x} e^{\pi t^2}\,.$$

We regard t as constant, and want to differentiate this with respect to the real variable x. We know how to compute the derivative of each factor; the only missing ingredient is to verify that $(fg)' = fg' + f'g$ when f and g are complex-valued. This is easily accomplished: If $f(x) = u + iv$ and $g(x) = r + is$, then

$$(10.41)\qquad \begin{aligned}(fg)' &= ((ur - vs) + i(us + vr))' = (ur - vs)' + i(us + vr)' \\ &= us' + u's + vr' + v'r + i(us' + u's + vr' + v'r),\end{aligned}$$

while

$$\begin{aligned} fg' &= (u + iv)(r' + is') = ur' - vs' + i(us' + vr'), \\ f'g &= (u' + iv')(r + is) = u'r - v's + i(u's + v'r). \end{aligned}$$

We sum these two identities, and compare the result with (10.41) to confirm that indeed $(fg)' = fg' + f'g$, even when f and g are complex-valued.

Example 10.10. Suppose that z is a positive real number. If $f(x) = e^{-\pi x^2 z}$, then by Theorem 10.1 (c) and Example 10.9 it follows that $\widehat{f}(t) = z^{-1/2}e^{-\pi t^2/z}$. Hence by the Poisson summation formula,

$$\sum_{n=-\infty}^{\infty} e^{-\pi n^2 z} = \frac{1}{z^{1/2}} \sum_{k=-\infty}^{\infty} e^{-\pi k^2/z}\,.$$

Put

$$(10.42)\qquad \vartheta(z) = \sum_{n=-\infty}^{\infty} e^{-\pi n^2 z}\,.$$

Thus we have shown that

$$(10.43)\qquad \vartheta(z) = \vartheta(1/z)/z^{1/2}$$

for all $z > 0$. With somewhat more work it can be shown that the identity (10.43) holds also for complex z, provided that $\operatorname{Re} z > 0$. Here ϑ is the Greek letter theta, and the function $\vartheta(z)$ is an example of a kind of function known as a *theta function*.

Exercises

1. (a) Show that if $z > 0$ and α is real, then
$$\sum_{n=-\infty}^{\infty} e^{-\pi(n+\alpha)^2 z} = \frac{1}{z^{1/2}} \sum_{k=-\infty}^{\infty} e(k\alpha)e^{-\pi k^2/z}\,.$$
(b) By taking $\alpha = 1/2$ and letting $z \to \infty$, show that
$$\lim_{r\to 1^-} \sum_{k=0}^{\infty}(-1)^k r^{k^2} = \frac{1}{2}\,.$$
2. Our object here is to show that equality is attained in Theorem 10.29 when $f(x)$ is of the form $f(x) = e^{-\pi x^2 z}$ for some $z > 0$. Throughout this exercise, f is defined in this way.
(a) By integration by parts, show that
$$\int_{-\infty}^{\infty} |xf(x)|^2\,dx = \frac{1}{4\pi\sqrt{2}z^{3/2}}\,.$$
(b) Show similarly that
$$\int_{-\infty}^{\infty} |t\widehat{f}(t)|^2\,dt = \frac{z^{1/2}}{4\pi\sqrt{2}}\,.$$
(c) Note that the product of these two integrals is $1/(32\pi^2 z)$.
(d) Show that
$$\int_{-\infty}^{\infty} |f(x)|^2\,dx = \frac{1}{\sqrt{2z}}\,.$$
(e) Deduce that in the present situation, equality holds in Theorem 10.29.
3. (a) Suppose that $\operatorname{Re} z > 0$, and that $\vartheta(z)$ is defined as in (10.42). Show that $\vartheta(z+2i) = \vartheta(z)$. (Thus, in the half-plane $\operatorname{Re} z > 0$, ϑ not only satisfies the functional equation (10.43), but also is periodic with period $2i$.)
(b) Show that if $\operatorname{Re} z = 0$, then $|e^{-\pi n^2 z}| = 1$. Deduce that the series (10.42) does not converge when $\operatorname{Re} z = 0$.
(c) Suppose that $\operatorname{Re} z < 0$. Show that $|e^{-\pi n^2 z}| \to \infty$ as $n \to \pm\infty$. Deduce that the series (10.42) does not converge when $\operatorname{Re} z < 0$.

10.5. Linear combinations of translates

Suppose that $f \in L^1(\mathbb{R})$, and consider functions $g(x)$ of the form
$$g(x) = \sum_{k=1}^{K} a_k f(x + d_k)\,. \tag{10.44}$$
The question is whether functions of this form are dense in $L^1(\mathbb{R})$. That is, if $h \in L^1(\mathbb{R})$, and $\varepsilon > 0$ is given, does it follow that there exists a K, and a set of a_k and displacements d_k such that $\|g - h\|_1 < \varepsilon$. Of course, if this can be done, then

by (10.2) it follows that $|\widehat{g}(t) - \widehat{h}(t)| < \varepsilon$ for all t. By K applications of Theorem 10.1 (a), we find that

$$\widehat{g}(t) = \widehat{f}(t)\Big(\sum_{k=1}^{K} a_k e(d_k t)\Big).$$

If there is a real number t_0 such that $\widehat{f}(t_0) = 0$, then $\widehat{g}(t_0) = 0$ no matter how the a_k and d_k are chosen. Thus if $\widehat{h}(t_0) \ne 0$, then $\|g - h\|_1$ cannot be made arbitrarily small—indeed, it cannot be made smaller than $|\widehat{h}(t_0)|$. For any given t_0 we know functions $h \in L^1(\mathbb{R})$ such that $\widehat{h}(t_0) \ne 0$. Thus if there is a t_0 for which $\widehat{f}(t_0) = 0$, then the functions $g(x)$ are certainly not dense in $L^1(\mathbb{R})$. This much is obvious. What lies much deeper is that the converse is true: Norbert Wiener proved that if $f \in L^1(\mathbb{R})$ and if the equation $\widehat{f}(t) = 0$ has no solution in real numbers t, then functions of the form (10.44) are dense in $L^1(\mathbb{R})$. In §.4.3 we discussed some Abelian and Tauberian theorems. Although it is not immediately evident, Wiener's theorem forms the basis of a further, and much more general, theory of Tauberian theorems.

Notes

For a more extended study of the Fourier transform, one might for example consult Goldberg (1970) or Titchmarsh (1986).

10.3. Rayleigh (1889) used the Plancherel formula in a study of the radiation of heat from a black body, and Plancherel's formula is often called "Rayleigh's formula" in the applied literature, but Plancherel (1910) was the first to give a rigorous proof.

The depiction of Young Heisenberg on p. 267 is due to cartoonist/anthropologist Emily Holt, and is printed here with her permission.

10.4. Although the points of view and derivations are quite different, it can be argued that the Poisson summation formula and the Euler–Maclaurin summation formula (9.29) are closely connected.

The cardinal series has been discovered and rediscovered many times. Cauchy (1841) had it for trigonometric polynomials (recall (6.2)). Certainly Whittaker (1915) had it for the Fourier transform, and perhaps there were others before him, but certainly there were many later rediscoveries. For more on the cardinal series, see Benedetto and Ferreira (2001).

Chapter 11

Higher Dimensions

We have considered the Discrete Fourier Transform, Fourier Series, and the Fourier Transform, each in dimension 1. Most of what we have done can be done in higher dimensions. We take an informal approach in which we just state formulas without giving much in the way of proofs. For the most part we restrict ourselves to dimension 2, just to keep the notation simple, but arbitrary dimensions can be done similarly.

11.1. Multiple Discrete Fourier Transforms

11.1.1. Suppose that $f(n_1, n_2)$ is defined for all pairs (n_1, n_2) of integers, and that f has period q_1 with respect to n_1, and period q_2 with respect to n_2. Thus

$$f(n_1 + m_1 q_1, n_2 + m_2 q_2) = f(n_1, n_2)$$

for arbitrary integers m_1, m_2. We can define a Discrete Fourier Transform for f,

$$\widehat{f}(k_1, k_2) = \sum_{n_1=1}^{q_1} \sum_{n_2=1}^{q_2} f(n_1, n_2) e\Big(-\frac{n_1 k_1}{q_1} - \frac{n_2 k_2}{q_2} \Big). \tag{11.1}$$

Here $\widehat{f}(k_1, k_2)$ has period q_1 with respect to k_1, and period q_2 with respect to k_2. Also, f has a Discrete Fourier expansion,

$$f(m_1, m_2) = \frac{1}{q_1 q_2} \sum_{k_1=1}^{q_1} \sum_{k_2=1}^{q_2} \widehat{f}(k_1, k_2) e\Big(\frac{k_1 m_1}{q_1} + \frac{k_2 m_2}{q_2} \Big) \tag{11.2}$$

for all pairs (m_1, m_2) of integers. In many situations we would have $q_1 = q_2$, but this is by no means necessary.

11.1.2. From (11.1) it follows by the triangle inequality that

$$|\widehat{f}(k_1, k_2)| \le \sum_{n_1=1}^{q_1} \sum_{n_2=1}^{q_2} |f(n_1, n_2)| \tag{11.3}$$

for all k_1, k_2, and similarly from (11.2),

$$|f(n_1, n_2)| \le \frac{1}{q_1 q_2} \sum_{k_1}^{q_1} \sum_{k_2=1}^{q_2} |\widehat{f}(k_1, k_2)| \,. \tag{11.4}$$

11.1.3. We have the manipulative properties that one would anticipate:

(a) If a_1, a_2 are integers and $g(n_1, n_2) = f(n_1 + a_1, n_2 + a_2)$, then $\widehat{g}(k_1, k_2) = \widehat{f}(k_1, k_2)e(k_1 a_1/q_1 + k_2 a_2/q_2)$ for all k_1, k_2.

(b) If b_1, b_2 are integers and $h(n_1, n_2) = f(n_1, n_2)e(b_1 n_1/q_1 + b_2 n_2/q_2)$, then $\widehat{h}(k_1, k_2) = \widehat{f}(k_1 - b_1, k_2 - b_2)$.

(c) If $j(n_1, n_2) = f(\varepsilon_1 n_1, \varepsilon_2 n_2)$ where ε_1 and ε_2 are independently ± 1, then $\widehat{j}(k_1, k_2) = \widehat{f}(\varepsilon_1 k_1, \varepsilon_2 k_2)$.

(d) If $\ell(n_1, n_2) = \overline{f(n_1, n_2)}$, then $\widehat{\ell}(k_1, k_2) = \overline{\widehat{f}(-k_1, -k_2)}$.

11.1.4. If f and g have periods q_1, q_2, then their *convolution* is

$$(f * g)(n_1, n_2) = \sum_{a_1=1}^{q_1} \sum_{a_2=1}^{q_2} f(n_1 - a_1, n_2 - a_2) g(a_1, a_2) \,. \tag{11.5}$$

We find that if $h = f * g$, then

$$\widehat{h}(k_1, k_2) = \widehat{f}(k_1, k_2)\widehat{g}(k_1, k_2) \,. \tag{11.6}$$

11.1.5. In the direction of a Parseval Identity, we note that if f and g have periods q_1, q_2, then

$$\sum_{n_1=1}^{q_1} \sum_{n_2=1}^{q_2} f(n_1, n_2)\overline{g(n_1, n_2)} = \frac{1}{q_1 q_2} \sum_{k_1=1}^{q_1} \sum_{k_2=1}^{q_2} \widehat{f}(k_1, k_2)\overline{\widehat{g}(k_1, k_2)} \,. \tag{11.7}$$

If $f = g$, then we have the symmetric form,

$$\sum_{n_1=1}^{q_1} \sum_{n_2=1}^{q_2} |f(n_1, n_2)|^2 = \frac{1}{q_1 q_2} \sum_{k_1=1}^{q_1} \sum_{k_2=1}^{q_2} |\widehat{f}(k_1, k_2)|^2 \,. \tag{11.8}$$

11.2. Multiple Fourier Series

11.2.1. We say that a real or complex valued function $f(x_1, x_2)$ of two real variables has period 1 with respect to both those variables if

$$f(x_1 + m_1, x_2 + m_2) = f(x_1, x_2)$$

for all integers m_1, m_2 and all real numbers x_1, x_2. Such a function can therefore be considered to be a function defined on $\mathbb{T}^2$. For such a function we put

$$\|f\|_1 = \int_0^1 \int_0^1 |f(x_1, x_2)| \, dx_1 \, dx_2. \tag{11.9}$$

We say that $f \in L^1(\mathbb{T}^2)$ if f has period 1 in both its variables, and if $\|f\|_1 < \infty$. For $f \in L^1(\mathbb{T}^2)$, we put

$$\widehat{f}(n_1, n_2) = \int_0^1 \int_0^1 f(x_1, x_2) e(-n_1 x_1 - n_2 x_2)\, dx_1\, dx_2, \tag{11.10}$$

and note that $|\widehat{f}(n_1, n_2)| \le \|f\|_1$ for all n_1, n_2. The Fourier Series of such an f is

$$\sum_{n_1=-\infty}^{\infty} \sum_{n_2=-\infty}^{\infty} \widehat{f}(n_1, n_2) e(n_1 x_1 + n_2 x_2)\,. \tag{11.11}$$

In dimension d we would have d-fold integrals and d-fold sums, which we would express using vector notation: For f defined over $\mathbb{T}^d$ we set

$$\|f\|_1 = \int_{\mathbb{T}^d} |f(\boldsymbol{x})|\, \boldsymbol{dx},$$

and then $L^1(\mathbb{T}^d)$ is the set of functions for which $\|f\|_1 < \infty$. Such an f has Fourier coefficients,

$$\widehat{f}(\boldsymbol{n}) = \int_{\mathbb{T}^d} f(\boldsymbol{x}) e(-\boldsymbol{n} \cdot \boldsymbol{x})\, \boldsymbol{dx},$$

and a Fourier Series,

$$\sum_{\boldsymbol{n} \in \mathbb{Z}^d} \widehat{f}(\boldsymbol{n}) e(\boldsymbol{n} \cdot \boldsymbol{x})\,.$$

The Fourier coefficients of a function $f \in L^1(\mathbb{T}^2)$ have the anticipated manipulative properties:

(a) If a_1 and a_2 are real numbers and $g(x_1, x_2) = f(x_1 + a_1, x_2 + a_2)$, then $\widehat{g}(n_1, n_2) = \widehat{f}(n_1, n_2) e(n_1 a_1 + n_2 a_2)$.

(b) If b_1 and b_2 are integers, and $h(x_1, x_2) = f(x_1, x_2) e(b_1 x_1 + b_2 x_2)$, then $\widehat{h}(n_1, n_2) = \widehat{f}(n_1 - b_1, n_2 - b_2)$.

(c) If $j(x_1, x_2) = f(\varepsilon_1 x_1, \varepsilon_2 x_2)$ where ε_1 and ε_2 are independently ± 1, then $\widehat{j}(n_1, n_2) = \widehat{f}(\varepsilon_1 n_1, \varepsilon_2 n_2)$.

(d) If m_1, m_2 are positive integers, and $p(x_1, x_2) = f(m_1 x_1, m_2 x_2)$, then $\widehat{p}(n_1, n_2) = \widehat{f}(n_1/m_1, n_2/m_2)$ if $m_1 | n_2$ and $m_2 | n_2$, and $\widehat{p}(n_1, n_2) = 0$ otherwise.

(e) If $q(x_1, x_2) = \overline{f(x_1, x_2)}$, then $\widehat{q}(n_1, n_2) = \overline{\widehat{f}(-n_1, -n_2)}$.

11.2.2. If f is an even function in the sense that $f(-x_1, -x_2) = f(x_1, x_2)$ for all x_1, x_2, then $\widehat{f}(-n_1, -n_2) = \widehat{f}(n_1, n_2)$. Similarly, if $f(-x_1, -x_2) = -f(x_1, x_2)$ for all x_1, x_2, then $\widehat{f}(-n_1, -n_2) = -\widehat{f}(n_1, n_2)$. There is also the possibility of a 4-fold symmetry in which $f(\pm x_1, \pm x_2) = f(x_1, x_2)$ for all x_1, x_2 and all independent choices of the signs. For such a function, $\widehat{f}(\pm n_1, \pm n_2) = \widehat{f}(n_1, n_2)$ for all integers n_1, n_2 and all independent choices of signs.

11.2.3. If f is real-valued, then $\widehat{f}(-n_1, -n_2) = \overline{\widehat{f}(n_1, n_2)}$, so that in the Fourier Series of f, the two terms $\widehat{f}(n_1, n_1) e(n_1 x_1 + n_2 x_1)$ and $\widehat{f}(-n_1, -n_2) e(-n_1 x_1 - n_2 x_2)$ are complex conjugates. Their combined contribution is therefore real, which makes sense.

11.2.4. From Lebesgue measure theory we find that if $f \in L^1(\mathbb{T}^2)$, then

$$\lim_{\substack{\delta_1\to 0\\ \delta_2\to 0}} \int_0^1 \int_0^1 |f(x_1+\delta_1, x_2+\delta_2) - f(x_1,x_2)|\, dx_1\, dx_2 = 0\,. \tag{11.12}$$

From this there follows a Riemann–Lebesgue lemma, namely that if $f \in L^1(\mathbb{T}^2)$, then

$$\lim_{|n_1|+|n_2|\to\infty} \widehat{f}(n_1,n_2) = 0\,. \tag{11.13}$$

11.2.5. If $f, g \in L^1(\mathbb{T}^2)$, then their *convolution* is

$$(f*g)(x_1,x_2) = \int_0^1 \int_0^1 f(u_1,u_2)g(x_1-u_1, x_2-u_2)\, dx_1\, dx_2\,. \tag{11.14}$$

For such functions, $\|f*g\|_1 \le \|f\|_1\|g\|_1$, and

$$\widehat{f*g}(n_1,n_2) = \widehat{f}(n_1,n_2)\widehat{g}(n_1,n_2)\,. \tag{11.15}$$

11.2.6. A *trigonometric polynomial* in two variables is a finite sum of the form

$$T(x_1,x_2) = \sum_{n_1=-N_1}^{N_1} \sum_{n=-N_2}^{N_2} t(n_1,n_1)e(n_1x_1+n_2x_2)\,. \tag{11.16}$$

Clearly,

$$\widehat{T}(n_1,n_2) = \begin{cases} t(n_1,n_2) & \text{if } -N_1 \le n_1 \le N_1 \text{ and } -N_2 \le n_2 \le N_2, \\ 0 & \text{otherwise.} \end{cases}$$

The *degree* of T is now the maximum of $|n_1|+|n_2|$ over those pairs n_1, n_2 for which $t(n_1,n_2) \ne 0$.

Example 11.1. Let

$$\begin{aligned} T(x_1,x_2) &= 4\sin\pi x_1\, \sin\pi x_2\, \sin\pi(x_1-x_2) \\ &= -\sin 2\pi x_1 + \sin 2\pi x_2 + \sin 2\pi(x_1-x_2)\,. \end{aligned} \tag{11.17}$$

This is a trigonometric polynomial of degree 2. Suppose that $0 \le x_1, x_2 < 1$. We note that $T(x_1,x_2) = 0$ when $x_1 = 0$, when $x_2 = 0$, and when $x_1 = x_2$. We also note that $T(x_1,x_2) > 0$ when $0 < x_2 < x_1 < 1$ and $T(x_1,x_2) < 0$ when $0 < x_1 < x_2 < 1$. Clearly

$$\begin{aligned} \frac{\partial T}{\partial x_1}(x_1,x_2) &= -2\pi\cos 2\pi x_1 + 2\pi\cos 2\pi(x_1-x_2), \\ \frac{\partial T}{\partial x_2}(x_1,x_2) &= 2\pi\cos 2\pi x_2 - 2\pi\cos 2\pi(x_1-x_2)\,. \end{aligned}$$

If these are both 0, then

$$\cos 2\pi x_1 = \cos 2\pi(x_1-x_2) = \cos 2\pi x_2\,.$$

Thus either $x_1 = x_2$ or $x_2 = 1 - x_1$. In the first case, the common value above is 1, so $x_1 = x_2 = 0$. In the second case, $\cos 2\pi x_1 = \cos 4\pi x_1 = 2\cos^2 2\pi x_1 - 1$, so $\cos 2\pi x_1$ is a root of $2u^2 - u - 1 = 0$. But $2u^2 - u - 1 = (2u+1)(u-1)$, so $\cos 2\pi x_1 = -1/2$, and hence $(x_1,x_2) = (1/3, 2/3)$ or $(2/3, 1/3)$. Thus T is maximum at $(1/3, 2/3)$, minimum at $(2/3, /1/3)$, and has only one other critical point, $(0,0)$.

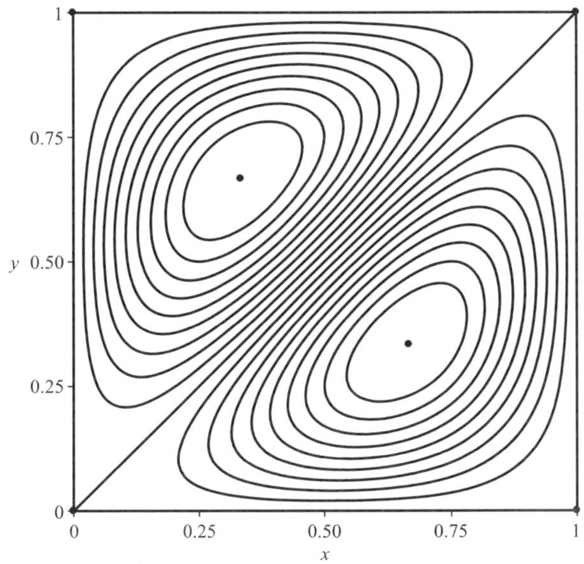

Figure 11.1. Level curves $T(x_1, x_2) = k/4$ for $k = -9, -8, \dots, 9$ and critical points for $T(x_1, x_2)$ as in (11.17).

This is unusual, since according to Morse Theory, a smooth real-valued function on the torus $\mathbb{T}^2$ will have at least 4 critical points. The explanation for this discrepancy is that Morse Theory refers to generic critical points. The critical point at $(0, 0)$ is not generic, since

$$\frac{\partial^2 T}{\partial x_1^2}(0,0) = \frac{\partial^2 T}{\partial x_1 \partial x_2}(0,0) = \frac{\partial^2 T}{\partial x_2^2}(0,0) = 0\,.$$

We know that if a trigonometric polynomial of degree N in one variable has more than $2N$ roots, then it is identically zero. In the above example we see that in two variables, the set of zeros of a trigonometric polynomial may contain a continuum. However, there are still limitations. In particular, if $x_1 < x_2 < \cdots < x_R < x_1 + 1$, $y_1 < y_2 < \cdots < y_S < y_1 + 1$, $R > 2N_1$, $S > 2N_2$, and if $T(x_r, y_s) = 0$ for $1 \le r \le R$, $1 \le s \le S$, then $\widehat{f}(n_1, n_2) = 0$ for all pairs n_1, n_2, and T is identically 0. Hence in particular, if there is a point (x_0, y_0) such that $T(x, y) = 0$ for all (x, y) in a neighborhood of (x_0, y_0) (say $\sqrt{(x - x_0)^2 + (y - y_0)^2} < \delta$ where $\delta > 0$), then T is identically 0.

If T is a trigonometric polynomial in two variables, and $f \in L^1(\mathbb{T}^2)$, then

$$(T * f)(x_1, x_2) = \sum_{n_1=-N_1}^{N_1} \sum_{n=-N_2}^{N_2} t(n_1, n_1)\widehat{f}(n_1, n_2)e(n_1x_1 + n_2x_2)\,.$$

With $T(x_1, x_2) = \Delta_{N_1}(x_1)\Delta_{N_2}(x_2)$ and $f \in L^1(\mathbb{T}^2)$, we set

$$\begin{aligned}\sigma_{N_1,N_2}(x_1, x_2) &= (T * f)(x_1, x_2) \\ &= \sum_{-N_1}^{N_1} \sum_{-N_2}^{N_2} \Big(1 - \frac{|n_1|}{N_1}\Big)\Big(1 - \frac{|n_2|}{N_2}\Big)\widehat{f}(n_1, n_2)e(n_1x_1 + n_2x_2)\,. \end{aligned} \tag{11.18}$$

11.2.7. We let $C(\mathbb{T}^2)$ denote the set of functions that are continuous on $\mathbb{T}^2$. For $f \in C(\mathbb{T}^2)$, this means that f is uniformly continuous on the closed unit square $[0,1] \times [0,1]$, of course with $f(x_1, 0) = f(x_1, 1)$ for all x_1 and $f(0, x_2) = f(1, x_2)$ for all x_2.

If $f(x_1, x_2)$ is a bounded function with period 1 in both arguments, and if $g \in L^1(\mathbb{T}^2)$, then $f * g \in C(\mathbb{T}^2)$.

If $f \in C(\mathbb{T}^2)$, then

$$\lim_{\substack{N_1 \to \infty \\ N_2 \to \infty}} \sigma_{N_1, N_2}(x_1, x_2) = f(x_1, x_2) \tag{11.19}$$

uniformly in x_1, x_2.

If $f \in C(\mathbb{T}^2)$ and $\widehat{f}(n_1, n_2) = 0$ for all pairs of integers n_1, n_2, then $f(x_1, x_2) = 0$ for all x_1, x_2.

If $f \in C(\mathbb{T}^2)$ and

$$\sum_{n_1=-\infty}^{\infty} \sum_{n_2=-\infty}^{\infty} |\widehat{f}(n_1, n_2)| < \infty,$$

then

$$f(x_1, x_2) = \sum_{n_1=-\infty}^{\infty} \sum_{n_2=-\infty}^{\infty} \widehat{f}(n_1, n_2) e(n_1 x_1 + n_2 x_2) \tag{11.20}$$

for all x_1, x_2.

11.2.8. If $f \in L^1(\mathbb{T}^2)$, then

$$\lim_{\substack{N_1 \to \infty \\ N_2 \to \infty}} \int_0^1 \int_0^1 |f(x_1, x_2) - \sigma_{N_1, N_2}(x_1, x_2)| \, dx_1 \, dx_2 = 0 . \tag{11.21}$$

As a consequence, if $f \in L^1(\mathbb{T}^2)$ and $\widehat{f}(n_1, n_2) = 0$ for all integers n_1, n_2, then $f(x_1, x_2) = 0$ for almost all $(x_1, x_2) \in \mathbb{T}^2$.

11.2.9. Suppose that $f \in L^1(\mathbb{T}^2)$. We put

$$\|f\|_2 = \left(\int_0^1 \int_0^1 |f(x_1, x_2)|^2 \, dx_1 \, dx_2 \right)^{1/2} . \tag{11.22}$$

We say that $f \in L^2(\mathbb{T}^2)$ if f has period 1 with respect to its two variables, and if $\|f\|_2 < \infty$.

By the Cauchy–Schwarz inequality it follows that $\|f\|_1 \le \|f\|_2$, so if $f \in L^2(\mathbb{T}^2)$, then $f \in L^1(\mathbb{T}^2)$. That is, $L^2(\mathbb{T}^2) \subset L^1(\mathbb{T}^2)$, and there are members of L^1 that are not in L^2, so L^2 is strictly smaller.

If $f \in L^2(\mathbb{T}^2)$, then

$$\lim_{\substack{\delta_1 \to 0 \\ \delta_2 \to 0}} \int_0^1 \int_0^1 |f(x_1 + \delta_1, x_2 + \delta_2) - f(x_1, x_2)|^2 \, dx_1 \, dx_2 = 0 . \tag{11.23}$$

The functions $e(n_1x_1 + n_2x_2)$ are orthonormal on $\mathbb{T}^2$, so we have a Bessel's inequality: If $f \in L^2(\mathbb{T}^2)$, then

$$\sum_{n_1=-\infty}^{\infty} \sum_{n_2=-\infty}^{\infty} |\widehat{f}(n_1, n_2)|^2 \le \|f\|_2^2. \tag{11.24}$$

If $f \in L^2(\mathbb{T}^2)$, then

$$\lim_{\substack{N_1\to\infty \\ N_2\to\infty}} \int_0^1 \int_0^1 |f(x_1, x_2) - \sigma_{N_1,N_2}(x_1, x_2)|^2 \, dx_1 \, dx_2 = 0 \,. \tag{11.25}$$

If f and g are members of $L^2(\mathbb{T}^2)$, then

$$\int_0^1 \int_0^1 f(x_1, x_2)\overline{g(x_1, x_2)} \, dx_1 \, dx_2 = \sum_{n_1=-\infty}^{\infty} \sum_{n_2=-\infty}^{\infty} \widehat{f}(n_1, n_2)\overline{\widehat{g}(n_1, n_2)} \,. \tag{11.26}$$

If $g = f$, then we have the symmetric form of Parseval's Identity: If $f \in L^2(\mathbb{T}^2)$, then

$$\int_0^1 \int_0^1 |f(x_1, x_2)|^2 \, dx_1 \, dx_2 = \sum_{n_1=-\infty}^{\infty} \sum_{n_2=-\infty}^{\infty} |\widehat{f}(n_1, n_2)|^2 \,. \tag{11.27}$$

Example 11.2. Suppose that $f(x_1, x_2) \in L^2(\mathbb{T}^2)$. Our object is to show that

$$\int_0^1 \Big| \int_0^1 f(x_1, x_2) \, dx_1 \Big|^2 dx_2 + \int_0^1 \Big| \int_0^1 f(x_1, x_2) \, dx_2 \Big|^2 dx_1$$
$$\le \Big| \int_0^1 \int_0^1 f(x_1, x_2) \, dx_1 \, dx_2 \Big|^2 + \int_0^1 \int_0^1 |f(x_1, x_2)|^2 \, dx_1 \, dx_2 \,.$$

By Parseval's Identity the left hand side is

$$\sum_{n_2=-\infty}^{\infty} |\widehat{f}(0, n_2)|^2 + \sum_{n_1=-\infty}^{\infty} |\widehat{f}(n_1, 0)|^2,$$

while the right hand side of the proposed inequality is

$$|\widehat{f}(0,0)|^2 + \sum_{n_1=-\infty}^{\infty} \sum_{n_2=-\infty}^{\infty} |\widehat{f}(n_1, n_2)|^2 \,.$$

The result is now clear, and we see that the inequality becomes an identity if f is of the form $f(x_1, x_2) = g(x_1) + h(x_2)$.

11.2.10. In Chapter 6 we discussed the Fejér–Riesz Theorem (Theorem 6.8), which asserts that a nonnegative trigonometric polynomial is the modulus squared of an exponential sum. This is false in 2 or more dimensions. In the first place, we see that $T(x_1, x_2) = 4 + 2\cos 2\pi x_1 + 2\cos 2\pi x_2$ is nonnegative, but is not of the form

$$\Big| \sum_{0\le k_1,k_2\le K} c(k_1, k_2) e(k_1x_1 + k_2x_k) \Big|^2 .$$

It is the sum of two squares, since $T(x_1, x_2) = |1 + e(x_1)|^2 + |1 + e(x_2)|^2$, but the situation is worse: There are nonnegative trigonometric polynomials in two variables that cannot be expressed as a sum of squares. The Fejér–Riesz Theorem can be extended to higher dimensions, but in a more complicated way. See Dumitrescu (2007).

11.2.11. If $f \in C(\mathbb{T}^2)$ and

$$\sum_{n_1=-\infty}^{\infty} \sum_{n_2=-\infty}^{\infty} |\widehat{f}(n_1, n_2)| < \infty,$$

then we say that $f \in A(\mathbb{T}^2)$. For such a function,

$$f(x_1, x_2) = \sum_{n_1=-\infty}^{\infty} \sum_{n_2=-\infty}^{\infty} \widehat{f}(n_1, n_2) e(n_1 x_1 + n_2 x_2)$$

for all x_1 and x_2. If $f \in A(\mathbb{T}^2)$ and the equation $f(x_1, x_2) = 0$ has no solution, then $1/f \in A(\mathbb{T}^2)$.

11.2.12. Weyl's criterion extends to $\mathbb{T}^d$. A sequence $\boldsymbol{u}_n$ of points in $\mathbb{T}^d$ is uniformly distributed if and only if

$$\sum_{n=1}^{N} e(\boldsymbol{k} \cdot \boldsymbol{u}_n) = o(N)$$

as $N \to \infty$ for every fixed nonzero lattice point $\boldsymbol{k} \in \mathbb{Z}^d$. Thus if $1, \alpha_1, \alpha_2, \ldots, \alpha_d$ are linearly independent over the field $\mathbb{Q}$ of rational numbers, then the sequence $\boldsymbol{u}_n = (n\alpha_1, n\alpha_2, \ldots, n\alpha_d) = n\boldsymbol{\alpha}$ is uniformly distributed in $\mathbb{T}^d$, because $\boldsymbol{k} \cdot \boldsymbol{\alpha}$ is never an integer. The weaker assertion that the points $n\boldsymbol{\alpha}$ are dense in $\mathbb{T}^d$ is known as Kronecker's Diophantine Approximation Theorem. If $1, \alpha_1, \ldots, \alpha_d$ are linearly independent over $\mathbb{Q}$, if $\mathcal{B}$ is a box in $\mathbb{T}^d$ with positive side-lengths, and if $0 < n_1 < n_2 < \cdots$ are the values of n for which $n\boldsymbol{\alpha} \in \mathcal{B}$, then the gaps $n_{j+1} - n_j$ are bounded. This is important in the study of almost periodic functions.

11.3. Multiple Fourier Transforms

11.3.1. For a function $f(x_1, x_2)$ of two real variables we set

$$\|f\|_1 = \int_{-\infty}^{\infty} \int_{-\infty}^{\infty} |f(x_1, x_2)| \, dx_1 \, dx_2,$$

and we let $L^1(\mathbb{R}^2)$ denote the set of all functions for which this norm is finite. For such functions we define a Fourier Transform:

$$\widehat{f}(t_1, t_2) = \int_{-\infty}^{\infty} \int_{-\infty}^{\infty} f(x_1, x_2) e(-t_1 x_1 - t_2 x_2) \, dx_1 \, dx_2 \, .$$

Our hope is that in suitable circumstances, we can express f in terms of its Fourier Transform, so that

$$f(x_1, x_2) = \int_{-\infty}^{\infty} \int_{-\infty}^{\infty} \widehat{f}(t_1, t_2) e(t_1 x_1 + t_2 x_2) \, dt_1 \, dt_2 \, .$$

The Fourier Transform of a function $f \in L^1(\mathbb{R}^2)$ has the anticipated manipulative properties:

(a) If a_1 and a_2 are real numbers and $g(x_1, x_2) = f(x_1 + a_1, x_2 + a_2)$, then $\widehat{g}(t_1, t_2) = \widehat{f}(t_1, t_2) e(t_1 a_1 + t_2 a_2)$.

(b) If b_1 and b_2 are real numbers, and $h(x_1,x_2) = f(x_1,x_2)e(b_1x_1 + b_2x_2)$, then $\widehat{h}(t_1,t_2) = \widehat{f}(t_1 - b_1, t_2 - b_2)$.

(c) If $j(\boldsymbol{x}) = f(A\boldsymbol{x})$ where A is an invertible 2×2 matrix, then

$$\widehat{j}(\boldsymbol{t}) = \frac{\widehat{f}((A^{-t})\boldsymbol{t})}{|\det A|}.$$

(Here A^{-t} denotes the transpose of the inverse of A.)

(d) If $p(x_1,x_2) = \overline{f(x_1,x_2)}$, then $\widehat{p}(t_1,t_2) = \overline{\widehat{f}(-t_1,-t_2)}$.

11.3.2. If $f \in L^1(\mathbb{R}^2)$, then $\widehat{f}(t_1,t_2)$ is uniformly continuous in the plane $\mathbb{R}^2$, and $|\widehat{f}(t_1,t_2)| \le \|f\|_1$ for all t_1, t_2.

11.3.3. From Lebesgue measure theory we find that if $f \in L^1(\mathbb{R}^2)$, then

$$\lim_{\substack{\delta_1\to 0\\ \delta_2\to 0}} \int_{-\infty}^{\infty}\int_{-\infty}^{\infty} |f(x_1+\delta_1, x_2+\delta_2) - f(x_1,x_2)|\,dx_1\,dx_2 = 0\,. \tag{11.28}$$

From this there follows a Riemann–Lebesgue lemma, namely that if $f \in L^1(\mathbb{R}^2)$, then

$$\lim_{|t_1|+|t_2|\to\infty} \widehat{f}(n_1,n_2) = 0\,. \tag{11.29}$$

11.3.4. If $f, g \in L^1(\mathbb{R}^2)$, then their *convolution* is

$$(f * g)(x_1,x_2) = \int_{-\infty}^{\infty}\int_{-\infty}^{\infty} f(u_1,u_2)g(x_1-u_1, x_2-u_2)\,dx_1\,dx_2\,. \tag{11.30}$$

For such functions, $\|f*g\|_1 \le \|f\|_1\|g\|_1$, and

$$\widehat{f*g}(t_1,t_2) = \widehat{f}(t_1,t_2)\widehat{g}(t_1,t_2)\,. \tag{11.31}$$

11.3.5. For $f \in L^1(\mathbb{R}^2)$, let

$$\sigma_{T_1,T_2}(x_1,x_2) = \int_{-T_1}^{T_1}(1-|t_1|/T_1)\int_{-T_2}^{T_2}(1-|t_2|/T_2)\widehat{f}(t_1,t_2)e(t_1x_1+t_2x_2)\,dt_2\,dt_1\,.$$

Then

$$\lim_{\substack{T_1\to\infty\\ T_2\to\infty}} \int_{-\infty}^{\infty}\int_{-\infty}^{\infty} |f(x_1,x_2) - \sigma_{T_1,T_2}(x_1,x_2)|\,dx_2\,dx_1 = 0\,.$$

Also, if f is continuous at (x_1,x_2), then

$$\lim_{\substack{T_1\to\infty\\ T_2\to\infty}} \sigma_{T_1,T_2}(x_1,x_2) = f(x_1,x_2)\,.$$

If $f \in L^1(\mathbb{R}^2)$, $\widehat{f} \in L^1(\mathbb{R}^2)$, and if f is continuous at (x_1,x_2), then

$$\int_{-\infty}^{\infty}\int_{-\infty}^{\infty} \widehat{f}(t_1,t_2)e(t_1x_1+t_2x_2)\,dx_2\,dx_1 = f(x_1,x_2)\,.$$

11.3.6. From Lebesgue measure theory we find that if $f \in L^2(\mathbb{R}^2)$, then

$$\lim_{\substack{\delta_1 \to 0 \\ \delta_2 \to 0}} \int_{-\infty}^{\infty} \int_{-\infty}^{\infty} |f(x_1 + \delta_1, x_2 + \delta_2) - f(x_1, x_2)|^2 \, dx_1 \, dx_2 = 0 \,. \tag{11.32}$$

If $f \in L^1(\mathbb{R}^2) \cap L^2(\mathbb{R}^2)$, then

$$\int_{-\infty}^{\infty} \int_{-\infty}^{\infty} |f(x_1, x_2)|^2 \, dx_2 \, dx_1 = \int_{-\infty}^{\infty} \int_{-\infty}^{\infty} |\widehat{f}(t_1, t_2)|^2 \, dt_2 \, dt_1 \,.$$

If $f \in L^1(\mathbb{R}^2) \cap L^2(\mathbb{R}^2)$ and $g \in L^1(\mathbb{R}^2) \cap L^2(\mathbb{R}^2)$, then

$$\int_{-\infty}^{\infty} \int_{-\infty}^{\infty} f(x_1, x_2)\overline{g(x_1, x_2)} \, dx_2 \, dx_1 = \int_{-\infty}^{\infty} \int_{-\infty}^{\infty} \widehat{f}(t_1, t_2)\overline{\widehat{g}(t_1, t_2)} \, dt_2 \, dt_1 \,.$$

11.3.7. The Poisson summation formula extends easily. If $f \in L^1(\mathbb{R}^d)$, then for $\boldsymbol{\alpha} \in \mathbb{T}^d$ we put

$$F(\boldsymbol{\alpha}) = \sum_{\boldsymbol{n} \in \mathbb{Z}^d} f(\boldsymbol{n} + \boldsymbol{\alpha}) \,.$$

As in the case $d = 1$ we find that $F \in L^1(\mathbb{T}^d)$, $\|F\|_1 \le \|f\|_1$, so that the sum $F(\boldsymbol{\alpha})$ is absolutely convergent for almost all $\boldsymbol{\alpha}$, and $\widehat{F}(\boldsymbol{k}) = \widehat{f}(\boldsymbol{k})$ for all $\boldsymbol{k} \in \mathbb{Z}^d$. As in the case $d = 1$, the question remains whether the Fourier Series of F converges to $F(\boldsymbol{\alpha})$. Certainly if F is continuous and

$$\sum_{\boldsymbol{k} \in \mathbb{Z}^d} |\widehat{f}(\boldsymbol{k})| < \infty,$$

then

$$\sum_{\boldsymbol{n} \in \mathbb{Z}^d} f(\boldsymbol{n} + \boldsymbol{\alpha}) = \sum_{\boldsymbol{k} \in \mathbb{Z}^d} \widehat{f}(\boldsymbol{k}) e(\boldsymbol{k} \cdot \boldsymbol{\alpha})$$

for all $\boldsymbol{\alpha} \in \mathbb{T}^d$. In particular, these conditions are fulfilled if

$$|f(\boldsymbol{x})| = O\Big(\frac{1}{1 + |\boldsymbol{x}|^\lambda}\Big) \quad \text{and} \quad |\widehat{f}(\boldsymbol{t})| = O\Big(\frac{1}{1 + |\boldsymbol{t}|^\lambda}\Big)$$

for all $\boldsymbol{x} \in \mathbb{R}^d$ and all $\boldsymbol{t} \in \mathbb{R}^d$ where $\lambda > d$.

We close with an interesting application of the machinery we have constructed.

Theorem 11.1. (Minkowski) *Let $\mathcal{C}$ be a convex body in $\mathbb{R}^d$ that is symmetric about the origin, and let V denote the volume (i. e., d-dimensional content) of $\mathcal{C}$. If $V > 2^d$, then $\mathcal{C}$ contains a nonzero lattice point.*

A set $\mathcal{C}$ in $\mathbb{R}^d$ is said to be *convex* if for every $\boldsymbol{x} \in \mathcal{C}$ and every $\boldsymbol{y} \in \mathcal{C}$, all points $\{(1-t)\boldsymbol{x} + t\boldsymbol{y} : 0 \le t \le 1\}$ of the line segment joining $\boldsymbol{x}$ to $\boldsymbol{y}$ lie in $\mathcal{C}$. A set $\mathcal{C}$ in $\mathbb{R}^d$ is said to be *symmetric about the origin* if $-\boldsymbol{x} \in \mathcal{C}$ if and only if $\boldsymbol{x} \in \mathcal{C}$. For our purposes a *lattice point* is simply a member of $\mathbb{Z}^d$. (There are more general lattices than $\mathbb{Z}^d$, but we do not need to go into that.)

Minkowski's theorem is sharp, for if $\mathcal{C} = (-1, 1)^d$, then $V = 2^d$ but $\mathcal{C}$ contains no nonzero lattice point.

Proof. Choose V' so that $2^d < V' < V$, and let $\mathcal{C}'$ denote the intersection of $\mathcal{C}$ with the ball $|\boldsymbol{x}| < R$, where R is taken large enough to ensure that the volume of $\mathcal{C}'$ is $> V'$. Thus $\mathcal{C}'$ is convex, symmetric about the origin, has volume nearly as large as that of $\mathcal{C}$, and has the advantage that it is compact. In other words, without loss of generality we may assume that $\mathcal{C}$ is compact. Let $\frac{1}{2}\mathcal{C}$ denote a scaled copy of $\mathcal{C}$, and

$$f(\boldsymbol{x}) = \chi_{\frac{1}{2}\mathcal{C}}(\boldsymbol{x})$$

be the characteristic function of $\frac{1}{2}\mathcal{C}$. Thus $\|f\|_1 = V/2^d < \infty$, so $f \in L^1(\mathbb{R}^d)$. Now set $j(\boldsymbol{x}) = f(-\boldsymbol{x})$. By taking $A = -I$ in §11.3.1 (c) we find that $\widehat{j}(\boldsymbol{t}) = \widehat{f}(-\boldsymbol{t})$. But $\frac{1}{2}\mathcal{C}$ is symmetric about the origin, so $j(\boldsymbol{x}) = f(\boldsymbol{x})$, and hence $\widehat{f}(-\boldsymbol{t}) = \widehat{f}(\boldsymbol{t})$. Thus

$$\widehat{f}(\boldsymbol{t}) = \frac{\widehat{f}(\boldsymbol{t}) + \widehat{f}(-\boldsymbol{t})}{2} = \int_{\frac{1}{2}\mathcal{C}} \cos 2\pi \boldsymbol{t}\cdot\boldsymbol{x}\,\boldsymbol{dx} = \frac{1}{2^d}\int_{\mathcal{C}} \cos \pi \boldsymbol{t}\cdot\boldsymbol{x}\,\boldsymbol{dx}$$

is a real number. (This is just an extension of the familiar fact that the Fourier coefficients of an even real-valued function are real.)

We now digress to discuss convolutions and convex sets. If $p * q = r$, $\operatorname{supp} p = \mathcal{P}$, and $\operatorname{supp} q = \mathcal{Q}$, then in order for $r(\boldsymbol{x})$ to be nonzero there must be a $\boldsymbol{u}$ for which $p(\boldsymbol{u}) \neq 0$ and $q(\boldsymbol{x} - \boldsymbol{u}) \neq 0$. Thus $\boldsymbol{x} = \boldsymbol{u} + (\boldsymbol{x} - \boldsymbol{u})$ is in the sum set $\mathcal{P} + \mathcal{Q} = \{\boldsymbol{p} + \boldsymbol{q} : \boldsymbol{p} \in \mathcal{P}, \boldsymbol{q} \in \mathcal{Q}\}$. In anticipation of setting $g = f * f$, we now show that $\frac{1}{2}\mathcal{C} + \frac{1}{2}\mathcal{C} = \mathcal{C}$. We recall that a convex set is determined by its supporting halfplanes (or halfhyperplanes). Let $\boldsymbol{u}$ be a unit vector, and choose $c = c(\boldsymbol{u})$ so that $\boldsymbol{u}\cdot\boldsymbol{x} \le c$ for all $\boldsymbol{x} \in \frac{1}{2}\mathcal{C}$. Thus the halfplane $\boldsymbol{u}\cdot\boldsymbol{x} \le c$ is a supporting halfplane of $\frac{1}{2}\mathcal{C}$. Let $\boldsymbol{x}$ and $\boldsymbol{y}$ be arbitrary members of $\frac{1}{2}\mathcal{C}$. Then $\boldsymbol{u}\cdot(\boldsymbol{x} + \boldsymbol{y}) \le 2c$. But $\boldsymbol{u}\cdot\boldsymbol{x} \le 2c$ is a supporting halfplane of $\mathcal{C}$, so $\frac{1}{2}\mathcal{C} + \frac{1}{2}\mathcal{C} \subseteq \mathcal{C}$. Conversely, if $\boldsymbol{x} \in \mathcal{C}$, then $\boldsymbol{x} = \frac{1}{2}\boldsymbol{x} + \frac{1}{2}\boldsymbol{x} \in \frac{1}{2}\mathcal{C} + \frac{1}{2}\mathcal{C}$, so $\mathcal{C} \subseteq \frac{1}{2}\mathcal{C} + \frac{1}{2}\mathcal{C}$.

Set $g = f * f$. Then $\operatorname{supp} g \subseteq \mathcal{C}$ and $\widehat{g}(\boldsymbol{t}) = \widehat{f}(\boldsymbol{t})^2 \ge 0$. Since g is the convolution of a bounded function with a function in L^1, it follows that g is continuous. Put

$$G(\boldsymbol{\alpha}) = \sum_{\boldsymbol{n}\in\mathbb{Z}^d} g(\boldsymbol{n} + \boldsymbol{\alpha})\,.$$

Since $\mathcal{C}$ is compact, there are only a bounded number of nonzero terms in the above sum when $\boldsymbol{\alpha} \in [0,1]^d$, so G is continuous.

Set

$$F(\boldsymbol{\alpha}) = \sum_{\boldsymbol{n}\in\mathbb{Z}^d} f(\boldsymbol{n} + \boldsymbol{\alpha})\,.$$

Since $\mathcal{C}$ is compact, the number of nonzero terms in the sum is bounded, and hence F is bounded. Also, $\widehat{F}(\boldsymbol{k}) = \widehat{f}(\boldsymbol{k})$, so by Parseval's Identity for multiple Fourier Series (as in §11.2.9),

$$\sum_{\boldsymbol{k}\in\mathbb{Z}^d} |\widehat{f}(\boldsymbol{k})|^2 = \int_{\mathbb{T}^d} |F(\boldsymbol{\alpha})|^2\,\boldsymbol{d\alpha} < \infty\,.$$

Hence the Fourier Series of G is absolutely convergent, and the Poisson summation formula applies, which is to say that

$$\sum_{\boldsymbol{n}\in\mathbb{Z}^d} g(\boldsymbol{n} + \boldsymbol{\alpha}) = \sum_{\boldsymbol{k}\in\mathbb{Z}^d} |\widehat{f}(\boldsymbol{k})|^2 e(\boldsymbol{k}\cdot\boldsymbol{\alpha})$$

for all $\boldsymbol{\alpha}$. In particular, for $\boldsymbol{\alpha} = \mathbf{0}$, we find that

$$\sum_{\boldsymbol{n}\in\mathbb{Z}^d} g(\boldsymbol{n}) = \sum_{\boldsymbol{k}\in\mathbb{Z}^d} |\widehat{f}(\boldsymbol{k})|^2 .$$

The contribution of $\boldsymbol{k} = \mathbf{0}$ on the right hand side is $V^2/2^{2d}$, so the right hand side above is

$$\geq \frac{V^2}{2^{2d}} .$$

On the other hand, the contribution of $\boldsymbol{n} = \mathbf{0}$ on the left hand side is $V/2^d$, which is smaller since $V > 2^d$. Thus on the left hand side there must be an $\boldsymbol{n} \neq \mathbf{0}$ for which $g(\boldsymbol{n}) > 0$. But then such an $\boldsymbol{n}$ must be a member of $\mathcal{C}$, so our proof is complete. □

Minkowski's theorem has many applications to Diophantine Approximation. In particular, we can derive Dirichlet's theorem (Corollary 9.20) from Minkowski's theorem.

Second Proof of Corollary 9.20. Let $\mathcal{C}$ be the parallelogram

$$|x| < N+1, \ |y - \theta x| < \frac{1+\varepsilon}{N+1}$$

in the plane. This is convex, symmetric about the origin, and has area $(1+\varepsilon)4 > 4$. Hence there exists a lattice point $(n_1, n_2) \neq (0,0)$ in $\mathcal{C}$. That is, $|n_1| < N+1$ and $|\theta n_1 + n_2| < (1+\varepsilon)/(N+1)$. If $n_1 = 0$, then the second inequality implies that $n_2 = 0$, so $n_1 \neq 0$. The second inequality asserts that $\|n_1\theta\| < (1+\varepsilon)/(N+1)$. If $n_1 < 0$ we replace n_1 by $-n_1$. There are only N possible values of n, and since there is an acceptable n for every $\varepsilon > 0$, there must be one that is acceptable for every such ε. For that n we have $1 \leq n \leq N$ and $\|n\theta\| \leq 1/(N+1)$. □

Notes

11.2. Example 11.1 was developed in 2008 by Edward Crane and Elmer Rees, and is included here with their permission. Rees also notes that the function $f(x_1, x_2) = \sin 2\pi x_1 \sin 2\pi x_2$ has period 1 with respect to x_1 and x_2, and is also invariant under the transformation $(x_1, x_2) \to (-x_1, x_2 + 1/2)$. This is a fixed-point free orientation reversing involution of the torus, so it is a map on the Klein bottle. It has eight critical points on the torus, and four critical points on the Klein bottle.

11.3. Minkowski's theorem can be proved in a much more elementary way. Our proof, which is due to Siegel (1935), gives additional information. For further thoughts of Siegel on the subject of the geometry of numbers see Siegel (1989).

Appendix B

The Binomial Theorem

B.1. Binomial coefficients

The *binomial coefficient* $\binom{n}{k}$ (orally "n-choose-k") is the number of subsets containing k elements of a set of n elements. We show that

$$\binom{n}{k} = \frac{n!}{k!(n-k)!} \tag{B.1}$$

for integers k, $0 \le k \le n$. Clearly $\binom{n}{k} = 0$ if $k < 0$ or $k > n$. To establish (B.1) we create an ordered list of our n elements, by two different procedures. First, we can take any one of the n elements to be the first member of our list, any one of the $n - 1$ remaining elements to be the second member, and so on. Thus there are $n!$ ordered lists that we can create. Once the list has been formed, we can take the first k members of the list to be the members of our subset of size k. Alternatively, we can choose a subset of k elements in $\binom{n}{k}$ ways, order these elements in $k!$ ways, then order the $n - k$ elements not in our subset in $(n - k)!$ ways, and append this second list to the first to create an ordered list of all n elements, in $\binom{n}{k}k!(n-k)!$ ways. Thus

$$\binom{n}{k} k!(n-k)! = n!,$$

so we have (B.1). Note that by convention, $0! = 1$, so that $\binom{n}{0} = \binom{n}{n} = 1$.

The binomial coefficients have many interesting properties. For example,

$$\binom{n+1}{k} = \binom{n}{k} + \binom{n}{k-1}. \tag{B.2}$$

We give two proofs of this. First, by two applications of (B.1) we see that

$$\begin{aligned}\binom{n}{k} + \binom{n}{k-1} &= \frac{n!}{k!(n-k)!} + \frac{n!}{(k-1)!(n-k+1)!}\\ &= \frac{n!}{(k-1)!(n-k)!}\Big(\frac{1}{k} + \frac{1}{n-k+1}\Big)\\ &= \frac{n!}{(k-1)!(n-k)!}\Big(\frac{n-k+1}{k(n-k+1)} + \frac{k}{k(n-k+1)}\Big)\\ &= \frac{n!}{(k-1)!(n-k)!}\,\frac{n+1}{k(n-k+1)} = \frac{(n+1)!}{k!(n-k+1)!},\end{aligned}$$

and this is $\binom{n+1}{k}$ by a third application of (B.1). Secondly, let $x_1, \ldots, x_n, x_{n+1}$ denote the $n+1$ elements from which we are going to choose a subset of size k. There are two cases: Either x_{n+1} is in our subset, or it is not. If x_{n+1} is in our subset, then we choose $k-1$ elements from among $x_1, \ldots, x_n$, and this can be done in $\binom{n}{k-1}$ ways. If x_{n+1} is not in our subset, then we choose k elements from among $x_1, \ldots, x_n$, and this can be done in $\binom{n}{k}$ ways. The sum of these two counts is the number of ways of choosing k elements from among $n+1$, so we again have (B.2).

The relation (B.2) allows the binomial coefficients to be generated recursively, and we obtain *Pascal's Triangle*:

$$\begin{array}{ccccccccc} & & & & 1 & & & & \\ & & & 1 & & 1 & & & \\ & & 1 & & 2 & & 1 & & \\ & 1 & & 3 & & 3 & & 1 & \\ 1 & & 4 & & 6 & & 4 & & 1 \\ & & & & \vdots & & & & \end{array}$$

From (B.1) it is immediate that

(B.3) $$\binom{n}{k} = \binom{n}{n-k}.$$

This can also be seen combinatorially: When we choose the k elements of our subset, we are simultaneously choosing the $n-k$ elements that are not in our subset. So instead of choosing k elements in $\binom{n}{k}$ ways, we could choose the complementary subset of $n-k$ elements in $\binom{n}{n-k}$ ways.

Consider an arbitrary subset of the n distinct elements $x_1, \ldots, x_n$. Either x_1 is in the subset, or it is not. Either x_2 is in the subset or it is not. For each element there are two possibilities, and so we see that the total number of subsets is 2^n. If we group these subsets according to size, we find that

(B.4) $$\sum_{k=0}^{n} \binom{n}{k} = 2^n.$$

Exercises

1. Consider the identity

(B.5) $$k\binom{n}{k} = n\binom{n-1}{k-1}.$$

   (a) Use (B.1) to derive the above.
   (b) Suppose that from among n people you are going to form a committee with k members, one of whom is designated as the chair of the committee. If you first choose the k members of the committee, and then choose the chair from among those k members, how many ways are there of completing the task? If you first choose the chair from among the n people, and then choose the $k-1$ other members of the committee from among the $n-1$ other people, how many ways are there to complete the task?

2. Consider the identity

(B.6) $$\sum_{k=1}^{n} k\binom{n}{k} = n2^{n-1}.$$

   (a) Prove the above by summing (B.5) over k.
   (b) From n people in how many ways can you choose a committee of arbitrary size, one of whom is the chair? Do this in two ways, to give a second proof of (B.6).

3. (a) Prove the identity (B.18) by induction on n.
   (b) From among the numbers $0, 1, \ldots, n$, how many subsets are there of size $r+1$ whose largest element is k? Give a further proof of (B.18).

4. Consider the identity (B.13). Suppose that S and T are disjoint sets, that S has m elements and T has n elements. How many subsets of $S \cup T$ are there, with k elements? How many subsets of $S \cup T$ are there, with r elements from S and $k-r$ elements from T? Deduce (B.13).

5. Use (B.13) to show that

(B.7) $$\sum_{r=0}^{n} \binom{n}{r}^2 = \binom{2n}{n}.$$

B.2. Binomial theorems

The simplest form of the binomial theorem states that

(B.8) $$(x+y)^n = \sum_{k=0}^{n} \binom{n}{k} x^k y^{n-k}.$$

This is easily proved by induction on n. When $n = 1$ it is trivial, and if it is true for n, then

$$\begin{aligned}(x+y)^{n+1} &= (x+y)\sum_{k=0}^{n}\binom{n}{k}x^k y^{n-k}\\ &= \sum_{k=0}^{n}\binom{n}{k}x^{k+1}y^{n-k} + \sum_{k=0}^{n}\binom{n}{k}x^k y^{n+1-k}.\end{aligned}$$

In the first sum we adjust the indexing, so that $k+1$ is replaced by ℓ. Thus the above is

$$\begin{aligned}&= \sum_{\ell=1}^{n+1}\binom{n}{\ell-1}x^{\ell}y^{n+1-\ell} + \sum_{k=0}^{n}\binom{n}{k}x^k y^{n+1-k}\\ &= \sum_{k=0}^{n+1}\left(\binom{n}{k-1}+\binom{n}{k}\right)x^k y^{n+1-k}\\ &= \sum_{k=0}^{n+1}\binom{n+1}{k}x^k y^{n+1-k}\end{aligned}$$

by (B.2). This is (B.8) with n replaced by $n+1$, so the proof is complete.

We can also arrive at (B.8) through combinatorial reasoning. When we expand $(x+y)^n$, we obtain 2^n monomials, each consisting of a product of n terms, with each factor being either x or y. Thus each monomial is of the form $x^k y^{n-k}$ for some k. For any particular k, the monomial $x^k y^{n-k}$ occurs $\binom{n}{k}$ times.

A second form of the binomial theorem asserts that

$$(1+z)^n = \sum_{k=0}^{n}\binom{n}{k}z^k \tag{B.9}$$

for all z. We can derive this from (B.8) by taking $x = z$, $y = 1$. Conversely, if we take $z = x/y$ in (B.9) and then multiply by y^n, then we obtain (B.8). This works when $y \ne 0$, and (B.8) is trivial when $y = 0$, so we see that (B.9) implies (B.8).

Let $P(z) = a_0 + a_1 z + \cdots + a_n z^n$ be an arbitrary polynomial. Then

$$\begin{aligned}P'(z) &= a_1 + 2a_2 z + 3a_3 z^2 + \cdots + na_n z^{n-1},\\ P''(z) &= 2a_2 + 3\cdot 2a_3 z + 4\cdot 3a_4 z^2 + \cdots n(n-1)a_n z^{n-2},\end{aligned}$$

and in general

$$P^{(r)}(z) = r!\sum_{k=r}^{n}\binom{k}{r}a_k z^{k-r}.$$

On setting $z = 0$, we deduce that

$$P^{(r)}(0) = r!a_r. \tag{B.10}$$

In the particular case that $P(z) = (1+z)^n$, we see that $P'(z) = n(1+z)^{n-1}$, $P''(z) = n(n-1)(1+z)^{n-2}$, and in general

$$P^{(r)}(z) = r!\binom{n}{r}(1+z)^{n-r},$$

so $P^{(r)}(0) = r!\binom{n}{r}$. On inserting this information into (B.10), we discover that the coefficient of z^r in the polynomial $(1+z)^n$ is $\binom{n}{r}$). That is, we have a proof of (B.9) that depends on calculus ideas.

The binomial theorem enables us to establish further identities involving binomial coefficients. For example, by taking $z = -1$ in (B.9), we discover that

$$\sum_{k=0}^{n}(-1)^n\binom{n}{k} = 0 \tag{B.11}$$

when $n > 0$. Also, when we multiply two polynomials, say

$$P(z)Q(z) = \left(\sum_{i=0}^{m} a_i z^i\right)\left(\sum_{j=0}^{n} b_j z^j\right) = \sum_{k=0}^{m+n} c_k z^k,$$

the coefficient c_k of z^k in the product is

$$c_k = \sum_{r=0}^{k} a_r b_{k-r}\,. \tag{B.12}$$

This is known as the *Cauchy convolution* of the two sequences. If $P(z) = (1+z)^m$ and $Q(z) = (1+z)^n$, then $P(z)Q(z) = (1+z)^{m+n}$, so by (B.12) and three applications of the binomial theorem in the form (B.9) we find that

$$\sum_{r=0}^{k}\binom{m}{r}\binom{n}{k-r} = \binom{m+n}{k}. \tag{B.13}$$

We observe that

$$\binom{n}{k} = \frac{n!}{k!(n-k)!} = \frac{n(n-1)(n-2)\cdots(n-k+1)}{k!}\,.$$

If on the right hand side we replace the positive integer n by an arbitrary complex number α, we obtain the expression

$$\frac{\alpha(\alpha-1)(\alpha-2)\cdots(\alpha-k+1)}{k!},$$

which is a polynomial in α of degree k, with rational coefficients. This allows us to widen our definition of binomial coefficients: We put

$$\binom{\alpha}{k} = \frac{\alpha(\alpha-1)\cdots(\alpha-k+1)}{k!} \tag{B.14}$$

for any real or complex number α. Of course, if α happens to be a positive integer, then this is consistent with our initial definition. These more general binomial coefficients $\binom{\alpha}{k}$ arise when we attempt to expand $(1+z)^\alpha$ in a power series. For simplicity we restrict ourselves to real α and real x.

Theorem B.1. *Let α be a real number. Then*

$$(1+x)^\alpha = \sum_{k=0}^{\infty}\binom{\alpha}{k}x^k \tag{B.15}$$

for $-1 < x < 1$.

Proof. Taylor's formula with the integral form of the remainder states that if f has continuous derivatives through the order $n+1$, then

$$f(x) = \sum_{k=0}^{n} \frac{f^{(k)}(0)}{k!}x^k + \frac{1}{n!}\int_0^x f^{(n+1)}(u)(x-u)^n\,du\,. \tag{B.16}$$

In the case that $f(x) = (1+x)^\alpha$, we have $f'(x) = \alpha(1+x)^{\alpha-1}$, $f''(x) = \alpha(\alpha-1)(1+x)^{\alpha-2}$, and in general

$$\frac{f^{(k)}(x)}{k!} = \binom{\alpha}{k}(1+x)^{\alpha-k}\,.$$

For a fixed x with $-1 < x < 1$, the integral in (B.16) is at most exponentially large as a function of n, but $n! > (n/e)^n$, so $n!$ tends to infinity faster than any exponential function. Thus the last term in (B.16) tends to 0 as $n \to \infty$, so we have the stated result. □

We observe that

$$\binom{\alpha}{k} = \frac{\alpha(\alpha-1)\cdots(\alpha-k+1)}{k!} = (-1)^k\frac{(k-1-\alpha)(k-2-\alpha)\cdots(1-\alpha)(-\alpha)}{k!}$$
$$= (-1)^k\binom{k-1-\alpha}{k}.$$

By setting $\beta = -\alpha - 1$ and replacing x by $-x$ in Theorem B.1, we obtain

Corollary B.2. *Let β be a real number. Then*

$$\frac{1}{(1-x)^{\beta+1}} = \sum_{k=0}^{\infty}\binom{k+\beta}{k}x^k \tag{B.17}$$

for $-1 < x < 1$.

This is known as the *negative binomial theorem.* In fact (B.17) is equivalent to (B.15), since the only difference between them is a change of notation. When $\beta = 0$, the above is simply the familiar formula for the sum (0.3) of a geometric series.

By convolving the power series coefficients of $(1-x)^{-r-1}$ with those of the power series $1/(1-x)$ we obtain the coefficients of the power series $(1-x)^{-r-2}$. Thus by two applications of Corollary B.2 we find that

$$\sum_{k=r}^{n}\binom{k}{r} = \binom{n+1}{r+1}. \tag{B.18}$$

Example B.1. Suppose we are to combine $n+1$ symbols with n applications of a binary operation, which we will speak of as multiplication. If the multiplication is neither commutative nor associative, then the outcome may depend on which order the multiplications are executed. For $n = 2$ we have two possibilities: $(ab)c$ and $a(bc)$. For $n = 3$ we have five possibilities: $a(b(cd))$, $a((bc)d)$, $(ab)(cd)$, $(a(bc))d$, $((ab)c)d$. Let C_n denote the number of possible ways of placing the parentheses. By convention we put $C_0 = C_1 = 1$. Consider C_{n+1}, which is to say the multiplication of $n+2$ symbols. The outermost parentheses divide the list into $k+1$ symbols on the left, and $n+1-k$ symbols on the right:

$$\underbrace{(\qquad\qquad)}_{k+1\text{ terms}}\underbrace{(\qquad\qquad)}_{n+1-k\text{ terms}}$$

Within the first grouping the interior parentheses can be inserted in C_k ways, and in the second one in C_{n-k} ways. Thus

$$C_{n+1} = \sum_{k=0}^{n} C_k C_{n-k}\,. \tag{B.19}$$

This suggests the Cauchy convolution of the coefficients of a power series with itself, and so we define

$$f(x) = \sum_{n=0}^{\infty} C_n x^n\,. \tag{B.20}$$

At this stage we have no guarantee that this power series has positive radius of convergence, but it seems that

$$f(x)^2 = \sum_{n=0}^{\infty} C_{n+1}x^n = \frac{f(x)-1}{x}\,.$$

That is, $xf^2 - f + 1 = 0$. We know how to solve for the roots of a quadratic polynomial, so we find that

$$f(x) = \frac{1 \pm \sqrt{1-4x}}{2x}\,.$$

If we take the plus sign, the numerator tends to 2 as $x \to 0$, while the denominator is tending to 0, which is inconsistent with the idea that f can be written as in (B.20). Thus we take the minus sign, and argue by Theorem B.1 that

(B.21)

$$f(x) = \frac{1-\sqrt{1-4x}}{2x} = \frac{1}{2x}\left(1 - \sum_{n=0}^{\infty}\binom{1/2}{n}(-4)^n x^n\right) = -\frac{1}{2}\sum_{n=1}^{\infty}\binom{1/2}{n}(-4)^n x^{n-1}\,.$$

But

$$\binom{1/2}{n} = \frac{\frac{1}{2}(\frac{-1}{2})\cdots(\frac{1}{2}-n+1)}{n!} = (-1)^{n-1}\frac{1\cdot 3\cdot 5\cdots(2n-3)}{2^n n!}\,.$$

We multiply and divide by $2\cdot 4\cdots(2n-2) = 2^{n-1}(n-1)!$ to see that the above is

$$= (-1)^{n-1}\frac{(2n-2)!}{2^{2n-1}n!(n-1)!}\,.$$

On inserting this in (B.20), we are led to believe that

$$f(x) = \sum_{n=1}^{\infty}\frac{(2n-2)!}{n!(n-1)!}x^{n-1} = \sum_{n=0}^{\infty}\frac{\binom{2n}{n}}{n+1}x^n\,.$$

This suggests that

$$C_n = \frac{\binom{2n}{n}}{n+1} \tag{B.22}$$

for all n. While this might seem a bit speculative, by Theorem B.1 we see that the right hand side above satisfies the same recurrence (B.19) as the C_n, and have the same initial conditions $C_0 = C_1 = 1$, so in fact we have a rigorous proof that (B.22) holds for all n.

The numbers C_n are known as the *Catalan numbers*. Their initial values are

$$1, 1, 2, 5, 14, 42, 132, 429, 1430, 4862, 16796, 58786, 208012, 742900, 2674440, \ldots$$

They have many combinatorial interpretations. Taken out of context, it would not be obvious that $n+1$ divides evenly into $\binom{2n}{n}$, but since we have a combinatorial interpretation of this quotient, we know that it is an integer.

Exercises

1. Let the binomial coefficient $\binom{\alpha}{k}$ be defined as in (B.14). Show that
$$\binom{\alpha}{k} + \binom{\alpha}{k-1} = \binom{\alpha+1}{k}$$
for positive integers k.

2. With $\binom{x}{k}$ defined as in (B.14), show that any polynomial $P(x)$ of degree at most n can be expressed in the form
$$P(x) = \sum_{k=0}^{n} c_k \binom{x}{k}. \tag{B.23}$$
Show also that the c_k are uniquely determined.

3. Let f be a function of a real variable, and let Δf be the function
$$\Delta f(x) = f(x+1) - f(x). \tag{B.24}$$
For $k > 1$ let $\Delta^{k+1} f = \Delta(\Delta^k f)$. Show that
$$\Delta^k f(x) = \sum_{j=0}^{k} (-1)^j \binom{k}{j} f(x+k-j).$$
This is called the kth *forward difference* of f.

4. With P defined as in (B.23) and the operator Δ defined as in (B.24), show that
$$\Delta P(x) = \sum_{k=1}^{n} c_k \binom{x}{k-1}.$$
Note the similarity to the formula for the derivative of a polynomial.

Chebyshev Polynomials

We observe that

$$\begin{aligned}
\cos 0\theta &= 1, \\
\cos 1\theta &= \cos\theta, \\
\cos 2\theta &= 2\cos^2\theta - 1, \\
\cos 3\theta &= 4\cos^3\theta - 3\cos\theta, \\
\cos 4\theta &= 8\cos^4\theta - 8\cos^2\theta + 1, \\
\cos 5\theta &= 16\cos^5\theta - 20\cos^3\theta + 5\cos\theta, \\
\cos 6\theta &= 32\cos^6\theta - 48\cos^4\theta + 18\cos^2\theta - 1.
\end{aligned}$$

These identities suggest that perhaps $\cos n\theta$ can be expressed as a polynomial in $\cos\theta$ for all positive integers n. To see how this can be established we recall the identity (T.28)

$$\cos(a \pm b) = \cos a \cos b \mp \sin a \sin b.$$

By substituting into this, we deduce that

(C.1) $$\cos(n+1)\theta = \cos n\theta \cos\theta - \sin n\theta \sin\theta,$$

(C.2) $$\cos(n-1)\theta = \cos n\theta \cos\theta + \sin n\theta \sin\theta.$$

On summing these two identities, we see that

$$\cos(n+1)\theta + \cos(n-1)\theta = 2\cos n\theta \cos\theta,$$

which is to say that

$$\cos(n+1)\theta = 2\cos\theta\cos n\theta - \cos(n-1)\theta.$$

Thus if $\cos(n-1)\theta$ is a polynomial in $\cos\theta$, say $\cos(n-1)\theta = T_{n-1}(\cos\theta)$, and if $\cos n\theta$ is a polynomial in $\cos\theta$, say $\cos n\theta = T_n(\cos\theta)$, then $\cos(n+1)\theta$ is a polynomial in $\cos\theta$, $\cos(n+1)\theta = T_{n+1}(\cos\theta)$, where $T_{n+1}(\cos\theta) = 2\cos\theta T_n(\cos\theta) - T_{n-1}(\cos\theta)$. Hence by induction it follows that

(C.3) $$\cos n\theta = T_n(\cos\theta)$$

for all nonnegative integers n, where the polynomials $T_n(x)$ are determined inductively by the identities

(C.4) $$T_0(x) = 1, \qquad T_1(x) = x, \qquad T_{n+1}(x) = 2xT_n(x) - T_{n-1}(x).$$

By taking $\theta = 0$ in (C.3) we see that

(C.5) $$T_n(1) = 1.$$

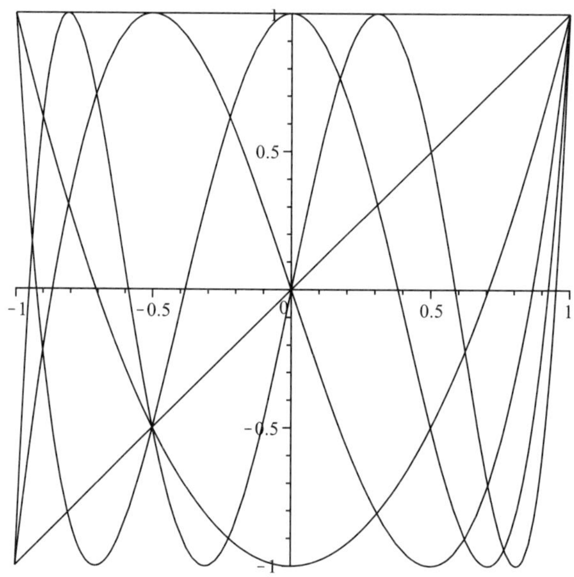

Figure C.1. Graphs of $T_n(x)$ for $-1 \le x \le 1$, $0 \le n \le 5$.

By putting $\theta = \pi$ in (C.3), we similarly find that

(C.6) $$T_n(-1) = (-1)^n.$$

By using (C.4) and arguing by induction we also see that $T_n(x)$ is a polynomial of degree n with leading coefficient 2^{n-1} when $n > 0$. Put

(C.7) $$r_k = \cos\frac{(2k-1)\pi}{2n} \qquad (1 \le k \le n).$$

By (C.3) we see that $T_n(r_k) = \cos\frac{(2k-1)\pi}{2} = 0$. Since $\cos\theta$ is strictly decreasing on $[0, \pi]$ with $\cos 0 = 1$, $\cos\pi = -1$, it follows that

(C.8) $$1 > r_1 > r_2 > \cdots > r_n > -1.$$

Thus $T_n(x)$ has n distinct zeros in the interval $(-1, 1)$, and hence the factored form of $T_n(x)$ is

(C.9) $$T_n(x) = 2^{n-1}(x - r_1)(x - r_2)\cdots(x - r_n)$$

when $n > 0$.

The Chebyshev polynomials are orthogonal in the following sense.

Theorem C.1. *If* $0 \le m \le n$, *then*

$$\int_{-1}^{1} T_m(x)T_n(x)(1-x^2)^{-1/2}\,dx = \begin{cases} \pi & (m=n=0), \\ \pi/2 & (m=n>0), \\ 0 & (m<n). \end{cases}$$

Proof. By (T.25) we know that $\cos m\theta \cos n\theta = \frac{1}{2}\cos(m+n)\theta + \frac{1}{2}\cos(n-m)\theta$. Hence

$$\int_{0}^{\pi} T_m(\cos\theta)T_n(\cos\theta)\,d\theta = \begin{cases} \pi & (m=n=0), \\ \pi/2 & (m=n>0), \\ 0 & (m<n). \end{cases}$$

We put $x = \cos\theta$, with the result that $d\theta = -dx/(1-x^2)^{1/2}$ to obtain the stated result. □

The Chebyshev polynomials have many important extremal properties, of which we give an important example. In passing we recall that a polynomial $P(x) = a_n x^n + \cdots + a_0$ is said to be *monic* if $a_n = 1$.

Theorem C.2. *Suppose that* $P(x)$ *is a monic polynomial of degree* $n > 0$. *Then*

$$\max_{-1\le x\le 1} |P(x)| \ge 2^{1-n},$$

and equality is achieved if we take $P(x) = 2^{1-n}T_n(x)$.

Proof. The second assertion is obvious. Suppose that the maximum in question is $< 2^{1-n}$, and put $R(x) = P(x) - 2^{1-n}T_n(x)$. Since $P(x)$ and $2^{1-n}T_n(x)$ are both

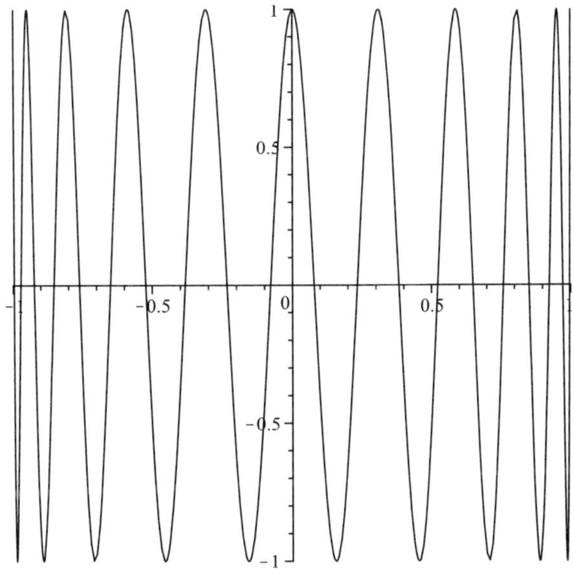

Figure C.2. Graph of $T_{20}(x)$ for $-1 \le x \le 1$.

monic of degree n, it follows that $R(x)$ has degree $< n$. Put

$$x_k = \cos\frac{k\pi}{n} \qquad (0 \le k \le n).$$

Thus

$$1 = x_0 > x_1 > \cdots > x_n = -1,$$

and $T_n(x_k) = \cos k\pi = (-1)^k$. Hence if k is even, then

$$R(x_k) = P(x_k) - 2^{1-n} < 0,$$

while if k is odd, then

$$R(x_k) = P(x_k) + 2^{1-n} > 0.$$

Since $R(x)$ changes sign in each of the intervals $[x_n, x_{n-1}]$, $[x_{n-1}, x_{n-2}]$, ..., $[x_1, x_0]$, by the intermediate value theorem it follows that $R(x)$ has a zero in each of these n intervals. But then $R(x)$ has more zeros than its degree, and that can happen only when $R(x)$ is identically 0. But in that case $P(x) = 2^{1-n}T_n(x)$. Since

$$\max_{-1\le x\le 1} 2^{1-n}|T_n(x)| = 2^{1-n},$$

we have a contradiction, and the proof is complete. □

Exercises

1. Use formula (C.3) to determine the value of $T_n(0)$.
2. Use the definition (C.4) and induction to show that $T_n(x)$ is an odd function if n is odd, and an even function if n is even.
3. (a) Show that if m and n are nonnegative integers, then $T_{mn}(x) = T_m(T_n(x))$ for $-1 \le x \le 1$.
   (b) Explain why the identity $T_{mn}(x) = T_m(T_n(x))$ must hold for all x.
4. Suppose that $a < b$, and put $I = [-1, 1]$, $J = [a, b]$. Let x and u be related by the identity

$$u = \frac{b-a}{2}x + \frac{a+b}{2}.$$

   (a) Show that if $x \in I$, then $u \in J$ and vice versa.
   (b) Let $P(u)$ be a polynomial of degree $n > 0$ with leading coefficient c. Put $Q(x) = P(\frac{b-a}{2}x + \frac{a+b}{2})$. Show that $Q(x)$ is of degree n with leading coefficient $c(b-a)^n/2^n$.
   (c) Show also that

$$\max_{x\in I}|Q(x)| = \max_{u\in J}|P(u)|.$$

   (d) Use Theorem C.2 to show that

$$\max_{a\le u\le b}|P(x)| \ge |c|(b-a)^n 2^{1-2n}$$

   when $n > 0$.
5. (a) By induction on n, or by using the methods of §.F.4, show that if $-1 \le x \le 1$, then

$$T_n(x) = \frac{1}{2}\big(x + i\sqrt{1-x^2}\big)^n + \frac{1}{2}\big(x - i\sqrt{1-x^2}\big)^n.$$

(b) In Figure C.1 it appears that $T_n(-1/2) = -1/2$ for $n = 1, 2, 4, 5$. Show that $T_n(-1/2) = -1/2$ when $3 \nmid n$, and that $T_n(-1/2) = 1$ when $3|n$.

6. In Figure C.2 it appears that the roots are closer together near the endpoints of the interval. Construct a quadratic approximation to $\cos x$ at $x = 0$, and use it to show that the root r_k given by (C.7) is approximately $1 - (2k-1)^2\pi^2/(8n^2)$ when k is small compared with n.

7. Use (C.4) and induction on n to show that

$$T_n(x) = \frac{n}{2} \sum_{0 \le m \le n/2} (-1)^m \frac{(n-m-1)!}{m!(n-2m)!}(2x)^{n-2m}.$$

8. (a) Use (T.28) to show that $2\cos m\theta \cos n\theta = \cos(m+n)\theta + \cos(n-m)\theta$.
(b) Show that if $0 \le m \le n$, then

$$2T_m(x)T_n(x) = T_{m+n}(x) + T_{n-m}(x).$$

9. (a) Use Exercise 3.4.3 (c) to show that

$$1 + 2\sum_{n=1}^{N} T_n(u)T_n(v) = \frac{T_{N+1}(u)T_N(v) - T_N(u)T_{N+1}(v)}{u - v}.$$

(b) Write the numerator above as $(T_{N+1}(u) - T_{N+1}(v))T_N(v) - (T_N(u) - T_N(v))T_{N+1}(v)$. Allow u to tend to v to show that

$$1 + 2\sum_{n=1}^{N} T_n^2(v) = T'_{N+1}(v)T_N(v) - T'_N(v)T_{N+1}(v).$$

(c) Conclude that $T'_{N+1}(v)T_N(v) - F'_N(v)T_{N+1}(v) > 0$ for all v.
(d) Show that if v is a number such that $T_N(v) = 0$, then the sign of $T'_N(v)$ is opposite that of $T_{N+1}(v)$.
(e) Show that if $T'_N(v) = 0$, then $T_N(v)$ and $T'_{N+1}(v)$ have the same sign.

10. (a) Show that $T'_n(\cos\theta)\sin\theta = n\sin n\theta$.
(b) Show that $2\sin\theta\sin n\theta = \cos(n-1)\theta - \cos(n+1)\theta$.
(c) Show that $T'_n(\cos\theta)\sin^2\theta = n(T_{n-1}(\cos\theta) - T_{n+1}(\cos\theta))/2$.
(d) Use (C.4) to deduce that

$$T'_n(\cos\theta)\sin^2\theta = nT_{n-1}(\cos\theta) - n\cos\theta T_{n+1}(\cos\theta).$$

(e) Show that $(1-x^2)T'_n(x) = nT_{n-1}(x) - nxT_n(x)$.
(f) Show that $y = T_n(x)$ is a solution of the differential equation

$$(1-x^2)y'' - xy' + n^2y = 0.$$

(g) Show that $(T'_n(x)\sqrt{1-x^2})' = -n^2T_n(x)/\sqrt{1-x^2}$.

11. (a) Use Exercise 10 (e) and (C.4) to show that

$$T_{n+1}(x) = xT_n(x) - \frac{1-x^2}{n}T'_n(x).$$

(b) Use Exercise 10(e) (with n replaced by $n+1$) in conjunction with part (a) to see that

$$T'_{n+1}(x) = (n+1)T_n(x) + \frac{n+1}{n}xT'_n(x).$$

Note that these identities enable one to find the pair $(T_{n+1}(x), T'_{n+1}(x))$ in terms of the pair $(T_n(x), T'_n(x))$. This is similar to (C.4), which allows us to determine the pair $(T_{n+1}(x), T_n(x))$ in terms of the pair $(T_n(x), T_{n-1}(x))$.

The Chebyshev polynomials that we have discussed are more properly called the *Chebyshev polynomials of the first kind.* There is also a second family of Chebyshev polynomials that arise from the identities

$$\begin{aligned} \sin\theta &= \sin\theta, \\ \sin 2\theta &= 2\cos\theta\sin\theta, \\ \sin 3\theta &= (4\cos^2\theta - 1)\sin\theta, \\ \sin 4\theta &= (8\cos^3\theta - 4\theta)\sin\theta, \end{aligned}$$

from which we might guess that for each positive n there is a polynomial $U_n(x)$ of degree n such that $\sin(n+1)\theta = U_n(\cos\theta)\sin\theta$. These are the *Chebyshev polynomials of the second kind.*

12. (a) Use the formula (T.13) to show that

$$\sin(n+1)\theta + \sin(n-1)\theta = 2\sin n\theta\cos\theta.$$

(b) Put $U_0(x) = 1$, $U_1(x) = 2x$. Thus $\sin\theta = U_0(\cos\theta)\sin\theta$ and $\sin 2\theta = U_1(\cos\theta)\sin\theta$. Supposing that $U_{n-1}(x)$ is a polynomial such that $\sin n\theta = U_{n-1}(\cos\theta)\sin\theta$, and that $U_{n-2}(x)$ is a polynomial such that $\sin(n-1)\theta = U_{n-2}(\cos\theta)\sin\theta$, show that there is a polynomial $U_n(x)$ such that

$$\sin(n+1)\theta = U_n(\cos\theta)\sin\theta,$$

and that

$$U_n(x) = 2xU_{n-1}(x) - U_{n-2}(x).$$

13. Let the $U_n(x)$ be defined as above.
(a) Show that $U_n(x)$ has degree n and leading coefficient 2^n.
(b) Show that $U_n(x)$ is an odd function if n is odd, and that it is an even function if n is even.
(c) Show that $U_n(1) = n + 1$.
(d) Show that $U_n(-1) = (-1)^n(n+1)$.

14. (a) Show that for positive integers m and n,

$$\int_0^\pi \sin m\theta \sin n\theta \, d\theta = \begin{cases} \pi/2 & (m = n), \\ 0 & (m \neq n). \end{cases}$$

(b) By the change of variable $x = \cos\theta$, deduce that

$$\int_{-1}^1 U_m(x)U_n(x)(1-x^2)^{1/2}\, dx = \begin{cases} \pi/2 & (m = n), \\ 0 & (m \neq n) \end{cases}$$

for nonnegative integers m and n.

15. (a) Suppose that $0 < \theta < \pi$. Explain why

$$|U_n(\cos\theta)| = \left|\frac{\sin(n+1)\theta}{\sin\theta}\right| \le \frac{1}{\sin\theta} = \frac{1}{\sqrt{1-\cos^2\theta}}.$$

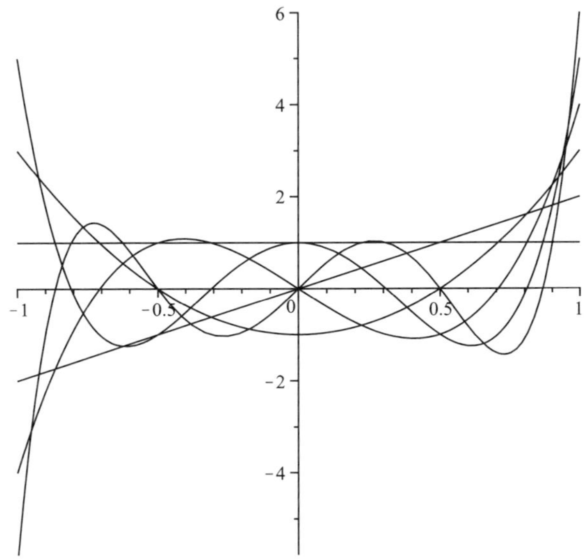

Figure C.3. Graphs of $U_n(x)$ for $-1 \le x \le 1$, $1 \le n \le 5$.

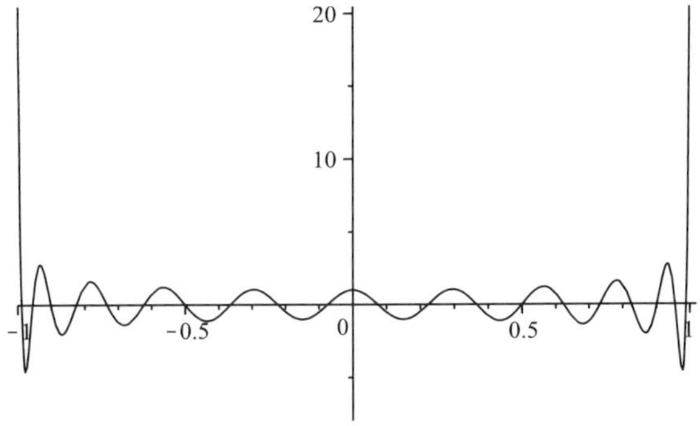

Figure C.4. Graph of $U_{20}(x)$ for $-1 \le x \le 1$.

(b) Show that if $-1 < x < 1$, then

$$|U_n(x)| \le \frac{1}{\sqrt{1-x^2}}.$$

(c) For $0 \le k \le n$ put $x_k = \cos\big((2k+1)\pi/(2(n+1))\big)$. Show that $1 > x_0 > x_1 > \cdots > x_n > -1$.

(d) Show that $U_n(x_k)\sqrt{1-x_k^2} = (-1)^k$ for $0 \le k \le n$.

(e) Show that if $P(x)$ is a monic polynomial of degree n, then

$$\max_{-1\le x\le 1} |P(x)|\sqrt{1-x^2} \ge 2^{-n},$$

and that equality is achieved by taking $P(x) = 2^{-n}U_n(x)$.

16. (a) Differentiate (C.3) with respect to θ to show that

$$T_n'(\cos\theta) = n\frac{\sin n\theta}{\sin\theta}.$$

(b) By letting θ tend to 0, deduce that

$$T_n'(1) = n^2.$$

(c) Show also that $T_n'(x) = nU_{n-1}(x)$ for all x.

(d) Show that

$$\frac{\sin n\theta}{\sin\theta} = e^{-(n-1)i\theta}\big(1 + e^{2i\theta} + e^{4i\theta} + \cdots + e^{2(n-1)i\theta}\big).$$

(e) Deduce that

$$\left|\frac{\sin n\theta}{\sin\theta}\right| \le n,$$

with equality only when θ is an integer multiple of π.

(f) Deduce that $|U_n(x)| \le n+1$ for $-1 \le x \le 1$, with equality only when $x = \pm 1$.

(g) Conclude that $|T'(x)| \le n^2$ for $-1 \le x \le 1$. Note: It is known that if $P(x)$ is a polynomial of degree n, then

$$\max_{-1\le x\le 1} |P'(x)| \le n^2 \max_{-1\le x\le 1} |P(x)|.$$

Thus by this exercise we see that the $T_n(x)$ are extremal also in this context.

17. The object of this exercise is to establish a further extremal property of the Chebyshev polynomials $T_n(x)$, namely that among polynomials of degree n (not necessarily monic), if $x > 1$, then the ratio

$$\frac{|P(x)|}{\max_{-1\le x\le 1} |P(x)|}$$

is maximized by taking $P(x) = T_n(x)$. To prove this, suppose that there is a polynomial $P(x)$ of degree n such that $|P(x)| \le 1$ for $-1 \le x \le 1$, and that there is an $x > 1$ for which $P(x) > T_n(x)$. We need to give this x a name, say x_{-1}. Choose c so that $T_n(x_{-1})/P(x_{-1}) < c < 1$, and put $R(x) = T_n(x) - cP(x)$.

(a) Show that $R(x_{-1}) < 0$.

(b) Let $x_0, x_1, \ldots, x_n$ be as in the proof of Theorem C.2. Show that $R(x_k) > 0$ if k is even, and that $R(x_k) < 0$ if k is odd.

(c) Explain why $R(x)$ has a zero in each of the intervals (x_0, x_{-1}), (x_1, x_0), (x_2, x_1), …, (x_n, x_{n-1}).

(d) Show that $R(x)$ is a polynomial of degree at most n that has at least $n+1$ zeros, and explain why this is a contradiction.

18. The object of this exercise is to prove *Rodrigue's formula*, which asserts that

$$T_n(x) = \frac{(-1)^n}{1{\cdot}3{\cdot}5\cdots(2n-1)}(1-x^2)^{1/2}\frac{d^n}{dx^n}(1-x^2)^{n-1/2} \tag{C.10}$$

for $-1 < x < 1$. We find it convenient to prove at the same time a second formula of this type, namely

$$T_n'(x) = \frac{(-1)^n n^2}{1{\cdot}3{\cdot}5\cdots(2n-1)}(1-x^2)^{-1/2}\frac{d^{n-1}}{dx^{n-1}}(1-x^2)^{n-1/2} \tag{C.11}$$

for $-1 < x < 1$. We argue by induction on n.

(a) Show that (C.10) and (C.11) both hold for $n = 1$. This is the basis of the induction.

(b) Suppose that $f(x) = p(x)(1-x^2)^\alpha$ where $p(x)$ is a polynomial. Show that there is a polynomial $q(x)$ such that $f'(x) = q(x)(1-x^2)^{\alpha-1}$.

(c) Deduce that the right hand side of (C.10) is a polynomial, and that the right hand side of (C.11) is a polynomial.

(d) Suppose that (C.10) and (C.11) both hold for $n-1$; we wish to show that they hold for n. Compute the first derivative of $(1-x^2)^{n-1/2}$.

(e) Recall Leibnitz's formula for the mth derivative of a product $f(x)g(x)$ of two functions.

(f) Use the formula of (e) with $m = n-1$, $f(x) = x$, $g(x) = (1-x^2)^{n-3/2}$ to show that

$$\begin{aligned}\frac{d^{n-1}}{dx^{n-1}}\, x(1-x^2)^{n-3/2} &= x\,\frac{d^{n-1}}{dx^{n-1}}\,(1-x^2)^{n-3/2} + (n-1)\,\frac{d^{n-2}}{dx^{n-2}}\,(1-x^2)^{n-3/2}\\ &= S_1 + S_2,\end{aligned}$$

say.

(g) Use the inductive hypothesis (C.10) (with n replaced by $n-1$) to show that

$$S_1 = (-1)^{n-1}1{\cdot}3\cdots(2n-3)(1-x^2)^{-1/2}xT_{n-1}(x).$$

(h) Use the inductive hypothesis (C.11) (with n replaced by $n-1$) to show that

$$S_2 = (-1)^n 1{\cdot}3\cdots(2n-3)\frac{(1-x^2)^{1/2}}{n-1}T'_{n-1}(x).$$

(i) Deduce that

$$S_1 + S_2 = (-1)^{n-1}1{\cdot}3\cdots(2n-3)(1-x^2)^{-1/2}\Big(xT_{n-1}(x) - \frac{1-x^2}{n-1}T'_{n-1}(x)\Big),$$

and use Exercise 11(a) to complete the proof of (C.10).

(j) It remains to derive (C.11) from (C.10), in order to complete the inductive step. Let

$$f(x) = (1-x^2)^{-1/2}\frac{d^{n-1}}{dx^{n-1}}\,(1-x^2)^{n-1/2}.$$

Use (C.10) to show that

$$\big(f(x)(1-x^2)^{1/2}\big)' = (-1)^n 1{\cdot}3\cdots(2n-1)(1-x^2)^{-1/2}T_n(x).$$

(k) Use 10(g) to show that that right hand side above is

$$= (-1)^{n-1}1{\cdot}3\cdots(2n-1)n^{-2}\big(T'_n(x)(1-x^2)^{1/2}\big)' = g'(x)$$

for $-1 < x < 1$, where $g(x) = (-1)^{n-1}1{\cdot}3\cdots(2n-1)n^{-2}T'_n(x)(1-x^2)^{1/2}$.

(l) Deduce that there is a constant c such that $f(x)(1-x^2)^{1/2} = g(x) + c$ for $-1 < x < 1$.

(m) Use part (c) above to show that $f(x)$ is a polynomial, and hence deduce that

$$\lim_{x\to 1^-} f(x)(1-x^2)^{1/2} = 0.$$

(n) Show that $\lim_{x\to 1^-} g(x) = 0$, and hence deduce that $c = 0$.

(o) Deduce (C.11). This completes the inductive step, so we have (C.10) and (C.11) for all n.

19. (a) Use (C.4) to show that all coefficients of $T_n(x)$ are integers.

(b) Use Exercise 6(a) to show that if $-1 \le x \le 1$, then

$$T_n(x) = \binom{n}{0}x^n - \binom{n}{2}x^{n-2}(1-x^2) + \binom{n}{4}x^{n-4}(1-x^2)^2 - \binom{n}{6}x^{n-6}(1-x^2)^3 + \cdots .$$

(c) Explain why this identity must hold for all x.

(d) Use the above identity to give a second proof that all coefficients of $T_n(x)$ are integers.

(e) Show that if $x \ge 1$ or $x \le -1$, then

$$T_n(x) = \frac{1}{2}\left(x + \sqrt{x^2-1}\right)^n + \frac{1}{2}\left(x - \sqrt{x^2-1}\right)^n.$$

(f) In Figure C.2 we see that $T_{20}(x)$ is increasing very rapidly when $x = 1$, and by Exercise 3(b) we find that $T'_{20}(1) = 400$. It stands to reason that $T_{20}(2)$ is very large. Explain why this number must be an integer, and use the formula above to calculate it exactly.

Applications of the Fundamental Theorem of Algebra

F.1. Zeros of the derivative of a polynomial

To prepare for our main topic, we first develop some machinery. If f is a real-valued function of a real variable, then we say that

$$\lim_{x\to a} f(x) = L$$

when for every $\varepsilon > 0$ there is a $\delta > 0$ such that $|f(x) - L| < \varepsilon$ whenever $0 < |x - a| < \delta$. The set of x for which $0 < |x - a| < \delta$ is an open interval $(a - \delta, a + \delta)$ with the point a removed. In symbols it is $(a - \delta, a + \delta) \setminus \{a\}$. We call this a *punctured neighborhood* of a. The function f may or may not be defined at a, but if it is, and if it happens that $f(a) = L$, then we say that f is *continuous* at a. Correspondingly, if f is a complex-valued function of a complex variable z, then we say that

$$\lim_{z\to a} f(z) = L$$

when for every $\varepsilon > 0$ there is a $\delta > 0$ such that $|f(z) - L| < \varepsilon$ whenever $0 < |z - a| < \delta$. In this situation, our punctured neighborhood of a is an open disc with the center deleted. The function f may or may not be defined at a, but if it is, and if $f(a) = L$, then we say that f is continuous at a. Here a and L are complex numbers (possibly real). Just as with real variables, in the complex situation the limit of a sum is the sum of the limits (provided that they exist), and the limit of a product is the product of the limits. In calculus we call

$$\frac{f(x) - f(a)}{x - a}$$

the *difference–quotient* of f. This is the slope of the chord from $(a, f(a))$ to $(x, f(x))$. The *derivative* of a function f is the limit of the difference–quotient as x tends to a,

$$f'(a) = \lim_{x\to a} \frac{f(x) - f(a)}{x - a}.$$

Here x can approach a from the left or the right, and it is essential that the limit is the same in both cases. For a complex-valued function f of a complex variable we set

$$f'(a) = \lim_{z\to a} \frac{f(z) - f(a)}{z - a}.$$

Suppose that $f(z) = z^n$ identically. From the factorization $z^n - a^n = (z-a)(z^{n-1} + z^{n-2}a + \cdots + za^{n-2} + a^{n-1})$ we see that

$$\frac{f(z) - f(a)}{z - a} = z^{n-1} + z^{n-2}a + \cdots + za^{n-2} + a^{n-1}.$$

Here there are n terms on the right hand side, each one tending to a^{n-1} as $z \to a$, and so we see that

$$\frac{d}{dz} z^n = nz^{n-1},$$

just as in the real-variable case. Thus in the complex domain, polynomials differentiate in the same way as in ordinary calculus.

If f is a real-valued differentiable function of the real variable x, and if $f(x) > 0$, then we can form $\log f(x)$. The derivative of this is $f'(x)/f(x)$, by the chain rule. This is called the *logarithmic derivative* of f. Since $\log f(x)g(x) = \log f(x) + \log g(x)$, it follows that that the logarithmic derivative of fg is the logarithmic derivative of f plus the logarithmic derivative of g. In the complex domain the logarithm is more complicated to define, but we can still form $f'(z)/f(z)$ provided that f is differentiable and that $f(z) \ne 0$, and we still call this the logarithmic derivative, even though there is no logarithm in sight. Also,

$$\frac{(f(z)g(z))'}{f(z)g(z)} = \frac{f'(z)g(z) + f(z)g'(z)}{f(z)g(z)} = \frac{f'(z)}{f(z)} + \frac{g'(z)}{g(z)}.$$

The logarithmic derivative of $z - a$ is $1/(z - a)$. Thus if f is a polynomial in the factored form

$$f(z) = a_n(z - z_1)\cdots(z - z_n),$$

then

$$\text{(F.1)} \qquad \frac{f'(z)}{f(z)} = \frac{1}{z - z_1} + \cdots + \frac{1}{z - z_n}.$$

The polynomial $f'(z)$ has degree $n - 1$, and hence has $n - 1$ zeros. We now use (F.1) to show that the zeros of f' can only lie in certain areas relative to the zeros z_j of f. Suppose that z is a complex number such that $\operatorname{Re} z > \operatorname{Re} z_j$ for all j. Then $\operatorname{Re}(z - z_j) > 0$ for all j. But the reciprocal of a number with positive real part also has positive real part. Thus $\operatorname{Re} 1/(z - z_j) > 0$ for all j. Since the sum of numbers with positive real part also has positive real part we see that the right hand side above has positive real part. Hence it is not 0. That is, $f'(z) \ne 0$. In this way we see that the half-plane $\operatorname{Re} z > \max \operatorname{Re} z_j$ is completely free of zeros of f'. But this works when we project the zeros z_j onto other lines, not just the real axis. Let w be a complex number with $|w| = 1$. Then the set $\{rw : -\infty < r < \infty\}$ is a line through the origin that passes through w. If $\operatorname{Re} z/w > \operatorname{Re} z_j/w$ then $\operatorname{Re}(z - z_j)/w > 0$. The reciprocal of this therefore also has positive real part: $\operatorname{Re} w/(z - z_j) > 0$. If this is true for all j, then we sum over j and by (3) we see that $\operatorname{Re} wf'(z)/f(z) > 0$. Hence $f'(z) \neq 0$. To better understand precisely which z are excluded by this argument, suppose that there is a line L_1 with the property that z lies on one side of the line and all the z_j lie on the other side of the line. We choose w so that the segment from 0 to w is perpendicular to L_1, and pointing toward the side that z lies on. Then $\operatorname{Re} z/w$ will be larger than any of the $\operatorname{Re} z_j/w$. Consider the line $\operatorname{Re} z = \max \operatorname{Re} z_i$. At least one of the z_j lies on this line. If there is more than one, choose the one with greatest imaginary part. Let's call that point z_1. Now rotate the line about z_1 in a counter-clockwise direction, until it hits another of the z_j. It may hit several z_j simultaneously; in that case let z_2 be the point farthest from z_1. Continue rotating until you are back at the the beginning. The z_j that are used as pivots are the vertices of a convex polygon, which we will call C. In general, if S is a set in the plane, then the *convex hull* of S is the smallest convex set that contains S as a subset. The polygon C that we constructed is the convex hull of the z_j. If z lies outside C, then we can separate z from C with a line L_1, and hence $f'(z) \neq 0$. That is, all zeros of $f'(z)$ must lie inside C.

Example F.1. Suppose that $f(z) = z^3+z^2+z+1 = (z-i)(z+1)(z+i)$. When we construct the polygon C as described above, we have $z_1 = i$, $z_2 = -1$, and $z_3 = -i$. The zeros of f' are $(-1 \pm i\sqrt{2})/3$, which lie inside the triangle C.

Example F.2. Suppose that $f(z) = z^3 - z = (z-1)z(z+1)$. When we construct C, we find that $z_1 = 1$, $z_2 = -1$, so C is a degenerate polygon that passes from 1 to -1 and back. The zeros of f' are at $\pm 1/\sqrt{3}$, so they lie within C.

Exercises

1. Let $f(z) = z^4 + 1$. Find the z_j, describe C, and find the zeros of f'.
2. Let $f(z) = a_n z^n + \cdots + a_0 = a_n(z-z_1)\cdots(z-z_n)$. The zeros of f' lie in the convex hull of the zeros of f, the zeros of f'' lie in the convex hull of the zeros of f', and so on. The linear polynomial $f^{(n-1)}(z)$ has one zero, which lies in all these convex hulls.
   (a) Show that $f^{(n-1)}(z) = a_n n! z + a_{n-1}(n-1)!$.
   (b) Show that $f^{(n-1)}(z) = 0$ precisely when $z = -a_{n-1}/(na_n)$.
   (c) Show that $z_1 + z_2 + \cdots + z_n = -a_{n-1}/a_n$.
   (d) Conclude that the root of $f^{(n-1)}(z) = 0$ is $(z_1 + z_2 + \cdots + z_n)/n$.
3. Let $f(z) = z^2 - z - 6$ as in Figure 1.4, and let $\mathcal{C}(r)$ denote the curve $\mathcal{C}(r) = \{f(r(\cos\theta + i\sin\theta)) : 0 \le \theta \le 2\pi\}$. Three examples of such curves are depicted in Figure 1.4.
   (a) When r is small, $\mathcal{C}(r)$ is a simple closed curve, approximately a circle centered at $f(0) = -6$. The curve grows with r, but at some point it ceases to be a simple closed curve. In Figure 1.4 (a) we see that this has happened before r reaches 1.5. At what r does the transition take place? How and why does this happen?
   (b) Describe, for a general polynomial of degree n, how such transitions take place.

4. Let f be a polynomial of degree > 1 with real or complex coefficients, and let A denote the average of its roots. Let A' denote the average of the roots of the derivative f'. Show that $A = A'$.

5. Suppose that f is a polynomial of degree n with n distinct real zeros $x_1 < x_2 < \cdots < x_n$. Show that f' has exactly one zero in each of the open intervals (x_j, x_{j+1}) for $j = 1, 2, \ldots, n-1$. (Hint: Recall Rolle's theorem.)

6. Let $f(z) = a_n z^n + \cdots + a_0 = a_n(z - z_1)\cdots(z - z_n)$ be a polynomial, let w be a non-zero complex number, and put $g(z) = f(wz) = b_n z^n + \cdots + b_0$.
   (a) Show that $b_k = a_k w^k$ for $k = 0, 1, \ldots, n$.
   (b) Show that the zeros of g are the numbers z_j/w.
   (c) Note that

$$f'(z) = na_n z^{n-1} + (n-1)a_{n-1}z^{n-2} + \cdots + a_1,$$
$$g'(z) = nb_n z^{n-1} + (n-1)b_{n-1}z^{n-2} + \cdots + b_1.$$

   Deduce that $g'(z) = wf'(wz)$. (This is consistent with the chain rule, but we do not use that rule here because we have not shown that it extends to the complex setting.)
   (d) Conclude that

$$\frac{g'(z)}{g(z)} = \sum_{j=1}^{n} \frac{1}{z - z_j/w} = w\sum_{j=1}^{n} \frac{1}{wz - z_j} = w\frac{f'(wz)}{f(wz)}\,.$$

7. In Example F.1, the zeros of f' fell in the interior of C, but in Example F.2 they were on the boundary. Show that if f is a polynomial whose zeros z_j do not all lie on the same line, then the zeros of f' are strictly inside C (with none on the boundary).

F.2. Linear differential equations with constant coefficients

We conclude this chapter with three brief sections that illustrate some of the uses we find for the Fundamental Theorem of Algebra. For brevity, some proofs are left as exercises. Suppose we wish to find functions f that satisfy a differential equation of the form

$$a_n f^{(n)}(x) + a_{n-1} f^{(n-1)}(x) + \cdots + a_0 f(x) = 0 \tag{F.2}$$

where $a_0, a_1, \ldots, a_n$ are constants and $a_n \neq 0$. We observe that this is linear and homogeneous, in the sense that if f_1 and f_2 are two functions with this property, then $\alpha f_1(x) + \beta f_2(x)$ is a further such function, for any α and β. We might try a function of the form $f(x) = e^{\lambda x}$ where λ is some real or complex number. By induction we see that the kth derivative of $e^{\lambda x}$ is $\lambda^k e^{\lambda x}$. Hence if $y = e^{\lambda x}$, then the left hand side of (F.2) is

$$a_n\lambda^n e^{\lambda x} + a_{n-1}\lambda^{n-1}e^{\lambda x} + \cdots + a_0 e^{\lambda x} = \big(a_n\lambda^n + a_{n-1}\lambda^{n-1} + \cdots + a_0\big)e^{\lambda x}\,.$$

This is 0 precisely when λ is a root of the polynomial

$$P(z) = a_n z^n + a_{n-1}z^{n-1} + \cdots + a_0\,.$$

In most cases there will be n distinct real or complex numbers that satisfy this equation, say $\lambda_1, \ldots, \lambda_n$. This gives n independent solutions of the form $e^{\lambda_j x}$, and on taking linear combinations we have the general solution,

$$f(x) = \alpha_1 e^{\lambda_1 x} + \cdots + \alpha_n e^{\lambda_n} x$$

where α_1, $\alpha_2, \ldots,$ α_n are arbitrary constant. With a little more work it can be shown that all solutions are of this form, and also that for any real number b and any given numbers $c_0, c_1, \ldots, c_{n-1}$ there is a unique choice of the α_j such that

$$f^{(k)}(b) = c_k \qquad (0 \leq k < n).$$

Suppose now that the roots λ_j are not all distinct, say in particular that λ_1 is a double root of the polynomial. Then $e^{\lambda_1 x}$ and $xe^{\lambda_1 x}$ are both solutions of the differential equation, and we can generate all solutions, as before. If λ_1 is a triple root of the polynomial, then $e^{\lambda_1 x}$, $xe^{\lambda_1 x}$, and $x^2 e^{\lambda_1 x}$ are roots of the differential equation, and so on for roots of higher multiplicity. We note that $e^{\lambda x} \to 0$ as $x \to +\infty$ if $\operatorname{Re} \lambda < 0$. Thus if all roots have negative real part, then every solution will tend to 0 as $x \to +\infty$. This would occur in a model of a physical situation with resistance or drag. The polynomial $P(z)$ may have complex roots even when the a_k are all real and we are expecting a real-valued solution. For example, if $-3 \pm 2i$ are roots, then we will have solutions of the form $ae^{(-3+2i)x} + be^{(-3-2i)x}$. If $b = \overline{a}$, then the combined contribution of these two terms is real, and will simplify to give a term of the sort $ce^{-3x} \sin(2x + \phi)$ where c and ϕ are real.

Exercises

1. (a) Find all solutions of the equation $y'' + y = 0$.
   (b) Find all real-valued solutions of the above.
2. Find all solutions of the equation $y'' - 2y' + y = 0$.

F.3. Partial fraction expansions

Let $f(z)/g(z)$ be a rational function. We can divide g into f to obtain a quotient polynomial $q(z)$ and a remainder $r(z)$, which is to say that

$$\frac{f(z)}{g(z)} = q(z) + \frac{r(z)}{g(z)}$$

where $\deg r < \deg g$. The rational function $r(z)/g(z)$ has a partial fraction expansion, which is a sum of terms of the form $c/(z-a)^m$ for various c, a, and m where c and a are complex numbers, $c \neq 0$, and m is a positive integer. We may suppose that $r(z)$ and $g(z)$ have no common roots, for $r(a) = g(a) = 0$, then we could divide them both by $z - a$, and we would still have the same rational function. Suppose that $g(a) = 0$, and that m is the exact power of $z - a$ in the factorization of g. Thus $(z-a)^m | g(z)$, but $(z-a)^{m+1} \nmid g(z)$. Write $g(z) = (z-a)^m h(z)$. Then $h(a) \neq 0$.

Take $c_1 = r(a)/h(a)$. Then

$$\frac{r(z)}{g(z)} - \frac{c_1}{(z-a)^m} = \frac{r(z) - c_1 h(z)}{(z-a)^m h(z)}.$$

From our choice of c_1 we see that the numerator on the right hand side vanishes when $z = a$. Thus it has a factor $z - a$, which can be divided from the numerator and denominator. Thus the right hand side is of the form $r_1(z)/\big((z-a)^{m-1}h(z)\big)$. We repeat this procedure until all factors of $z - a$ in the denominator have been removed:

$$\frac{r(z)}{g(z)} = \frac{c_1}{(z-a)^m} + \frac{c_2}{(z-a)^{m-1}} + \cdots + \frac{c_m}{z-a} + \frac{r_m(z)}{h(z)}.$$

At this point we start over, with $r(z)/g(z)$ replaced by $r_m(z)/h(z)$. Eventually we reach a representation of $r(z)/g(z)$ of the desired sort. Suppose we expand $g(z)$ in powers of $(z-a)^k$,

$$g(z) = \sum_{k=0}^{n} \frac{g^{(k)}(a)}{k!}(z-a)^k.$$

The assertion that $(z-a)^m$ exactly divides $g(z)$ is equivalent to $g^{(k)}(a) = 0$ for $0 \le k < m$, and $g^{(m)}(a) \ne 0$. We deduce that $h(a) = g^{(m)}(a)/m!$.

Example F.3. We show that

$$\frac{1}{z(z+1)(z+2)\cdots(z+n)} = \frac{1}{n!}\sum_{k=0}^{n}\binom{n}{k}\frac{(-1)^k}{z+k}.$$

We compute the number c_k so that $c_k/(z+k)$ is one of the terms of the expansion. When z is near $-k$, we see that

$$\frac{1}{z(z+1)\cdots(z+k-1)(z+k+1)\cdots(z+n)}$$

is approximately $1/\big((-1)^k k!(n-k)!\big)$. Hence we should take $c_k = (-1)^k\binom{n}{k}/n!$. When we subtract $c_k/(z+k)$ from the left hand side, the result is a rational function that has no pole at $-k$. Thus the difference between the two sides has no pole, and hence is a polynomial. But each side tends to 0 as $z \to \infty$, so the polynomial in question is identically zero, and the two sides are equal.

Exercises

1. Write $z/(z^2-1)$ in partial fractions.
2. Write $(1+2z+3z^2)/z^3$ in partial fractions.
3. (a) Determine the partial fraction expansion of

$$\frac{1}{(z^2+1)(z^2+4)}.$$

(b) Find constants a, b, c, d so that

$$\frac{1}{(z^2+1)(z^2+4)} = \frac{az+b}{z^2+1} + \frac{cz+d}{z^2+4}.$$

(c) Find

$$\int \frac{1}{(x^2+1)(x^2+4)}\,dx\,.$$

F.4. Linear recurrences

We say that a sequence u_k of real or complex numbers satisfies a linear recurrence of order n if

$$u_{k+n} = a_{n-1}u_{k+n-1} + \cdots + a_0 u_k \tag{F.3}$$

for $k = 0, 1, 2, \ldots$. Here $a_0, a_1, \ldots, a_{n-1}$ are constants with $a_0 \neq 0$. (If a_0 were to vanish, then the above would be a linear recurrence, but its order would be $< n$.) Once values of $u_0, \ldots, u_{n-1}$ have been selected, the value of u_k for $k \geq n$ is uniquely determined by the above recurrence. The theory that unfolds here is very similar to that of linear differential equations, as in §F.2.

Let

$$P(z) = z^k - a_{k-1}z^{k-1} - \ldots - a_1 z - a_0. \tag{F.4}$$

Suppose that λ is a root of $P(z)$. That is,

$$\lambda^n = a_{n-1}\lambda^{n-1} + a_{n-2}\lambda^{n-2} + \cdots + a_1\lambda + a_0\,.$$

On multiplying through by λ^k, we have

$$\lambda^{k+n} = a_{k-1}\lambda^{k+n-1} + \cdots + a_0\lambda^k\,.$$

Thus the sequence $u_k = \lambda^k$ satisfies the linear recurrence (F.3). If $\lambda_1, \ldots, \lambda_n$ are the roots of $P(z)$, then every sequence u_k satisfying (14) can be written in the form

$$u_k = \sum_{j=1}^{n} \alpha_j \lambda_j^k \tag{F.5}$$

for suitable α_j. Here we are assuming that the roots are distinct. With a little work it can be shown that for any given collection of numbers $c_0, c_1, \ldots, c_{n-1}$ there will be a unique choice of the α_j so that $u_k = c_k$ for $0 \leq k < n$. We note that the product of the λ_j is $\pm a_0 \neq 0$. Hence the λ_j are all non-zero.

Now suppose that the zeros of $P(z)$ are not all distinct. If λ is a zero of $P(z)$ of multiplicity m, then the m sequences λ^k, $k\lambda^k$, $\ldots$, $k^{m-1}\lambda^k$ satisfy (14), and these sequences take the place of m sequences arising from m distinct λ's. On taking linear combinations of these basic solutions, we obtain the general solution. Thus if the factored form of $P(z)$ is

$$P(z) = (z-\lambda_1)^{m_1}(z-\lambda_2)^{m_2}\cdots(z-\lambda_r)^{m_r} \tag{F.6}$$

where $m_1 + m_2 + \cdots + m_r = n$, then the general solution of the recurrence is

$$u_k = \sum_{j=1}^{r} p_j(k)\lambda_j^k \tag{F.7}$$

where $p_j(u)$ is a polynomial of degree not exceeding $m_j - 1$. For given initial values $u_0 = c_0, \ldots u_{n-1} = c_{n-1}$ there is a unique choice of the coefficients of the polynomials p_j so that the initial conditions are satisfied. Since p_j has potentially m_j coefficients, and $m_1 + \cdots + m_j = n$, we have n coefficients to choose so that the n initial conditions are satisfied.

If we define a sequence u_k by the formula (F.7), then it will satisfy the linear recurrence (F.3). Thus a sequence satisfies a linear recurrence if and only if it can be written in the form (F.7).

Example F.4. The *Fibonacci numbers* F_k are integers defined by the relations $F_0 = 0$, $F_1 = 1$, and $F_{k+2} = F_{k+1} + F_k$ for $k \geq 0$. This is a linear recurrence of order 2, and the associated polynomial is $P(z) = z^2 - z - 1$, whose zeros are $(1 \pm \sqrt{5})/2$. Thus

$$F_k = \alpha_1 \Big(\frac{1+\sqrt{5}}{2}\Big)^k + \alpha_2 \Big(\frac{1-\sqrt{5}}{2}\Big)^k$$

for some with choice of α_1 and α_2. These constants must be chosen so that the initial conditions are met. That is, we want

$$\alpha_1 + \alpha_2 = 0,$$
$$\alpha_1 \frac{1+\sqrt{5}}{2} + \alpha_2 \frac{1-\sqrt{5}}{2} = 1\,.$$

With a little calculation we find that we should take $\alpha_1 = 1/\sqrt{5}$ and $\alpha_2 = -1/\sqrt{5}$. Thus

$$F_k = \frac{1}{\sqrt{5}}\Big(\frac{1+\sqrt{5}}{2}\Big)^k - \frac{1}{\sqrt{5}}\Big(\frac{1-\sqrt{5}}{2}\Big)^k. \tag{F.8}$$

We note that $(1+\sqrt{5})/2 = 1.61803\ldots$ and that $(1-\sqrt{5})/2 = -0.61803\ldots$. Thus the kth power of this second root is exponentially small, and its sign depends on whether k is odd or even. The kth power of the first root is exponentially large, and the first term is exponentially close F_k, which is an integer. Thus if one uses this formula to calculate F_k, it would suffice to compute the first term, and then round to the nearest integer. For example,

$$\frac{1}{\sqrt{5}}\Big(\frac{1+\sqrt{5}}{2}\Big)^{100} = 354224848179261915074.999999876\ldots,$$

so $F_{100} = 354224848179261915075$.

Suppose that u_k satisfies the linear recurrence (F.3), and put

$$F(z) = \sum_{k=0}^{\infty} u_k z^k\,. \tag{F.9}$$

From (F.7) we know that $|u_k| < r^{(1+\varepsilon)k}$ for all large k, where $r = \max_j |\lambda_j|$. Thus the power series (F.9) has radius of convergence at least $1/r$. Write

$$F(z) = u_0 + u_1 z + \cdots + u_{n-1} z^{n-1} + \sum_{k=n}^{\infty} u_k z^k = S(z) + \sum_{k=0}^{\infty} u_{n+k} z^{n+k}\,.$$

If u_k satisfies (F.3), then the sum on the right is

$$F(z) - S(z) = \sum_{k=0}^{\infty}\Big(\sum_{j=0}^{n-1} a_j u_{j+k}\Big) z^{n+k}$$
$$= \sum_{j=0}^{n-1} a_j z^{n-j} \sum_{k=0}^{\infty} u_{j+k} z^{j+k}$$
$$= \sum_{j=0}^{n-1} a_j z^{n-j} \sum_{k=j}^{\infty} u_k z^k$$
$$= \sum_{j=0}^{n-1} a_j z^{n-j} \Big(F(z) - \sum_{k=0}^{j-1} u_j z^j\Big). \tag{F.10}$$

Let

$$Q(z) = 1 - a_{n-1}z - a_{n-2}z^2 - \cdots - a_0 z^n. \tag{F.11}$$

On rearranging (F.10), we find that

$$Q(z)F(z) = S(z) - \sum_{j=0}^{n-1} a_j z^{n-j} \sum_{k=0}^{j-1} u_j z^j = \sum_{h=0}^{n-1}\Big(u_h - \sum_{\ell=0}^{h-1} a_{\ell+n-h} u_\ell\Big) z^h = R(z),$$

say. Here $R(z)$ is a polynomial of degree at most $n-1$. Thus $F(z)$ is a rational function,

$$F(z) = \frac{R(z)}{Q(z)}. \tag{F.12}$$

The coefficients of Q are the same as those of P except that they are in reverse order. That is, $Q(z) = z^n P(1/z)$. Hence

$$Q(z) = (1-\lambda_1 z)^{m_1}(1-\lambda_2 z)^{m_2}\cdots(1-\lambda_r z)^{m_r}. \tag{F.13}$$

Thus F has poles at points of the form $1/\lambda_j$.

Suppose we define $F(z)$ as in (F.9), but instead of using (F.3) we invoke the representation (F.7). Then we find that

$$F(z) = \sum_{j=1}^{r}\sum_{k=0}^{\infty} p_j(k)(\lambda_j z)^k.$$

With the negative binomial theorem in mind (recall (B.17)), we express the polynomial $p_j(u)$ as a linear combination of the basic polynomials $\binom{u+h}{h}$ for $h = 0, 1, \ldots, m_j - 1$, say

$$p_j(u) = \sum_{h=0}^{m_j-1} c_{h\,j}\binom{u+h}{h}.$$

Then

$$\sum_{k=0}^{\infty} p_j(k)(\lambda_j z)^k = \sum_{h=0}^{m_j-1} c_{h\,j}\sum_{k=0}^{\infty}\binom{k+h}{h}(\lambda_j z)^k = \sum_{h=0}^{m_j-1}\frac{c_{h\,j}}{(1-\lambda_j z)^{h+1}} \tag{F.14}$$

by (B.17). Here we again see that $F(z)$ is a rational function, this time expressed in its partial fraction expansion. Conversely, suppose we start with a rational function $F(z)$, say of the form (F.12) with Q of degree n and R of degree at most $n-1$, and assume further that $Q(0) \neq 0$. By dividing both the numerator and the denominator of (F.12) by $Q(0)$ we may assume that $Q(0) = 1$. Then Q can be expressed in the form (F.13) for suitable λ_j and m_j. The partial fraction expansion of $F(z)$ will then be of the form (F.14), and by reversing the steps of our reasoning above, we find that the coefficients u_k of the power series expansion of $F(z)$ are given by a formula of the shape (F.7). That is, they satisfy a linear recurrence.

In summary, we have shown that the following three assertions concerning a sequence u_k are equivalent:

(1) The sequence u_k satisfies a linear recurrence of order n, such as (F.3).

(2) The sequence u_k is given by a formula of the shape (F.7).

(3) The power series (F.9) is a rational function, as in (F.12), with Q of degree n, P of degree $< n$, and $Q(0) \neq 0$.

Example F.4, continued. Let $\omega = (1+\sqrt{5})/2$ denote the golden ratio. Thus (F.8) asserts that $F_k = \omega^k/\sqrt{5} - (-1/\omega)^k/\sqrt{5}$. Hence

$$\begin{aligned}\sum_{k=0}^{\infty} F_k z^k &= \frac{1}{\sqrt{5}}\sum_{k=0}^{\infty}(\omega z)^k - \frac{1}{\sqrt{5}}\sum_{k=0}^{\infty}(-z/\omega)^k = \frac{1}{\sqrt{5}(1-\omega z)} - \frac{1}{\sqrt{5}(1+z/\omega)}\\ &= \frac{z}{1-z-z^2}\end{aligned}$$

for $|z| < 1/\omega$. In particular,

$$\sum_{k=0}^{\infty} \frac{F_k}{2^k} = 2\,.$$

Exercises

1. The *Lucas numbers* L_k are defined by the initial conditions $L_0 = 2$, $L_1 = 1$ and the recurrence $L_{k+2} = L_{k+1} + L_k$.
   (a) Find a formula for L_k similar to the one for F_k.
   (b) Express $\sum_{k=0}^{\infty} L_k z^k$ as a rational function.
2. Let u_k be a sequence that satisfies a linear recurrence of order m, and v_k be a sequence that satisfies a linear recurrence of order n.
   (a) Show that $u_k + v_k$ satisfies a linear recurrence of order at most $m+n$.
   (b) Show that $u_k v_k$ satisfies a linear recurrence of order at most mn.

Inequalities

I.1. The Arithmetic–Geometric Mean Inequality

In its simplest form, the inequality in question asserts that if a and b are nonnegative real numbers, then

$$\sqrt{ab} \le \frac{a+b}{2}. \tag{I.1}$$

Here the left hand side is the geometric mean of a and b, and the right hand side is the geometric mean of a and b. This inequality is easy to prove: By squaring both sides we see that the above is equivalent to the inequality

$$ab \le \left(\frac{a+b}{2}\right)^2 = \frac{(a+b)^2}{4}.$$

On multiplying by 4, we see that this is equivalent to the assertion that

$$4ab \le (a+b)^2 = a^2 + 2ab + b^2.$$

On subtracting $4ab$ from both sides we see that this is equivalent to the inequality

$$0 \le a^2 - 2ab + b^2 = (a-b)^2.$$

Since the square of a real number is always nonnegative, this last inequality is obviously true. Moreover, we see that it holds with equality if and only if $a = b$. Since all our steps are reversible, we deduce that (I.1) holds for all nonnegative real numbers a and b, and that equality holds precisely when $a = b$.

The inequality (I.1) can also be seen geometrically, since we can place four $a \times b$ rectangles in a square of side length $a + b$. Hence the sum of the areas of the rectangles, $4ab$, does not exceed the area of the square, $(a+b)^2$. In this packing, the wasted area, $(a-b)^2$, arose also in our algebraic proof. In three dimensions we may similarly ask whether it is possible to pack twenty-seven $a \times b \times c$ blocks into a cube of side length $a + b + c$. This forms the basis of a moderately challenging puzzle. (Suggestion: Make 27 blocks of $4 \times 5 \times 6$ units to place in a $15 \times 15 \times 15$ cube.)

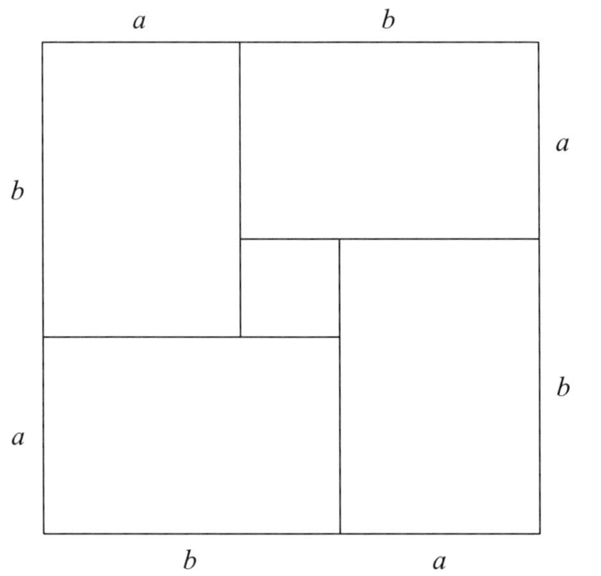

Figure I.1. Visual AGM.

The inequality (I.1) is case $n = 2$ of the Arithmetic–Geometric Mean Inequality (abbreviated AGM), which asserts that if $a_1, a_2, \ldots, a_n$ are nonnegative real numbers, then

$$(a_1 a_2 \cdots a_n)^{1/n} \le \frac{a_1 + a_2 + \cdots + a_n}{n}. \tag{I.2}$$

Here the left hand side is the *geometric mean* of the a_j, while the right hand side is their *arithmetic mean.*

The case $n = 4$ of (I.2) can be derived by three applications of (I.1):

$$(a_1 a_2 a_3 a_4)^{1/4} = \left((a_1 a_2)^{1/2}(a_3 a_4)^{1/2}\right)^{1/2} \le \frac{(a_1 a_2)^{1/2} + (a_3 a_4)^{1/2}}{2},$$

and by two further applications of (I.1) we see that the above is

$$\le \frac{\dfrac{a_1 + a_2}{2} + \dfrac{a_3 + a_4}{2}}{2} = \frac{a_1 + a_2 + a_3 + a_4}{4}.$$

By repeating this operation, we can establish (I.2) when $n = 2^k$, but for general n we adopt a different approach—based on convexity.

Lemma I.1. *For all real numbers x,*

$$e^x \ge 1 + x, \tag{I.3}$$

with equality if and only if $x = 0$.

Proof. Let $f(x) = e^x$. Since $f(0) = f'(0) = 1$, the linear approximation $L(x)$ to $f(x)$ at $x = 0$ is $L(x) = 1 + x$. But $f''(x) = e^x > 0$ for all x, and so $f(x)$ is strictly concave upwards, which is to say that its graph lies above its tangent lines, with contact only at the point of tangency. That is, $e^x > 1 + x$ except when $x = 0$. □

Theorem I.2. (The Arithmetic–Geometric Inequality) *If* $a_1, a_2, \ldots, a_n$ *are nonnegative real numbers, then*

$$(a_1 a_2 \cdots a_n)^{1/n} \le \frac{a_1 + a_2 + \cdots + a_n}{n}$$

with equality only when all the a_i *are equal.*

Proof. Let G denote the left hand side, and A denote the right hand side. For each j put $x_j = \log(a_j/G)$. Then

$$x_1 + x_2 + \cdots + x_n = \log\frac{a_1}{G} + \log\frac{a_2}{G} + \cdots + \log\frac{a_n}{G} = \log\frac{a_1 a_2 \cdots a_n}{G^n} = \log 1 = 0.$$

On the other hand, since $a_j = Ge^{x_j}$, it follows that

$$A = \frac{a_1 + a_2 + \cdots + a_n}{n} = \frac{Ge^{x_1} + Ge^{x_2} + \cdots + Ge^{x_n}}{n} = G\frac{e^{x_1} + e^{x_2} + \cdots + e^{x_n}}{n}.$$

By Lemma I.1 we see that this last expression is

$$\ge G\frac{(1 + x_1) + (1 + x_2) + \cdots + (1 + x_n)}{n} = G\Big(1 + \frac{x_1 + x_2 + \cdots + x_n}{n}\Big) = G.$$

This gives the inequality, and we see that equality occurs only when all the x_j are 0, which is to say that all the a_j are equal. □

Exercises

1. Suppose that $a_1, a_2, \ldots, a_n$ are the edge lengths of a box in n-dimensional space $\mathbb{R}^n$. We say that the box has volume (more properly n-content) $a_1 a_2 \cdots a_n$. Suppose that you wish to form a cube (hypercube) in n dimensions whose volume is the same as that of the box. What is the edge length of the cube?
2. (a) Show that $\int ue^u \, du = xe^x - e^x + C$.
   (b) Show that
   $$e^x = 1 + x + \int_0^x (x - u)e^u \, du.$$
   (c) Show that the integral above is positive when $x > 0$.
   (d) Show that the integral above is positive when $x < 0$. (Thus we have a second proof of Lemma I.1.)
3. For a vector $\boldsymbol{a} \in \mathbb{R}^n$, let $A(\boldsymbol{a})$ denote the arithmetic mean of its coordinates. If the coordinates a_i are all nonnegative, let $G(\boldsymbol{a})$ denote the geometric mean of the coordinates.
   (a) Show that if $a_1 = a_2 = \cdots = a_n = a$, then $A(\boldsymbol{a}) = a$.
   (b) Show that if $a_1 = a_2 = \cdots = a_n = a > 0$, then $G(\boldsymbol{a}) = a$.
   (c) Show that for any real number c, $A(c\boldsymbol{a}) = cA(\boldsymbol{a})$.
   (d) Show that for any nonnegative real number c, $G(c\boldsymbol{a}) = cG(\boldsymbol{A})$.
   (e) Show that if $a_i \le b_i$ for all i, then $A(\boldsymbol{a}) \le A(\boldsymbol{b})$.
   (f) Deduce that for any $\boldsymbol{a}$, $\min_i a_i \le A(\boldsymbol{a}) \le \max_i a_i$.
   (g) Show that if $0 \le a_i \le b_i$ for all i, then $G(\boldsymbol{a}) \le G(\boldsymbol{b})$.

(h) Deduce that $\min_i a_i \le G(\boldsymbol{a}) \le \max_i a_i$.

4. Let $a_1, a_2, \ldots, a_n$ be the integers $1, 2, \ldots, n$ in any order.
   (a) Use the AGM to show that
   $$\frac{a_1}{1} + \frac{a_2}{2} + \cdots + \frac{a_n}{n} \ge n.$$
   (b) When does equality hold?

5. (a) Suppose that $a_1 \ge a_2$ and that $b_1 \ge b_2$. Show that $a_1b_1 + a_2b_2 \ge a_1b_2 + a_2b_1$.
   (b) Let S_n denote the set of all $n!$ permutations of the numbers $1, 2, \ldots, n$. (This set is called the *symmetric group.*) For $\pi \in S_n$, let $s(\pi) = a_1b_{\pi(1)} + a_2b_{\pi(2)} + \cdots + a_nb_{\pi(n)}$. Suppose that the a_j and b_j are real numbers such that $a_1 \ge a_2 \ge \cdots \ge a_n$ and $b_1 \ge b_2 \ge \cdots \ge b_n$. Show that
   $$\max_{\pi \in S_n} s(\pi) = a_1b_1 + a_2b_2 + \cdots + a_nb_n.$$
   (c) Show that
   $$\min_{\pi \in S_n} s(\pi) = a_1b_n + a_2b_{n-1} + \cdots + a_nb_1.$$
   (d) Use the above to give a second solution (one that does not use the AGM) of Exercise 4.

6. Suppose that $a_1, a_2, \ldots, a_n$ and $b_1, b_2, \ldots, b_n$ are nonnegative real numbers. Show that
   $$\big((a_1+b_1)(a_2+b_2)\cdots(a_n+b_n)\big)^{1/n} \ge (a_1a_2\cdots a_n)^{1/n} + (b_1b_2\cdots b_n)^{1/n}.$$

7. Let p be a positive real number. Show that if $a_1, a_2, \ldots, a_n$ are nonnegative real numbers, then
   $$(a_1a_2\cdots a_n)^{1/n} \le \left(\frac{a_1^p + a_2^p + \cdots a_n^p}{n}\right)^{1/p}$$
   with equality only when all the a_i's are equal. (Here the right hand side is called the pth *power mean* of the a_i.)

8. Suppose that $a_1, a_2, \ldots, a_n$ are positive real numbers, and let G denote their geometric mean. Let H denote their *harmonic mean*,
   $$H = \frac{n}{\dfrac{1}{a_1} + \dfrac{1}{a_2} + \cdots + \dfrac{1}{a_n}}.$$
   (This is the pth power mean with $p = -1$.) Show that $G \ge H$, with equality only when all the a_i are equal.

9. Let p be a negative real number. Show that if $a_1, a_2, \ldots, a_n$ are positive real numbers, then
   $$(a_1a_2\cdots a_n)^{1/n} \ge \left(\frac{a_1^p + a_2^p + \cdots a_n^p}{n}\right)^{1/p}.$$

10. Let $a_1 = 1$, $a_2 = 2$, $a_3 = 3$.
    (a) Compute the arithmetic, geometric, and harmonic means of these three numbers, and verify that $H < G < A$ in this case.
    (b) Compute the pth power mean of these numbers when $p = 100$. What number is this near?

(c) Compute the pth power mean of these numbers when $p = -100$. What number is this near?

11. For $x > -1$ let $f(x) = \log(1+x)$.
   (a) Find $f(0)$ and $f'(0)$, and thus determine the linear approximation $L(x)$ to $f(x)$ at $x = 0$.
   (b) Find a formula for $f''(x)$, and use it to prove that $f(x)$ is strictly concave downwards.
   (c) Combine your results to show that

$$\log(1+x) \le x \tag{I.4}$$

   whenever $x > -1$, with equality only when $x = 0$. (This inequality can also be derived by simply taking logarithms of both sides of the inequality (I.3) in Lemma I.1.)

12. Suppose that $a_1, a_2, \ldots, a_n$ are positive real numbers, and let A and G denote their arithmetic and geometric means, respectively. Put $\delta_i = a_i/A - 1$.
   (a) Show that $\delta_1 + \delta_2 + \cdots \delta_n = 0$.
   (b) Use Lemma I.1 to show that

$$G^n = A^n(1+\delta_1)(1+\delta_2)\cdots(1+\delta_n) \le A^n e^{\delta_1} e^{\delta_2} \cdots e^{\delta_n}\,.$$

   (c) Combine the above results to show that $G \le A$, with equality only when all the a_i are equal.

13. Let $w_1, w_2, \ldots, w_n$ be non-negative real numbers such that $w_1+w_2+\cdots+w_n = 1$. We refer to these numbers as *weights*, and define the corresponding *weighted arithmetic mean* A_w of numbers $a_1, a_2, \ldots, a_n$ to be

$$A_w = w_1a_1 + w_2a_2 + \cdots + w_na_n.$$

(The ordinary unweighted average A is the special case $w_1 = w_2 = \cdots = w_n = 1/n$.) Similarly, form the *weighted geometric mean*,

$$G_w = a_1^{w_1} a_2^{w_2} \cdots a_n^{w_n}$$

provided that the a_i are nonnegative. Show that the proof of Theorem I.2 generalizes to this situation, and thus prove that

$$G_w \le A_w.$$

14. Let $w_1, w_2, \ldots, w_n$ be a system of weights, as in the preceding exercise.
   (a) Show that if $a_1 = a_2 = \cdots = a_n = a$, then $A_w(\boldsymbol{a}) = a$.
   (b) Show that if $a_1 = a_2 = \cdots = a_n = a > 0$, then $G_w(\boldsymbol{a}) = a$.
   (c) Show that for any real number c, $A_w(c\boldsymbol{a}) = cA_w(\boldsymbol{a})$.
   (d) Show that for any nonnegative real number c, $G_w(c\boldsymbol{a}) = cG_w(\boldsymbol{A})$.
   (e) Show that if $a_j \le b_j$ for all j, then $A_w(\boldsymbol{a}) \le A_w(\boldsymbol{b})$.
   (f) Deduce that for any a, $\min_j a_j \le A_w(\boldsymbol{a}) \le \max_j a_j$.
   (g) Show that if $0 \le a_j \le b_j$ for all j, then $G_w(\boldsymbol{a}) \le G_w(\boldsymbol{b})$.
   (h) Deduce that $\min_j a_j \le G_w(\boldsymbol{a}) \le \max_j a_j$.

15. Suppose that $f(x)$ is continuous on the interval $[a, b]$, and that $f(x) > 0$ when $a \le x \le b$. Show that

$$\exp\Big(\frac{1}{n}\sum_{i=1}^{n} \log f(x_i)\Big) \le \frac{1}{n}\sum_{i=1}^{n} f(x_i),$$

and hence deduce that

$$\exp\left(\frac{1}{b-a}\int_a^b \log f(x)\,dx\right) \le \frac{1}{b-a}\int_a^b f(x)\,dx.$$

16. Show that if $x > 0$, then $x + 1/x \ge 2$ with equality only when $x = 1$.
17. Let A and P be the area and perimeter of a rectangle.
    (a) Show that $A \le P^2/16$ with equality only when the rectangle is a square.
    (b) Is there any inequality in the opposite direction?
18. A rectangular box has volume V, surface area S, and sum of edge lengths E.
    (a) Show that $S \ge 6V^{2/3}$ with equality only when the box is a cube. Is there an inequality in the opposite direction?
    (b) Show that $E \ge 12V^{1/3}$, with equality only when the box is a cube. Is there an inequality in the opposite direction?
19. Let a, b, c denote the side-lengths of a triangle. Heron's formula asserts that the area of the triangle is given by the formula

$$A = \sqrt{s(s-a)(s-b)(s-c)}$$

where s is the "semiperimeter", $s = (a+b+c)/2$.
    (a) Show that $\big((s-a)(s-b)(s-c)\big)^{1/3} \le (a+b+c)/6$.
    (b) Deduce that $\big((s-a)(s-b)(s-c)\big)^{1/2} \le \big((a+b+c)/6\big)^{3/2}$.
    (c) Show that if a triangle has area A and perimeter P, then

$$A \le \frac{P^2}{12\sqrt{3}}$$

with equality only when the triangle is equilateral.
20. Let P and Q be two points on the unit circle $x^2 + y^2 = 1$, and let θ denote the measure of the angle $\angle POQ$.
    (a) Show that the triangle $\triangle POQ$ has area $\frac{1}{2}\sin\theta$. Hence an n-gon with vertices on the unit circle has area

$$A = \frac{1}{2}\sum_{i=1}^{n} \sin\theta_i$$

where the θ_i are the angles between the successive vertices. (We are assuming, for simplicity, that the origin lies in the interior of the n-gon.)
    (b) Find the linear approximation $L(x)$ to $\sin x$ at the point a.
    (c) Show that if $0 \le a \le \pi$ and $0 \le x \le \pi$, then $\sin x \le L(x)$.
    (d) Suppose that $\theta_1, \theta_2, \ldots, \theta_n$ lie in the interval $[0, \pi]$, and let a denote their average. Put $x_i = \theta_i - a$. Show that

$$\sum_{i=1}^{n} \sin\theta_i \le \sum_{i=1}^{n} L(a + x_i).$$

    (e) Show that $\sum_{i=1}^{n} x_i = 0$.
    (f) Deduce that

$$\frac{1}{n}\sum_{i=1}^{n} \sin\theta_i \le \sin\left(\frac{1}{n}\sum_{i=1}^{n}\theta_i\right).$$

(g) Conclude that if an n-gon is inscribed in a unit circle, then its area A satisfies

$$A \le \frac{n}{2}\sin\frac{2\pi}{n}$$

with equality only when the n-gon is regular.

(h) What if the origin does not lie within the n-gon?

21. Suppose that $a_0 \ge b_0 > 0$, and suppose that two sequences a_n and b_n are generated by the recurrences

$$a_{n+1} = \frac{a_n + b_n}{2}, \qquad b_{n+1} = \sqrt{a_n b_n}.$$

(a) Supposing that $a_n \ge b_n > 0$, show that $b_n \le b_{n+1} \le a_{n+1} \le a_n$.

(b) Show that

$$0 \le a_{n+1} - b_{n+1} = \frac{(a_n - b_n)^2}{2(\sqrt{a_n} + \sqrt{b_n})^2}.$$

(c) Deduce that the decreasing sequence a_n and the increasing sequence b_n have a common limit; call it $M(a_0, b_0)$.

(d) How many iterations are required before the sequences are apparently constant, when you calculate $M(\sqrt{2}, 1)$?

(e) Using your numerical value calculated above, compare $1/M(\sqrt{2}, 1)$ with the quantity

$$\frac{2}{\pi}\int_0^1 \frac{dt}{\sqrt{1-t^4}} = 0.83462684167407318628\ldots$$

I.2. Hölder's Inequality

We proved Cauchy's inequality first in §.0.1, and will prove it again in §L.1 in Appendix L. We now establish an important generalization of Cauchy's inequality. We begin with a useful lemma.

Lemma I.3. (Young's Inequality) *Let a and b be nonnegative real numbers, and let p and q be real numbers such that $1 < p < \infty$, $1 < q < \infty$, and $1/p + 1/q = 1$. Then*

$$ab \le \frac{a^p}{p} + \frac{b^q}{q}. \tag{I.5}$$

Numbers p and q as above are known as *conjugate Hölder exponents.* When $p = q = 2$, the above is the simplest case of the Arithmetic Geometric Mean inequality.

Proof. If $a = 0$ or $b = 0$, then the inequality is trivial, so we assume that $a > 0$ and $b > 0$. We make the change of variable $A = a^p$, $B = b^q$. Thus the desired inequality is equivalent to the inequality

$$A^{1/p}B^{1/q} \le \frac{A}{p} + \frac{B}{q}. \tag{I.6}$$

We make a further change of variable by setting $x = 1/p$. Thus the above is equivalent to

$$A^x B^{1-x} \le Ax + B(1-x)\,. \tag{I.7}$$

Let $f(x)$ denote the function on the left and $g(x)$ the function on the right. We note that $f(0) = g(0) = B$ and that $f(1) = g(1) = A$. Now $f(x) = B(A/B)^x$, which is a positive multiple of an exponential function. Hence it is convex. Indeed, $f''(x) = B(A/B)^x(\log A/B)^2 \ge 0$. The graph of g is the chord joining $(0, B)$ to $(1, A)$. Since f is convex, it follows that $f(x) \le g(x)$ for $0 \le x \le 1$. Thus we have (I.7), hence (I.6), and hence (I.5), so the proof is complete. □

From the proof just completed we see that equality holds if and only if $A = B$, which is to say that $a^p = b^q$.

Theorem I.4. (Hölder's Inequality) *Let p and q be real numbers, $1 < p < \infty$, $1 < q < \infty$, such that*

$$\frac{1}{p} + \frac{1}{q} = 1.$$

Then for any nonnegative real numbers $a_1, a_2, \ldots, a_n$ and $b_1, b_2, \ldots, b_n$ we have

$$\sum_{j=1}^{n} a_j b_j \le \Big(\sum_{j=1}^{n} a_j^p\Big)^{1/p} \Big(\sum_{j=1}^{n} b_j^q\Big)^{1/q} \tag{I.8}$$

with equality only if the a_i^p are proportional to the b_i^q.

Proof. Let $\|\boldsymbol{a}\|_p$ denote the first factor on the right hand side and $\|\boldsymbol{b}\|_q$ the second factor. If $\|\boldsymbol{a}\|_p = 0$, then $a_j = 0$ for all j, and then the inequality is trivial. Similarly the inequality is trivial if $\|\boldsymbol{b}\|_q = 0$. Thus we may assume that $\|\boldsymbol{a}\|_p > 0$ and $\|\boldsymbol{b}\|_q > 0$. Put $A_j = a_j/\|\boldsymbol{a}\|_p$ and $B_j = b_j/\|\boldsymbol{b}\|_q$. On dividing both sides of (I.8) by the right hand side, we see that Hölder's inequality is equivalent to the assertion that

$$\sum_{j=1}^{n} A_j B_j \le 1\,.$$

To prove this, we apply Young's inequality to each term on the left. Thus the left hand side above is

$$\begin{aligned} &\le \sum_{j=1}^{n} \Big(\frac{A_j^p}{p} + \frac{B_j^q}{q}\Big) = \frac{1}{p}\sum_{j=1}^{n} \frac{a_j^p}{\|\boldsymbol{a}\|_p^p} + \frac{1}{q}\sum_{j=1}^{n} \frac{b_j^q}{\|\boldsymbol{b}\|_q^q} \\ &= \frac{1}{p\|\boldsymbol{a}\|_p^p}\sum_{j=1}^{n} a_j^p + \frac{1}{q\|\boldsymbol{b}\|_q^q}\sum_{j=1}^{n} b_j^q\,. \end{aligned}$$

Here the first sum is $\|\boldsymbol{a}\|_p^p$ and the second sum is $\|\boldsymbol{b}\|_q^q$, so the above is

$$= \frac{1}{p} + \frac{1}{q} = 1\,.$$

Thus the proof is complete. □

Equality holds in Hölder's inequality when the numbers a_j^p are proportional to the b_j^q.

Suppose that $a_1, a_2, \ldots, a_n$ are positive real numbers, and for $p \neq 0$ let $M_p(\boldsymbol{a})$ denote their p^{th} power mean,

$$M_p(\boldsymbol{a}) = \Big(\frac{a_1^p + a_2^p + \cdots + a_n^p}{n}\Big)^{1/p}. \tag{I.9}$$

Hölder's inequality can be used to describe the behavior of these means as p varies.

Theorem I.5. *If $0 < p_1 < p_2$, then $M_{p_1}(\boldsymbol{a}) \le M_{p_2}(\boldsymbol{a})$, with equality if and only if all the a_j are equal.*

Proof. By Hölder's inequality with $p = p_2/p_1$ we see that

$$\sum_{j=1}^n a_j^{p_1} \le \Big(\sum_{j=1}^n a_j^{p_2}\Big)^{p_1/p_2} \Big(\sum_{j=1}^n 1^q\Big)^{(p_2-p_1)/p_2}$$

with equality if and only if the numbers $a_j^{p_2}$ are proportional to the numbers 1^q, i.e., if the a_j are all equal. Since the second sum on the right hand side above is n, on dividing both sides by n we see that

$$\frac{1}{n}\sum_{j=1}^n a_j^{p_1} \le \Big(\frac{1}{n}\sum_{j=1}^n a_j^{p_2}\Big)^{p_1/p_2}.$$

We raise both sides to the power $1/p_1$ to obtain the desired result. □

By applying Theorem I.5 to the numbers $1/a_i$ and then taking reciprocals we find that the inequality $M_{p_1}(\boldsymbol{a}) \le M_{p_2}(\boldsymbol{a})$ also holds when $p_1 < p_2 < 0$. By the arithmetic mean inequality we know that $M_p(\boldsymbol{a}) \ge G(\boldsymbol{a})$ when $p > 0$ and that $M_p(\boldsymbol{a}) \le G(\boldsymbol{a})$ when $p < 0$. To complete the discussion of $M_p(\boldsymbol{a})$ when p is near 0, we prove

Theorem I.6. $\lim_{p\to 0} M_p(\boldsymbol{a}) = G(\boldsymbol{a})$.

It is now natural to define $M_p(\boldsymbol{a})$ also when $p = 0$; we set $M_0(\boldsymbol{a}) = G(\boldsymbol{a})$.

Proof. We know that $e^x \ge 1 + x$ for all real x. In the opposite direction we note that $e^x \le 1 + x + x^2$ for $x \le 1$. (To prove this put $f(x) = e^x - 1 - x - x^2$. Then $f(0) = 0$ and f is decreasing for $0 \le x \le 1$.) If p is close enough to 0, then $p \log a_j < 1$ for all j, and hence

$$1 + p\log a_j \le a_j^p \le 1 + p\log a_j + p^2(\log a_j)^2.$$

We sum this over j and divide by n to see that

$$1 + p\log G(\boldsymbol{a}) \le \frac{1}{n}\sum_{j=1}^n a_j^p \le 1 + p\log G(\boldsymbol{a}) + Cp^2$$

where $C = \frac{1}{n}\sum_{j=1}^n (\log a_j)^2$. For the moment we restrict our attention to the limit as p tends to 0 from above. If $p > 0$, then by taking p^{th} roots we find that

$$(1 + p\log G(\boldsymbol{a}))^{1/p} \le M_p(\boldsymbol{a}) \le (1 + p\log G(\boldsymbol{a}) + Cp^2)^{1/p}.$$

As $p \to 0^+$, the expression on the left tends to $e^{\log G(\boldsymbol{a})} = G(\boldsymbol{a})$. The expression on the right has the same limit, so by the squeeze theorem we see that $M_p(\boldsymbol{a})$ tends to $G(\boldsymbol{a})$ as $p \to 0^+$. If $p < 0$, then taking p^{th} roots reverses the inequalities, but the limits are the same as before, so we obtain the same result when $p \to 0^-$. □

The situation when p tends to infinity is comparatively easy.

Theorem I.7. *For positive real numbers a_j,*

$$\lim_{p\to+\infty} M_p(\boldsymbol{a}) = \max_{1\le j\le n} a_j, \qquad \lim_{p\to-\infty} M_p(\boldsymbol{a}) = \min_{1\le j\le n} a_j.$$

With this result in mind, it is sometimes convenient to put $M_\infty(\boldsymbol{a}) = \max a_j$ and $M_{-\infty}(\boldsymbol{a}) = \min a_j$.

Proof. Suppose that $a_1 = \max a_j$. Then

$$a_1^p \le \sum_{j=1}^n a_j^p \le n a_1^p.$$

On dividing by n and taking p^{th} roots, we deduce that

$$\frac{a_1}{n^{1/p}} \le M_p(\boldsymbol{a}) \le a_1.$$

Since $\lim_{p\to+\infty} n^{1/p} = n^0 = 1$, the first result follows by the squeeze theorem. To obtain the result for $p \to -\infty$, it suffices to apply the first result to the numbers $1/a_i$, and then take reciprocals. □

We now show that Hölder's inequality can be used to obtain some further information concerning the relative sizes of the means $M_p(\boldsymbol{a})$.

Theorem I.8. *Let $f(p) = \log M_p(\boldsymbol{a})^p = \log \frac{1}{n}\sum_{j=1}^n a_j^p$. If p_1, p_2, p_3 are real numbers such that $p_1 < p_2 < p_3$, then*

$$f(p_2) \le \frac{p_3 - p_2}{p_3 - p_1} f(p_1) + \frac{p_2 - p_1}{p_3 - p_1} f(p_3).$$

In geometric language, this asserts that the point $(p_2, f(p_2))$ lies below the chord joining the points $(p_1, f(p_1))$, $(p_3, f(p_3))$. In other words, the function $f(p)$ is convex. Sometimes this is expressed by saying that $M_p(\boldsymbol{a})^p$ is *logarithmically convex.*

Proof. We apply Hölder's inequality with

$$p = \frac{p_3 - p_1}{p_3 - p_2}, \qquad q = \frac{p_3 - p_1}{p_2 - p_1},$$

and we note that $p_1/p + p_3/q = p_2$. Hence

$$\sum_{j=1}^n a_j^{p_2} = \sum_{j=1}^n a_j^{p_1/p} a_j^{p_3/q} \le \Big(\sum_{j=1}^n a_j^{p_1}\Big)^{1/p} \Big(\sum_{j=1}^n a_j^{p_3}\Big)^{1/q}.$$

To complete the proof it suffices to divide both sides by n and then take logarithms. □

We next use Hölder's inequality to generalize the triangle inequality to the p-norm defined by

$$\|\boldsymbol{x}\|_p = \Big(\sum_{j=1}^{n} |x_j|^p\Big)^{1/p}. \tag{I.10}$$

Theorem I.9. (Minkowski's Inequality) *Let p be fixed, $1 \le p < \infty$. Then for any nonnegative real numbers x_j and y_j,*

$$\Big(\sum_{j=1}^{n}(x_j + y_j)^p\Big)^{1/p} \le \Big(\sum_{j=1}^{n} x_j^p\Big)^{1/p} + \Big(\sum_{j=1}^{n} y_j^p\Big)^{1/p}.$$

Equality occurs when the x_j and y_j are proportional. If x_j and y_j are complex numbers, then $|x_j + y_j| \le |x_j| + |y_j|$, so by applying Minkowski's Inequality to $|x_j|$ and $|y_j|$ we find that

$$\Big(\sum_{j=1}^{n} |x_j + y_j|^p\Big)^{1/p} \le \Big(\sum_{j=1}^{n} |x_j|^p\Big)^{1/p} + \Big(\sum_{j=1}^{n} |y_j|^p\Big)^{1/p}.$$

For vectors $\boldsymbol{x} \in \mathbb{C}^n$ we put

$$\|\boldsymbol{x}\|_p = \Big(\sum_{j=1}^{n} |x_j|^p\Big)^{1/p},$$

and we call this the ℓ_p norm of $\boldsymbol{x}$ (orally, "little ℓ p norm"). Thus $\|\boldsymbol{x} + \boldsymbol{y}\|_p \le \|\boldsymbol{x}\|_p + \|\boldsymbol{y}\|_p$. Although this generalizes the triangle inequality for the 2-norm, the proof is now more difficult.

Proof. We may suppose that $p > 1$, since the case $p = 1$ is trivial. Let $z_j = (x_j + y_j)^{p-1}$. Then

$$\|\boldsymbol{x} + \boldsymbol{y}\|_p^p = \sum_{j=1}^{n}(x_j + y_j)^p = \sum_{j=1}^{n}(x_j + y_j)z_j = \sum_{j=1}^{n} x_j z_j + \sum_{j=1}^{n} y_j z_j.$$

We apply Hölder's inequality to these last two sums to see that the above is

$$\le \Big(\sum_{j=1}^{n} x_j^p\Big)^{1/p}\Big(\sum_{i=j}^{n} z_j^q\Big)^{1/q} + \Big(\sum_{j=1}^{n} y_j^p\Big)^{1/p}\Big(\sum_{j=1}^{n} z_j^q\Big)^{1/q}.$$

Here $q = p/(p-1)$, so that $z_j^q = (x_j + y_j)^p$. Hence the above is

$$\begin{aligned} &= \|\boldsymbol{x}\|_p\|\boldsymbol{x} + \boldsymbol{y}\|_p^{p-1} + \|\boldsymbol{y}\|_p\|\boldsymbol{x} + \boldsymbol{y}\|_p^{p-1} \\ &= (\|\boldsymbol{x}\|_p + \|\boldsymbol{y}\|_p)\|\boldsymbol{x} + \boldsymbol{y}\|_p^{p-1}. \end{aligned}$$

We may suppose that $\|\boldsymbol{x} + \boldsymbol{y}\|_p > 0$, since otherwise $x_j + y_j = 0$ for all j, and the relation is trivial. On dividing both sides above by $\|\boldsymbol{x} + \boldsymbol{y}\|_p^{p-1}$, we obtain the stated result. □

The p-norms are also monotonic in p, but in the reverse direction.

Theorem I.10. *If* $0 < p_1 < p_2$, *then for any nonnegative numbers* a_j,

$$\Big(\sum_{j=1}^{n} a_j^{p_1}\Big)^{1/p_1} \ge \Big(\sum_{j=1}^{n} a_j^{p_2}\Big)^{1/p_2}$$

with equality when only one of the a_j *is positive.*

Here the case of equality is quite decidedly different than in our former results. This result is also markedly less sophisticated, since it depends only on the fact that e^x is an increasing function, whereas our former results depended on the fact that e^x is convex.

Proof. Put $p = p_2/p_1$ and set $A_j = a_j^{p_1}$. Then the inequality to be proved is equivalent to the inequality

$$\sum_{j=1}^{n} A_j^p \le \Big(\sum_{j=1}^{n} A_j\Big)^p.$$

Put $A = \sum_{j=1}^{n} A_j$ and set $\alpha_j = A_j/A$. On dividing both sides of the above by A^p we obtain the equivalent inequality

$$\sum_{j=1}^{n} \alpha_j^p \le 1$$

where $\sum_{j=1}^{n} \alpha_j = 1$. However, this last inequality is obvious, since the inequalities $0 \le \alpha_j \le 1$, $p > 1$ imply that $\alpha_j^p \le \alpha_j$ with equality if and only if $\alpha_j = 0$ or 1. Hence

$$\sum_{j=1}^{n} \alpha_j^p \le \sum_{j=1}^{n} \alpha_j = 1,$$

and the proof is complete. □

Hölder's inequality for Riemann integrals can be established easily by forming a Riemann sum, applying Hölder's inequality to the sum, and then letting the mesh of the partition tend to 0. However, the passage from Riemann integrals to the Lebesgue integral is a little complicated, so we find it easier to derive Hölder's inequality for integrals by the same method that we used for sums.

Theorem I.11. (Hölder's Inequality for integrals) *Suppose that* $1 < p < \infty$, $1 < q < \infty$, *and that* $1/p + 1/q = 1$. *Then*

$$\Big|\int_a^b f(x)g(x)\,dx\Big| \le \Big(\int_a^b |f(x)|^p\,dx\Big)^{1/p}\Big(\int_a^b |g(x)|^q\,dx\Big)^{1/q}. \tag{I.11}$$

Proof. By the triangle inequality,

$$\Big|\int_a^b f(x)g(x)\,dx\Big| \le \int_a^b |f(x)||g(x)|\,dx,$$

so without loss of generality we may assume that f and g take only nonnegative values. Let $\|f\|_p$ denote the first factor on the right hand side in (I.11), and $\|g\|_q$ the second. If $\|f\|_p = 0$, then $f(x) = 0$ for almost all x, so then the left hand side in Hölder's inequality is 0, and the inequality is trivial. Thus we may assume that $\|f\|_p > 0$. Similarly we may assume that $\|g\|_q > 0$. If $\|f\|_p = \infty$, then the inequality asserts nothing, so we may assume that $\|f\|_p < \infty$, and similarly that $\|g\|_q < \infty$. Put

$$F(x) = \frac{f(x)}{\|f\|_p}, \quad G(x) = \frac{g(x)}{\|g\|_q}.$$

On dividing the left hand side of Hölder's inequality by the right hand side, we see that Hölder's inequality is equivalent to the inequality

$$\int_a^b F(x)G(x)\,dx \le 1.$$

To establish this we first apply Young's Inequality (Lemma I.3) to the integrand to see that the left hand side above is

$$\begin{aligned}&\le \int_a^b \frac{F(x)^p}{p} + \frac{G(x)^q}{q}\,dx = \frac{1}{p\|f\|_p^p}\int_a^b f(x)^p\,dx + \frac{1}{q\|g\|_q^q}\int_a^b g(x)^q\,dx\\ &= \frac{1}{p} + \frac{1}{q} = 1.\end{aligned}$$

This completes the proof. □

Equality holds in Hölder's inequality for integrals when f and g are nonnegative and $f(x)^p$ is proportional to $g(x)^q$ for almost all x.

We derive Minkowski's inequality for the Riemann or Lebesgue integral from Hölder's inequality for integrals by the same method we used to derive Minkowski's inequality for vectors from Hölder's inequality for sums.

Theorem I.12. (Minkowski's inequality for integrals) *Suppose that* $1 < p < \infty$. *Then*

$$\text{(I.12)}\qquad \Big(\int_a^b |f(x)+g(x)|^p\,dx\Big)^{1/p} \le \Big(\int_a^b |f(x)|^p\,dx\Big)^{1/p} + \Big(\int_a^b |g(x)|^p\,dx\Big)^{1/p}.$$

Equality holds in the above when $f(x)$ and $g(x)$ are proportional except possibly for a set of x of measure zero.

Proof. Since $|f(x)+g(x)| \le |f(x)|+|g(x)|$, we may assume that f and g take only nonnegative values. Also, if $\int_a^b (f(x)+g(x))^p\,dx = 0$, then there is nothing to prove, so we may assume that this integral is positive. Put $h(x) = (f(x)+g(x))^{p-1}$. Then

$$\int_a^b (f(x)+g(x))^p\,dx = \int_a^b (f(x)+g(x))h(x)\,dx = \int_a^b f(x)h(x)\,dx + \int_a^b g(x)h(x)\,dx,$$

which by two applications of Hölder's inequality for integrals is

$$\le \Big(\int_a^b f(x)^p\,dx\Big)^{1/p}\Big(\int_a^b h(x)^q\,dx\Big)^{1/q} + \Big(\int_a^b g(x)^p\,dx\Big)^{1/p}\Big(\int_a^b h(x)^q\,dx\Big)^{1/q}.$$

Now $q = p/(p-1)$, so $h(x)^q = (f(x)+g(x))^p$. On dividing both sides by $\|h\|_q$ we obtain the desired inequality. □

We have now established the main classical inequalities (AGM, Cauchy, Young, Hölder, Minkowski), from which many other useful inequalities can be derived. We conclude with a few remarks concerning the lessons we learn from these inequalities. Suppose that $p > 0$ and we know that

$$\int_0^1 |f(x)|^p \, dx \le C_p. \tag{I.13}$$

This gives us some information concerning the size of the subset of $[0, 1]$ on which $|f(x)|$ is large. Let V be a chosen value, and let $L(V)$ be the size of the subset on which $|f(x)| \ge V$. Put

$$g(x) = \begin{cases} V & (|f(x)| \ge V), \\ 0 & (|f(x)| < V). \end{cases}$$

Since $0 \le g(x) \le |f(x)|$ for all x, it follows that

$$\int_0^1 g(x)^p \, dx \le \int_0^1 |f(x)|^p \, dx \le C_p.$$

Since the integral on the left is precisely $V^p L(V)$, we deduce that

$$L(V) \le \frac{C_p}{V^p}. \tag{I.14}$$

Thus $|f(x)|$ can be large only on a small set of x. If we are interested in an upper bound of this kind that tends to 0 rapidly as V becomes large, then we are interested in larger values of p. If we have bounds of the kind (I.14) for several p, then the one that is least will depend on the size of V.

Example I.1. Suppose we know that

$$\int_0^1 |f(x)| \, dx \le 1, \qquad \int_0^1 |f(x)|^3 \, dx \le 4. \tag{I.15}$$

Then by (I.14) we know that $L(V) \le 1/V$, and also that $L(V) \le 4/V^3$. In addition, we always have the trivial bound $L(V) \le 1$. Thus altogether, $L(V) \le \min(1, 1/V, 4/V^3)$. More explicitly,

$$L(V) \le \begin{cases} 1 & (0 \le V \le 1), \\ 1/V & (1 \le V \le 2), \\ 4/V^3 & (2 \le V < \infty). \end{cases} \tag{I.16}$$

This is graphed as the heavy curve in Figure I.2. Now by Hölder's inequality we can combine the bounds (I.15) to obtain a bound for the integral $\int_0^1 |f(x)|^p \, dx$ for $1 < p < 3$. In particular, by Cauchy's inequality we see that

$$\int_0^1 |f(x)|^2 \, dx \le \Big(\int_0^1 |f(x)| \, dx\Big)^{1/2} \Big(\int_0^1 |f(x)|^3 \, dx\Big)^{1/2} \le 1^{1/2} 4^{1/2} = 2.$$

From (I.14) it therefore follows that $L(V) \le 2/V^2$. This bound is indicated by the dotted curve in Figure I.2, and we see that for no value of V does this provide new information concerning the size of $L(V)$. Indeed, $2/V^2$ is the geometric mean of $1/V$ and $4/V^3$, so it will always lie between these two quantities. This is representative of the general situation: When one interpolates using Hölder's inequality, any upper

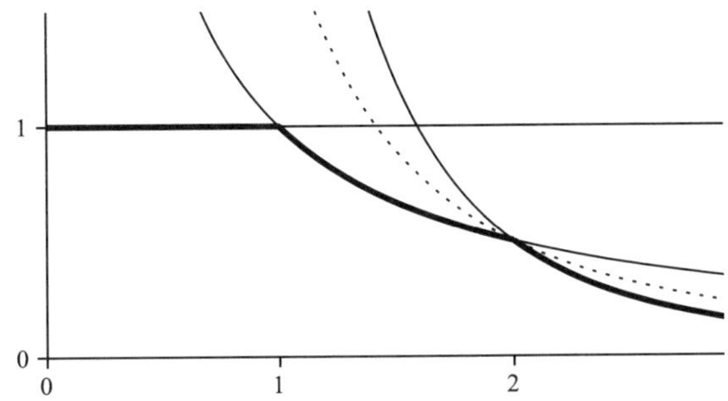

Figure I.2. Bounds for $L(V)$.

bound for $L(V)$ that is then derived by (I.14) could have been derived from the initial estimates on which the application of Hölder's inequality was based.

The inequality (I.13) implies (I.14), but the converse is false, since we can construct examples of functions for which (I.14) holds but for which the integral in (I.13) is as large as we please (cf. Exercise 17 below). Although the information in (I.14) is weaker than that in (I.13), it is only slightly weaker. For example, suppose that we know that

$$\int_0^1 |f(x)|^{p_1}\,dx \le C_{p_1}, \qquad \int_0^1 |f(x)|^{p_2}\,dx \le C_{p_2}.$$

By Hölder's inequality it follows that

$$\int_0^1 |f(x)|^p\,dx \le C_{p_1}^{\frac{p_3-p}{p_3-p_1}} C_{p_2}^{\frac{p-p_1}{p_3-p_1}}$$

for $p_1 \le p \le p_3$. On the other hand, suppose we start from the weaker information that

$$L(V) \le \frac{C_{p_1}}{V^{p_1}}, \qquad L(V) \le \frac{C_{p_2}}{V^{p_2}}.$$

With a little care it can be shown that if $p_1 < p < p_2$, then

$$\int_0^1 |f(x)|^p\,dx \le K C_{p_1}^{\frac{p_3-p}{p_3-p_1}} C_{p_2}^{\frac{p-p_1}{p_3-p_1}}$$

where K is a constant that depends only on the exponents p_1, p, p_2. Thus the weaker information still yields the same bound for the interpolated integral, though with a slightly inferior constant.

Example I.2. Suppose that f_1, f_2, and f_3 have period 1, and that we wish to estimate

$$\int_0^1\int_0^1\int_0^1 |f_1(x_2-x_3)f_2(x_3-x_1)f_3(x_1-x_2)|\,dx_1\,dx_2\,dx_3\,.$$

By Hölder's inequality we see that the above is

$$\le \Big(\int_0^1 |f_1(x)|^3\,dx\Big)^{1/3}\Big(\int_0^1 |f_2(x)|^3\,dx\Big)^{1/3}\Big(\int_0^1 |f_3(x)|^3\,dx\Big)^{1/3} = \|f_1\|_3\|f_2\|_3\|f_3\|_3\,.$$

Suppose that

$$f_1(x) = f_2(x) = f_3(x) = \begin{cases} 1 & (\|x\| \le \delta), \\ 0 & (\text{otherwise}). \end{cases} \tag{I.17}$$

Then the original expression is $\asymp \delta^2$, but our upper bound is $\asymp \delta$, which is much worse if δ is small. Set

$$\begin{aligned} g_1(x_1, x_2, x_3) &= |f_1(x_2 - x_3) f_2(x_3 - x_1)|^{1/2}, \\ g_2(x_1, x_2, x_3) &= |f_2(x_3 - x_1) f_3(x_1 - x_2)|^{1/2}, \\ g_3(x_1, x_2, x_3) &= |f_3(x_1 - x_2) f_1(x_2 - x_3)|^{1/2}. \end{aligned}$$

Thus our original expression is

$$\int_0^1 \int_0^1 \int_0^1 |g_1(x_1, x_2, x_3) g_2(x_1, x_2, x_3) g_3(x_1, x_2, x_3)| \, dx_1 \, dx_2 \, dx_3,$$

which by Hölder's inequality is

$$\begin{aligned} &\le \prod_{j=1}^{3} \Big(\int_0^1 \int_0^1 \int_0^1 |g_j(x_1, x_2, x_3)|^3 \, dx_1 \, dx_2 \, dx_3 \Big)^{1/3} \\ &= \prod_{j=1}^{3} \Big(\int_0^1 |f_j(x)|^{3/2} \, dx \Big)^{2/3} = \|f_1\|_{3/2} \|f_2\|_{3/2} \|f_3\|_{3/2}. \end{aligned}$$

In the case of the functions (I.17) already considered, this new upper bound is $4\delta^2$, which is the correct order of magnitude.

Exercises

1. If x, y, and z are positive real numbers such that $x^3 + y^3 + z^3 = 1$, then how large can $9x + 16y + 25z$ be? What choice of the variables gives the maximum?
2. If x, y, and z are real numbers such that $|x|^3 + |y|^3 + |z|^3 = 1$, then how large can $9x - 16y - 25z$ be? What choice of the variables gives the maximum?
3. If x, y, z, w are real numbers such that $x - y + 25z - 36w = 49$, then how small can $|x|^3 + |y|^3 + |z|^3 + |w|^3$ be?
4. Use Hölder's inequality twice to show that if $a_i > 0$, $b_i > 0$ and $c_i > 0$ for $1 \le i \le n$ and $p > 1$, $q > 1$, $r > 1$ with

   $$\frac{1}{p} + \frac{1}{q} + \frac{1}{r} = 1,$$

   then

   $$\sum_{j=1}^{n} a_j b_j c_j \le \Big(\sum_{j=1}^{n} a_j^p \Big)^{1/p} \Big(\sum_{j=1}^{n} b_j^q \Big)^{1/q} \Big(\sum_{j=1}^{n} c_j^r \Big)^{1/r}.$$

5. Let $w_1, w_2, \ldots, w_n$ be a system of weights, so that $w_i \geq 0$ for all i and $w_1+w_2+\cdots+w_n = 1$. For positive numbers a_j let $M_{p,w}(\boldsymbol{a})$ be the weighted p^{th}-power mean,

$$M_{p,w}(\boldsymbol{a}) = \Big(\sum_{j=1}^{n} w_j a_j^p\Big)^{1/p}.$$

Use Hölder's inequality with $p = p_2/p_1$ to show that if $0 < p_1 < p_2$, then

$$\sum_{j=1}^{n} w_j a_j^{p_1} \leq \Big(\sum_{j=1}^{n} w_j a_j^{p_2}\Big)^{p_1/p_2}.$$

Hence show that $M_{p_1,w}(\boldsymbol{a}) \leq M_{p_2,w}(\boldsymbol{a})$.

6. Let $f(p) = \log M_p(\boldsymbol{a})^p$ where the numbers a_i are positive. In Theorem I.8 we showed that $f(p)$ is convex. In this exercise we give a new proof of this by showing that $f''(p) > 0$.
   (a) Show that

$$f'(p) = \frac{\sum_{j=1}^n a_j^p \log a_j}{\sum_{j=1}^n a_j^p}.$$

   (b) Show that

$$f''(p) = \frac{\Big(\sum_{j=1}^n a_j^p\Big)\Big(\sum_{j=1}^n a_j^p (\log a_j)^2\Big) - \Big(\sum_{j=1}^n a_j^p \log a_j\Big)^2}{\Big(\sum_{j=1}^n a_j^p\Big)^2}.$$

   (c) Use Cauchy's inequality to show that the numerator above is ≥ 0, with equality only if all the a_j are equal. Thus $f''(p) > 0$ unless all the variables are equal.

7. Let $A = [a_{ij}]$ be an $m \times n$ matrix whose entries are nonnegative.
   (a) Suppose that $1 \leq p < \infty$, and consider the following inequality:

$$\Big(\sum_{i=1}^{m}\Big(\sum_{j=1}^{n} a_{ij}\Big)^p\Big)^{1/p} \leq \sum_{j=1}^{n}\Big(\sum_{i=1}^{m} a_{ij}{}^p\Big)^{1/p}.$$

   When $n = 2$, this is Minkowski's inequality. By $n-1$ applications of Minkowski's inequality (or by induction), show that the above is true for general n.
   (b) We can take the p_1 norm of each row of A, and then the p_2 norm of the resulting m numbers, or we could take the p_2 norm of each column of A, followed by the p_1 norm of the resulting n numbers. Show that

$$\Big(\sum_{i=1}^{m}\Big(\sum_{j=1}^{n} a_{ij}{}^{p_1}\Big)^{p_2/p_1}\Big)^{1/p_2} \leq \Big(\sum_{j=1}^{n}\Big(\sum_{i=1}^{m} a_{ij}{}^{p_2}\Big)^{p_1/p_2}\Big)^{1/p_1}.$$

   (c) By suitably rescaling the above, show that this inequality among norms yield a corresponding inequality among means:

$$\Big(\frac{1}{m}\sum_{i=1}^{m}\Big(\frac{1}{n}\sum_{j=1}^{n} a_{ij}{}^{p_1}\Big)^{p_2/p_1}\Big)^{1/p_2} \leq \Big(\frac{1}{n}\sum_{j=1}^{n}\Big(\frac{1}{m}\sum_{i=1}^{m} a_{ij}{}^{p_2}\Big)^{p_1/p_2}\Big)^{1/p_1}.$$

8. Let $f(x)$ and $g(x)$ be continuous functions on the interval $[a,b]$, and suppose that $p>1$, $q>1$, $1/p+1/q=1$.
   (a) Show that if $x_1, x_2, \dots, x_n$ are any points in the interval $[a,b]$, then

$$\frac{b-a}{n}\sum_{j=1}^{n}|f(x_j)g(x_j)| \le \Big(\frac{b-a}{n}\sum_{j=1}^{n}|f(x_j)|^p\Big)^{1/p}\Big(\frac{b-a}{n}\sum_{j=1}^{n}|g(x_j)|^q\Big)^{1/q}.$$

   (b) Take the x_j to be equally spaced, and deduce the continuous version of Hölder's inequality for the Riemann integral:

$$\int_a^b |f(x)g(x)|\,dx \le \Big(\int_a^b |f(x)|^p\,dx\Big)^{1/p}\Big(\int_a^b |g(x)|^q\,dx\Big)^{1/q}.$$

   (c) Suppose that $w(x)$ is continuous and non-negative for $a \le x \le b$. Use (b) to show that

$$\int_a^b w(x)|f(x)g(x)|\,dx \le \Big(\int_a^b w(x)|f(x)|^p\,dx\Big)^{1/p}\Big(\int_a^b w(x)|g(x)|^q\,dx\Big)^{1/q}.$$

9. For functions f continuous on an interval $[a,b]$, put

$$M_p(f) = \Big(\frac{1}{b-a}\int_a^b |f(x)|^p\,dx\Big)^{1/p}.$$

Show that if $0<p_1<p_2$, then $M_{p_1}(f) \le M_{p_2}(f)$.

10. Suppose that f is continuous on $[a,b]$, and that w is continuous and non-negative on $[a,b]$, and that $\int_a^b w(x)\,dx = 1$. Show that if $0<p_1<p_2$, then

$$\Big(\int_a^b w(x)|f(x)|^{p_1}\,dx\Big)^{1/p_1} \le \Big(\int_a^b w(x)|f(x)|^{p_2}\,dx\Big)^{1/p_2}.$$

11. Suppose that f is continuous on $[a,b]$, and that w is continuous and non-negative on $[a,b]$. Show that if $p_1<p_2<p_3$, then

$$\int_a^b |f(x)|^{p_2}\,dx \le \Big(\int_a^b |f(x)|^{p_1}\,dx\Big)^{\frac{p_3-p_2}{p_3-p_1}}\Big(\int_a^b |f(x)|^{p_3}\,dx\Big)^{\frac{p_2-p_1}{p_3-p_1}}.$$

12. Would you buy a used car from someone who tells you that they have a function f for which $\int_0^1 |f(x)|\,dx = 1$, $\int_0^1 |f(x)|^2 = 3$, and $\int_0^1 |f(x)|^3\,dx = 4$? Explain why not.

13. Suppose that $f(x,y)$ is nonnegative and integrable on the rectangle $[a,b]\times[c,d]$, and that $1\le p<\infty$. Prove the following analogue of Minkowski's inequality:

$$\Big(\int_a^b \Big|\int_c^d f(x,y)\,dy\Big|^p\,dx\Big)^{1/p} \le \int_c^d \Big(\int_a^b |f(x,y)|^p\,dx\Big)^{1/p}\,dy\,.$$

As in the discrete case, this can be rescaled to yield an inequality among means.

14. (a) Suppose that $\int_0^1 A(x)^4\,dx = 41$. Show that $\max_{0\le x\le 1}|A(x)| \ge 41^{1/4} = 2.5304\ldots$.
   (b) Give an example of a function $A(x)$ for which $\int_0^1 A(x)^4\,dx = 41$ and $\max_{0\le x\le 1}|A(x)| = 41^{1/4}$.
   (c) For the function found in (b), show that $\int_0^1 A(x)^2\,dx = 41^{1/2} = 6.40312\ldots$.

15. (a) Explain why the following inequality holds:

$$\int_0^1 A(x)^4\,dx \le \Big(\max_{0\le x\le 1}|A(x)|\Big)^2 \int_0^1 A(x)^2\,dx.$$

(b) Show that if $\int_0^1 A(x)^4\,dx = 41$ and $\int_0^1 A(x)^2\,dx = 5$, then

$$\max_{0\le x\le 1}|A(x)| \ge \sqrt{41/5} = 2.86356\ldots.$$

(c) Find a number p, $0 \le p \le 1$ such that if

$$A(x) = \begin{cases} \sqrt{41/5} & (0 \le x \le p), \\ 0 & (p \le x \le 1), \end{cases}$$

then $\int_0^1 A(x)^4\,dx = 41$ and $\int_0^1 A(x)^2\,dx = 5$.

(d) Show that for this function, $\int_0^1 A(x)\,dx = 5\sqrt{5/41} = 1.7460757\ldots$.

16. We now consider functions $A(x)$ for which

$$\text{(I.18)} \qquad \int_0^1 |A(x)|\,dx = 2, \qquad \int_0^1 |A(x)|^2\,dx = 5, \qquad \int_0^1 |A(x)|^4\,dx = 41,$$

and we seek a lower bound for $\max_{0\le x\le 1}|A(x)|$. Using only the last two constraints we already have the lower bound of 15 (b) above. The question is whether the first constraint can be used to derive a still better lower bound.

(a) Suppose that $P(u)$ is a polynomial of the special form $P(u) = a_4u^4 + a_2u^2 + a_1u + a_0$. Show that if $A(x)$ satisfies (I.18), then

$$\int_0^1 P(|A(x)|)\,dx = 41a_4 + 5a_2 + 2a_1 + a_0.$$

(b) Let $P(x) = (x-1)^2(x-3)(x+5)$. Show that this polynomial is of the above type, and determine the coefficients a_0, a_1, a_2, a_4.

(c) Show that for the polynomial selected, $41a_4 + 5a_2 + 2a_1 + a_0 = 0$.

(d) Draw a graph of $P(x)$, and explain why $P(x) \le 0$ for $0 \le x \le 3$. At what points in this interval does equality occur?

(e) Show that if $|A(x)| < 3$ for all x and if $\int_0^1 P(|A(x)|)\,dx = 0$, then $|A(x)| = 1$ for (almost) all x.

(f) Show that if $|A(x)| = 1$ for almost all x, then $\int_0^1 |A(x)|\,dx = 1$, and hence $A(x)$ does not satisfy (I.18). Deduce that if $A(x)$ satisfies (I.18) then $\max|A(x)| \ge 3$.

(g) Consider functions $A(x)$ of the following shape:

$$A(x) = \begin{cases} 1 & (0 \le x \le p), \\ 3 & (p < x \le 1). \end{cases}$$

Find a value of p so that such an $A(x)$ satisfies (I.18).

17. Let K be a large parameter, put

$$f(x) = \begin{cases} K & (0 \le x \le 1/K), \\ 1/x & (1/K \le x \le 1), \end{cases}$$

and let $L(V)$ be the size of the subset of $[0,1]$ on which $f(x) \ge V$.

(a) Show that $L(V) = 1$ for $0 \le V \le 1$.

(b) Show that $L(V) = 1/V$ for $1 \leq V \leq K$.
(c) Show that $L(V) = 1/K$ for $V > K$.
(d) Deduce that $L(V) \leq 1/V$ for all $V > 0$.
(e) Show that $\int_0^1 f(x)\,dx = 1 + \log K$. (Hence the inequality $L(V) \leq 1/V$ does not imply an upper bound for $\int_0^1 |f(x)|\,dx$.)

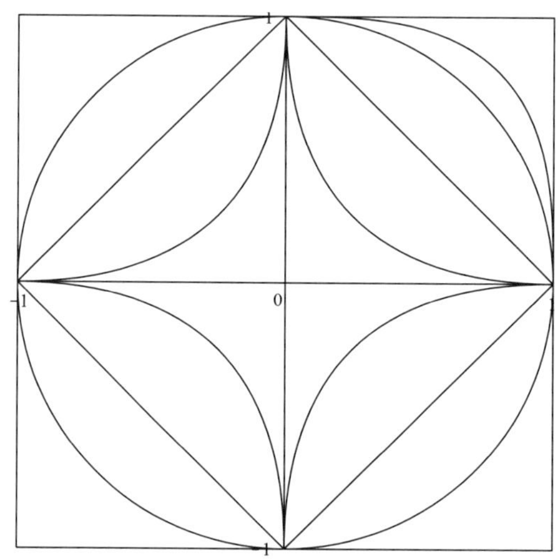

Figure I.3. The unit "circle" $\|\boldsymbol{x}\|_p = 1$ in $\mathbb{R}^2$ for $p = 1/2, 1, 2, 3, \infty$.

18. Suppose that $p \geq 1$, that $\|\boldsymbol{x}\|_p \leq 1$, that $\|\boldsymbol{y}\|_p \leq 1$, and that $0 \leq \lambda \leq 1$. Use Minkowski's inequality to show that

$$\|\lambda \boldsymbol{x} + (1 - \lambda)\boldsymbol{y}\|_p \leq 1.$$

Hence deduce that the "disc" $\|\boldsymbol{x}\|_p \leq 1$ is a convex set in $\mathbb{R}^n$.

19. Suppose that $0 < p < 1$.
(a) Show that $\|(0, 1)\|_p = 1$, and that $\|(1, 0)\|_p = 1$.
(b) Show that $\|(1/2, 1/2)\|_p = 2^{1/p-1} > 1$.
(c) Deduce that the "disc" $\|\boldsymbol{x}\|_p \leq 1$ is not a convex set when $0 < p < 1$.

Notes

I.1. Exercise I.1.21 hints at work of Gauss; see Borwein and Borwein (1987).

I.2. The reader will note that practically all our inequalities are derived using convexity.

Hölder's inequality originates with Hölder (1889). Young's inequality, which provides an easy path to Hölder, originates with Young (1912). Minkowski's inequality originates with Minkowski (1910, pp. 115–117).

The classical resource on the subject of inequalities is Hardy, Littlewood and Pólya (1952). The reader might also enjoy Steele (2004).

Appendix L

Topics in Linear Algebra

L.1. Familiar vector spaces

We have some past experience with the vector space $\mathbb{R}^n$, which consists of vectors $\boldsymbol{v} = (v_1, v_2, \ldots, v_n)$ with n real coordinates. Such a vector can be multiplied by a scalar to form $c\boldsymbol{v} = (cv_1, cv_2, \ldots, cv_n)$. Two vectors may be added, so that if $\boldsymbol{w} = (w_1, w_2, \ldots, w_n)$, then $\boldsymbol{v} + \boldsymbol{w} = (v_1 + w_1, v_2 + w_2, \ldots, v_n + w_n)$. By combining these operations we may form a linear combination $c_1\boldsymbol{v}_1 + c_2\boldsymbol{v}_2 + \cdots + c_k\boldsymbol{v}_k$ of several vectors. We say that the vectors $\boldsymbol{v}_1, \ldots, \boldsymbol{v}_k$ are *linearly independent* if the equation

$$c_1\boldsymbol{v}_1 + c_2\boldsymbol{v}_2 + \cdots + c_k\boldsymbol{v}_k = \boldsymbol{0}$$

holds only when all the c_i are 0. A subset V of $\mathbb{R}^n$ is called a *subspace* if it is closed scalar multiplication and vector addition. Such a subspace has a certain *dimension*, which is the maximal size of a collection of linearly independent vectors in the subspace. If $\dim V = k$, then V has a *basis* consisting of k vectors, and any basis for V consists of exactly k vectors.

In addition to forming a vector space, $\mathbb{R}^n$ has the further property that we can define an inner (or dot) product of two vectors,

$$\langle \boldsymbol{v}, \boldsymbol{w} \rangle = \sum_{j=1}^{n} v_j w_j\,. \tag{L.1}$$

Several systems of notation are in use for this quantity. Some authors write $(\boldsymbol{v}, \boldsymbol{w})$, while others write $\boldsymbol{v}\cdot\boldsymbol{w}$. This function takes an ordered pair of vectors, and produces a real number. In set-theoretic language, the inner product maps $\mathbb{R}^n \times \mathbb{R}^n$ to $\mathbb{R}$. If we think of our vectors as being matrices with 1 column and n rows, then we can use matrix multiplication to write the inner product, $\langle \boldsymbol{v}, \boldsymbol{w} \rangle = \boldsymbol{w}^t\boldsymbol{v}$ where $\boldsymbol{A}^t$ denotes the transpose of a matrix A. The dot product is both *linear* and *symmetric*, by which we mean that

$$\langle \alpha\boldsymbol{a} + \beta\boldsymbol{b}, \boldsymbol{c} \rangle = \alpha\langle \boldsymbol{a}, \boldsymbol{c} \rangle + \beta\langle \boldsymbol{b}, \boldsymbol{c} \rangle \tag{L.2}$$

for any vectors $\boldsymbol{a}, \boldsymbol{b}, \boldsymbol{c}$ and any scalars α, β, and

$$\langle \boldsymbol{v}, \boldsymbol{w} \rangle = \langle \boldsymbol{w}, \boldsymbol{v} \rangle \,. \tag{L.3}$$

By using the symmetry, then the linearity, and finally the symmetry a second time, we may deduce that the inner product is also linear in its second argument:

$$\langle \boldsymbol{a}, \beta\boldsymbol{b} + \gamma\boldsymbol{c} \rangle = \beta\langle \boldsymbol{a}, \boldsymbol{b} \rangle + \gamma\langle \boldsymbol{a}, \boldsymbol{c} \rangle \tag{L.4}$$

for any vectors $\boldsymbol{a}, \boldsymbol{b}, \boldsymbol{c}$, and any scalars β, γ.

We use the inner product to define the *norm* of a vector,

$$\|\boldsymbol{v}\| = \sqrt{\langle \boldsymbol{v}, \boldsymbol{v} \rangle} = \Big(\sum_{j=1}^{n} v_j^2 \Big)^{1/2} \,. \tag{L.5}$$

This norm satisfies a triangle inequality, which is to say that $\|\boldsymbol{v} + \boldsymbol{w}\| \le \|\boldsymbol{v}\| + \|\boldsymbol{w}\|$ for any vectors $\boldsymbol{v}, \boldsymbol{w}$.

We say that two vectors $\boldsymbol{v}$, $\boldsymbol{w}$ are *perpendicular*, or *orthogonal* if $\langle \boldsymbol{v}, \boldsymbol{w} \rangle = 0$. If $\boldsymbol{v}_1, \boldsymbol{v}_2, \ldots, \boldsymbol{v}_k$ are mutually orthogonal, then

$$\begin{aligned} \|\boldsymbol{v}_1 + \boldsymbol{v}_2 + \cdots + \boldsymbol{v}_m\|^2 &= \Big\langle \sum_{j=1}^{m} \boldsymbol{v}_j, \sum_{k=1}^{m} \boldsymbol{v}_k \Big\rangle \\ &= \sum_{j=1}^{m}\sum_{k=1}^{m} \langle \boldsymbol{v}_j, \boldsymbol{v}_k \rangle = \sum_{j=1}^{m} \langle \boldsymbol{v}_j, \boldsymbol{v}_j \rangle = \sum_{j=1}^{m} \|\boldsymbol{v}_j\|^2 \,. \end{aligned}$$

This is Pythagoras's Theorem for $\mathbb{R}^n$. In addition to defining orthogonality, we can also determine the angle ϕ between two vectors $\boldsymbol{a}$, $\boldsymbol{b}$ by the identity

$$\cos\phi = \frac{\langle \boldsymbol{a}, \boldsymbol{b} \rangle}{\|\boldsymbol{a}\|\,\|\boldsymbol{b}\|} \,.$$

That there is such an angle is assured by the important fact that the right hand side above always lies in the interval $[-1, 1]$. That is,

$$|\langle \boldsymbol{a}, \boldsymbol{b} \rangle| \le \|\boldsymbol{a}\|\,\|\boldsymbol{b}\| \tag{L.6}$$

for any vectors $\boldsymbol{a}$, $\boldsymbol{b}$. In elementary terms, this asserts that

$$\Big| \sum_{j=1}^{n} a_j b_j \Big| \le \Big(\sum_{j=1}^{n} a_j^2 \Big)^{1/2} \Big(\sum_{j=1}^{n} b_j^2 \Big)^{1/2} \,. \tag{L.7}$$

This is *Cauchy's Inequality*, which was already proved in Chapter 0. However, the following two additional proofs are instructive.

Second Proof. We observe that

$$\begin{aligned} 0 \le \sum_{j=1}^{n} (\lambda a_j - b_j)^2 &= \lambda^2 \Big(\sum_{j=1}^{n} a_j^2 \Big) - 2\lambda \Big(\sum_{j=1}^{n} a_j b_j \Big) + \Big(\sum_{j=1}^{n} b_j^2 \Big) \\ &= a\lambda^2 + b\lambda + c \end{aligned}$$

where

$$a = \sum_{j=1}^{n} a_j^2, \qquad b = -2\sum_{j=1}^{n} a_j b_j, \qquad c = \sum_{j=1}^{n} b_j^2 \,.$$

If $a = 0$, then $a_j = 0$ for all j, and the inequality is true in this case. Suppose that $a > 0$. A quadratic polynomial is nonnegative for all real λ only when its discriminant $b^2 - 4ac$ is ≤ 0. That is,

$$4\Big(\sum_{j=1}^{n} a_j b_j\Big)^2 - 4\Big(\sum_{j=1}^{n} a_j^2\Big)\Big(\sum_{j=1}^{n} b_j^2\Big) \leq 0,$$

which is the desired inequality.

Instead of quoting knowledge of the discriminant of a quadratic polynomial, we could alternatively note that the derivative of $a\lambda^2 + b\lambda + c$ is $2a\lambda + b$, which is 0 when $\lambda = -b/(2a)$. For this λ,

$$a\lambda^2 + b\lambda + c = -\frac{b^2}{4a} + c\,.$$

Since the polynomial is everywhere ≥ 0, it follows that the above is ≥ 0, so we have the same result as before. □

Third Proof. Consider the b_j to be fixed, and the a_j to be variables. If each a_j is replaced by ca_j, then the left hand side of the proposed inequality is multiplied by $|c|$, and so is the right hand side. By choosing c suitably, we may rescale the a_j so that $\sum_{j=1}^{n} a_j^2 = 1$. For such a_j, the issue is to determine the maximum of $\sum_{j=1}^{n} a_j b_j$. Put

$$f(\boldsymbol{a}) = \sum_{j=1}^{n} a_j b_j, \qquad g(\boldsymbol{a}) = \sum_{j=1}^{n} a_j^2\,.$$

We observe that these functions have the gradients

$$\nabla f = (b_1, b_2, \ldots, b_n), \qquad \nabla g = (2a_1, 2a_2, \ldots, 2a_n)\,.$$

By the method of Lagrange multipliers, we recognize that if $f(\boldsymbol{a})$ is maximal subject to the condition $g(\boldsymbol{a}) = 1$, then ∇f is parallel to ∇g, which is to say that there is a λ such that $b_j = 2\lambda a_j$ for all j. For such a_j we obtain the same result as before.

In this situation it may appear to be excessive to use Lagrange multipliers, but the approach is useful in more difficult circumstances, as it aids in identifying the shape of the extremal configurations. □

We now consider the question of how all this changes when we work with complex numbers, instead of real numbers. Certainly the procedures of solving systems of linear equations can be executed as before, as they involve nothing more than addition, subtraction, multiplication, and division. However, a difference arises when we consider the inner product. For example, if we consider $\boldsymbol{v} = (1, i)$ as a member of $\mathbb{C}^2$, then for the inner product as we have defined it, we would have $\|\boldsymbol{v}\|^2 = 1^2 + i^2 = 0$. However, the idea of the norm is to measure the distance from a given vector from the origin, so obviously this fails.

For vectors in $\mathbb{C}^n$, we define the inner product $\langle \boldsymbol{u}, \boldsymbol{v}\rangle$ to be

$$\langle \boldsymbol{u}, \boldsymbol{v}\rangle = \sum_{k=1}^{n} u_j \overline{v_j} = \overline{\boldsymbol{v}}^t \boldsymbol{u}\,.$$

From this definition it is clear that

$$\langle \alpha \boldsymbol{a} + \beta \boldsymbol{b}, \boldsymbol{c}\rangle = \alpha\langle \boldsymbol{a}, \boldsymbol{c}\rangle + \beta\langle \boldsymbol{b}, \boldsymbol{c}\rangle .$$

However, an asymmetry develops, since $\langle \boldsymbol{v}, \boldsymbol{u}\rangle = \overline{\langle \boldsymbol{u}, \boldsymbol{v}\rangle}$. Thus in particular,

$$\langle \boldsymbol{a}, \beta \boldsymbol{b} + \gamma \boldsymbol{c}\rangle = \overline{\beta}\langle \boldsymbol{a}, \boldsymbol{c}\rangle + \overline{\gamma}\langle \boldsymbol{a}, \boldsymbol{c}\rangle .$$

Although our definition of inner product has been redefined, we still take the norm of $\boldsymbol{v}$ to be

$$\|\boldsymbol{v}\| = \sqrt{\langle \boldsymbol{v}, \boldsymbol{v}\rangle} = \sum_{j=1}^{n} |v_j|^2 .$$

Additionally, we still have

$$|\langle \boldsymbol{a}, \boldsymbol{b}\rangle| \le \|\boldsymbol{a}\|\,\|\boldsymbol{b}\|,$$

since

$$|\langle \boldsymbol{a}, \boldsymbol{b}\rangle| = \Big|\sum_{j=1}^{n} a_j\overline{b_j}\Big| \le \sum_{j=1}^{n} |a_j\overline{b_j}| = \sum_{j=1}^{n} |a_j|\,|b_j|,$$

which by Cauchy's inequality is

$$\begin{aligned} &\le \Big(\sum_{j=1}^{n} |a_j|^2\Big)^{1/2}\Big(\sum_{j=1}^{n} |b_j|^2\Big)^{1/2} \\ &= \|\boldsymbol{a}\|\,\|\boldsymbol{b}\| . \end{aligned}$$

Note that the triangle inequality is applied first, and then Cauchy's inequality for the numbers $|a_j|$, $|b_j|$. Thus Cauchy's inequality is fundamentally an inequality concerning nonnegative real numbers.

In the more advanced theory of linear algebra, further differences emerge between the theory over the real numbers $\mathbb{R}$ versus that over the complex numbers $\mathbb{C}$. To set the stage, we recall some definitions and results concerning linear algebra over the real numbers $\mathbb{R}$. A square matrix Q is said to be *orthogonal* if $Q^tQ = I$. That is to say, the columns of Q form an orthonormal family of vectors. Indeed, the following are equivalent:

Q is an orthogonal matrix;

$Q^tQ = I$;

The columns of Q form an orthonormal family;

The rows of Q form an orthonormal family;

$\|Q\boldsymbol{v}\| = \|\boldsymbol{v}\|$ for all $\boldsymbol{v} \in \mathbb{R}^n$.

Since the linear transformation $\boldsymbol{v} \mapsto Q\boldsymbol{v}$ preserves distances, we may regard this linear transformation as consisting of a rotation, possibly with a reflection. Orthogonal matrices arise in the discussion of diaganalization of a symmetric matrix S. We say that S is *symmetric* if $S^t = S$, which is to say that $s_{ij} = s_{ji}$ for all i and j. For any such matrix, there is an orthogonal matrix Q such that

$$S = QDQ^t$$

where D is a diagonal matrix. The diagonal elements of D are the eigenvalues of S, which are all real. The columns of Q are the associated eigenvectors; they form an orthonormal basis for $\mathbb{R}^n$. From the distance-preserving property of Q it is clear that all eigenvalues of Q lie on the unit circle $|z| = 1$ in the complex plane. These eigenvalues are in general non-real, as are the coordinates of the associated eigenvectors, even though the elements of Q are real.

Over the field $\mathbb{C}$ of complex numbers, if $A = [a_{ij}]$ is a square matrix, then set $A^* = \overline{A^t}$, that is, $A^* = [\overline{a_{ji}}]$. This matrix is the *adjoint* of A. A matrix is said to be *Hermitian* if $A = A^*$. That is, $a_{ji} = \overline{a_{ij}}$ for all i and j. In particular, the diagonal elements a_{ii} are real. Hermitian matrices over $\mathbb{C}$ correspond to symmetric matrices over $\mathbb{R}$. Like a symmetric matrix, the eigenvalues of a Hermitian matrix are all real. An $n \times n$ matrix U is *unitary* if $U^*U = I$. Indeed, the following are equivalent:

U is a unitary matrix;

$U^*U = I$;

The columns of U form an orthonormal family;

The rows of U form an orthonormal family;

$\|U\boldsymbol{v}\| = \|\boldsymbol{v}\|$ for all $\boldsymbol{v} \in \mathbb{C}^n$.

For any Hermitian matrix A there exists a unitary matrix U such that

$$A = UDU^*$$

where D is a diagonal matrix whose diagonal elements are the eigenvalues of A (all of which are real). Moreover, the columns of U are the corresponding eigenvectors; they form an orthonormal basis for $\mathbb{C}^n$. From the distance-preserving property of U it is clear that all eigenvalues of U lie on the unit circle $|z| = 1$ in the complex plane.

Exercises

1. If x, y, and z are real numbers such that $x^2 + y^2 + z^2 = 1$, then how large can $2x + 3y + 6z$ be? What choice of the variables gives the maximum?
2. Let $a_1, a_2, \dots, a_n$ be any real numbers. Show that
$$\frac{a_1 + a_2 + \cdots + a_n}{n} \le \left(\frac{a_1^2 + a_2^2 + \cdots + a_n^2}{n}\right)^{1/2}.$$
3. Let
$$f(x) = \frac{1}{n}\sum_{j=1}^{n}(a_j - x)^2$$
be the mean square distance of the numbers a_i from x.
   (a) Find $f'(x)$ and all critical points of f.
   (b) Find $f''(x)$, and show that $f''(x) > 0$.

(c) Let A denote the average of the a_i. Show that

$$\min_x f(x) = f(A) = \frac{1}{n}\sum_{j=1}^{n} a_j^2 - A^2.$$

(From the definition of $f(x)$ it is clear that $f(x) \geq 0$ for all x. The right hand side above is not so obviously nonnegative, but that fact can be seen by the preceding exercise.)

4. Let $a_1, a_2, \ldots$ be an infinite sequence of positive real numbers. Show that the inequalities

$$\sum_{n=1}^{\infty} \frac{a_n}{n} < \infty \qquad \sum_{n=1}^{\infty} \frac{1}{a_n n} < \infty$$

cannot both hold.

L.2. Abstract vector spaces

A set V is called a vector space over $\mathbb{R}$ if the following hold:

Table L.1. Definition of a vector space over $\mathbb{R}$

(1) For every pair of vectors $\boldsymbol{u}, \boldsymbol{v}$ in V, there is a vector in V called $\boldsymbol{u} + \boldsymbol{v}$.
(2) $\boldsymbol{u} + \boldsymbol{v} = \boldsymbol{v} + \boldsymbol{u}$ for all $\boldsymbol{u}, \boldsymbol{v}$ in V.
(3) $(\boldsymbol{u} + \boldsymbol{v}) + \boldsymbol{w} = \boldsymbol{u} + (\boldsymbol{v} + \boldsymbol{w})$ for all $\boldsymbol{u}, \boldsymbol{v}, \boldsymbol{w}$ in V.
(4) There is a vector in V called $\boldsymbol{0}$ such that $\boldsymbol{0} + \boldsymbol{v} = \boldsymbol{v}$ for all $\boldsymbol{v}$ in V.
(5) For every vector $\boldsymbol{v}$ in V there is a vector in V called $-\boldsymbol{v}$ such that $\boldsymbol{v} + (-\boldsymbol{v}) = \boldsymbol{0}$.
(6) For every real number c and every vector $\boldsymbol{v}$ in V there is a vector in V called $c\boldsymbol{v}$.
(7) $c(\boldsymbol{u} + \boldsymbol{v}) = c\boldsymbol{u} + c\boldsymbol{v}$ for all real numbers c and all vectors $\boldsymbol{u}, \boldsymbol{v}$ in V.
(8) $(a + b)\boldsymbol{v} = a\boldsymbol{v} + b\boldsymbol{v}$ for all real numbers a, b and all vectors $\boldsymbol{v}$ in V.
(9) $a(b\boldsymbol{v}) = (ab)\boldsymbol{v}$ for all real numbers a, b and all vectors $\boldsymbol{v}$ in V.
(10) $1\boldsymbol{v} = \boldsymbol{v}$ for all $\boldsymbol{v}$ in V.

A vector space over $\mathbb{C}$ is defined similarly, with all references to real numbers changed to complex numbers. Indeed, the scalars can come from an arbitrary field F, and then we speak of a vector space V over F.

In the case of a vector space over $\mathbb{R}$ we sometimes have additional structure. We say that the vector space V is an *inner product space* when the properties in Table L.2 hold.

By using both properties (2) and (3) we see that the inner product is linear in its first argument, in the sense that

$$\langle a\boldsymbol{u} + b\boldsymbol{v}, \boldsymbol{w}\rangle = a\langle \boldsymbol{u}, \boldsymbol{w}\rangle + b\langle \boldsymbol{v}, \boldsymbol{w}\rangle\,. \tag{L.8}$$

By using property (4), then (L.8), and finally (4) again, we deduce that the inner product is also linear in its second argument:

$$\langle \boldsymbol{u}, a\boldsymbol{v} + b\boldsymbol{w}\rangle = a\langle \boldsymbol{u}, \boldsymbol{v}\rangle + b\langle \boldsymbol{u}, \boldsymbol{w}\rangle\,. \tag{L.9}$$

Table L.2. Definition of an inner product space over $\mathbb{R}$

(1) For all vectors $\boldsymbol{u}, \boldsymbol{v}$ in V, there is a real number called $\langle \boldsymbol{u}, \boldsymbol{v} \rangle$.
(2) $\langle a\boldsymbol{u}, \boldsymbol{v} \rangle = a\langle \boldsymbol{u}, \boldsymbol{v} \rangle$ for all real numbers a and all vectors $\boldsymbol{u}, \boldsymbol{v}$ in V.
(3) $\langle \boldsymbol{u} + \boldsymbol{v}, \boldsymbol{w} \rangle = \langle \boldsymbol{u}, \boldsymbol{w} \rangle + \langle \boldsymbol{v}, \boldsymbol{w} \rangle$ for all vectors $\boldsymbol{u}, \boldsymbol{v}, \boldsymbol{w}$ in V.
(4) $\langle \boldsymbol{u}, \boldsymbol{v} \rangle = \langle \boldsymbol{v}, \boldsymbol{u} \rangle$ for all vectors $\boldsymbol{u}, \boldsymbol{v}$ in V.
(5) $\langle \boldsymbol{v}, \boldsymbol{v} \rangle \ge 0$ for all vectors $\boldsymbol{v}$ in V.
(6) $\langle \boldsymbol{v}, \boldsymbol{v} \rangle = 0$ if and only if $\boldsymbol{v} = \boldsymbol{0}$.

Of course we can use both (L.8) and (L.9) to obtain a still more general identity, namely

$$\text{(L.10)} \quad \begin{aligned} &\langle a_1\boldsymbol{u}_1 + a_2\boldsymbol{u}_2, b_1\boldsymbol{v}_1 + b_2\boldsymbol{v}_2 \rangle \\ &\qquad = a_1b_1\langle \boldsymbol{u}_1, \boldsymbol{v}_1 \rangle + a_1b_2\langle \boldsymbol{u}_1, \boldsymbol{v}_2 \rangle + a_2b_1\langle \boldsymbol{u}_2, \boldsymbol{v}_1 \rangle + a_2b_2\langle \boldsymbol{u}_2, \boldsymbol{v}_2 \rangle \,. \end{aligned}$$

In the case of a vector space V over $\mathbb{C}$, we may also be able to define an inner product that should satisfy the same axioms as above, except that (4) should be replaced by

(4′) $\langle \boldsymbol{u}, \boldsymbol{v} \rangle = \overline{\langle \boldsymbol{v}, \boldsymbol{u} \rangle}$ for all vectors $\boldsymbol{u}, \boldsymbol{v}$ in V.

In this situation, (L.8) holds as before, but (L.9) becomes

$$\text{(L.9}'\text{)} \qquad \langle \boldsymbol{u}, a\boldsymbol{v} + b\boldsymbol{w} \rangle = \overline{a}\langle \boldsymbol{u}, \boldsymbol{v} \rangle + \overline{b}\langle \boldsymbol{u}, \boldsymbol{w} \rangle,$$

and hence (L.10) is replaced by

$$\text{(L.10}'\text{)} \quad \begin{aligned} &\langle a_1\boldsymbol{u}_1 + a_2\boldsymbol{u}_2, b_1\boldsymbol{v}_1 + b_2\boldsymbol{v}_2 \rangle \\ &\qquad = a_1\overline{b_1}\langle \boldsymbol{u}_1, \boldsymbol{v}_1 \rangle + a_1\overline{b_2}\langle \boldsymbol{u}_1, \boldsymbol{v}_2 \rangle + a_2\overline{b_1}\langle \boldsymbol{u}_2, \boldsymbol{v}_1 \rangle + a_2\overline{b_2}\langle \boldsymbol{u}_2, \boldsymbol{v}_2 \rangle \,. \end{aligned}$$

From these axioms we shall derive several important results. One advantage of the axiomatic approach is that these results need to be proved only once, rather than repeatedly in each instance. Also, by reasoning in this way, we can see more clearly what characteristics of a situation yield various consequences. We concentrate on inner product spaces over $\mathbb{C}$. The corresponding proofs for an inner product space over $\mathbb{R}$ are similar, but simpler, and are left as exercises.

Theorem L.1. (The Schwarz inequality) *Let $\boldsymbol{u}$ and $\boldsymbol{v}$ be any vectors in an inner product space over $\mathbb{C}$. Then*

$$|\langle \boldsymbol{u}, \boldsymbol{v} \rangle|^2 \;\le\; \langle \boldsymbol{u}, \boldsymbol{u} \rangle \langle \boldsymbol{v}, \boldsymbol{v} \rangle \,.$$

We define the *norm* of a vector to be $\|\boldsymbol{v}\| = \sqrt{\langle \boldsymbol{v}, \boldsymbol{v} \rangle}$. Thus the above may be expressed equivalently as the inequality

$$|\langle \boldsymbol{u}, \boldsymbol{v} \rangle| \;\le\; \|\boldsymbol{u}\| \, \|\boldsymbol{v}\| \,.$$

Proof. We assume first that $\boldsymbol{u} \ne \boldsymbol{0}$. By property (5) we know that

$$\langle \lambda\boldsymbol{u} + \boldsymbol{v}, \lambda\boldsymbol{u} + \boldsymbol{v} \rangle \ge 0$$

for any complex number λ. By (L.10′) we know that the left hand side above is

$$= |\lambda|^2\langle \boldsymbol{u}, \boldsymbol{u} \rangle + \lambda\langle \boldsymbol{u}, \boldsymbol{v} \rangle + \overline{\lambda}\langle \boldsymbol{v}, \boldsymbol{u} \rangle + \langle \boldsymbol{v}, \boldsymbol{v} \rangle \,.$$

By (4′) we see further that this is

$$(\mathrm{L}.11) \quad = |\lambda|^2\langle \boldsymbol{u},\boldsymbol{u}\rangle + \lambda\langle \boldsymbol{u},\boldsymbol{v}\rangle + \bar{\lambda}\,\overline{\langle \boldsymbol{u},\boldsymbol{v}\rangle} + \langle \boldsymbol{v},\boldsymbol{v}\rangle = |\lambda|^2\langle \boldsymbol{u},\boldsymbol{u}\rangle + 2\,\mathrm{Re}\,\big(\lambda\langle \boldsymbol{u},\boldsymbol{v}\rangle\big) + \langle \boldsymbol{v},\boldsymbol{v}\rangle\,.$$

From our assumption that $\boldsymbol{u} \neq \boldsymbol{0}$ it follows by properties (5) and (6) that $\langle \boldsymbol{u},\boldsymbol{u}\rangle > 0$. Thus we may take

$$\lambda = \frac{-\overline{\langle \boldsymbol{u},\boldsymbol{v}\rangle}}{\langle \boldsymbol{u},\boldsymbol{u}\rangle}\,.$$

With this choice of λ, we find that the expression (L.11) is

$$= \frac{|\langle \boldsymbol{u},\boldsymbol{v}\rangle|^2}{\langle \boldsymbol{u},\boldsymbol{u}\rangle} - 2\frac{|\langle \boldsymbol{u},\boldsymbol{v}\rangle|^2}{\langle \boldsymbol{u},\boldsymbol{u}\rangle} + \langle \boldsymbol{v},\boldsymbol{v}\rangle = -\frac{|\langle \boldsymbol{u},\boldsymbol{v}\rangle|^2}{\langle \boldsymbol{u},\boldsymbol{u}\rangle} + \langle \boldsymbol{v},\boldsymbol{v}\rangle\,.$$

Since this last quantity is ≥ 0, we obtain the desired inequality when $\boldsymbol{u} \neq \boldsymbol{0}$. If $\boldsymbol{u} = \boldsymbol{0}$ but $\boldsymbol{v} \neq \boldsymbol{0}$, we may repeat the same argument with the roles of $\boldsymbol{u}$ and $\boldsymbol{v}$ interchanged, to obtain the result again. Finally, if $\boldsymbol{u} = \boldsymbol{v} = \boldsymbol{0}$, then the inequality is still true, in view of property (6) of the inner product. □

Theorem L.2. (The triangle inequality) *Let* $\boldsymbol{u}$ *and* $\boldsymbol{v}$ *be any vectors in an inner product space over* $\mathbb{C}$. *Then*

$$\|\boldsymbol{u}+\boldsymbol{v}\| \leq \|\boldsymbol{u}\| + \|\boldsymbol{v}\|\,.$$

Proof. The square of the left hand side is

$$(\mathrm{L}.12) \qquad \|\boldsymbol{u}+\boldsymbol{v}\|^2 = \langle \boldsymbol{u}+\boldsymbol{v}, \boldsymbol{u}+\boldsymbol{v}\rangle\,.$$

By (L.10′) this is

$$(\mathrm{L}.13) \qquad = \langle \boldsymbol{u},\boldsymbol{u}\rangle + \langle \boldsymbol{u},\boldsymbol{v}\rangle + \langle \boldsymbol{v},\boldsymbol{u}\rangle + \langle \boldsymbol{v},\boldsymbol{v}\rangle$$

$$(\mathrm{L}.14) \qquad = \langle \boldsymbol{u},\boldsymbol{u}\rangle + 2\,\mathrm{Re}\langle \boldsymbol{u},\boldsymbol{v}\rangle + \langle \boldsymbol{v},\boldsymbol{v}\rangle$$

by (4′). But $\mathrm{Re}\langle \boldsymbol{u},\boldsymbol{v}\rangle \leq |\langle \boldsymbol{u},\boldsymbol{v}\rangle|$, which by the Schwarz Inequality is $\leq \|\boldsymbol{u}\|\,\|\boldsymbol{v}\|$. Thus the above is

$$(\mathrm{L}.15) \qquad \leq \|\boldsymbol{u}\|^2 + 2\|\boldsymbol{u}\|\,\|\boldsymbol{v}\| + \|\boldsymbol{v}\|^2$$

$$(\mathrm{L}.16) \qquad = (\|\boldsymbol{u}\| + \|\boldsymbol{v}\|)^2\,.$$

This is the square of the right hand side of the proposed inequality, so we have the stated result. □

If V is a subspace of $\mathbb{R}^n$, then there is a complementary subspace $V^\perp$ called "V-perp", which consists of all vectors $\boldsymbol{w}$ such that $\langle \boldsymbol{v},\boldsymbol{w}\rangle = 0$ for all $\boldsymbol{v} \in V$. Every vector $\boldsymbol{v}$ in $\mathbb{R}^n$ can be uniquely expressed as a sum of a member of V and a member of $V^\perp$. Indeed, if $\varphi_1, \varphi_2, \ldots, \varphi_R$ is an orthonormal basis for V, and we set

$$\begin{aligned} \boldsymbol{v}_1 &= \langle \boldsymbol{v},\varphi_1\rangle\varphi_1 + \langle \boldsymbol{v},\varphi_2\rangle\varphi_2 + \cdots + \langle \boldsymbol{v},\varphi_R\rangle\varphi_R \\ \boldsymbol{v}_2 &= \boldsymbol{v} - \boldsymbol{v}_1, \end{aligned}$$

then $\boldsymbol{v} = \boldsymbol{v}_1 + \boldsymbol{v}_2$, $\boldsymbol{v}_1 \in V$, $\boldsymbol{v}_2 \in V^\perp$ and $\|\boldsymbol{v}\|^2 = \|\boldsymbol{v}_1\|^2 + \|\boldsymbol{v}_2\|^2$. This line of reasoning serves as a guide in the derivation of

Theorem L.3. (Bessel's Inequality) *Let V be an inner product space over $\mathbb{C}$, and suppose that $\varphi_1, \varphi_2, \ldots, \varphi_R$ is an orthonormal collection of vectors in V. Then for any $\boldsymbol{v} \in V$, we have*

$$\sum_{r=1}^{R} |\langle \boldsymbol{v}, \boldsymbol{\varphi}_r \rangle|^2 \le \|\boldsymbol{v}\|^2 .$$

Proof. For brevity we put $c_r = \langle \boldsymbol{v}, \boldsymbol{\varphi}_r \rangle$. By property (5) of an inner product we know that

$$0 \le \left\| \boldsymbol{v} - \sum_{r=1}^{R} c_r \boldsymbol{\varphi}_r \right\|^2 .$$

We note that in general, $\|\boldsymbol{a} - \boldsymbol{b}\|^2 = \|\boldsymbol{a}\|^2 - 2\,\mathrm{Re}\langle \boldsymbol{a}, \boldsymbol{b} \rangle + \|\boldsymbol{b}\|^2$. Hence the above is

$$= \|\boldsymbol{v}\|^2 - 2\,\mathrm{Re}\left\langle \boldsymbol{v}, \sum_{r=1}^{R} c_r \boldsymbol{\varphi}_r \right\rangle + \left\| \sum_{r=1}^{R} c_r \boldsymbol{\varphi}_r \right\|^2 . \tag{L.17}$$

By (L.9′) we see that

$$\left\langle \boldsymbol{v}, \sum_{r=1}^{R} c_r \boldsymbol{\varphi}_r \right\rangle = \sum_{r=1}^{R} \overline{c_r} \langle \boldsymbol{v}, \boldsymbol{\varphi}_r \rangle = \sum_{r=1}^{R} |c_r|^2$$

and that

$$\begin{aligned} \left\| \sum_{r=1}^{R} c_r \boldsymbol{\varphi}_r \right\|^2 &= \left\langle \sum_{r=1}^{R} c_r \boldsymbol{\varphi}_r, \sum_{r=1}^{R} c_r \boldsymbol{\varphi}_r \right\rangle \\ &= \sum_{r=1}^{R} \sum_{s=1}^{R} c_r \overline{c_s} \langle \boldsymbol{\varphi}_r, \boldsymbol{\varphi}_s \rangle . \end{aligned}$$

But $\langle \boldsymbol{\varphi}_r, \boldsymbol{\varphi}_s \rangle = 1$ if $r = s$, and is 0 otherwise. Hence the above is

$$= \sum_{r=1}^{R} |c_r|^2 .$$

On inserting these findings in (L.17), we discover that

$$0 \le \|\boldsymbol{v}\|^2 - \sum_{r=1}^{R} |c_r|^2 .$$

This is the desired inequality. □

L.3. Circulant matrices

Let $C = [c_{mn}]$ be a $q \times q$ matrix. We say that C is a *circulant matrix* if there is a real or complex-valued arithmetic function $a(n)$ of period q such that $c_{mn} = a(n - m)$ for all m and n. Thus when $q = 3$,

$$C = \begin{bmatrix} a(0) & a(1) & a(2) \\ a(2) & a(0) & a(1) \\ a(1) & a(2) & a(0) \end{bmatrix} .$$

Let $\boldsymbol{v}_k \in \mathbb{C}^q$ be a column vector whose n^{th} coordinate is $e(-kn/q)/\sqrt{q}$. Then $C\boldsymbol{v}_k$ is a column vector whose m^{th} coordinate is

$$\begin{aligned}\sum_{n=1}^{n} a(n-m)e(-kn/q)/\sqrt{q} &= \frac{e(-mk/q)}{\sqrt{q}} \sum_{n=1}^{q} a(n-m)e(-k(n-m)/q) \\ &= \widehat{a}(k)\frac{e(-mk/q)}{\sqrt{q}}\,.\end{aligned}$$

That is, $C\boldsymbol{v}_k = \widehat{a}(k)\boldsymbol{v}_k$. Thus $\widehat{a}(k)$ is an eigenvalue of C, and $\boldsymbol{v}_k$ is the associated eigenvector. Let $U = [u_{nk}]$ be the $q \times q$ matrix whose columns are the $\boldsymbol{v}_k$. Thus $u_{nk} = e(-nk/q)/\sqrt{q}$. By taking inner products we find that the columns of U are orthonormal, which is to say that U is unitary. Let $CU = B = [b_{mk}]$. Thus $b_{mk} = \widehat{a}(k)e(-km/q)/\sqrt{q}$. Let $V = [v_{jm}] = U^*$ be the adjoint of U. That is, $v_{jm} = e(jm/q)/\sqrt{q}$. Put $U^*B = D = [d_{jk}]$. Thus

$$d_{jk} = \sum_{m=1}^{q} \frac{e(jm/q)}{\sqrt{q}}\widehat{a}(k)\frac{e(-km/q)}{\sqrt{q}} = \begin{cases} \widehat{a}(k) & (j=k), \\ 0 & (\text{otherwise}). \end{cases}$$

Hence

$$U^*CU = D, \tag{L.18}$$

and so the circulant matrix C is unitarily similar to the diagonal matrix $D = \operatorname{Diag}\big(\widehat{a}(k)\big)$. Of course it follows that

$$\det C = \prod_{k=1}^{q} \widehat{a}(k)\,. \tag{L.19}$$

Notes

L.1. Although our notation is quite widely used, some authors call a different matrix the adjoint. Also, some write A^{H} for our A^*, and call it the "Hermitian transpose".

Appendix O

Orders of Magnitude

The terms and notations we define here are fairly standard throughout pure mathematics and most of applied mathematics, but in other sciences (e.g., physics, astronomy, chemistry, etc.) may have significantly differently meanings.

We say that $f(x)$ is asymptotic to $g(x)$ as x tends to some limiting value a, and write $f(x) \sim g(x)$ as $x \to a$, if

$$\lim_{x \to a} \frac{f(x)}{g(x)} = 1 .$$

For example,

$$\cot x \sim \frac{1}{x}$$

as $x \to 0$. We say that $f(x)$ is big-oh of $g(x)$, and write $f(x) = O(g(x))$ when there is a constant C such that

$$|f(x)| \leq Cg(x)$$

for all x under consideration. Note that in this situation, the function $g(x)$ must be nonnegative. A constant C for which the above holds is referred to as the *implicit constant*. For example,

$$\sin x = O(1),$$

which is just a way of saying that $\sin x$ is a bounded function. Also,

$$\sin x = O(|x|).$$

Both of these assertions are true for all x, and in both cases the implicit constant can be taken to be 1. Often we use the big-oh notation to indicate a bound on the degree of precision of an estimate. For example, we might write

$$\cos x = 1 - \frac{1}{2}x^2 + O(x^4) .$$

Here we are making an assertion about an implicitly defined function, namely the difference between $\cos x$ and $1 - \frac{1}{2}x^2$. Thus the above asserts that if

$$f(x) = \cos x - \left(1 - \frac{1}{2}x^2\right),$$

then $f(x) = O(x^4)$.

In the above examples the implicit constants are absolute. However, this notation is also used a little more vaguely. For example, if $f(x)$ is continuous on the interval $[0,1]$, then $|f(x)|$ is bounded. So we might write $f(x) = O(1)$, which is true, but the implicit constant depends on f. Whether the implicit constant is absolute, or depends on the functions in play is generally not stated explicitly, but is more often left as an implicit understanding between the author and the reader. Sometimes it is useful to admit dependence of constants on functions or parameters in play, in which case one can indicate the dependence with a subscript. In the above situation we might write $f = O_f(1)$. As another example, we note that we can derive an asymptotic formula for the logarithmic integral by repeated integration by parts:

$$\mathrm{li}\,(x) := \int_2^x \frac{1}{\log u}\,du = x\sum_{k=1}^{K-1} \frac{(k-1)!}{(\log x)^k} + O_K\big(x(\log x)^{-K}\big).$$

We say that $f(x) = o(g(x))$ as x tends to some number a if

$$\lim_{x\to a} \frac{f(x)}{g(x)} = 0\,.$$

For example, if $f(x) = 1/x$, then $f(x) = o(1)$ as $x \to \infty$.

We say that two functions f and g are comparable, and write $f(x) \asymp g(x)$ if there exist constants $C > c > 0$ such that

$$cf(x) \le g(x) \le Cf(x)$$

for all x under consideration. In other words, $f = O(g)$ and $g = O(f)$.

Appendix T

Trigonometry

We generally measure angles in radians, and convert degrees to radians by multiplying by $\pi/180$. Similarly, we can convert back from radians to degrees by multiplying by $180/\pi$. Thus a full rotation of $360°$ is 2π radians, the a straight angle of $180°$ is π radians, and a right angle of $90°$ is $\pi/2$ radians. The sine function is denoted $\sin\theta$, the cosine function $\cos\theta$, the tangent function $\tan\theta$, the cotangent function $\cot\theta$, the secant function $\sec\theta$, and the cosecant function $\csc\theta$ or $\operatorname{cosec}\theta$. In §T.1 below we define these functions in terms of triangles, and in §T.2 we give more general definitions in terms of power series.

T.1. Trigonometric functions in plane geometry

In the right triangle of Figure T.1 (a), we have

$$\text{(T.1)} \qquad \sin\theta = \frac{b}{c} \qquad \cos\theta = \frac{a}{c} \qquad \tan\theta = \frac{b}{a} = \frac{\sin\theta}{\cos\theta}$$

$$\text{(T.2)} \qquad \sec\theta = \frac{c}{a} = \frac{1}{\cos\theta} \qquad \csc\theta = \frac{c}{b} = \frac{1}{\sin\theta} \qquad \cot\theta = \frac{a}{b} = \frac{1}{\tan\theta}$$

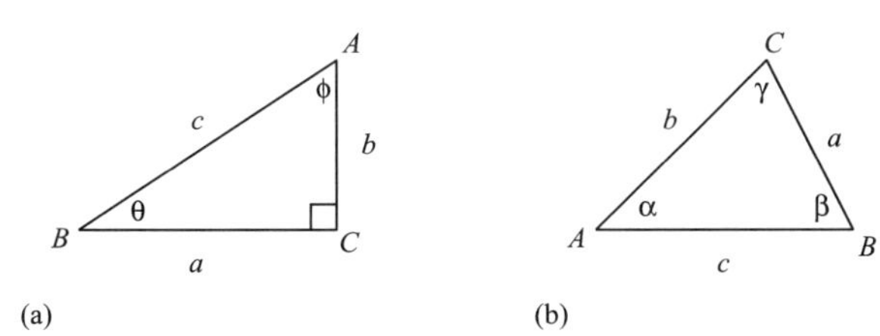

Figure T.1. (a) A right triangle; (b) A general triangle.

Pythagoreas's theorem asserts that $a^2 + b^2 = c^2$, so

$$\text{(T.3)} \qquad \sin^2\theta + \cos^2\theta = 1 \qquad \tan^2\theta + 1 = \sec^2\theta \qquad 1 + \cot^2\theta = \csc^2\theta\,.$$

For the more general triangle of Figure T.1 (b), with angles α, β, and γ, and opposite sides of lengths a, b, and c, the *law of sines* asserts that

$$\frac{a}{\sin\alpha} = \frac{b}{\sin\beta} = \frac{c}{\sin\gamma} = d \tag{T.4}$$

where d is the diameter of the circumscribed circle. For this same triangle, the *law of cosines* states that

$$c^2 = a^2 + b^2 - 2ab\cos\gamma\,. \tag{T.5}$$

If $\gamma = \pi/2$, then this is again the Pythagorean theorem. If γ is acute (i.e., $\gamma < \pi/2$), then $c^2 < a^2 + b^2$, while if γ is obtuse (i.e., $\gamma > \pi/2$), then $c^2 > a^2 + b^2$.

For a general real number θ, the point $(\cos\theta, \sin\theta)$ lies on the circle $x^2+y^2 = 1$, and makes an angle θ with the positive x-axis. If we allow the angle to increase from θ to $\theta + 2\pi$, then we are back where we started, which is to say that

$$\sin(\theta + 2\pi) = \sin\theta, \tag{T.6}$$

$$\cos(\theta + 2\pi) = \cos\theta\,. \tag{T.7}$$

That is, the sine and cosine functions are periodic with period 2π. From (T.2) it follows that the secant and cosecant functions also have period 2π. If we allow the angle to increase from θ to $\theta + \pi$, then we move halfway around the circle to the point $(-\cos\theta, -\sin\theta)$. Thus

$$\sin(\theta + \pi) = -\sin\theta, \tag{T.8}$$

$$\cos(\theta + \pi) = -\cos\theta\,. \tag{T.9}$$

On combining this with the definition (T.1) of $\tan\theta$, we deduce that the tangent and cotangent functions have period π.

Let T_1 be the right triangle with vertices $(0,0)$, $(\cos\theta, 0)$, and $(\cos\theta, \sin\theta)$. If we reflect this triangle about the x-axis, we obtain a triangle T_2 with vertices $(0,0)$, $(\cos\theta, 0)$, $(\cos\theta, -\sin\theta)$. But this new triangle makes an angle $-\theta$ at the origin, and so we deduce that

$$\sin(-\theta) = -\sin\theta, \tag{T.10}$$

$$\cos(-\theta) = \cos\theta, \tag{T.11}$$

$$\tan(-\theta) = -\tan\theta\,. \tag{T.12}$$

That is, sine and tangent are odd functions, while cosine is even.

For sums of two angles we have

$$\sin(\theta \pm \phi) = \sin\theta\cos\phi \pm \cos\theta\sin\phi, \tag{T.13}$$

$$\cos(\theta \pm \phi) = \cos\theta\cos\phi \mp \sin\theta\sin\phi, \tag{T.14}$$

$$\tan(\theta \pm \phi) = \frac{\tan\theta \pm \tan\phi}{1 \mp \tan\theta\tan\phi}, \tag{T.15}$$

$$\cot(\theta \pm \phi) = \frac{\pm\cot\theta\,\cot\phi - 1}{\cot\theta \pm \cot\phi}\,. \tag{T.16}$$

The first two of these formulas can be proved by geometric reasoning, while the others follow by division. The identities (T.8) and (T.9) can be derived from (T.13) and (T.14), using values found in Table T.1. Similarly, from (T.13), (T.14), and values in Table T.1 we see that

(T.17) $$\sin(\theta + \pi/2) = \cos\theta,$$

(T.18) $$\cos(\theta + \pi/2) = -\sin\theta,$$

(T.19) $$\tan(\theta + \pi/2) = -\cot\theta\,.$$

In general, a function of the form $\sin(x + c)$ is a linear combination of $\sin x$ and $\cos x$. Conversely, a function of the form

(T.20) $$f(x) = a\cos x + b\sin x$$

can be expressed in the form

(T.21) $$f(x) = C\sin(x + c)$$

where $C = \sqrt{a^2 + b^2}$ and c is chosen so that

(T.22) $$\sin c = \frac{b}{\sqrt{a^2 + b^2}}, \qquad \cos c = \frac{a}{\sqrt{a^2 + b^2}}\,.$$

From (T.13) and (T.14) we easily find that

(T.23) $$\sin\theta\sin\phi = \frac{-1}{2}\cos(\theta + \phi) + \frac{1}{2}\cos(\theta - \phi),$$

(T.24) $$\sin\theta\cos\phi = \frac{1}{2}\sin(\theta + \phi) + \frac{1}{2}\sin(\theta - \phi),$$

(T.25) $$\cos\theta\cos\phi = \frac{1}{2}\cos(\theta + \phi) + \frac{1}{2}\cos(\theta - \phi)\,.$$

By taking $\alpha = \theta + \phi$ and $\beta = \theta - \phi$ we obtain a set of formulæ that run in the opposite direction, namely

(T.26) $$\sin\alpha + \sin\beta = 2\sin\frac{\alpha + \beta}{2}\cdot\cos\frac{\alpha - \beta}{2},$$

(T.27) $$\sin\alpha - \sin\beta = 2\cos\frac{\alpha + \beta}{2}\cdot\sin\frac{\alpha - \beta}{2},$$

(T.28) $$\cos\alpha + \cos\beta = 2\cos\frac{\alpha + \beta}{2}\cdot\cos\frac{\alpha - \beta}{2},$$

(T.29) $$\cos\alpha - \cos\beta = -2\sin\frac{\alpha + \beta}{2}\cdot\sin\frac{\alpha - \beta}{2}\,.$$

For tangent and cotangent we have

(T.30) $$\tan\theta \pm \tan\phi = \frac{\sin(\theta \pm \phi)}{\cos\theta\cos\phi}$$

(T.31) $$\cot\theta \pm \cot\phi = \pm\frac{\sin(\phi \pm \theta)}{\sin\theta\sin\phi}$$

By taking $\phi = \theta$ in (T.13)–(T.15) we find in particular that

(T.32) $$\sin 2\theta = 2\sin\theta\cos\theta,$$

(T.33) $$\cos 2\theta = \cos^2\theta - \sin^2\theta = 2\cos^2\theta - 1 = 1 - 2\sin^2\theta,$$

(T.34) $$\tan 2\theta = \frac{2\tan\theta}{1 - \tan^2\theta}\,.$$

Table T.1. Trigonometric functions at familiar angles.

degrees	radians	$\sin\theta$	$\cos\theta$	$\tan\theta$
0	0	0	1	0
30	$\pi/6$	$1/2$	$\sqrt{3}/2$	$1/\sqrt{3}$
45	$\pi/4$	$1/\sqrt{2}$	$1/\sqrt{2}$	1
60	$\pi/3$	$\sqrt{3}/2$	$1/2$	$\sqrt{3}$
90	$\pi/2$	1	0	∞
120	$2\pi/3$	$\sqrt{3}/2$	$-1/2$	$-\sqrt{3}$
135	$3\pi/4$	$1/\sqrt{2}$	$-1/\sqrt{2}$	-1
150	$5\pi/6$	$1/2$	$-\sqrt{3}/2$	$-1/\sqrt{3}$
180	π	0	-1	0

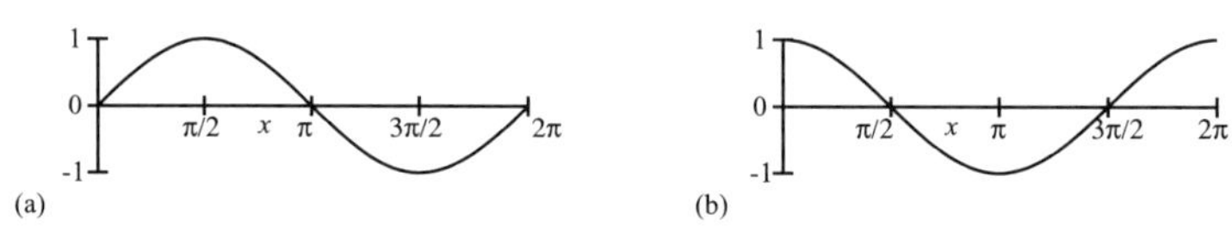

Figure T.2. (a) Graph of $\sin x$ for $0 \leq x \leq 2\pi$; (b) Graph of $\cos x$ for $0 \leq x \leq 2\pi$.

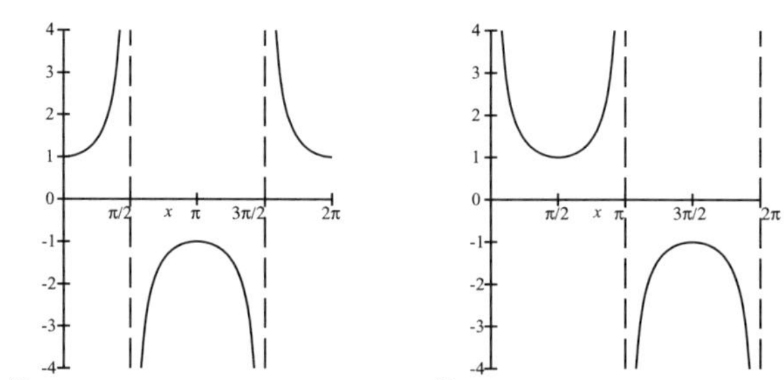

Figure T.3. (a) Graph of $\sec x$ for $0 \leq x \leq 2\pi$; (b) Graph of $\csc x$ for $0 \leq x \leq 2\pi$.

More generally, $\cos n\theta$ can be written as a polynomial in $\cos\theta$,

$$\cos n\theta = T_n(\cos\theta), \tag{T.35}$$

and $\sin n\theta$ is $\sin\theta$ times a polynomial in $\cos\theta$,

$$\sin n\theta = U_n(\cos\theta)\sin\theta\,. \tag{T.36}$$

Here the T_n and U_n are the Chebyshev polynomials of the first and second kinds, as discussed in Appendix C.

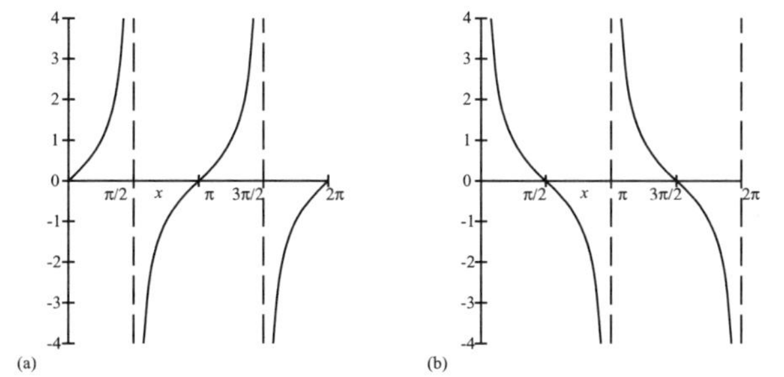

Figure T.4. (a) Graph of $\tan x$ for $0 \le x \le 2\pi$; (b) Graph of $\cot x$ for $0 \le x \le 2\pi$.

From (T.33) we obtain the half angle formulas

$$\sin \frac{1}{2}\theta = \pm\sqrt{\frac{1-\cos\theta}{2}}, \tag{T.37}$$

$$\cos \frac{1}{2}\theta = \pm\sqrt{\frac{1+\cos\theta}{2}}, \tag{T.38}$$

$$\tan \frac{1}{2}\theta = \pm\sqrt{\frac{1-\cos\theta}{1+\cos\theta}} = \frac{1-\cos\theta}{\sin\theta} = \frac{\sin\theta}{1+\cos\theta}. \tag{T.39}$$

For the differences between squares of sines and cosines we have

$$\sin^2\theta - \sin^2\phi = \sin(\theta+\phi)\,\sin(\theta-\phi), \tag{T.40}$$

$$\cos^2\theta - \cos^2\phi = -\sin(\theta+\phi)\,\sin(\theta-\phi), \tag{T.41}$$

$$\cos^2\theta - \sin^2\phi = \cos(\theta+\phi)\,\cos(\theta-\phi)\,. \tag{T.42}$$

Exercises

1. (a) In the triangle of Figure T.1 (b), show that $c = b\cos\alpha + a\cos\beta$, and note that this is still true if α or β is obtuse.
   (b) Multiply both sides of the identity from (a) by c to see that

$$c^2 = bc\cos\alpha + ac\cos\beta\,. \tag{T.43}$$

   Note that similarly,

$$a^2 = ab\cos\gamma + ac\cos\beta, \tag{T.44}$$

$$b^2 = ab\cos\gamma + bc\cos\alpha\,. \tag{T.45}$$

   (c) Add the identities (T.44), (T.45), and subtract (T.43) from this. The resulting left hand side is $a^2 + b^2 - c^2$. What is the right hand side?
   (d) Deduce the law of cosines (T.5).

2. In this exercise we use complex numbers to derive the law of cosines. For the triangle $\triangle ABC$ we take $C = 0$, $A = b$, and $B = a(\cos\gamma + i\sin\gamma)$. Here a and b are positive real numbers. The side CA has length b, the side CB has length a,

and the angle between these two sides is γ. The third side has length $c = |B-A|$. Use the identity $|x+iy|^2 = x^2+y^2$ to deduce that $c^2 = (a\cos\gamma - b)^2 + (a\sin\gamma)^2$. Simplify to obtain (T.5).

3. (a) In the triangle of Figure T.1 (b), show that the height of the triangle is $b\sin\alpha$.
(b) In the triangle of Figure T.1 (b), show that the height of the triangle is $a\sin\beta$.
(c) Conclude that

$$\frac{a}{\sin\alpha} = \frac{b}{\sin\beta}\,.$$

4. For the triangle of Figure T.1 (b), the *law of tangents* asserts that

$$\frac{a-b}{a+b} = \frac{\tan\frac{1}{2}(\alpha-\beta)}{\tan\frac{1}{2}(\alpha+\beta)}\,. \tag{T.46}$$

(a) Use (T.26) and (T.27) to show that

$$\frac{\tan\frac{1}{2}(\alpha-\beta)}{\tan\frac{1}{2}(\alpha+\beta)} = \frac{\sin\alpha - \sin\beta}{\sin\alpha + \sin\beta}\,.$$

(b) Let d be defined as in (T.4), so that $a = d\sin\alpha$ and $b = d\sin\beta$. Show that

$$\frac{a-b}{a+b} = \frac{\sin\alpha - \sin\beta}{\sin\alpha + \sin\beta}\,.$$

5. Use (T.26)–(T.28) to show that

$$\tan\frac{1}{2}(\alpha\pm\beta) = \frac{\sin\alpha \pm \sin\beta}{\cos\alpha + \cos\beta}\,. \tag{T.47}$$

6. (a) Show that

$$\prod_{j=0}^{k-1} \cos 2^j x = \frac{\sin 2^k x}{2^k \sin x} \tag{T.48}$$

(b) Deduce that

$$\prod_{j=1}^{k} \cos\frac{x}{2^j} = \frac{\sin x}{2^k \sin x/2^k}\,. \tag{T.49}$$

(c) Show that $\lim_{k\to\infty} 2^k \sin x/2^k = x$.
(d) Deduce the following formula of Euler:

$$\prod_{j=1}^{\infty} \cos\frac{x}{2^j} = \frac{\sin x}{x}\,. \tag{T.50}$$

(e) Note the special case, due to Viète:

$$\prod_{j=1}^{\infty} \cos\frac{\pi}{2^{j+1}} = \frac{2}{\pi}\,. \tag{T.51}$$

(f) This product converges with remarkable speed. Show that

$$\prod_{j=1}^{k} \cos \frac{\pi}{2^{j+1}} = \frac{2}{\pi} + O(2^{-2k}) . \tag{T.52}$$

(g) Let $a_1 = \sqrt{2}$, and $a_j = \sqrt{2 + a_{j-1}}$ for $j > 1$. Show that

$$\cos \frac{\pi}{2^{j+1}} = \frac{a_j}{2} . \tag{T.53}$$

(h) Conclude that

$$\frac{\sqrt{2}}{2} \cdot \frac{\sqrt{2+\sqrt{2}}}{2} \cdot \frac{\sqrt{2+\sqrt{2+\sqrt{2}}}}{2} \cdots = \frac{2}{\pi} . \tag{T.54}$$

7. Let α, β, and γ be the angles in a triangle, as in Figure T.1(b). That is, $\alpha + \beta + \gamma = \pi$. Show that

$$\tan \alpha + \tan \beta + \tan \gamma = \tan \alpha \tan \beta \tan \gamma . \tag{T.55}$$

T.2. Trigonometric functions in calculus

We define the exponential function, sine function, and cosine function by their power series:

$$e^z = \sum_{k=0}^{\infty} \frac{z^k}{k!}, \tag{T.56}$$

$$\sin z = \sum_{k=0}^{\infty} (-1)^k \frac{z^{2k+1}}{(2k+1)!}, \tag{T.57}$$

$$\cos z = \sum_{k=0}^{\infty} (-1)^k \frac{z^{2k}}{(2k)!} . \tag{T.58}$$

These formulas hold for all real and complex values of z. If we replace z by $-z$ in (T.57), then we obtain $-\sin z$. Thus the sine function is odd, not just for real arguments θ, but for all complex arguments. Similarly, $\cos z$ is even.

From (T.56) we see that

$$e^{iz} = \sum_{k=0}^{\infty} i^k \frac{z^k}{k!} .$$

But $i^2 = -1$, $i^3 = -i$, and $i^4 = 1$, so we separate the contributions of even k from odd k to see that the above is

$$= \sum_{k=0}^{\infty} (-1)^k \frac{z^{2k}}{(2k)!} + i \sum_{k=0}^{\infty} (-1)^k \frac{z^{2k+1}}{(2k+1)!},$$

so

$$e^{iz} = \cos z \; + \; i \sin z . \tag{T.59}$$

On replacing z by $-z$ we find also that

(T.60) $$e^{-iz} = \cos z - i \sin z\,.$$

On adding (T.59) and (T.60) and dividing by 2 we discover that

(T.61) $$\cos z = \frac{e^{iz} + e^{-iz}}{2}\,.$$

Similarly, by subtracting (T.60) from (T.59) and dividing by $2i$ we find that

(T.62) $$\sin z = \frac{e^{iz} - e^{-iz}}{2i}\,.$$

Indeed, these formulas could be taken to be the definitions of sine and cosine, in which case we would recover (T.57) and (T.58) from (T.56). We note that

$$\Big(\frac{e^{iz} - e^{-iz}}{2i}\Big)^2 + \Big(\frac{e^{iz} + e^{-iz}}{2}\Big)^2 = \frac{e^{2iz} - 2 + e^{-2iz}}{-4} + \frac{e^{2iz} + 2 + e^{-2iz}}{4} = 1,$$

and hence

(T.63) $$\sin^2 z + \cos^2 z = 1$$

for all real or complex z.

As was shown in Chapter 1, the familiar relation $e^{a+b} = e^a e^b$ holds not just for real numbers a and b, but also for complex numbers. Thus in particular,

(T.64)
$$\begin{aligned}\cos(\theta + \phi) + i \sin(\theta + \phi) &= e^{i(\theta+\phi)} \\ &= e^{i\theta} e^{i\phi} \\ &= (\cos\theta + i \sin\theta)(\cos\phi + i\sin\phi) \\ &= (\cos\theta\cos\phi - \sin\theta\sin\phi) + i(\sin\theta\cos\phi + \cos\theta\sin\phi)\,.\end{aligned}$$

On equating the real and imaginary parts of the two sides, we obtain the addition formulas (T.13) and (T.14), which were stated earlier without proof.

In (T.57) and (T.58) we replace the complex variable z by a real variable x, and differentiate term–by–term to see that

(T.65) $$\frac{d}{dx}\sin x = \cos x, \qquad \frac{d}{dx}\cos x = -\sin x\,.$$

By using formulas for the derivative of a quotient or reciprocal, it follows that

(T.66) $$\frac{d}{dx}\tan x = \sec^2 x, \qquad \frac{d}{dx}\cot x = -\csc^2 x,$$

(T.67) $$\frac{d}{dx}\sec x = \tan x \sec x, \qquad \frac{d}{dx}\csc x = -\cot x \csc x\,.$$

Thus $\sin x$ and $\cos x$ are both solutions of the homogeneous differential equation $y + y'' = 0$; sine with initial conditions $y(0) = 0$, $y'(0) = 1$, and cosine with $y(0) = 1$, $y'(0) = 0$. The most general solution of this differential equation is therefore $y(x) = a \sin x + b \cos x$.

Let $\boldsymbol{s} = \boldsymbol{s}(t) = (x(t), y(t))$ be a position vector in the plane, and suppose that $x(t) = \cos t$, $y(t) = \sin t$. This point moves with velocity

$$\boldsymbol{v} = \Big(\frac{dx}{dt}, \frac{dy}{dt}\Big) = (-\sin t, \cos t),$$

by (T.65). The speed of this motion is

$$|\boldsymbol{v}| = \sqrt{\sin^2 t + \cos^2 t} = 1\,.$$

Thus the point $\boldsymbol{s}(t) = (\cos t, \sin t)$ moves around the unit circle, and always at unit speed in the positive sense (i.e., counter–clockwise). For this reason, cosine and sine are sometimes referred to as *circular functions*. Since the unit circle has arc-length 2π, it follows that $(\sin t, \cos t) = (1, 0)$ precisely when $t = 2\pi k$ for some integer k. In addition, $\sin t$ and $\cos t$ have period 2π.

As for integration, we note that

(T.68) $$\int \sin x\, dx = -\cos x + C, \qquad \int \cos x\, dx = \sin x + C,$$

(T.69) $$\int \tan x\, dx = -\log \cos x + C, \qquad \int \cot x\, dx = \log \sin x + C,$$

(T.70) $$\int \sec x\, dx = \log(\sec x + \tan x) + C, \qquad \int \csc x\, dx = \log(\csc x - \cot x) + C\,.$$

If x and y are rational numbers such that $x^2 + y^2 = 1$, then the slope t of the line through $(-1, 0)$ and (x, y) is rational, $t = y/(x+1)$. Conversely, if t is rational, then the simultaneous equations

(T.71) $$t(x+1) = y,$$

(T.72) $$x^2 + y^2 = 1$$

have two solutions. When we substitute (T.71) into (T.72) so as to eliminate y, we obtain a quadratic equation in x with rational coefficients. Since one root of this quadratic is $x = -1$, the other root must also be rational. The same thing happens when we substitute to eliminate x. Specifically, we find that

$$0 = (1+t^2)x^2 + 2t^2x + t^2 - 1 = (x+1)((1+t^2)x + t^2 - 1),$$
$$0 = (1+t^2)y^2 - 2ty = y((1+t^2)y - 2t)\,.$$

Hence the second point of intersection of the line with the circle occurs at the point $P = \big((1-t^2)/(1+t^2), 2t/(1+t^2)\big)$. Thus rational points t on the real line are in one-to-one correspondence with rational points (x, y) on the unit circle, and for real t we have a rational parameterization of the unit circle. If $x = \cos\theta$ and $y = \sin\theta$ as in Figure T.5, then the line $y = t(x+1)$ makes an angle $\theta/2$ with the x-axis. Consequently, when we are presented with the task of determining the value of an integral of the form

$$\int R(\sin\theta, \cos\theta)\, d\theta$$

where R is a rational function, we make the *Weierstrass change of variable*

(T.73) $$t = \tan\frac{\theta}{2},$$

which yields

$$\text{(T.74)}\qquad \sin\theta = \frac{2t}{1+t^2}, \qquad \cos\theta = \frac{1-t^2}{1+t^2}, \qquad d\theta = \frac{2}{1+t^2}\,dt,$$

$$\text{(T.75)}\qquad \tan\theta = \frac{2t}{1-t^2}, \qquad \cot\theta = \frac{1-t^2}{2t}, \qquad e^{i\theta} = \frac{1+it}{1-it},$$

$$\text{(T.76)}\qquad \sec\theta = \frac{1+t^2}{1-t^2}, \qquad \csc\theta = \frac{1+t^2}{2t}, \qquad e^{-i\theta} = \frac{1-it}{1+it}.$$

Thus

$$\text{(T.77)}\qquad \int R(\sin\theta,\cos\theta)\,d\theta = \int R\Big(\frac{2t}{1+t^2},\frac{1-t^2}{1+t^2}\Big)\frac{2}{1+t^2}\,dt\,.$$

The integrand on the right is simply a rational function of t.

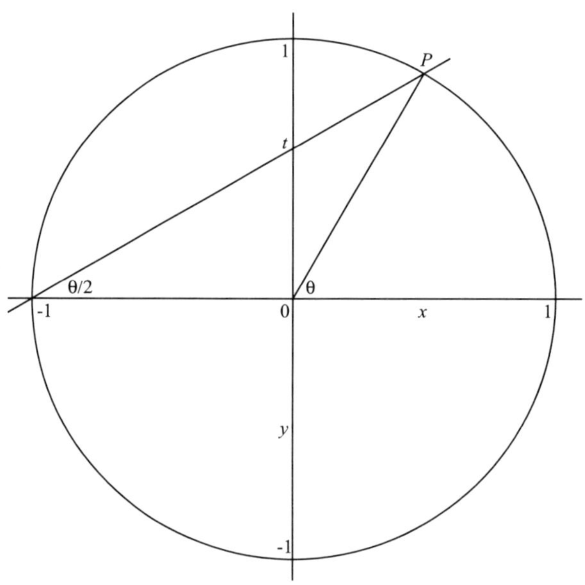

Figure T.5. Line with slope t passing through $(-1, 0)$ and $P = (\cos\theta, \sin\theta)$.

By using the result of Exercise 9.5.5(d), it can be shown that

$$\text{(T.78)}\qquad \tan z = z + \frac{z^3}{3} + \frac{2z^5}{15} + \frac{17z^7}{315} + \cdots$$

$$\text{(T.79)}\qquad = \sum_{k=1}^{\infty} \frac{(-1)^{k-1}(2^{4k}-2^{2k})B_{2k}}{(2k)!} z^{2k-1} \qquad (|z| < \pi/2)$$

$$\text{(T.80)}\qquad \cot z = \frac{1}{z} - \frac{z}{3} - \frac{z^3}{45} - \frac{2z^5}{945} - \cdots$$

$$\text{(T.81)}\qquad = \sum_{k=0}^{\infty} (-1)^k \frac{2^{2k}B_{2k}}{(2k)!} z^{2k-1} \qquad /2(|z| < \pi)$$

$$\sec z = 1 + \frac{z^2}{2} + \frac{5z^4}{24} + \frac{61z^6}{720} + \cdots \tag{T.82}$$

$$= \sum_{k=0}^{\infty} (-1)^{k+1} 4^{2k+1} \frac{B_{2k+1}(1/4)}{(2k+1)!} z^{2k} \qquad (|z| < \pi/2) \tag{T.83}$$

$$\csc z = \frac{1}{z} + \frac{z}{6} + \frac{7z^3}{360} + \frac{31z^5}{15120} + \cdots \tag{T.84}$$

$$= \sum_{k=0}^{\infty} \frac{(-1)^{k-1}(2^{2k}-2)B_{2k}}{(2k)!} z^{2k-1} \qquad (|z| < \pi). \tag{T.85}$$

Here B_k is a Bernoulli number and $B_k(x)$ is a Bernoulli polynomial, as discussed in §8.3.

In much the same way that a polynomial can be expressed as a product over its zeros, the sine function can be written as a product over its zeros. In order to ensure that the product converges, we group the contribution of the zero at πk with that at $-\pi k$:

$$\sin z = z \prod_{k=1}^{\infty} \left(1 - \frac{z^2}{\pi^2 k^2}\right). \tag{T.86}$$

Similarly,

$$\cos z = \prod_{k=1}^{\infty} \left(1 - \frac{4z^2}{\pi^2 (2k-1)^2}\right). \tag{T.87}$$

In the same way that a rational function can be written in partial fraction form, we have

$$\cot z = \frac{1}{z} + \sum_{k=1}^{\infty} \left(\frac{1}{z - \pi k} + \frac{1}{z + \pi k}\right), \tag{T.88}$$

$$\csc z = \frac{1}{z} + \sum_{k=1}^{\infty} (-1)^k \left(\frac{1}{z - \pi k} + \frac{1}{z + \pi k}\right), \tag{T.89}$$

$$\csc^2 z = \sum_{k=-\infty}^{\infty} \frac{1}{(z - \pi k)^2}. \tag{T.90}$$

For proofs of these formulas see Exercise 5. Similar partial fraction formulas for $\tan z$, $\sec z$, and $\sec^2 z$ can be derived by the translation identities (T.17)–(T.19).

We now construct upper and lower bounds for sine and cosine. We start by observing that

$$\cos x \le 1 \tag{T.91}$$

for all $x \ge 0$. Hence

$$\sin x = \int_0^x \cos u \, du \le \int_0^x 1 \, du = x \tag{T.92}$$

for all $x \geq 0$. Hence

$$1 - \cos x = \int_0^x \sin u\, du \leq \int_0^x u\, du = \frac{x^2}{2}$$

so

$$\cos x \geq 1 - \frac{x^2}{2} \tag{T.93}$$

for all $x \geq 0$. Hence

$$\sin x = \int_0^x \cos u\, du \geq \int_0^x 1 - \frac{u^2}{2}\, du = x - \frac{x^3}{3!} \tag{T.94}$$

for all $x \geq 0$. Hence

$$1 - \cos x = \int_0^x \sin u\, du \geq \int_0^x u - \frac{u^3}{3!}\, du = \frac{x^2}{2} - \frac{x^4}{4!}$$

so

$$\cos x \leq 1 - \frac{x^2}{2!} + \frac{x^4}{4!} \tag{T.95}$$

for all $x \geq 0$. This procedure can be continued indefinitely, and we find that

$$\sum_{n=0}^{2N+1} (-1)^n \frac{x^{2n}}{(2n)!} \leq \cos x \leq \sum_{n=0}^{2N} (-1)^n \frac{x^{2n}}{(2n)!}, \tag{T.96}$$

$$\sum_{n=0}^{2N+1} (-1)^n \frac{x^{2n+1}}{(2n+1)!} \leq \sin x \leq \sum_{n=0}^{2N} (-1)^n \frac{x^{2n+1}}{(2n+1)!} \tag{T.97}$$

for all nonnegative integers N and nonnegative real x. These inequalities are reminiscent of the alternating series test, but that test does not apply here, since the terms are not monotonically decreasing in absolute value when x is large. The cosine function and its approximations are even, and so the inequalities (T.96) hold without change when x is negative. The sine function and its approximations are odd, so the direction of the inequalities is reversed when x is negative.

We note that $(\tan x)' = \sec^2 x \geq 1$, so

$$\tan x \geq x \tag{T.98}$$

for $0 \leq x < \pi/2$ with equality only at $x = 0$.

Since $\sin x$ is concave downwards in the interval $[0, \pi/2]$, $\sin 0 = 0$, and $\sin \pi/2 = 1$, it follows that

$$\sin x \geq \frac{2}{\pi} x \tag{T.99}$$

for $0 \leq x \leq \pi/2$, with equality only at the endpoints.

Exercises

1. (a) Show that for all θ,

(T.100) $$\sin\theta = \cos(\pi/2-\theta) = \sin(\pi - x)\,.$$

(b) Show that for all θ,

(T.101) $$\cos\theta = \sin(\pi/2-\theta) = -\cos(\pi-\theta)\,.$$

(c) Show that for all θ,

(T.102) $$\tan\theta = \cot(\pi-\theta) = -\tan(\pi-\theta)\,.$$

2. Show that for all θ,

(T.103) $$\csc\theta = \cot\theta/2 - \cot\theta\,.$$

3. The identities of (T.64) still hold if θ and ϕ are complex, but when we equate real and imaginary parts to obtain (T.13) and (T.14), the reasoning requires that θ and ϕ are real.
(a) Use (T.61) and (T.62) to express $\cos z\cos w - \sin z\sin w$ in terms of exponentials. Expand, simplify, and use (T.61) again to show that
$$\cos(z+w) = \cos z\cos w - \sin z\sin w$$
for all complex z and w.
(b) Similarly, show that
$$\sin(z+w) = \sin z\cos w + \cos z\sin w$$
for all complex z and w.

4. (a) Show that if $\operatorname{Im} z > 0$, then $|e^{iz}| < |e^{-iz}|$.
(b) Show that if $\operatorname{Im} z < 0$, then $|e^{iz}| > |e^{-iz}|$.
(c) Deduce that if $\sin z = 0$, then z must be real.
(d) Describe the real zeros of $\sin z$.
(e) Note that in (T.86), the product is over all zeros of $\sin z$.

5. (a) Derive (T.88) from (T.86) by taking logarithmic derivatives.
(b) Show that
$$\csc z = \frac{1}{2}\cot\frac{z}{2} - \frac{1}{2}\cot\frac{z+\pi}{2}\,.$$
(c) Derive (T.89) from (T.88).
(d) Derive (T.90) by differentiating both sides of (T.88).

6. (a) Let $f(x) = \sin x - x\cos x$. Show that $f'(x) = -x\sin x$.
(b) Deduce that f is decreasing in the interval $[0,\pi]$.
(c) Note that $f(0) = 0$. Deduce that
$$x\cos x \le \sin x$$
for $0 \le x \le \pi$, with equality only when $x = 0$.

7. (a) By mimicking the proof of (T.99), show that
$$\cos\pi x \ge 1 - 2x$$
for $0 \le x \le 1/2$, with equality only at the endpoints.
(b) Put $g(x) = \sin\pi x - \pi x(1-x)$. Compute $g'(x)$, and show that $g'(x) \ge 0$ for

$0 \le x \le 1/2$, with equality only at the endpoints.
(c) Show that $g(1-x) = g(x)$ for all real x.
(d) Conclude that

$$\sin \pi x \ge \pi x(1-x)$$

for $0 \le x \le 1$, with equality only at the endpoints.

8. (a) Let $h(x) = \sin \pi x - 4x(1-x)$. Use (T.92) to show that $h(x) < 0$ for $0 < x < 1 - \pi/4 = 0.2146$.
(b) Show that $h'(1/6) = \pi\sqrt{3}/2 - 8/3 = 0.054 > 0$.
(c) Show that $h'(1/2) = 0$.
(d) Show that $h'(x)$ is concave downward for $0 \le x \le 1/2$.
(e) Use (b)–(d) to show that $h'(x) > 0$ for $1/6 \le x < 1/2$.
(f) Show that $h(1/2) = 0$. Deduce that $h(x) < 0$ for $1/6 \le x < 1/2$.
(g) Show that $h(1-x) = h(x)$ for all real x.
(h) Conclude that

$$\sin \pi x \le 4x(1-x)$$

for $0 \le x \le 1$, with equality only when $x = 0$, $x = 1/2$, or $x = 1$.

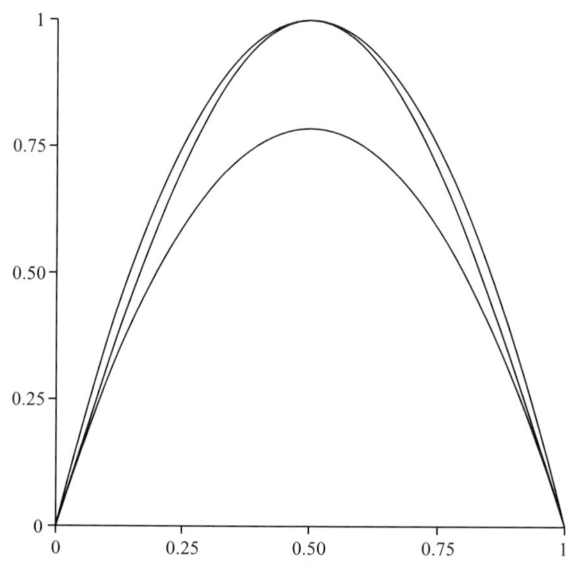

Figure T.6. Graphs of $\pi x(1-x)$, $\sin \pi x$, and $4x(1-x)$ for $0 \le x \le 1$.

T.3. Inverse trigonometric functions

The trigonometric functions are not one-to-one, and therefore do not have inverses when the domain is taken to be the real line. However, by restricting the domain, we can obtain a well-defined inverse. The functions $\sin x$, $\tan x$, and $\csc x$ are one-to-one on the interval $(-\pi/2, \pi/2)$, while $\cos x$, $\cot x$, and $\sec x$ are one-to-one on

$(0, \pi)$. Conventions vary as to which branch of the inverse function should be taken as the principal one. For the first three functions we take the principal branch to be the one taking values in $[-\pi/2, \pi/2]$, and for the latter three we take the branch with values in $[0, \pi]$. That is, for $-1 \le x \le 1$ we define $\operatorname{Arcsin} x$ to be the unique number $\theta \in [-\pi/2, \pi/2]$ such that $\sin\theta = x$; for real x we define $\operatorname{Arctan} x$ to be the unique $\theta \in (-\pi/2, \pi/2)$ such that $\tan\theta = x$; for $x \le -1$ or $x \ge 1$ we define $\operatorname{Arccsc} x$ to be the unique number $\theta \in [-\theta/2, \theta/2]$ such that $\csc\theta = x$; for $-1 \le x \le 1$ we define $\operatorname{Arccos} x$ to be the unique $\theta \in [0, \pi]$ such that $\cos\theta = x$; for real x we define $\operatorname{Arccot} x$ to be the unique $\theta \in (0, \pi)$ such that $\cot\theta = x$; and for $x \le -1$ or $x \ge 1$ we define $\operatorname{Arcsec} x$ to be the unique $\theta \in [0, \pi]$ such that $\sec\theta = x$.

We let $\theta = \arcsin x$ denote any real number such that $\sin\theta = x$. Thus $\arcsin x$ is a multiple-valued function, and we find that

$$\arcsin x = (-1)^n \operatorname{Arcsin} x + n\pi \tag{T.104}$$

$$\arctan x = \operatorname{Arctan} x + n\pi, \tag{T.105}$$

$$\operatorname{arccsc} x = (-1)^n \operatorname{Arccsc} x + n\pi, \tag{T.106}$$

$$\arccos x = \pm \operatorname{Arccos} x + 2n\pi, \tag{T.107}$$

$$\operatorname{arccot} x = \operatorname{Arccot} x + n\pi, \tag{T.108}$$

$$\operatorname{arcsec} x = \pm \operatorname{Arcsec} x + 2n\pi \tag{T.109}$$

for all integers n. Since cosine and secant are even while sine, cosecant, tangent and cotangent are odd, it follows that

(T.110)

$$\begin{aligned} &\arcsin(-x) = -\arcsin x, &&\arctan(-x) = -\arctan x, &&\operatorname{arcsec}(-x) = \operatorname{arcsec} x, \\ &\arccos(-x) = \arccos x, &&\operatorname{arccot}(-x) = -\operatorname{arccot} x, &&\operatorname{arccsc}(-x) = \operatorname{arccsc} x\,. \end{aligned}$$

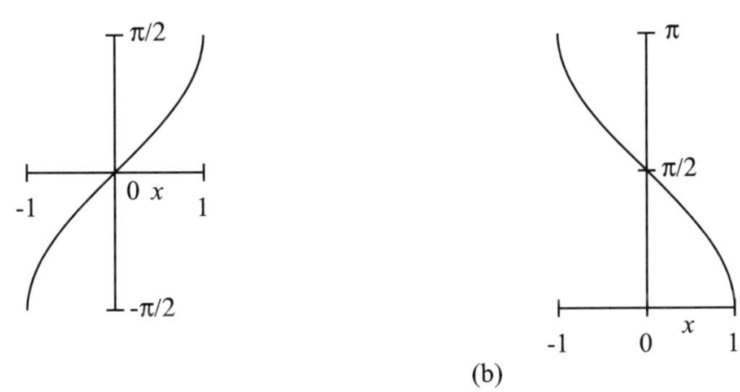

Figure T.7. (a) Graph of $\operatorname{Arcsin} x$ for $-1 \le x \le 1$. (b) Graph of $\operatorname{Arccos} x$ for $-1 \le x \le 1$.

We recall from calculus that if g is the inverse function of f, and $f(x) = y$, then $g'(y) = 1/f'(x)$. Thus from (T.63) and (T.65)–(T.67) we find that

$$\frac{d}{dx}\operatorname{Arcsin} x = \frac{1}{\sqrt{1-x^2}}, \tag{T.111}$$

$$\frac{d}{dx}\operatorname{Arctan} x = \frac{1}{1+x^2}, \tag{T.112}$$

$$\frac{d}{dx}\operatorname{Arccsc} x = \frac{-1}{|x|\sqrt{x^2-1}}, \tag{T.113}$$

$$\frac{d}{dx}\operatorname{Arccos} x = -\frac{1}{\sqrt{1-x^2}}, \tag{T.114}$$

$$\frac{d}{dx}\operatorname{Arccot} x = \frac{-1}{x^2+1}, \tag{T.115}$$

$$\frac{d}{dx}\operatorname{Arcsec} x = \frac{1}{|x|\sqrt{x^2-1}}. \tag{T.116}$$

These formulas allow us to express the inverse functions as definite integrals:

$$\operatorname{Arcsin} x = \int_0^x \frac{du}{\sqrt{1-u^2}} \qquad (-1 \le x \le 1), \tag{T.117}$$

$$\operatorname{Arctan} x = \int_0^x \frac{du}{u^2+1} \qquad (-\infty < x < \infty), \tag{T.118}$$

$$\operatorname{Arccsc} x = \int_x^\infty \frac{du}{u\sqrt{u^2-1}} \qquad (1 \le x < \infty), \tag{T.119}$$

$$\operatorname{Arccos} x = \int_x^1 \frac{du}{\sqrt{1-u^2}} \qquad (-1 \le x \le 1), \tag{T.120}$$

$$\operatorname{Arccot} x = \int_x^\infty \frac{du}{u^2+1} \qquad (-\infty < x < \infty), \tag{T.121}$$

$$\operatorname{Arcsec} x = \int_1^x \frac{du}{u\sqrt{u^2-1}} \qquad (1 \le x < \infty). \tag{T.122}$$

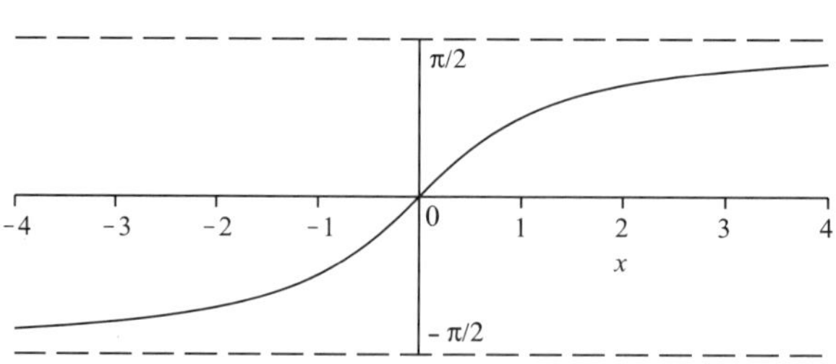

Figure T.8. Graph of $\operatorname{Arctan} x$ for $-4 \le x \le 4$.

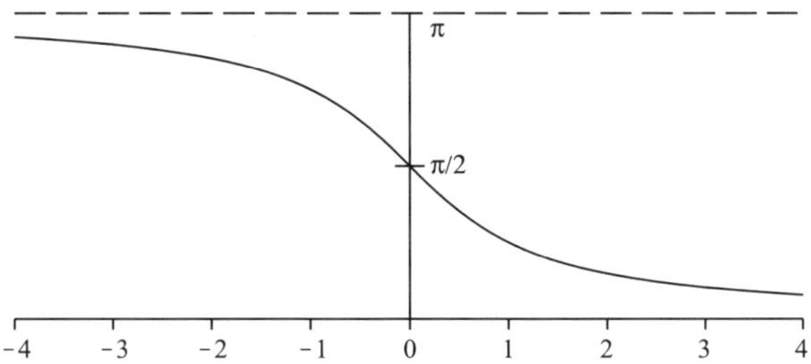

Figure T.9. Graph of Arccot x for $-4 \leq x \leq 4$.

From these identities we deduce that

$$\text{(T.123)} \qquad \operatorname{Arcsin} x + \operatorname{Arccos} x = \pi/2 \qquad (-1 \leq x \leq 1),$$

$$\text{(T.124)} \qquad \operatorname{Arctan} x + \operatorname{Arccot} x = \pi/2 \qquad (-\infty < x < \infty),$$

$$\text{(T.125)} \qquad \operatorname{Arcsec} x + \operatorname{Arccsc} x = \pi/2 \qquad (-\infty < x \leq -1 \text{ or } 1 \leq x < \infty)\,.$$

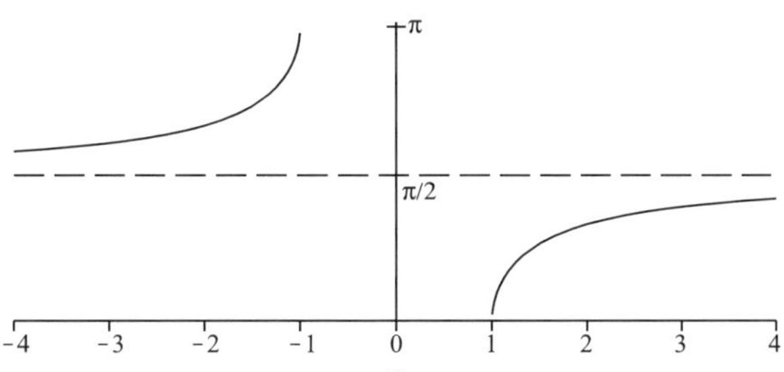

Figure T.10. Graph of Arcsec x for $-4 \leq x \leq 4$.

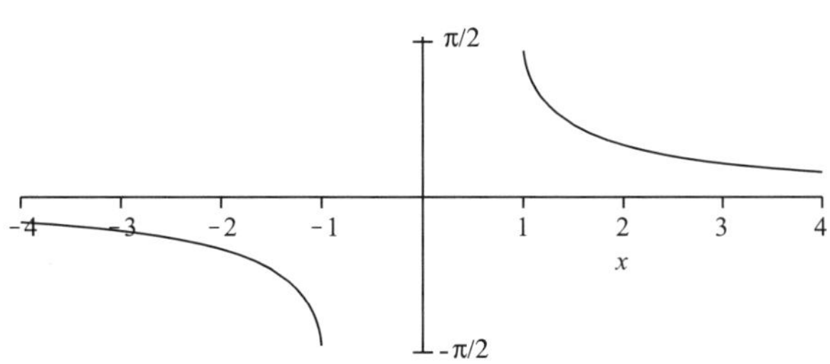

Figure T.11. Graph of Arccsc x for $-4 \leq x \leq 4$.

This is what we would expect, based on trigonometry, for if $x = b/c$ in Figure T.1 (a), then $\theta = \operatorname{Arcsin} x$, $\phi = \operatorname{Arccos} x$, and we know that $\theta + \phi = \pi/2$.

Since $\csc\theta = 1/\sin\theta$, $\sec\theta = 1/\cos\theta$, and $\cot\theta = 1/\tan\theta$, it is immediate that

(T.126) $$\operatorname{Arccsc} x = \operatorname{Arcsin} 1/x,$$

(T.127) $$\operatorname{Arcsec} x = \operatorname{Arccos} 1/x,$$

(T.128) $$\operatorname{Arccot} x = \begin{cases} \operatorname{Arctan} 1/x & (x > 0), \\ \pi + \operatorname{Arctan} 1/x & (x < 0). \end{cases}$$

By the binomial theorem in the form expressed in Corollary B.2 we find that

$$\frac{1}{\sqrt{1-u^2}} = \sum_{k=0}^{\infty} \binom{k-1/2}{k} u^{2k}$$

for $-1 < u < 1$. Now

$$\binom{k-1/2}{k} = \frac{(k-1/2)(k-3/2)\cdots(3/2)(1/2)}{k!} = \frac{1\cdot 3\cdot 5\cdots(2k-1)}{2^k k!} = \frac{(2k)!}{2^{2k}k!^2},$$

so from (T.117) we deduce that

(T.129)
$$\operatorname{Arcsin} x = x + \frac{1}{2\cdot 3}x^3 + \frac{1\cdot 3}{2\cdot 4\cdot 5}x^5 + \frac{1\cdot 3\cdot 5}{2\cdot 4\cdot 6\cdot 7}x^7 + \cdots = \sum_{k=0}^{\infty} \frac{(2k)!}{2^{2k}k!^2(2k+1)}x^{2k+1}$$

for $-1 \le x \le 1$. Similarly, since

$$\frac{1}{u^2+1} = \sum_{k=0}^{\infty} (-1)^k u^{2k}$$

for $-1 < u < 1$, it follows from (T.118) that

(T.130) $$\operatorname{Arctan} x = x - \frac{x^3}{3} + \frac{x^5}{5} - \frac{x^7}{7} + \cdots = \sum_{k=0}^{\infty} \frac{(-1)^k}{2k+1}x^{2k+1}$$

for $-1 \le x \le 1$. From (T.124), (T.128) and the above it follows that

(T.131) $$\operatorname{Arctan} x = \frac{\pi}{2} - \frac{1}{x} + \frac{1}{3x^3} - \frac{1}{5x^5} + \frac{1}{7x^7} - \cdots = \frac{\pi}{2} - \sum_{k=0}^{\infty} \frac{(-1)^k}{(2k+1)x^{2k+1}}$$

for $x \ge 1$ while

(T.132) $$\operatorname{Arctan} x = -\frac{\pi}{2} - \frac{1}{x} + \frac{1}{3x^3} - \frac{1}{5x^5} + \frac{1}{7x^7} - \cdots = -\frac{\pi}{2} - \sum_{k=0}^{\infty} \frac{(-1)^k}{(2k+1)x^{2k+1}}$$

for $x \le -1$.

We note the integration formulas

$$\int \operatorname{Arcsin} x\, dx = x \operatorname{Arcsin} x + \sqrt{1-x^2} + C, \tag{T.133}$$

$$\int \operatorname{Arccos} x\, dx = x \operatorname{Arccos} x - \sqrt{1-x^2} + C, \tag{T.134}$$

$$\int \operatorname{Arctan} x\, dx = x \operatorname{Arctan} x - \frac{1}{2}\log(1+x^2) + C, \tag{T.135}$$

$$\int \operatorname{Arccot} x\, dx = x \operatorname{Arccot} x + \frac{1}{2}\log(1+x^2) + C, \tag{T.136}$$

$$\int \operatorname{Arcsec} x\, dx = x \operatorname{Arcsec} x - \log(|x| + \sqrt{x^2-1}) + C, \tag{T.137}$$

$$\int \operatorname{Arccsc} x\, dx = x \operatorname{Arccsc} x + \log(|x| + \sqrt{x^2-1}) + C\,. \tag{T.138}$$

T.4. Hyperbolic functions

We set

$$\sinh z = \frac{e^z - e^{-z}}{2} = \sum_{k=0}^{\infty} \frac{z^{2k+1}}{(2k+1)!}, \tag{T.139}$$

$$\cosh z = \frac{e^z + e^{-z}}{2} = \sum_{k=0}^{\infty} \frac{z^{2k}}{(2k)!}\,. \tag{T.140}$$

Orally, these functions are pronounced, "sinch" and "cosh". In the reverse direction,

$$e^z = \cosh z + \sinh z, \qquad e^{-z} = \cosh z - \sinh z\,. \tag{T.141}$$

We further define

$$\tanh z = \frac{\sinh z}{\cosh z}, \qquad \coth z = \frac{\cosh z}{\sinh z} = \frac{1}{\tanh z}, \tag{T.142}$$

$$\operatorname{sech} z = \frac{1}{\cosh z}, \qquad \operatorname{csch} z = \frac{1}{\sinh z}\,. \tag{T.143}$$

Since these functions are defined in a manner similar to the trigonometric functions, it is not surprising that they satisfy many identities similar to trigonometric ones. By an easy calculation we find that

$$\cosh^2 z - \sinh^2 z = 1, \quad \tanh^2 z + \operatorname{sech}^2 z = 1, \quad \coth^2 z - \operatorname{csch}^2 z = 1\,. \tag{T.144}$$

Thus $(\cosh t, \sinh t)$ parameterizes the hyperbola $x^2 - y^2 = 1$, but the speed of the motion is not constant, unlike the case of $(\cos t, \sin t)$. The hyperbolic functions are related to the circular functions by a change of variable:

$$\sinh z = -i \sin iz, \qquad \cosh z = \cos iz, \qquad \tanh z = -i \tan iz, \tag{T.145}$$

$$\operatorname{csch} z = i \csc iz, \qquad \operatorname{sech} z = \sec iz, \qquad \coth z = i \cot iz\,. \tag{T.146}$$

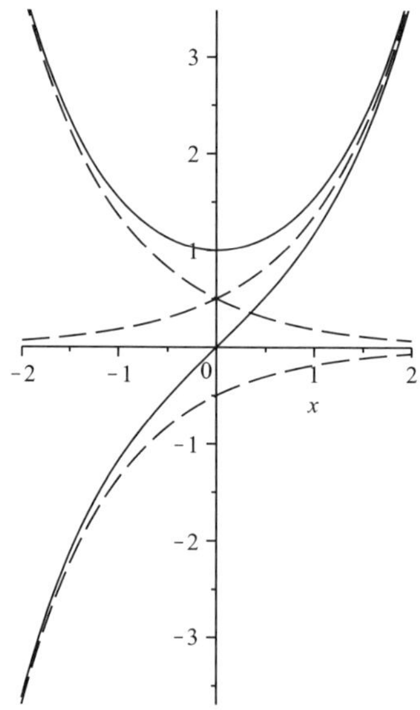

Figure T.12. Graphs of $\cosh x$ and $\sinh x$ for $-2 \le x \le 2$ (solid), with the asymptotic curves $\frac{1}{2}e^{\pm x}$ and $-\frac{1}{2}e^{-x}$.

By a further easy calculation we obtain addition formulas,

$$\sinh(z \pm w) = \sinh z \cosh w \pm \cosh z \sinh w, \tag{T.147}$$

$$\cosh(z \pm w) = \cosh z \cosh w \pm \sinh z \sinh w, \tag{T.148}$$

$$\tanh(z \pm w) = \frac{\tanh z \pm \tanh w}{1 \pm \tanh z \tanh w}, \tag{T.149}$$

$$\coth(z \pm w) = \frac{1 \pm \coth z \coth w}{\coth z \pm \coth w}. \tag{T.150}$$

Hence sinh and cosh are periodic with period $2\pi i$, while tanh is periodic with period πi. By differentiating term-by-term in (T.139) and (T.140) we obtain two differentiation formulas, and then corresponding formulas for the other four hyperbolic functions follow:

$$\frac{d}{dx}\sinh x = \cosh x, \qquad \frac{d}{dx}\operatorname{sech} x = -\operatorname{sech} x \tanh x, \tag{T.151}$$

$$\frac{d}{dx}\cosh x = \sinh x, \qquad \frac{d}{dx}\operatorname{csch} x = -\operatorname{csch} x \coth x, \tag{T.152}$$

$$\frac{d}{dx}\tanh x = \operatorname{sech}^2 x, \qquad \frac{d}{dx}\coth x = -\operatorname{csch}^2 x\,. \tag{T.153}$$

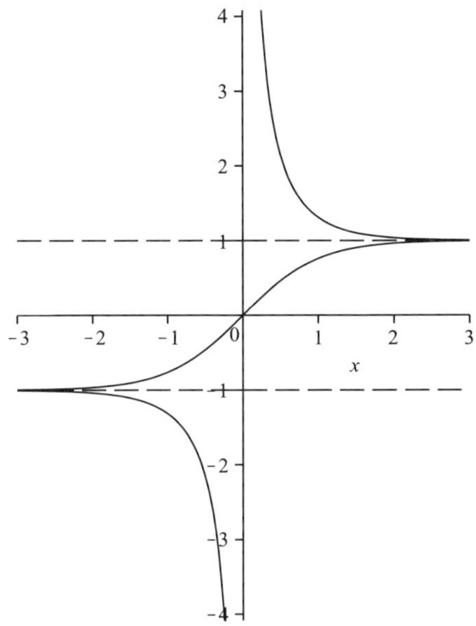

Figure T.13. Graphs of $\tanh x$ and $\coth x$ for $-3 \leq x \leq 3$ (solid), distinguished by $|\tanh x| < 1 < |\coth x|$, with the asymptotes $y = \pm 1$.

We note that $\cosh z$ and $\operatorname{sech} z$ are even functions, while $\sinh z$, $\tanh z$, $\coth z$, and $\operatorname{csch} z$ are odd. Analogously to (T.23)–(T.25) we have

(T.154) $$\sinh z \sinh w = \frac{1}{2}\cosh(z+w) - \frac{1}{2}\cosh(z-w)$$

(T.155) $$\sinh z \cosh w = \frac{1}{2}\sinh(z+w) + \frac{1}{2}\sinh(z-w)$$

(T.156) $$\cosh z \cosh w = \frac{1}{2}\cosh(z+w) + \frac{1}{2}\cosh(z-w)$$

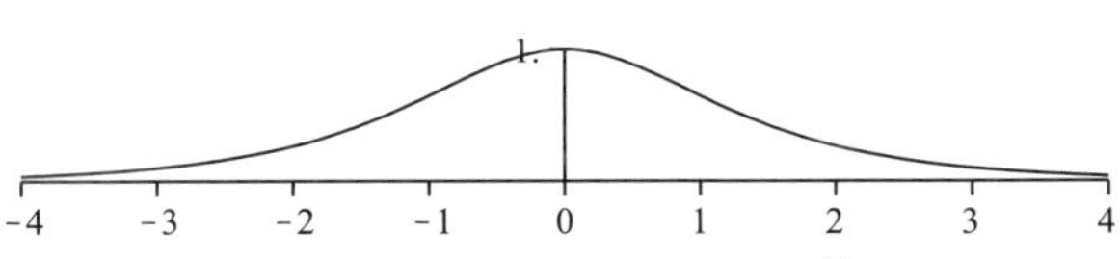

Figure T.14. Graph of $\operatorname{sech} x$ for $-3 \leq x \leq 3$.

Corresponding to (T.26)–(T.31) we have

(T.157) $$\sinh z + \sinh w = 2\sinh\frac{z+w}{2}\cosh\frac{z-w}{2},$$

(T.158) $$\sinh z - \sinh w = 2\cosh\frac{z+w}{2}\sinh\frac{z-w}{2},$$

(T.159) $$\cosh z + \cosh w = 2\cosh\frac{z+w}{2}\cosh\frac{z-w}{2},$$

(T.160) $$\cosh z - \cosh w = 2\sinh\frac{z+w}{2}\sinh\frac{z-w}{2},$$

(T.161) $$\tanh z \pm \tanh w = \frac{\sinh(z\pm w)}{\cosh z\cosh w},$$

(T.162) $$\coth z \pm \coth w = \frac{\sinh(w\pm z)}{\sinh z\cosh w}.$$

For doubled arguments we have

(T.163) $$\sinh 2z = 2\sinh z\cosh z,$$

(T.164) $$\cosh 2z = 2\cosh^2 z - 1 = 2\sinh^2 z + 1 = \cosh^2 z + \sinh^2 z,$$

(T.165) $$\tanh 2z = \frac{2\tanh z}{1+\tanh^2 z}.$$

For halved arguments,

(T.166) $$\sinh\frac{1}{2}z = \pm\sqrt{\frac{\cosh z - 1}{2}},$$

(T.167) $$\cosh\frac{1}{2}z = \sqrt{\frac{\cosh z + 1}{2}},$$

(T.168) $$\tanh\frac{1}{2}z = \pm\sqrt{\frac{\cosh z - 1}{\cosh z + 1}} = \frac{\cosh z - 1}{\sinh z} = \frac{\sinh z}{1+\cosh z}.$$

For sums and differences of squares,

(T.169) $$\sinh^2 z - \sinh^2 w = \cosh^2 z - \cosh^2 w = \sinh(z+w)\sinh(z-w),$$

(T.170) $$\sinh^2 z + \cosh^2 w = \cosh^2 z + \cosh^2 w = \cosh(z+w)\cosh(z-w).$$

If $z = x + iy$ with x and y real, then

(T.171) $$\sinh z = \sinh x\cos y + i\cosh x\sin y,$$

(T.172) $$\cosh z = \cosh x\cos y + i\sinh x\sin y,$$

(T.173) $$\tanh z = \frac{\sinh 2x + i\sin 2y}{\cosh 2x + \cos 2y},$$

(T.174) $$\coth z = \frac{\sinh 2x - i\sin 2y}{\cosh 2x - \cos 2y}.$$

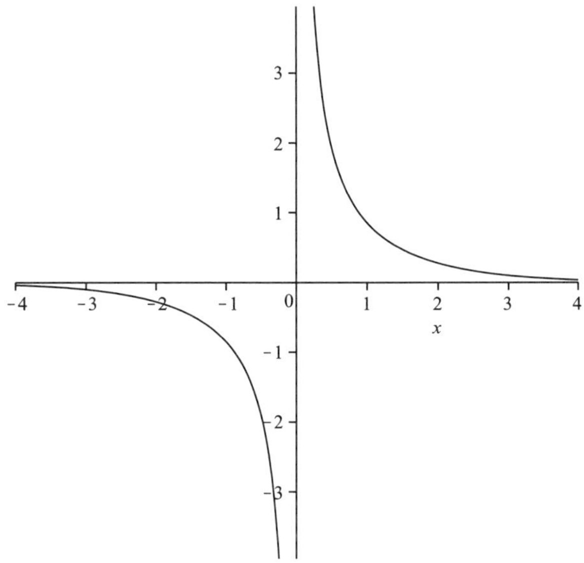

Figure T.15. Graph of $\operatorname{csch} x$ for $-3 \leq x \leq 3$.

We note the integration formulas

(T.175) $$\int \sinh x \, dx = \cosh x + C,$$

(T.176) $$\int \cosh x \, dx = \sinh x + C,$$

(T.177) $$\int \tanh x \, dx = \log \cosh x + C,$$

(T.178) $$\int \coth x \, dx = \log \sinh x + C,$$

(T.179) $$\int \operatorname{sech} x \, dx = \operatorname{Arctan}(\sinh x) + C,$$

(T.180) $$\int \operatorname{csch} x \, dx = \log \tanh(x/2) + C\,.$$

By combining (T.145), (T.146) with (T.78)–(T.84) we find that

(T.181) $$\tanh z = z - \frac{z^3}{3} + \frac{2z^5}{15} - \frac{17z^7}{315} + \cdots$$

(T.182) $$= \sum_{k=1}^{\infty} \frac{(2^{4k} - 2^{2k})B_{2k}}{(2k)!} z^{2k-1} \qquad (|z| < \pi/2)\,.$$

$$\coth z = \frac{1}{z} + \frac{z}{3} - \frac{z^3}{45} + \frac{2z^5}{945} - \cdots \tag{T.183}$$

$$= \sum_{k=0}^{\infty} \frac{2^{2k} B_{2k}}{(2k)!} z^{2k-1} \qquad (|z| < \pi)\,. \tag{T.184}$$

$$\operatorname{sech} z = 1 - \frac{z^2}{2} + \frac{5z^4}{24} - \frac{61z^6}{720} + \cdots \tag{T.185}$$

$$= -\sum_{k=0}^{\infty} 4^{2k+1} \frac{B_{2k+1}(1/4)}{(2k+1)!} z^{2k} \qquad (|z| < \pi/2)\,. \tag{T.186}$$

$$\operatorname{csch} z = \frac{1}{z} - \frac{z}{6} + \frac{7z^3}{360} - \frac{31z^5}{15120} + \cdots \tag{T.187}$$

$$= -\sum_{k=0}^{\infty} \frac{(2^{2k}-2)B_{2k}}{(2k)!} z^{2k-1} \qquad (|z| < \pi)\,. \tag{T.188}$$

Similarly, from (T.86) and (T.87) we find that

$$\sinh z = z \prod_{k=1}^{\infty} \Big(1 + \frac{z^2}{\pi^2 k^2}\Big), \tag{T.189}$$

$$\cosh z = \prod_{k=1}^{\infty} \Big(1 + \frac{4z^2}{\pi^2 (2k-1)^2}\Big)\,. \tag{T.190}$$

Exercises

1. (a) Explain why $\overline{e^z} = e^{\overline{z}}$.
   (b) Show that if $z = x + iy$, then
   $$|\sin z|^2 + |\cos z|^2 = \cosh 2y$$
   for all real or complex z. Note that this agrees with (T.63) when z is real.
2. (a) For $-1 < t < 1$, consider the line $y = t(x+1)$. This intersects the hyperbola $x^2 - y^2 = 1$ at the point $(-1, 0)$. Show that the second point of intersection is at $P = ((1+t^2)/(1-t^2), 2t/(1-t^2))$.
   (b) Choose z so that $(\cosh z, \sinh z) = P$. Deduce that $t = \sinh z/(1 + \cosh z)$.
   (c) Deduce that

$$t = \tanh \frac{z}{2}\,. \tag{T.191}$$

(d) Show that

$$\sinh z = \frac{2t}{1-t^2}, \qquad \cosh z = \frac{1+t^2}{1-t^2}, \qquad dz = \frac{2}{1-t^2}\,dt, \tag{T.192}$$

$$\tanh z = \frac{2t}{1+t^2}, \qquad \coth z = \frac{1+t^2}{2t}, \qquad e^z = \frac{1+t}{1-t}, \tag{T.193}$$

$$\operatorname{sech} z = \frac{1-t^2}{1+t^2}, \qquad \operatorname{csch} z = \frac{1-t^2}{2t}, \qquad e^{-z} = \frac{1-t}{1+t}. \tag{T.194}$$

(e) Conclude that if $R(u,v)$ is a rational function, then

$$\int R(\sinh z, \cosh z)\,dz = \int R\Big(\frac{2t}{1-t^2}, \frac{1+t^2}{1-t^2}\Big)\frac{2}{1-t^2}\,dt\,. \tag{T.195}$$

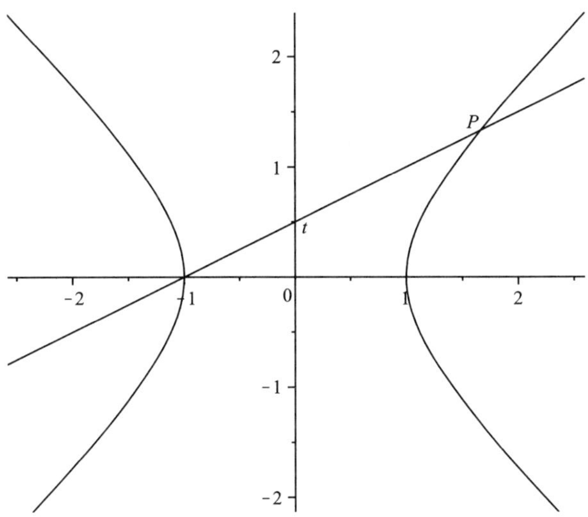

Figure T.16. Graph of the hyperbola $x^2 - y^2 = 1$ and the line $y = t(x+1)$, with the intersections $(-1,0)$ and $P = (\cosh z, \sinh z) = (\frac{1+t^2}{1-t^2}, \frac{2t}{1-t^2})$.

References

N. I. Achieser (1956). *Theory of Approximation* (Charles J. Hyman, translator), F. Unger Publ. Co., New York, x+307 pp.; 1992 reprint: Dover Publ., New York.

A. Baker (1990). *Transcendental number theory*, Second Edition, Cambridge Mathematical Library, Cambridge University Press, Cambridge, x+165 pp.

J. J. Benedetto (1997). *Harmonic analysis and applications*, CRC Press, Boca Raton, xix+336 pp.

J. J. Benedetto and P. J. S. G. Ferreira, Eds., (2001). *Modern Sampling Theory*, Springer Science+Business Media, New York, xvi+417 pp.

S. N. Bernstein (1912). "Sur l'ordre de la meilleure approximation des fonctions continues par des polynômes de degré donné", *Mém. Acad. Roy. Belgique* (2) **4**, 1–104.

R. P. Boas (1969). "Inequalities for the derivatives of polynomials", Math. Mag. **42**, 165–174.

M. Bôcher (1905–1906). "Introduction to the theory of Fourier's series", *Ann. of Math.* (2) **7**, 81–152.

P. D. G. du Bois-Reymond (1873). "Über die Fourierschen Reihen", *Nachr. Kön. Gesel. Wiss. Göttingen* (21), 571–584.

J. M. Borwein and P. B. Borwein (1987). *Pi and the AGM*, Canadian Math. Soc. Advanced Texts, Wiley, New York, xvi+414 pp.

P. B. Borwein and R. A. Ferguson (2004). "A complete description of Golay pairs for lengths up to 100", *Math. Comp.* **73**, 297–985.

P. B. Borwein and M. J. Mossinghoff (2009). "Wieferich primes and Barker sequences", *Des. Codes Cryptogr.* **53**, 149–163.

E. B. Burger and R. Tubbs (2004). *Making transcendence transparent*, Springer-Verlag, New York, x+263 pp.

G. Cantor (1874). "Über eine Eigenschaft des Inbegriffe aller reellen algebraischen Zahlen", *Jour. Math.* **77**, 258–262; *Gesammelte abhandlungen*, Springer-Verlag, Berlin, 1932, pp. 116–118.

G. Cantor (1884). "De la puissance des ansembles parfaits de points", *Acta Math.* **4**, 381–392.

L. Carleson (1966). "On convergence and growth of partial sums of Fourier series", *Acta Math.* **116**, 135–157.

A. L. Cauchy (1841). "Memoire sur diverses formulaes de analyse", *C. R. Paris* **12**, 283–296.

J. W. Cooley and J. W. Tukey (1965). "An algorithm for the machine computation of complex Fourier series", *Math. Comp.* **19**, 297–301.

R. B. Darst (1972). "Some Cantor sets and Cantor functions", *Mathematics Magazine* **45**, 2–7.

U. Dini (1880). *Serie di Fourier e altre rappresentazioni analitiche delle funzioni di una variabile reale*, Nistri, Pisa, vi+329 pp.

P. Lejeune Dirichlet (1829). "Sur la convergence des séries trigonométriques qui servent à représenter une fonction arbitraire entre des limites données", *J. Reine Angew. Math.* **4** (1829), 157–169.

B. Dumitrescu (2007). *Positive trigonometric polynomials and signal processing applications*, Signals and Communication Technology, Springer, Dordrecht.

P. Duren (2012). *Invitation to Classical Analysis*, Pure and Applied Undergraduate Texts 17, American Mathematical Society, Providence RI, ix+392 pp.

P. J. L. Fatou (1906). "Séries trigonométriques et séries de Taylor", *Acta Math.* **30**, 335–400.

L. Fejér (1916). "Über trigonometrische Polynome", *J. Reine Angew. Math.* **146**, 53–82.

E. S. Fischer (1907). "Sur la convergence en moyenne", *C. R. Acad. Sci. Paris* **144**, 1022–1024.

J. B. J. Fourier (1822). *Théorie analytique de la chaleur*, Fimin Didot Père et Fils, Paris, xxii+639 pp.

J. W. Gibbs (1898–1899). Letter in *Nature* **59**, 606.

M. J. E. Golay (1949). "Multislit spectrometry", *J. Opt. Soc. Amer.* **39**, 437–444.

R. R. Goldberg (1970). *Fourier Transforms*, Cambridge Tract 52, Cambridge University Press, Cambridge, viii+76 pp.

S. W. Graham and J. D. Vaaler (1981). "A class of extremal functions for the Fourier transform", *Trans. Amer. Math. Soc.* **265**, 283–302.

L.-S. Hahn (1994). *Complex Numbers and Geometry*, Math. Assoc. Amer., Washington, x+192 pp.

G. H. Hardy, J. E. Littlewood, and G. Pólya (1952). *Inequalities*, Second Edition, Cambridge University Press, Cambridge.

G. H. Hardy and W. W. Rogosinski (2013). *Fourier Series*, Dover, New York, x+100 pp.

Ch. Hermite (1873). "Sur la fonction exponentielle", *Comptes Rendus Paris*, 18–24; 74–79; 226–233; 285–293.

O. Hölder (1889). "Ueber einen Mittelwerthssatz", *Nachr. Königl. Gesell. Wiss. Göttingen*, 38–47.

J. Horváth (2004). "On the Riesz–Fischer theorem", *Studia Sci. Math. Hungar.* **41**, 467–478.

R. A. Hunt (1968). "On the convergence of Fourier series", Orthogonal Expansions and their Continuous Analogues (Proc. Conf., Edwardsville, IL, 1967), Southern Illinois Univ. Press, Carbondale, IL, pp. 235–255.

D. Jackson (1911). "Über eine trigonometrische Summe", *Rend. Circ. Mat. Palermo* **32**, 257–262.

D. Jackson (1912). *Über die Genauigkeit der Annäherung stetiger Funktionen durch rationale Funktionen gegebenen Grades und trigonometrische Summen gegebenen Ordnung*, doctoral thesis, University of Göttingen.

D. Jackson (1930). *The Theory of Approximations*, Amer. Math. Soc. Colloq. Publ. **XI**, American Mathematical Society, Providence, RI.

J.-P. Kahane (1980). "Sur les polynômes à coefficients unimodulaires", *Bull. London Math. Soc.* **12**, 321–342.

Y. Katznelson (2004). *An introduction to Harmonic Analysis*, Third Edition, Cambridge Math. Library, Cambridge University Press, Cambridge, xv+314 pp.

A. N. Kolmogorov (1926). "Une série de Fourier–Lebesgue divergente partout", *Comptes Rendus Paris* **183**, 1327–1328.

H. L. Lebesgue (1902). "Intégrale, longeur, aire", *Ann. Mat. Pura Appl.* (3) **7**, 231–359.

P. Lévy (1934). "Sur le convergence absolute des séries de Fourier", *Compositio Math.* **1**, 1–14.

F. Lindemann (1882). "Über die Zahl π", *Math. Ann.* **20**, 213–215.

J. Liouville (1840). "Sur l'irrationalité du nombre e", *J. Math. Pures Appl.* **5**, 192–193.

H. Mehta (2014). "The $L1$ norms of the de la Vallée Poussin kernels", *J. Math. Anal. Appl.*, to appear.

H. Minkowski (1910). *Geometrie der Zahlen*, Teubner, Leipzip–Berlin, ix+256 pp.

H. L. Montgomery and R. C. Vaughan (2007). *Multiplicative Number Theory I*, Cambridge University Press, Cambridge, xvii+552 pp.

D. J. Newman (1975). "A simple proof of Wiener's $1/f$ theorem", *Proc. Amer. Math. Soc.* **48**, 264–265.

I. Niven (1956). *Irrational Numbers*, Carus Monograph 11, Math. Assoc. Amer., Wiley, New York, xii+164 pp.

M.-A. Parseval (1799). *Mémoire sur les séries et sur l'intégration complète d'une équation aux différences partielles linéaires du second ordre, à coefficients constants.* Presented before the Académie des Sciences (Paris) on April 5, 1799.

M. Plancherel (1910). "Contribution à l'étude de la représention d'une fonction arbitraire par les intégrales définies", *Rend. Circ. Mat. Palermo* **30**, 298–335.

J. W. S. Rayleigh (1889). "On the character of the complete radiation at a given temperature", *Philosophical Magazine* **27**, 460–469.

F. Riesz (1907). "Sur les systèmes orthogonaux de fonctions", *C. R. Acad. Sci. Paris* **144**, 615–619.

F. Riesz (1916). "Über ein Problem des Herrn Carathéodory", *J. Reine Angew. Math.* **146**, 83–87.

M. Riesz (1928). "Sur les fonctions conjugé", *Math. Z.* **27**, 218–244.

W. Rudin (1959). "Some theorems on Fourier coefficients", *Proc. Amer. Math. Soc.* **10**, 855–859.

A. C. Schaeffer (1941). "Inequalities of A. Markoff and S. Bernstein for polynomials and related functions", *Bull. Amer. Math. Soc.* **61**, 934–940.

L. Scheeffer (1884). Allgemeine Untersuchungen über Rectification der Curven, *Acta Math.* **5**, 49–82.

H. S. Shapiro (1951). *Extremal problems for polynomials*, M. S. Thesis, MIT.

C. L. Siegel (1935). "Über Gitterpunkte in convexen Körpern und ein Damit Zussamendhängendes Extremalproblem", *Acta Math.* **65**, 307–323.

C. L. Siegel (1989). *Lectures on the Geometry of Numbers*, Springer-Verlag, Berlin, x+160 pp.

C. L. Siegel (1949), *Transcendental Numbers*, Annals of Mathematics Studies, No. 16, Princeton University Press, Princeton, NJ, viii+102 pp.

J. M. Steele (2004). *The Cauchy–Schwarz master class*, Cambridge University Press, Cambridge.

E. C. Titchmarsh (1986). *Introduction to the Theory of Fourier Integrals*, Third Edition, Chelsea Publ. Co., New York, x+394 pp.

R. Turyn (1974). "Hadamard matrices, Beumert–Hall units, four-symbol sequences, pulse compression, and surface wave encodings", *J. Comb. Theory Ser. A* **16**, 313–333.

S. Wagon (1987), "Fourteen proofs of a result about tiling a rectangle", *Amer. Math. Monthly* **94**, 601–617.

H. Weyl (1914). "Über ein Problem aus dem Gebeite der diophantischen Approximationen", *Nachr. Ges. Wiss. Göttingen Math.-phys. Kl.*, 234–244; *Gessammelte Abhandlungen*, Band 1, Springer-Verlag, Berlin, 1968, pp. 487–497.

H. Weyl (1916). "Über die Gleichverteilung von Zahlen mod Eins", *Math. Ann.* **77**, 313–352; *Gessammelte Abhandlungen*, Band 1, Springer-Verlag, Berlin, 1968, pp. 563–599; *Selecta Hermann Weyl*, Birkhäuser-Verlag, Basel–Stuttgart, 1956, 111–147.

E. T. Whittaker (1915). *On the functions which are represented by the expansions of the interpolation theory*, Proc. Roy. Soc. Edinburgh **35**, 181–194.

N. Wiener (1932). "Tauberian theorems", *Annals of Math.* (2) **33**, 1–100.

H. Wilbraham (1848). "On a certain periodic function", *Cambridge and Dublin Math. J.* **3**, 198–201.

W. H. Young (1912). "On classes of summable functions and their Fourier series", *Proc. Roy. Soc. Ser. A* **87**, 225–229.

A. Zygmund (1932). "A remark on conjugate series", *Proc. London Math. Soc.* **34**, 392–400.

A. Zygmund (1968). *Trigonometric Series*, Cambridge University Press, London, Vols. I, II. Second Edition, reprinted with corrections and some additions, xiv+383 pp., viii+364 pp.

Notation

Some symbols are used in more than one way. The intended interpretation should be clear from the context in which it arises.

Symbol	Meaning
$\mathbb{C}$	The set of complex numbers. See page 1.
$\mathbb{Q}$	The set of rational numbers.
$\mathbb{R}$	The set of real numbers.
$\mathbb{T}$	$\mathbb{R}/\mathbb{Z}$, i.e., the real numbers modulo 1. See page 55.
$\mathbb{Z}$	The set of rational integers.
$A(\mathbb{T})$	Space of absolutely convergent Fourier Series. See page 183.
$C(\mathbb{T})$	Space of continuous functions with period 1. See page 95.
$L^p(\mathbb{R})$	Space of functions with $\int_{-\infty}^{\infty} \lvert f\rvert^p < \infty$. See page 263.
$L^p(\mathbb{T})$	Space of functions with period 1 and $\int_0^1 \lvert f\rvert^p < \infty$. See page 55.
$\mathrm{Lip}(\mathbb{T})$	Lipschitz class. $\lvert f(x) - f(y)\rvert \le C\lvert x - y\rvert$. See page 184.
$\mathrm{Lip}_\alpha(\mathbb{T})$	$\lvert f(x) - f(y)\rvert \le C\lvert x - y\rvert^\alpha$. See page 184.
$B_k(x)$	A Bernoulli polynomial. See page 220.
$B_N(x)$	A Beurling trigonometric polynomial. See page 152.
$D_N(x)$	The Dirichlet kernel. See page 76.
$e(x)$	$e^{2\pi i x}$; the complex exponential with period 1. See page 34.
$J_N(x)$	The Jackson kernel. See 176.
$M(P)$	The Mahler measure of the polynomial P. See page 242.
$P_k(x)$	A Shapiro polynomial. See page 167.
$P_k(x)$	The mod-1 restriction of the Bernoulli polynomial $B_k(x)$. See page 222.
$P_r(x)$	The Poisson kernel. See page 132.
$p_n(x)$	The de la Vallée Poussin power kernel. See page 133.
$Q_k(x)$	A Shapiro polynomial of the second kind. See page 167.
$s(x)$	The sawtooth function. See page 58.
$s_N(x)$	The Nth symmetric partial sum of a Fourier Series. See page 76.

Symbol	Meaning
$S_K^{\pm}(x)$	Selberg trigonometric polynomials. See page 235.
$\mathcal{T}_N$	The set of trigonometric polynomials of degree not exceeding N. See pages 76, 149.
$T_n(x)$	A Chebyshev polynomial of the first kind. See page 299.
$U_n(x)$	A Chebyshev polynomial of the second kind. See page 304.
$V_N(x)$	The de la Vallée Pousin kernel. See page 133.
$\mathrm{Var}(f)$	Total variation of f. Also called $\\|f\\|_{BV}$. See page 198.
$\Delta_N(x)$	The Fejér kernel. See page 92.
$\Delta_T(x)$	Analogue of the Fejér kernel for $\mathbb{R}$. See page 257.
σ_k	The kth elementary symmetric function. See page 34.
$\sigma_N(x)$	The Nth Cesàro partial sum of a Fourier Series. See page 92.
$\sigma_T(x)$	Analogue of Cesàro partial sum for Fourier transforms. See page 261.
χ_S	The characteristic function of the set S. See page 126.
$\omega(\delta)$	The modulus of continuity of a function. See 175.
$[x]$	The integer part of the real number x: $[x] \in \mathbb{Z}$ and $[x] \le x < [x] + 1$.
$\{x\}$	The fractional part of the real number x: $\{x\} = x - [x]$.
$\\|x\\|$	The distance from the real number x to the nearest integer. See page 55.
$\\|\boldsymbol{v}\\|$	Length of the vector $\boldsymbol{v}$. See page 340.
$\\|f\\|_p$	The norm of the function f in $L^p(\mathbb{T})$ or $L^p(\mathbb{R})$. See pages 70, 249.
$\\|f\\|_{BV}$	Total variation of f. Also called $\mathrm{Var}(f)$. See page 198.
$\\|P\\|_1$	The ℓ^1 norm of the coefficients of the polynomial P. See page 241.
$\langle f, g\rangle$	Inner product of two functions. See page 135.
$\langle \boldsymbol{u}, \boldsymbol{v}\rangle$	The inner product of two vectors. See pages 37, 339.
$x \equiv y \pmod 1$	$\{x\} = \{y\}$; equivalently, $x - y \in \mathbb{Z}$. See page 55.
$m \equiv n \pmod q$	$q \mid (m-n)$. See page 36.
$d \mid n$	For integers n and d, n/d is an integer. See page 24.
$f(x) \sim g(x)$	$\lim f(x)/g(x) = 1$. See 349.
$f(x) = O(g(x))$	$\lvert f(x)\rvert \le Cg(x)$ for some constant C. See page 349.
$f(x) = o(g(x))$	$\lim f(x)/g(x) = 0$. See page 350.
$f(x) \asymp g(x)$	$cf(x) \le g(x) \le Cf(x)$ for constants c, C. See page 350.
$\operatorname{supp} f$	The support of a function: $\operatorname{supp} f = \{x : f(x) \ne 0\}$.

Index